THE HAEMODYNAMIC EFFECTS OF NITRIC OXIDE

THE HAEMODYNAMIC EFFECTS OF NITRIC OXIDE

Editors

Robert T. Mathie

Division of Surgery, Anaesthetics & Intensive Care,
Imperial College School of Medicine,
Hammersmith Hospital, London, UK

Tudor M. Griffith

Department of Diagnostic Radiology,
University of Wales College of Medicine, Cardiff, UK

ICP Imperial College Press

Published by

Imperial College Press
57 Shelton Street
Covent Garden
London WC2H 9HE

Distributed by

World Scientific Publishing Co. Pte. Ltd.
P O Box 128, Farrer Road, Singapore 912805
USA office: Suite 1B, 1060 Main Street, River Edge, NJ 07661
UK office: 57 Shelton Street, Covent Garden, London WC2H 9HE

British Library Cataloguing-in-Publication Data
A catalogue record for this book is available from the British Library.

THE HAEMODYNAMIC EFFECTS OF NITRIC OXIDE

ISBN 1-86094-081-1

Printed in Singapore.

To Hannah and Kirstie, Kate, Richard and Robert

Foreword

Nitric oxide (NO) is a ubiquitous signalling molecule with fundamental roles now apparent not only in vertebrates and invertebrates but also in insects and plants. Thus, NO is likely to have been one of the earliest regulatory molecules to appear during biological evolution. In the cardiovascular system, the haemodynamic forces of pulsation and shear regulate NO synthesis by the endothelial monolayer that lines the vascular tree on a moment-to-moment basis, thus dynamically regulating perfusion at a local level. This simple but elegant method of control is likely to have antedated the role of the autonomic nervous system in the regulation of the circulation.

In the mid-1970s, NO was shown to be a potent vasodilator and inhibitor of platelet aggregation by activating the enzyme guanylate cyclase, and to mediate the pharmacological actions of nitroglycerine, the principal drug used in the treatment of angina for over 100 years. Furchgott's recognition of the pivotal role of the endothelium in mediating vascular relaxations to acetylcholine then led to a new era in vascular biology. This culminated in the identification of Furchgott's endothelium-derived relaxing factor (EDRF) as NO. Since then, an explosion of research has shown that NO contributes to signalling in virtually all mammalian cell types. Indeed, a simple Internet search reveals that there are some 30,000 publications in the field, over 7,000 of which deal with the cardiovascular system. Under normal physiological conditions NO helps to maintain cardiovascular efficiency in its broadest sense.

This volume is the first to address in depth the specific role of NO in cardiovascular haemodynamics. The editors have assembled authoritative contributions from leading groups in a number of areas ranging from the role of NO in the heart itself to its role in the peripheral microcirculation. The initial chapters of the book summarize current knowledge about the synthesis, chemistry and metabolic fate of NO, the cellular mechanisms that underlie its mode of action in vascular smooth muscle and the effects of mechanical forces on the endothelial cell which underpin its haemodynamic role. Molecular biological and physiological approaches have both provided important information about the control of the circulation by NO. This volume exemplifies how valuable such complementary insights may be. Later chapters examine the role of NO in the regulation of perfusion in specific organ systems, focusing on the brain, heart, skeletal muscle, liver, intestine, kidney, and lung. These chapters highlight similarities, differences and organ-specific features which may be of particular physiological importance. Several chapters discuss the interaction of NO with other important vascular control systems including the prostaglandin pathway and mediators such as the endothelium-derived hyperpolarizing factor.

As well as beneficial effects, NO also has a "dark side" in that an inducible isoform of NO synthase can produce large amounts of NO in a variety of pathological situations. These include septic shock, where excessive NO production may result in profound vasodilatation and circulatory collapse. While the chemical structure of NO may be deceptively simple, its biological chemistry is extremely complex, particularly in pathophysiological states where reactions with other free radicals such as the superoxide anion generate toxic compounds that cause cell damage. In the cardiovascular system such free radical interactions may contribute to the genesis of atheroma, ischaemia-perfusion injury and neuronal death following cerebral ischaemia. All these important aspects are covered here, together with a discussion of possible therapeutic strategies and their likely limitations.

The comprehensive coverage of the role of NO in the cardiovascular system provided in this volume serves as a springboard for those wishing to assess the role of NO in the regulation of blood flow. As many basic principles have now been established, it also serves as an important source of reference even though new twists to the tale are continuously being revealed as research progresses. In addition, this volume will be an important resource to workers in a wide range of medical disciplines, for blood flow is central to the maintenance of tissue viability in all organs. The editors are to be congratulated on assembling such an important volume, with contributions from many of the leaders in the field.

Sir John Vane
The William Harvey Research Institute, London

Preface

Since the identification of endothelium-derived relaxing factor as NO in 1987, the elucidation of its many biological effects and mechanisms of action has developed at an extraordinary rate. NO is now implicated in many physiological and pathological processes, including not just the control of vascular resistance, but also autonomic nerve transmission, cell proliferation and the response to inflammation. Many scientific symposia and review articles during the last decade have helped the international research community to keep abreast of developments in the ever diversifying branches of NO investigation, and a journal dedicated to NO research in general has recently been launched.

Given the crucial importance of NO in the cardiovascular system, it seemed to us particularly relevant to introduce a text encompassing this particular aspect of NO. Nearly twenty years after the discovery of the pivotal role of the endothelium in blood flow regulation, we therefore consider it opportune to concentrate on the field of NO and haemodynamics, and provide a comprehensive framework of information from within this active field of research. The need for such a book is clear. It is very difficult for those new or peripheral to the field to extract the relevant details from the vast literature. Even for those actually involved in cardiovascular NO research, it is almost impossible to maintain a close grasp of associated specialised topics.

We believe that the presentation and critical evaluation of existing knowledge by internationally recognised specialists has created an important reference work. In an area of research moving at such a rapid pace, we naturally recognise that fresh discoveries can swiftly supersede what seemed to be established facts. However, we and our contributing authors have done our utmost to ensure that the literature cited throughout the text reflects current awareness of the subject. Thus, we hope the book will be of lasting benefit to those new to the topic as much as to those actively engaged in NO research, including physicians who have begun to apply basic discoveries to the management of the many disease processes now known to be linked in some way to the action of this intriguing molecule.

We would like to record our sincere gratitude to all the authors who contributed to the book — without such expert and diverse input, an up-to-date volume on the haemodynamics of NO could not have been produced. It is also a pleasure to acknowledge the enthusiasm and skills provided by Aileen Parlane of Imperial College Press and Yugarani Thanabalasingam of World Scientific Publishing during the editorial and production process. In addition, our thanks go to Linda

Theobald, Imperial College School of Medicine, and Wendy Simons, University of Wales College of Medicine, for their assistance with the detailed refinement of the manuscript. Finally, we thank both our families for their considerable patience and support during the preparation of the book.

Robert T Mathie
Tudor M Griffith

List of Contributors

SL Archer
Division of Cardiology,
University of Alberta,
Edmonton,
Alberta T6G 2B7,
Canada

JE Barker
Bernard O'Brien Institute of Microsurgery,
St Vincent's Hospital,
University of Melbourne,
Parkville,
Victoria 3065,
Australia

N Benjamin
Department of Clinical Pharmacology,
St Bartholomew's and the Royal London School of Medicine and Dentistry,
Charterhouse Square,
London EC1M 6BQ,
UK

VM Darley-Usmar
Center for Free Radical Biology and Department of Pathology,
University of Alabama,
Birmingham,
Alabama 35294,
USA

AM Dart
Department of Physiology,
University of Melbourne,
Parkville,
Victoria 3052,
Australia

AT Dinh-Xuan
Department de Physiologie et de Biologie Cellulaire,
Hôpital Cochin,
75679 Paris cedex 14,
France

K Djaballah
Department de Physiologie et de Biologie Cellulaire,
Hôpital Cochin,
75679 Paris cedex 14,
France

GJ Dusting
Howard Florey Institute of Experimental Physiology and Medicine,
University of Melbourne,
Parkville,
Victoria 3052,
Australia

TJ Evans
Department of Infectious Diseases,
Imperial College School of Medicine,
Hammersmith Hospital,
London W12 ONN,
UK

P Forte
Department of Clinical Pharmacology,
St Bartholomew's and the Royal London School of Medicine and Dentistry,
Charterhouse Square,
London EC1M 6BQ,
UK

TM Griffith
Department of Diagnostic Radiology,
University of Wales College of Medicine,
Heath Park,
Cardiff CF4 4XN,
UK

PH Groves
Department of Cardiology,
University Hospital of Wales,
Heath Park,
Cardiff CF4 4XW,
UK

H Hayakawa
Nephrology Section,
Department of Medicine,
VA Medical Center,
University of Minnesota,
Minneapolis,
Minnesota 55417,
USA

TW Higenbottam
Department of Medicine and Pharmacology,
Section of Respiratory Medicine,
The Medical School,
University of Sheffield,
Sheffield S10 2RX,
UK

AJ Hobbs
The Cruciform Project,
Rayne Institute,
University College London,
London WC1E 6JJ
UK

C Iadecola
Department of Neurology,
University of Minnesota,
Minneapolis,
Minnesota 55455,
USA

LJ Ignarro
Department of Molecular & Medical Pharmacology,
UCLA School of Medicine,
Los Angeles,
California 90095,
USA

EA Jaimes
Nephrology Section,
Department of Medicine,
VA Medical Center,
University of Minnesota,
Minneapolis,
Minnesota 55417,
USA

H Jo
Center for Free Radical Biology and Department of Pathology,
University of Alabama,
Birmingham,
Alabama 35294,
USA

L Jungersten
Department of Clinical Physiology,
Sahlgrenska University Hospital,
University of Gothenburg,
S-41345 Gothenburg,
Sweden

G Kaley
Department of Physiology,
New York Medical College,
Valhalla,
New York 10595,
USA

A Koller
Department of Physiology,
New York Medical College,
Valhalla,
New York 10595,
USA

P Kubes
Department of Medical Physiology,
Health Science Centre,
University of Calgary,
Calgary,
Alberta T2N 4N1,
Canada

D Lang
Department of Pharmacology, Therapeutics & Toxicology,
University of Wales College of Medicine,
Heath Park,
Cardiff CF4 4XN,
UK

MJ Lewis
Department of Pharmacology, Therapeutics & Toxicology,
University of Wales College of Medicine,
Heath Park,
Cardiff CF4 4XN,
UK

MC Maland
Center for Free Radical Biology and Department of Pathology,
University of Alabama,
Birmingham,
Alabama 35294,
USA

RT Mathie
Division of Surgery, Anaesthetics and Intensive Care,
Imperial College School of Medicine,
Hammersmith Hospital,
London W12 ONN,
UK

J McAndrew
Center for Free Radical Biology and Department of Pathology,
University of Alabama,
Birmingham,
Alabama 35294,
USA

D-M McCafferty
Department of Medical Physiology,
Health Science Centre,
University of Calgary,
Calgary,
Alberta T2N 4N1,
Canada

ED Michelakis
Division of Cardiology,
University of Alberta,
Edmonton,
Alberta T6G 2B7,
Canada

A Papapetropoulos
Department of Pharmacology,
Boyer Center for Molecular Medicine,
Yale University School of Medicine,
New Haven,
Connecticut 06536,
USA

RP Patel
Center for Free Radical Biology and Department of Pathology,
University of Alabama,
Birmingham,
Alabama 35294,
USA

U Pohl
Institute of Physiology and Pathophysiology,
Johannes Gutenberg — University of Mainz,
D-55099 Mainz,
Germany

BD Prendergast
Department of Cardiology,
University of Wales College of Medicine,
Heath Park,
Cardiff CF4 4XN,
UK

N Purcell
Center for Free Radical Biology and Department of Pathology,
University of Alabama,
Birmingham,
Alabama 35294,
USA

L Raij
Nephrology Section,
Department of Medicine,
VA Medical Center,
University of Minnesota,
Minneapolis,
Minnesota 55417,
USA

CM Rembold
Departments of Internal Medicine and Physiology,
University of Virginia Health Sciences Center,
Charlottesville,
Virginia 22908,
USA

WC Sessa
Department of Pharmacology,
Boyer Center for Molecular Medicine,
Yale University School of Medicine,
New Haven,
Connecticut 06536,
USA

AM Shah
Department of Cardiology,
Guy's, King's & St. Thomas' School of Medicine,
Denmark Hill Campus,
London SE5 9PJ,
UK

T Siddons
Department of Medicine and Pharmacology,
Section of Respiratory Medicine,
The Medical School,
University of Sheffield,
Sheffield S10 2RX,
UK

E Souil
Department de Physiologie et de Biologie Cellulaire,
Hôpital Cochin,
75679 Paris cedex 14,
France

GJ Southan
Inotek Inc.,
Cincinnati,
Ohio 45219,
USA

AG Stewart
Department of Pharmacology,
University of Melbourne,
Parkville,
Victoria 3052,
Australia

C Szabó
Division of Critical Care,
Children's Hospital Medical Center,
Cincinnati,
Ohio 45229,
USA

C Thiemermann
The William Harvey Research Institute,
St. Bartholomew's & The Royal London School of Medicine & Dentistry,
Charterhouse Square,
London EC1M 6BQ,
UK

Å Wennmalm
Department of Clinical Physiology,
Sahlgrenska University Hospital,
University of Gothenburg,
S-41345 Gothenburg,
Sweden

CR White
Molecular and Cellular Division and Department of Medicine,
University of Alabama,
Birmingham,
Alabama 35294,
USA

V Zachariou
Department of Pharmacology,
Boyer Center for Molecular Medicine,
Yale University School of Medicine,
New Haven,
Connecticut 06536,
USA

Contents

Part 1

Physiology and Biochemistry of Nitric Oxide

Chapter 1

Biochemical Effects of Nitric Oxide on Vascular Smooth Muscle

Derek Lang and Malcolm J Lewis

Introduction

It has long been known that the nitrovasodilator drugs, such as glyceryl trinitrate, can be used clinically for the treatment of angina and congestive heart failure. The mechanism of action of these compounds remained unclear for nearly a century, however. In the 1970s it was discovered that they could relax vascular smooth muscle by elevating intracellular guanosine 3',5'-cyclic monophosphate (cGMP) levels *via* the release of NO and the activation of soluble guanylate cyclase (sGC). With the identification of the 'endothelium-derived relaxing factor' (EDRF) of Furchgott and Zawadzki (1980)[1] as NO in 1987,[2] endothelial cells, *via* the production and release of NO, were themselves shown to be modulators of vascular smooth muscle tone. A whole new science has now evolved, illustrating the complex physiological significance of the NO/cGMP signal transduction system.

Nitric Oxide Formation in the Vasculature

The topic of NO biosynthesis is covered elsewhere in this book (see Chapter 3), and therefore will be only briefly summarized here. A family of enzymes has now been shown to be responsible for the production of NO in mammalian tissues. These NO synthases (NOS) fall into two general classes, known as the constitutive or Ca^{2+}-dependent (cNOS) and the inducible or Ca^{2+}-independent (iNOS) forms. They both synthesize NO from one of the terminal guanidino nitrogen atoms of L-arginine (Figure 1). The endothelial cNOS (eNOS) produces NO in response to receptor-dependent agents such as acetylcholine (ACh), bradykinin and adenosine 5'-triphosphate (ATP) and to shear stress.[3] Another form of cNOS (nNOS) is found in neural tissues, but may play a role in the control of blood vessel tone due to its presence in peri-adventitial nerves. Inflammatory stimuli, such as endotoxin, lipopolysaccharide and cytokines such as interleukin (IL)-1 and tumour necrosis factor (TNF), can induce the expression of the iNOS form in a wide variety of cells.[4]

Nitric Oxide and the Control of Vascular Tone

Activation of Soluble Guanylate Cyclase

Almost ten years before the identification of the biological activity of EDRF as being due to NO, it was demonstrated that NO could activate sGC[5] in vascular smooth muscle, leading to increases in cGMP and to the eventual relaxation of contracted arterial and bronchiolar smooth muscle strips. The highly lipophilic NO molecule diffuses readily across cell membranes, and binds to the haem component of sGC,[6] thereby activating the catalytic site on the enzyme. This reaction is not as straightforward as would at first seem, however, since it has been reported

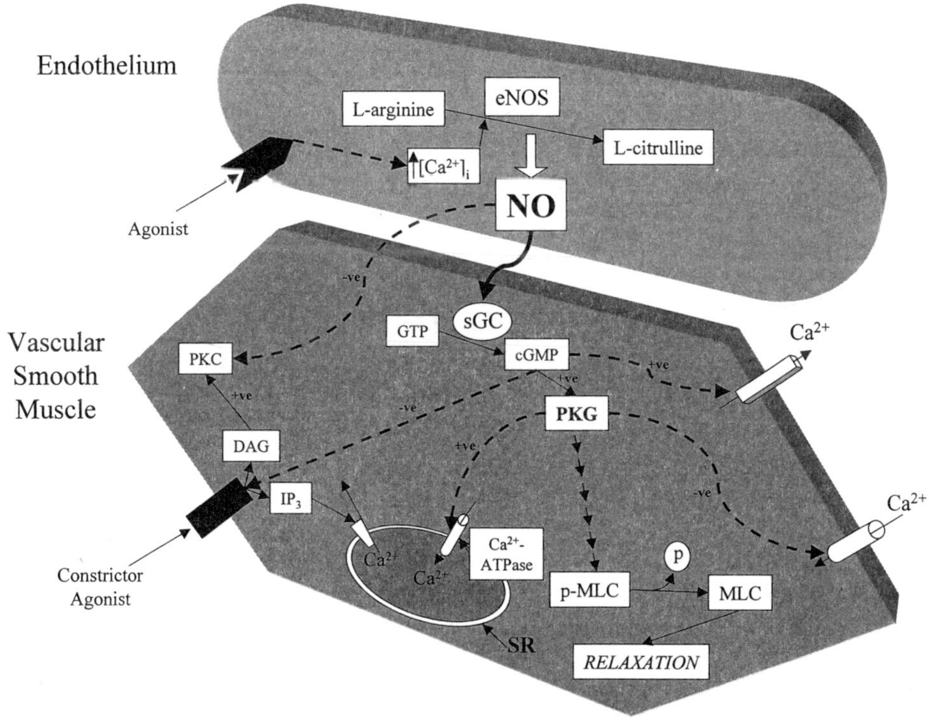

Figure 1. The major actions of NO and guanosine 3',5'-cyclic monophosphate (cGMP) in the control of vascular tone. Endothelial cell-derived NO, derived from L-arginine in the presence of essential cofactors O_2, nicotinamide adenine dinucleotide phosphate (NADPH) and tetrahydrobiopterin (BH_4), activates soluble guanylate cyclase (sGC) in the vascular smooth muscle cell. This results in the formation of cGMP and the activation of cGMP-dependent protein kinase (PKG). Together, NO, cGMP and PKG mediate various processes resulting in the decrease in smooth muscle cell calcium (Ca^{2+}) concentration, the dephosphorylation of contractile proteins such as myosin light chain (p-MLC) to MLC, and relaxation. eNOS, endothelial cell NO synthase; IP_3, inositol 1,4,5-trisphosphate; DAG, diacylglycerol; PKC, protein kinase C; GTP, guanosine 5'-triphosphate; SR, sarcoplasmic reticulum.

that the presence of sulphydryl groups is also necessary, both for the sequestration of NO by vascular smooth muscle cells (VSMC), and the subsequent activation of sGC.[7] Furthermore, it has also now been postulated that other haem-containing proteins are closely involved in the control of sGC activity. It appears that these proteins can reverse sGC from the 'activated' to the 'native' state by a direct effect on the enzyme itself, rather than preventing the action of NO by simple binding to the haem moiety.[8] It has also been suggested that optimal activation of sGC, *via* its haem group, requires NO to be in its oxidized (NO^+, nitrosonium) form.[9]

NO itself is known to be highly reactive and short-lived. The minimum concentration of NO needed to activate sGC is 5nM[10], showing it indeed to be a very reactive molecule. However, the fact that at this concentration the half-life of NO is 70 hours[10] shows that it is not as short-lived as once thought. NO can combine very readily with superoxide anions ($O_2^{\bullet-}$) to form peroxynitrite[11] (see below), a compound which is both a powerful and toxic oxidising agent.[10] It would appear, however, that the composition of the cellular environment strongly influences the reactivity of this compound. For example, in the presence of plasma, proteins, glucose or glutathione, peroxynitrite forms intermediates that can act as NO donors.[12] It has also been demonstrated that peroxynitrite can activate sGC in VSMC,[13] an effect that is probably mediated by peroxynitrite-derived NO.[14] Because of its reactivity, toxicity and modest efficacy as a relaxing agent, it is unlikely that peroxynitrite plays a major role in the physiological control of blood vessel tone.

cGMP, cGMP-dependent Protein Kinase and Intracellular Calcium

The increases in cGMP associated with VSMC relaxation are concomitant with decreases in the intracellular Ca^{2+} concentration ($[Ca^{2+}]_i$).[15–17] It is likely that cGMP has multiple sites of action to reduce $[Ca^{2+}]_i$ in VSMC, given that the mechanisms by which $[Ca^{2+}]_i$ is *increased* to elicit *constriction* are known to depend on the type of vasoconstrictor agent used, i.e. depolarization or G protein-coupled agonist.[18] The major receptor protein for cGMP in smooth muscle is cGMP-dependent protein kinase (PKG).[17] Furthermore, PKG has been shown to be both necessary and sufficient to account for the cGMP-induced changes in $[Ca^{2+}]_i$.[17] It is thought that cGMP-induced activation of PKG, together with the consequent phosphorylation of numerous other cellular proteins, constitutes a cascade of phosphorylation/dephosphorylation events that eventually leads to the dephosphorylation of myosin light chain and therefore to relaxation.[19] The decrease in the $[Ca^{2+}]_i$ in VSMC may produce a dephosphorylation of myosin light chain due to the inactivation of the Ca^{2+}/calmodulin-dependent myosin light chain kinase and subsequently facilitate relaxation (Figure 1). These events are discussed more fully in Chapter 4.

Constrictions to receptor-dependent agonists are associated with phospholipase C-mediated increases in inositol 1,4,5-trisphosphate (IP_3) formation[20] and protein kinase C (PKC) activity.[21] These agonist-induced changes are inhibited by elevations in cGMP levels,[22,23] although the mechanism by which cGMP inhibits IP_3 production and PKC activation is unknown. One possible site of action is at the G protein itself.[24] Indeed, activation of PKG has been shown to be associated with the phosphorylation of the alpha subunit of the inhibitory heterotrimeric G protein, G_i.[25] However, most other heterotrimeric G proteins are not especially good substrates for PKG;[25,26] therefore, the evidence for a direct phosphorylation of G proteins by PKG is not convincing. Interestingly, with regard to PKC itself, interaction of NO with the sulphydryl groups of this enzyme leads to an inactivation of the enzyme, and an impairment of its phorbol ester binding capacity.[27]

The evidence for other cGMP-induced effects on Ca^{2+} homeostasis is more convincing. For instance, the influx and intracellular release of Ca^{2+} in VSMC has been shown to be inhibited by cGMP.[15] The prevention of Ca^{2+} entry may be due to the phosphorylation of the Ca^{2+} channel (or associated regulatory proteins) by PKG.[28,29] The mechanism for the inhibition of intracellular Ca^{2+} release, however, is still unclear. Evidence suggests that this effect is not due to a direct inhibition of the IP_3 receptor on the sarcoplasmic reticulum.[28]

It has been demonstrated that cGMP can also actively control $[Ca^{2+}]_i$ by other means. It has been shown, for example, to activate the Ca^{2+} pumps which regulate the sequestration of Ca^{2+} into the sarcoplasmic reticulum,[30] and also the Na^+/Ca^{2+} exchange mechanism.[31] Phosphorylation of phospholamban, the Ca^{2+}-ATPase regulatory protein, is another mechanism whereby cGMP could activate intracellular Ca^{2+} sequestration.[32,33]

Hyperpolarization

Receptor-mediated hyperpolarization (and thus relaxation) of VSMC occurs as a result of activation of K^+ channels by a substance termed 'endothelium-derived hyperpolarizing factor' (EDHF), tentatively identified as a cytochrome P_{450}-derived arachidonic acid metabolite.[34] Some of the stimuli that release EDHF also release NO from the endothelium (e.g. ACh and substance P),[35] but the role of NO in endothelium-dependent hyperpolarization of VSMC is by no means clear (see also Chapter 2). Inhibition of Ca^{2+} entry into VSMC could be enhanced by an indirect effect on voltage-gated Ca^{2+} channels mediated by an NO-dependent activation of the Ca^{2+}-dependent K^+ channels that cause hyperpolarization.[36] To complicate the issue further, the involvement of Na^+-K^+-ATPase in VSMC relaxation has also been proposed by the activation of this enzyme by NO in intact arterial tissues.[37] It remains uncertain, however, if this latter effect is direct or secondary, as recent evidence suggests that NO inhibits the activity of Na^+-K^+-ATPase purified from cerebral cortex.[38] Some workers have also postulated that the effect of NO on Ca^{2+}-dependent K^+ channels is not cGMP-dependent, but due instead to the reaction of NO with protein sulphydryl groups contained within these channels.[39] Reports are conflicting, however, since new evidence suggests that the activation of K^+ channels may be mediated *via* the phosphorylating action of PKG.[40]

A direct NO-induced hyperpolarization has clearly been demonstrated in the guinea-pig uterine artery,[41] rat mesenteric resistance arteries[42] and in the coronary arteries from several species.[35] However, these NO-mediated effects seem only to occur under particular conditions. In the coronary artery, for instance, an NO-induced change in membrane potential was observed only when the artery segments were stretched to a tension at least equivalent to a transmural pressure of 50 mmHg.[35] Alternatively, saturating concentrations of NO failed to have any effect on the membrane potential of similarly prepared cerebral arteries.[43] From this evidence, therefore, it would seem that NO-induced hyperpolarization is an event that is artery-specific and highly dependent on experimental conditions.

Hyperpolarization is thus a phenomenon that has largely only been investigated *in vitro*, and it seems that concentrations of NO in the range of at least 0.5 mM are needed to produce this effect.[44] Given that the minimum concentration of NO needed to activate sGC is 5 nM,[10] it would seem that a direct effect of NO on membrane potential may be of little physiological importance, the cGMP-dependent effects of NO being much more relevant.

Nitric Oxide and Cytotoxicity

The cytotoxic effects of iNOS-derived NO have been implicated in various pathological conditions involving immunostimulation, inflammation and various forms of circulatory shock.[45] These effects result in a fall in blood pressure, a decrease in peripheral vascular resistance, and a hyporesponsiveness of arteries and veins to endogenous and exogenous vasoconstrictors.[45] The mechanism of this toxicity is not fully understood, however. Several possibilities exist, such as the inhibition of respiratory enzymes, the formation of peroxynitrite, and the induction of apoptosis.

Cellular Respiration

One important mechanism by which NO exerts cytotoxicity is the inhibition of cellular respiration secondary to the inactivation of mitochondrial aconitase and components of the electron transport chain.[46–48] NO may exert its physiological and pathophysiological effects by directly de-energising mitochondria.[49] In particular, NO at physiological concentrations has been shown to bind to and inhibit the activity of cytochrome c oxidase, the terminal member of the mitochondrial respiratory chain.[50] This latter reaction is reversible, however, and competitive with oxygen.[51] It has been suggested that the rapid mitochondrial sequestration of NO by cytochrome c oxidase may be a mechanism for NO breakdown within the cell.[52]

On the other hand, the presence of eNOS within the mitochondria of several mammalian tissues has now been demonstrated.[53] It is therefore possible that this enzyme, *via* the production of NO and its effect on cytochrome c oxidase activity, acts as a ubiquitous regulator of mitochondrial oxidative phosphorylation. It is possible, however, that a proportion of the NO produced by mitochondrial eNOS does not bind to cytochrome c oxidase, but reacts with superoxide anions produced by the electron transport chain[54] to form peroxynitrite (see below).

Peroxynitrite Formation

As described above, peroxynitrite is formed by the very rapid combination of NO with superoxide anion.[11] Once formed, it acts as a potent oxidising agent.[10] *In vivo*, under 'normal' conditions, the constant removal of NO by haemoglobin and other reactions allows superoxide dismutase (SOD) to reduce peroxynitrite formation by limiting the accumulation of superoxide anions. However, in certain pathological conditions (see below), where the levels of NO are vastly increased, the formation of peroxynitrite is also increased, due to the much greater affinity of superoxide anion for NO than for SOD.[10]

It has now been clearly demonstrated that peroxynitrite, as well as NO, can inhibit mitochondrial respiration,[55] though possibly by a different mechanism. For instance, the mechanism by which NO and superoxide anions inactivate mitochondrial aconitase may not simply be due to a direct action on the iron-sulphur cluster of the enzyme. It may instead rely on the formation of peroxynitrite, since aconitase is readily inactivated by peroxynitrite but only poorly by NO.[56] Furthermore, peroxynitrite has been shown to have little or no effect on cytochrome c oxidase, while still causing the inhibition of other respiratory complexes.[57] Interestingly, although it has been demonstrated that NO can cause a direct but transient inhibition of mitochondrial respiration, the pattern of this inhibition more closely resembles that observed when peroxynitrite is formed subsequently to NO exposure.[58] This strongly suggests that peroxynitrite is the ultimate reactive intermediate accounting for NO-dependent inactivation of mitochondrial respiration.

A further direct role for peroxynitrite in the disruption of mitochondrial function has also been described, in that it can induce cyclosporin-A-sensitive mitochondrial Ca^{2+} efflux,[59] which may in turn lead to cell damage and death. These observations would strongly suggest that peroxynitrite is the ultimate reactive intermediate accounting for NO-dependent inactivation of electron transport components and ATPase in living cells and tissues.

The formation of peroxynitrite is often linked with sites of iNOS induction and NO overproduction. Examples of this include immunostimulated macrophages,[60] various inflammatory disorders,[61,62] atherosclerosis[63] and circulatory shock.[64,65] Furthermore, peroxynitrite has been proposed as an important mediator of cell damage in inflammation and oxidative stress.[63,65,66] One mechanism whereby peroxynitrite is thought to play a central role in this cellular toxicity is *via* a direct action that initiates DNA strand breakage.[67] This mechanism does not involve the action of NO *per se*, since peroxynitrite is at least 100 times more potent in its ability to trigger this DNA strand breakage than NO.[68]

DNA strand breakage leads to the activation of the nuclear enzyme poly-ADP ribosyl synthase (PARS), the synthesis of cyclic ADP ribose and the depletion of intracellular stores of nicotinamide adenine dinucleotide (NAD^+), the substrate for PARS. Furthermore, there is a concomitant slowing of the rate of glycolysis, electron transport and ATP formation, which results in cell dysfunction and cell death.[67] Interestingly, a recent report suggests that PARS activity is actually required for the expression of the iNOS gene.[69] This could possibly lead to a vicious cycle of events, leading to cell death.

Peroxynitrite also has other detrimental actions. These include the inactivation of sarcoplasmic reticulum Ca^{2+} ATPase,[70] inhibition of endothelial prostacyclin synthase,[71] inhibition of NOS activity,[72] oxidation of lipids and thiols, stimulation of iron release and nitration of protein tyrosine residues.[55] The mechanism of this peroxynitrite-mediated toxicity may depend on a direct action of peroxynitrite on these cellular constituents, thereby altering their functional properties.[73] Indeed, the stable product nitrotyrosine, formed by the nitrosylation of free or protein-associated tyrosine by peroxynitrite, inhibits the haemodynamic responses of VSMC to adrenoceptor agonists[74] and to angiotensin II.[75] It is possible therefore that nitrotyrosine formation may also be involved in the pathogenesis of the haemodynamic disturbances associated with inflammatory conditions where peroxynitrite formation is favoured, such as ischaemia-reperfusion injury, atherosclerosis and sepsis (see also Chapters 17–19, and 22). Using antibodies to nitrotyrosine, tissue sections may be stained immunohistochemically for evidence of peroxynitrite-mediated damage (see Chapter 8).

Alternatively, the peroxynitrite-mediated toxicity may be a result of the heterolytic or homolytic cleavage of peroxynitrite and the subsequent generation of secondary oxidants including nitronium ions and an intermediate ion with hydroxyl radical character.[55] Whatever mechanisms are responsible for these peroxynitrite-mediated effects, they will cause serious disruption of important signalling pathways, leading to the onset of pathological changes and possibly cell death.

Apoptosis

Necrotic cell death is a pathological pathway that leads to uncontrolled cell loss and tissue damage. Apoptosis, on the other hand, is a physiological suicide pathway to maintain tissue homeostasis, with the caveat that a massive induction of this process, under certain conditions, could lead to tissue destruction. The cells of the cardiovascular system exist under a state of oxidative siege, in which survival requires an appropriate balance of oxidants and antioxidants. An imbalance in this sensitive equilibrium may lead to the induction of oxidative stress as a common mediator of apoptosis.

Apoptosis of VSMC has been shown to occur in coronary atherosclerotic plaques[76] and is thought to contribute significantly to the disease process. Although the exact mechanism of this induction of programmed cell death in VSMC is unknown, it may involve the expression of specific gene products and locally generated cytokines.[76] The stimulated expression of iNOS has been demonstrated at the sites of atherosclerotic lesions,[63] and there are now several reports suggesting a role for NO in the induction of apoptosis.[77–79] The exact mechanism of this NO-mediated effect is not clear, but it would appear that sustained exposure or low concentrations of NO is responsible for the induction of apoptosis, whereas sudden exposure to high concentrations of NO results in cell necrosis.[67]

It has been suggested that NO induces apoptosis in VSMC *via* a cGMP dependent mechanism, but this remains undefined.[78] Cyclic GMP-independent mechanisms have also been described, which involve direct effects of NO on the Na^+-H^+ exchange system[79] and upregulation of the type 1 membrane protein, Fas.[77] Fas belongs to the TNF family, and is involved in the mediation of the cell death signal.[80] Overexpression of Fas may therefore induce massive apoptosis, thus leading to tissue destruction such as plaque rupture.

Expression of the proto-oncogene bcl-2 in VSMC is also linked to the control of apoptosis.[76] It is possible that bcl-2 acts as an anti-apoptotic gene by shifting NAD^+ to NADH, thereby preventing the formation of cyclic ADP ribose.[81] As already described above, peroxynitrite is associated with cell toxicity, DNA strand breakage and the synthesis of cyclic ADP ribose. In some cell types, peroxynitrite has been shown to induce apoptosis in a time- and concentration-dependent manner.[82] It is therefore possible that the formation of peroxynitrite also plays a role in the NO-mediated induction of apoptosis in VSMC.

S-Nitrosothiols

The stability of NO in the physiological environment depends on the pH, the presence of reactive oxygen-derived free radicals (ODFR) and the availability of thiol groups. By reducing the amount of ODFR, thiols can have a stabilising effect on NO. Furthermore, and perhaps more importantly, thiols react with NO (*via* the intermediary products NO^+ and N_2O_3) to form thionitrites or *S*-nitrosothiols (RSNOs).[83] In blood and tissues, such as the lungs, NO is found predominantly as *S*-nitrosothiol adducts of proteins and peptides, such as *S*-nitrosoglutathione and *S*-nitrosocysteine.[84,85] The most common adduct in the plasma is thought to be *S*-nitroso serum albumin,[86] though a recent report suggests a major role for *S*-nitrosohaemoglobin in the transduction of NO-related activities.[87] Therefore, the formation of these RSNOs may serve both to stabilize NO during delivery to its target, and to palliate the toxicity of NO at its site of production.

Naturally occurring RSNOs have been discovered *in vivo,* where they have been shown to activate sGC and, as a result, cause relaxation of VSMC as described above.[88] RSNOs have been shown to release free NO,[89] though interestingly this is not a prerequisite for the demonstration of RSNO-mediated functional effects.[90] Questions still remain, therefore, as to both the availability and the mechanism by which the extracellular RSNO pool leads to the activation of sGC.

It is possible that RSNO activation of sGC may occur through a signal transduction event on the cell surface, *via* thiol-*S*-nitrosothiol exchange that may occur across the cell membrane. Alternatively, the intracellular formation of *S*-nitroso proteins as a mechanism of post-translational modification may itself serve as a signal transduction event.[84] The thiol/*S*-nitrosothiol exchange reaction has been demonstrated *in vivo*,[91] where low molecular weight *S*-nitrosothiols formed from *S*-nitroso serum albumin are transported across the cell membrane of VSMC to their site of action, or exchanged at the cell membrane itself. The latter process is not diffusion-limited since it occurs despite the presence of a significant NO gradient across VSMC membranes.[91]

Nitric Oxide and the Inhibition of Vascular Smooth Muscle Cell Growth

Cyclic Nucleotide-dependent Effects

As documented elsewhere in this book, NO is a multifunctional molecule that mediates a wide variety of physiological processes.[4] One such role of endothelial cell-derived NO is the maintenance of mitogenic quiescence of the underlying medial VSMC. Increases in cGMP in response to NO generation or in response to cGMP analogues[92–94] have been associated with an inhibition of VSMC proliferation. Furthermore, NO has also been shown to inhibit total protein and collagen synthesis[95] as well as migration[96] of cultured VSMC. The mechanisms underlying these anti-proliferative and anti-migratory effects are still largely unknown, however.

Before the anti-proliferative effects of cGMP were demonstrated, many workers had shown that cyclic adenosine 3',5' monophosphate (cAMP) could inhibit VSMC growth. In studies using cyclic nucleotide analogues, it was clearly shown that cAMP analogues were more potent inhibitors of the growth of VSMC than were cGMP analogues.[97] A possible explanation for this observation is that prolonged activation of sGC results in a 'cross-over' effect, whereby cGMP can also activate cAMP-dependent processes.[98] A long period of NO exposure may result in increases of cGMP concentrations to levels high enough to activate cAMP-dependent processes. Activation of cAMP-dependent kinase (PKA) by cGMP was shown in one study to be responsible for the inhibition of VSMC growth *in vitro*.[98]

c-myc

The growth arrest of VSMC by cyclic nucleotides may involve the down-regulation of genes required to maintain proliferation. Among such genes, the oncoprotein *c-myc* seems to be important. The *c-myc* protein and its mRNA are induced after mitogenic stimulation of VSMC[99] *in vitro* and also following arterial injury after balloon catheter.[100] Furthermore, the continuous expression of *c-myc* in proliferating VSMC suggests that *c-myc* may be important in the maintenance as well as the initiation of VSMC proliferation. It has now been demonstrated that prolonged exposure of VSMC to cGMP and cAMP analogues causes a downregulation of *c-myc* protein and mRNA.[101] Thus, the cGMP-mediated downregulation of *c-myc* provides a possible mechanism for the antiproliferative effect of NO observed *in vitro*,[102] and following vascular injury by balloon angioplasty *in vivo*.[94]

Nuclear Factor Kappa B

The involvement of cytokines in atherogenesis is well established, though the exact mechanism of VSMC activation and proliferation in response to these agents remains unclear. The oxidant-sensitive transcription factor, nuclear factor kappa B (NFκB), is a basal constituent in the cytoplasm of VSMC[103] and is thought to play an important role in VSMC proliferation. It has now been shown, for example, that cytokines (e.g. IL-1 or TNF) can activate the production and nuclear entry of NFκB, leading to VSMC activation and proliferation, through binding to DNA control elements and thereby inducing the synthesis of mRNA.[104]

A recent report has shown that NO can inhibit cytokine-induced expression of proinflammatory genes, and also the intercellular and vascular cell adhesion molecules ICAM-1 and VCAM-1.[105] The mechanism for this NO-mediated effect is cGMP-independent, occurs at the transcriptional level, and involves the inhibition of both basal constitutive and cytokine-induced NFκB activity.[105] Clearly, further studies are needed to elucidate the relationship between NO and NFκB, but these early reports provide some insight into the role of NO in the inflammatory response associated with atherosclerosis.

Summary

NO exerts many biochemical effects that influence VSMC function. Some of these can clearly be beneficial, like the reduction of blood vessel tone and the inhibition of VSMC growth/proliferation. Other effects are without doubt harmful, as evidenced by the tissue damage associated with excess NO production following iNOS induction with subsequent formation of peroxynitrite. Though the mechanisms of many of these effects are cGMP-dependent, more recent studies show that NO can have direct intracellular effects, extending down to the transcriptional level. For such a simple molecule, the effects of NO on the cardiovascular system are both profound and far-reaching.

References

1. Furchgott RF, Zawadzki JV. The obligatory role of endothelial cells in the relaxation of arterial smooth muscle by acetylcholine. *Nature* 1980; **288**: 373–376.
2. Palmer RMJ, Ferrige AG, Moncada S. Nitric oxide release accounts for the biological activity of endothelium-derived relaxing factor. *Nature* 1987; **327**: 524–525.
3. Moncada S, Palmer RMJ, Higgs EA. Biosynthesis of nitric oxide from L-arginine: a pathway for the regulation of cell function and communication. *Biochem Pharmacol* 1989; **38**: 1709–1715.
4. Moncada S, Palmer RMJ, Higgs EA. Nitric oxide: physiology, pathophysiology and pharmacology. *Pharmacol Rev* 1991; **43**: 109–142.
5. Katsuki S, Arnold N, Mittal C, Murad F. Stimulation of guanylate cyclase by sodium nitroprusside, nitroglycerin and nitric oxide in various tissue preparations and comparison to the effects of sodium azide and hydroxylamine. *J Cyclic Nucleotide Res* 1977; **3**: 23–35.
6. Moncada S, Higgs EA. The L-arginine: nitric oxide pathway. *N Engl J Med* 1993; **329**: 2002–2012.
7. Liu ZG, McLaughlin BE, Marks GS *et al*. Sulfhydryl involvement in nitric oxide sequestration and nitric oxide induced guanylyl cyclase activation in vascular smooth muscle. *Can J Physiol Pharmacol* 1995; **73**: 1144–1148.
8. Mittal CK. Oxygen-radical/nitric oxide mediate calcium-dependent hormone action on cyclic GMP system: A novel concept in signal transduction mechanisms. *Mol Cell Biochem* 1995; **149/150**: 257–262.
9. Severina IS, Bussygina OG, Grigorjev NB. Effect of nitroso complexes of some transition metals on the activity of soluble guanylate cyclase. *Biochem Int* 1992; **26**: 695–705.
10. Beckman JS, Koppenol WH. Nitric oxide, superoxide, and peroxynitrite: the good, the bad and the ugly. *Am J Physiol* 1996; **271**: C1424–C1437.
11. Beckman JS, Beckman TW, Chen J *et al*. Apparent hydroxyl radical production by peroxynitrite: implications for endothelial injury from nitric oxide and superoxide. *Proc Natl Acad Sci USA* 1990; **87**: 1620–1624.
12. Moro MA, Darley-Usmar VM, Lizasoain I *et al*. The formation of nitric oxide donors from peroxynitrite. *Br J Pharmacol* 1995; **116**: 1999–2004.
13. Tarpey MM, Beckman JS, Ischiropoulos H *et al*. Peroxynitrite stimulates vascular smooth muscle cell cyclic GMP synthesis. *FEBS Lett* 1995; **364**: 314–318.
14. Mayer B, Schrammel A, Klatt P *et al*. Peroxynitrite-induced accumulation of cyclic GMP in endothelial cells and stimulation of purified soluble guanylyl cyclase. Dependence on glutathione and possible role of S-nitrosation. *J Biol Chem* 1995; **270**: 17355–17360.

15. Collins P, Griffith TM, Henderson AH, Lewis MJ. Endothelium-derived relaxing factor alters calcium fluxes in rabbit aorta: a cyclic guanosine monophosphate-mediated effect. *J Physiol (Lond)* 1986; **381**: 427–437.
16. Rashatwar SS, Cornwell TL, Lincoln TM. Effects of 8-bromo-cGMP on Ca^{2+} levels in vascular smooth muscle cells: possible regulation of Ca^{2+}-ATPase by cGMP-dependent protein kinase. *Proc Natl Acad Sci USA* 1987; **84**: 5686–5689.
17. Cornwell TL, Lincoln TM. Regulation of intracellular Ca^{2+} levels in cultured vascular smooth muscle cells: Reduction of Ca^{2+} by atriopeptin and 8-bromo-cyclic GMP is mediated by cyclic GMP-dependent protein kinase. *J Biol Chem* 1989; **264**: 1146–1155.
18. Collins P, Henderson AH, Lang D, Lewis MJ. Endothelium-derived relaxing factor and nitroprusside compared in noradrenaline- and K^+-contracted rabbit and rat aortae. *J Physiol (Lond)* 1988; **400**: 395–404.
19. Murad F. Cyclic guanosine monophosphate as a mediator of vasodilation. *J Clin Invest* 1986; **78**: 1–5.
20. Hashimoto T, Hirata M, Itoh T *et al.* Inositol 1,4,5-trisphosphate activates pharmacomechanical coupling in smooth muscle of the rabbit mesenteric artery. *J Physiol (Lond)* 1986; **370**: 605–618.
21. Haller H, Smallwood JI, Rasmussen H. Protein kinase C translocation in intact vascular smooth muscle strips. *Biochem J* 1990; **270**: 375–381.
22. Lang D, Lewis MJ. Endothelium-derived relaxing factor inhibits the formation of inositol trisphosphate by rabbit aorta. *J Physiol (Lond)* 1989; **411**: 45–52.
23. Lang D, Lewis MJ. Endothelium-derived relaxing factor inhibits the endothelin-1-induced increase in protein kinase C activity in rat aorta. *Br J Pharmacol* 1991; **104**: 139–144.
24. Hirata M, Kohse KP, Chang C-H *et al.* Mechanism of cyclic GMP inhibition of inositol phosphate formation in rat aorta segments and cultured bovine aortic smooth muscle cells. *J Biol Chem* 1990; **265**: 1268–1273.
25. Pfeifer A, Nurnberg B, Kamm M *et al.* Cyclic GMP-dependent protein kinase blocks pertussis toxin-sensitive hormone receptor signalling pathways in Chinese hamster ovary cells. *J Biol Chem* 1995; **270**: 9052–9059.
26. Lincoln TM. Pertussis toxin-sensitive and insensitive guanine nucleotide binding proteins (G-proteins) are not phosphorylated by cGMP-dependent protein kinase. *Second Messengers Phosphoproteins* 1991; **13**: 99–109.
27. Gopalakrishna R, Chen ZH, Gundimenda U. Nitric oxide and nitric oxide-generating agents induce a reversible inactivation of protein kinase C activity and phorbol ester binding. *J Biol Chem* 1993; **268**: 27180–27185.
28. Blatter LA, Wier WG. Nitric oxide decreases $[Ca^{2+}]_i$ in vascular smooth muscle by inhibition of the calcium current. *Cell Calcium* 1994; **15**: 122–131.

29. Xiong Z, Sperelakis N, Fenoglio-Preiser C. Regulation of L-type calcium channels by cyclic nucleotides and phosphorylation in smooth muscle cells from rabbit portal vein. *J Vasc Res* 1994; **31**: 271–279.
30. Vrolix M, Raeymaekers L, Wuytack F *et al.* Cyclic GMP-dependent protein kinase stimulates the plasmalemmal Ca^{2+} pump of smooth muscle *via* phosphorylation of phosphatidylinositol. *Biochem J* 1988; **255**: 855–863.
31. Furakawa K-I, Ohshima N, Tawada-Iwata Y, Shigekawa M. Cyclic GMP stimulates Na^+/Ca^{2+} exchange in vascular smooth muscle cells in primary culture. *J Biol Chem* 1991; **266**: 12337–12341.
32. Lincoln TM, Cornwell TL. Towards an understanding of the mechanism of action of cyclic AMP and cyclic GMP in smooth muscle relaxation. *Blood Vessels* 1991; **28**: 129–137.
33. Lincoln TM, Cornwell TL, Komalavilas P, Boerth N. Cyclic GMP-dependent protein kinase in nitric oxide signalling. *Methods Enzymol* 1996; **269**: 149–166.
34. Popp R, Bauersachs J, Hecker M *et al.* A transferable, β-naphthoflavone-inducible, hyperpolarizing factor is synthesized by native and cultured porcine coronary endothelial cells. *J Physiol* (*Lond*) 1996; **497**: 699–709.
35. Parkington HC, Tare M, Tonta MA, Coleman HA. Stretch revealed three components in the hyperpolarization of guinea-pig coronary artery in response to acetylcholine. *J Physiol* (*Lond*) 1993; **465**: 459–476.
36. Kauser K, Rubanyi GM. Endothelium-dependent hyperpolarisation of vascular smooth muscle. In: *Endothelial Regulation of Vascular Tone*, Ryan US, Rubanyi GM (Eds). Marcel Decker, New York 1992: 341–354.
37. Gupta S, McArthur C, Grady C, Ruderman NB. Stimulation of vascular Na^+-K^+-ATPase activity by nitric oxide: a cGMP-independent effect. *Am J Physiol* 1994; **266**: H2146–H2151.
38. Sato T, Kamata Y, Irifune M, Nishikawa T. Inhibitory effect of several nitric oxide-generating compounds on purified Na^+,K^+-ATPase activity from porcine cerebral cortex. *J Neurochem* 1997; **68**: 1312–1318.
39. Bolotina VM, Najibi S, Palacino JJ *et al.* Nitric oxide directly activates calcium-dependent potassium channels in vascular smooth muscle. *Nature* 1994; **368**: 850–853.
40. Peng W, Hoidal JR, Farrukh I. Regulation of Ca^{2+}-activated K^+ channels in pulmonary vascular smooth muscle cells: role of nitric oxide. *J Appl Physiol* 1996; **81**: 1264–1272.
41. Tare M, Parkington HC, Neild TO, Dusting GJ. Hyperpolarisation and relaxation of arterial smooth muscle caused by nitric oxide derived from the endothelium. *Nature* 1990; **346**: 69–71.
42. Garland CJ, McPherson GA. Evidence that nitric oxide does not modulate the hyperpolarisation and relaxation to acetylcholine in the rat small mesenteric artery. *Br J Pharmacol* 1992; **105**: 429–435.

43. Plane F, Garland CJ. Differential effects of acetylcholine, nitric oxide and levcromakalim on smooth muscle membrane potential and tone in the rabbit basilar artery. *Br J Pharmacol* 1993; **110**: 651–656.
44. Cohen RA, Vanhoutte PM. Endothelium-dependent hyperpolarisation. Beyond nitric oxide and cyclic GMP. *Circulation* 1995; **92**: 3337–3349.
45. Szabó C. Alterations in nitric oxide production in various forms of circulatory shock. *New Horizons* 1995; **3**: 2–32.
46. Drapier JC, Hibbs JB. Murine cytotoxic activated macrophages inhibit aconitase in tumour cells. Inhibition involves the iron-sulfur prosthetic group and is reversible. *J Clin Invest* 1986; **78**: 790–797.
47. Wharton M, Granger DL, Durck DT. Mitochondrial iron loss from leukemia cells injured by macrophages. A possible mechanism for electron transport chain defects. *J Immunol* 1988; **141**: 1311–1317.
48. Stadler J, Curran RD, Ochoa JB *et al.* Effect of endogenous nitric oxide on mitochondrial respiration of rat hepatocytes in vitro and in vivo. *Arch Surg* 1991; **126**: 186–191.
49. Schweizer M, Richter C. Nitric oxide potently and reversibly deenergises mitochondria at low oxygen tension. *Biochem Biophys Res Commun* 1994; **204**: 169–175.
50. Cleeter MWJ, Cooper JM, Darley-Usmar VM *et al.* Reversible inhibition of cytochrome c oxidase the terminal enzyme in the mitochondrial respiratory chain by nitric oxide. *FEBS Lett* 1994; **345**: 50–54.
51. Torres J, Darley-Usmar V, Wilson MT. Inhibition of cytochrome c oxidase in turnover by nitric oxide: mechanism and implications for control of respiration. *Biochem J* 1995; **312**: 169–173.
52. Borutaite V, Brown GC. Rapid reduction of nitric oxide by mitochondria, and reversible inhibition of mitochondrial respiration by nitric oxide. *Biochem J* 1996; **315**: 295–299.
53. Bates TE, Loesch A, Burnstock G, Clark JB. Mitochondrial nitric oxide synthase: a ubiquitous regulator of oxidative phosphorylation. *Biochem Biophys Res Commun* 1996; **218**: 40–44.
54. Halliwell B. Reactive oxygen species and the central nervous system. *J Neurochem* 1992; **59**: 1609–1623.
55. Pryor W, Squadrito GL. The chemistry of peroxynitrite: a product from the reaction of nitric oxide with superoxide. *Am J Physiol* 1995; **268**: L699–L722.
56. Hausladen A, Fridovich I. Superoxide and peroxynitrite inactivate aconitases, but nitric oxide does not. *J Biol Chem* 1994; **269**: 29405–29408.
57. Radi R, Rodriguez M, Castro L, Telleri R. Inhibition of mitochondrial electron transport by peroxynitrite. *Arch Biochem Biophys* 1994; **308**: 89–95.
58. Cassina A, Radi R. Differential inhibitory action of nitric oxide and peroxynitrite on mitochondrial electron transport. *Arch Biochem Biophys* 1996; **328**: 309–316.

59. Packer MA, Murphy MP. Peroxynitrite causes calcium efflux from mitochondria which is prevented by cyclosporin A. *FEBS Lett* 1994; **345**: 237–240.
60. Ischiropoulos H, Zhu L, Beckman J. Peroxynitrite formation from macrophage-derived nitric oxide. *Arch Biochem Biophys* 1992; **298**: 446–451.
61. Halliwell B. Oxygen radicals, nitric oxide and human inflammatory joint disease. *Ann Rheumat Dis* 1995; **54**: 505–510.
62. Al-Mufti RA, Williamson RCN, Mathie RT. Increased nitric oxide activity in a rat model of acute pancreatitis. *Gut* 1998; **43**: 564–570.
63. Buttery LDK, Springall DR, Chester AH *et al.* Inducible nitric oxide synthase is present within human atherosclerotic lesions and promotes the formation and activity of peroxynitrite. *Lab Invest* 1996; **75**: 77–85.
64. Wizemann T, Gardner C, Laskin J *et al.* Production of nitric oxide and peroxynitrite in the lung during acute endotoxemia. *J Leukoc Biol* 1994; **56**: 759–768.
65. Szabó C. The role of peroxynitrite in the pathophysiology of shock, inflammation and ischemia-reperfusion injury. *Shock* 1996; **6**: 79–88.
66. Crow JP, Beckman JS. The role of peroxynitrite in nitric oxide-mediated toxicity. *Current Topic Microbiol Immunol* 1995; **196**: 57–73.
67. Szabó C. DNA strand breakage and activation of poly-ADP ribosyltransferase: a cytotoxic pathway triggered by peroxynitrite. *Free Radic Biol Med* 1996; **21**: 855–869.
68. Szabó C, Zingarelli B, O'Connor M, Salzman AL. DNA strand breakage, activation of poly-ADP ribosyl synthetase, and cellular energy depletion are involved in the cytotoxicity in macrophages and smooth muscle cells exposed to peroxynitrite. *Proc Natl Acad Sci USA* 1996; **93**: 1753–1758.
69. Pellat-Seceunyk K, Wietzerbin J, Drapier JC. Nicotinamide inhibits nitric oxide synthase mRNA induction in activated macrophages. *Biochem J* 1994; **297**: 53–58.
70. Viner RI, Huhmer AFR, Bigelow DJ, Schoneich C. The oxidative inactivation of sarcoplasmic reticulum Ca^{2+}-ATPase by peroxynitrite. *Free Radic Res* 1996; **24**: 243–259.
71. Zou MH, Ullrich V. Peroxynitrite formed by simultaneous generation of nitric oxide and superoxide selectively inhibits bovine aortic prostacyclin synthase. *FEBS Lett* 1996; **382**: 101–104.
72. Pasquet JPEE, Zou MH, Ullrich V. Peroxynitrite inhibition of nitric oxide synthases. *Biochimie* 1996; **78**: 785–791.
73. Moro MA, Darley-Usmar VM, Goodwin DA *et al.* Paradoxical fate and biological action of peroxynitrite on human platelets. *Proc Natl Acad Sci USA* 1994; **91**: 6702–6706.
74. Kooy NW, Lewis SJ. Nitrotyrosine attenuates the hemodynamic effects of adrenoceptor agonists in vivo: relevance to the pathophysiology of peroxynitrite. *Eur J Pharmacol* 1996; **310**: 155–161.

75. Kooy NW, Lewis SJ. The peroxynitrite product 3-nitro-L-tyrosine attenuates the hemodynamic responses to angiotensin II in vivo. *Eur J Pharmacol* 1996; **315**: 165–170.
76. Bennett MR, Evan GI, Schwartz SM. Apoptosis of human vascular smooth muscle derived from normal vessels and coronary atherosclerotic plaques. *J Clin Invest* 1995; **95**: 2266–2274.
77. Fukuo K, Hata S, Suhara T *et al.* Nitric oxide induces upregulation of fas and apoptosis in vascular smooth muscle. *Hypertension* 1996; **27**: 823–826.
78. Pollman MJ, Yamada T, Horiuchi M, Gibbons GH. Vasoactive substances regulate vascular smooth muscle cell apoptosis. Countervailing influences of nitric oxide and angiotensin II. *Circ Res* 1996; **79**: 748–756.
79. Nishio E, Fukushima K, Shiozaki M, Watanabe Y. Nitric oxide donor SNAP induces apoptosis in smooth muscle cells through cGMP-independent mechanism. *Biochem Biophys Res Commun* 1996; **221**: 163–168.
80. Nagata S, Golstein P. The Fas death factor. *Science* 1995; **267**: 1449–1456.
81. Richter C, Schweizer M, Cossarizza A, Franceschi C. Control of apoptosis by the cellular ATP level. *FEBS Lett* 1996; **378**: 107–110.
82. Lin KT, Xue JY, Nomen M *et al.* Peroxynitrite-induced apoptosis in HL-60 cells. *J Biol Chem* 1995; **270**: 16487–16490.
83. Ignarro LJ, Gruetter CA. Requirement of thiols for activation of coronary arterial guanylate cyclase by glyceryl trinitrate and sodium nitrite: Possible involvement of S-nitrosothiols. *Biochim Biophys Acta* 1980; **631**: 221–231.
84. Stamler JS, Simon DI, Osborne JA *et al.* S-nitrosylation of proteins with nitric oxide: synthesis and characterisation of biologically active compounds. *Proc Natl Acad Sci USA* 1992; **89**: 444–448.
85. Gaston B, Reilly J, Drazen JM *et al.* Endogenous nitrogen oxides and bronchodilator S-nitrosothiols in human airways. *Proc Natl Acad Sci USA* 1993; **90**: 10957–10961.
86. Stamler JS, Jaraki O, Osborne JA *et al.* Nitric oxide circulates in mammalian plasma primarily as an S-nitroso adduct of serum albumin. *Proc Natl Acad Sci USA* 1992; **89**: 7674–7677.
87. Li J, Bonaventura C, Bonaventura J, Stamler JS. S-nitrosohaemoglobin: a dynamic activity of blood involved in vascular control. *Nature* 1996; **380**: 221–226.
88. Ignarro LJ, Kadowitz PJ, Baricos WH. Evidence that regulation of hepatic guanylate cyclase activity involves interactions between catalytic site SH groups and both substrate and activator. *Arch Biochem Biophys* 1981; **208**: 75–86.
89. Ignarro LJ, Lipton H, Edwards JC *et al.* Mechanism of vascular smooth muscle relaxation by organic nitrates, nitrites, nitroprusside and nitric oxide: evidence for the involvement of S-nitrosothiols as active intermediates. *J Pharmacol Exp Ther* 1981; **218**: 739–749.

90. Ignarro LJ. Biosynthesis and metabolism of endothelium-derived nitric oxide. *Ann Rev Pharmacol Toxicol* 1990; **30**: 535–560.
91. Scharfstein JS, Keaney J, Slivka A *et al.* In vivo transfer of nitric oxide between a plasma protein-bound reservoir and low molecular weight thiols. *J Clin Invest* 1994; **94**: 1432–1439.
92. Garg UC, Hassid A. Nitric oxide-generating vasodilators and 8-bromo-cyclic guanosine monophosphate inhibit mitogenesis and proliferation of cultured rat vascular smooth muscle cells. *J Clin Invest* 1989; **83**: 1774–1777.
93. Assender JW, Southgate KM, Newby AC. Does nitric oxide inhibit smooth muscle proliferation? *J Cardiovasc Pharmacol* 1991; **17 (Suppl 3)**: S104–S107.
94. Groves PH, Banning AP, Penny WJ *et al.* The effects of exogenous nitric oxide on smooth muscle cell proliferation following porcine carotid angioplasty. *Cardiovasc Res* 1995; **30**: 87–96.
95. Kolpakov V, Gordon D, Kulik TJ. Nitric oxide-generating compounds inhibit total protein and collagen synthesis in cultured vascular smooth muscle cells. *Circ Res* 1995; **76**: 305–309.
96. Sarkar R, Meinberg EG, Stanley JC *et al.* Nitric oxide reversibly inhibits the migration of cultured vascular smooth muscle cells. *Circ Res* 1996; **78**: 225–230.
97. Southgate KM, Newby AC. Serum-induced proliferation of rabbit aortic smooth muscle cells from the contractile state is inhibited by 8-Br-cAMP but not 8-Br-cGMP. *Atherosclerosis* 1990; **82**: 113–123.
98. Cornwell TL, Arnold E, Boerth NJ, Lincoln TM. Inhibition of smooth muscle cell growth by nitric oxide and activation of cAMP-dependent protein kinase by cGMP. *Am J Physiol* 1994; **267**: C1405–C1413.
99. Kindy MS, Sonenshein GE. Regulation of oncogene expression in cultured aortic smooth muscle cells: post-transcriptional control of *c-myc* mRNA. *J Biol Chem* 1986; **261**: 12865–12868.
100. Bauters C, de Groote P, Adamantidis M *et al.* Proto-oncogene expression in rabbit aorta after wall injury: first marker of the cellular process leading to restenosis after angioplasty. *Eur Heart J* 1992; **13**: 556–559.
101. Bennett MR, Littlewood TD, Hancock DC *et al.* Down-regulation of the *c-myc* proto-oncogene in inhibition of vascular smooth-muscle cell proliferation: a signal for growth arrest? *Biochem J* 1994; **302**: 701–708.
102. Mooradian DL, Hutsell TC, Keefer LK. Nitric oxide (NO) donor molecules: effect of NO release rate on vascular smooth muscle cell proliferation in vitro. *J Cardiovasc Pharmacol* 1995; **25**: 674–678.
103. Bellas RE, Lee JS, Sonenshein GE. Expression of a constitutive NF-κB-like activity is essential for proliferation of cultured bovine vascular smooth muscle. *J Clin Invest* 1995; **96**: 2521–2527.

104. Schreck R, Albermann K, Baeuerle PA. Nuclear factor kappa B: an oxidative stress-responsive transcription factor of eukaryotic cells. *Free Radic Res Commun* 1992; **17**: 221–237.
105. Shin WS, Hong YH, Peng HB *et al.* Nitric oxide attenuates vascular smooth muscle cell activation by interferon-γ. The role of constitutive NF-κB activity. *J Biol Chem* 1996; **271**: 11317–11324.

Chapter 2

Shear Stress and Nitric Oxide Release: Physiological Integration of Cellular Mechanisms, Physical Forces and Flow Regulation

Tudor M Griffith

Introduction

The discovery of endothelium-dependent relaxation, first described in detail by Furchgott and Zawadzki in 1980 as a "paradoxical" response to acetylcholine (ACh), has led to the characterization of an endogenous relaxing factor as a diffusible nitrovasodilator now thought to be the NO radical.[1–4] NO release is also stimulated by longitudinal shear forces acting on the endothelium and is therefore modulated by flow velocity and perfusate viscosity.[5–10] It thereby contributes to the locally-mediated flow-dependent dilatation that occurs in conduit arteries, resistance arteries, arterioles and venules following increases in blood flow,[10–15] a phenomenon that permits the coupling of changes in resistance in different parts of the vascular bed through the integrating signal of shear stress. Longitudinal gradients in control mechanisms may thus be coordinated by the endothelium through sequential interactions in space and time, with the smallest arterioles being most responsive to metabolic factors and myogenic adaptation to changes in pressure being most pronounced in arterioles of intermediate size.[16]

Mechanical Forces and Nitric Oxide Release

Acute Regulation

In vivo, the mechanotransduction pathway stimulated by flow is activated within seconds and dilatation maximal within ~ 2 minutes.[17–21] Cascade bioassay experiments with isolated vessels have shown that NO release reaches a plateau when the time-averaged shear stress in the donor is ~ 10 dynes/cm^2, which compares with physiological values of 5–20 dynes/cm^2 in small vessels and ~ 30 dynes/cm^2 or greater in conduit arteries.[9,22] NO release is also sensitive to flow pulsatility, being maximal at pulse frequencies between 4 and 6 Hz in arteries from a variety of species (Figure 1).[6,8,9,23] Analogously, NO release is stimulated by the periodic compression imposed by contraction in the isolated rabbit heart,[24] and flow-dependent epicardial coronary vasodilatation to changes in pulse frequency are NO-dependent in conscious dogs.[25] These findings suggest that changes in heart rate could directly modulate peripheral resistance *via* effects on the endothelium.

In isolated arterioles the magnitude of flow-dependent dilatation is maximal at transmural pressures equivalent to those found under physiological conditions when myogenic tone is at an intermediate level,[13] but is blunted at high pressures in the femoral artery of the dog *in vivo*.[17] This could at least in part be explained by observations that acute increases in transmural pressure attenuate basal and agonist-evoked NO in cascade bioassay, and that NO release from conduit arteries subjected to pulsatile flow at a constant pulse frequency varies inversely with the amplitude of the pressure pulse.[8,26] Agonist-stimulated NO release from cultured endothelial cells grown on a flat surface (assessed by the accumulation of

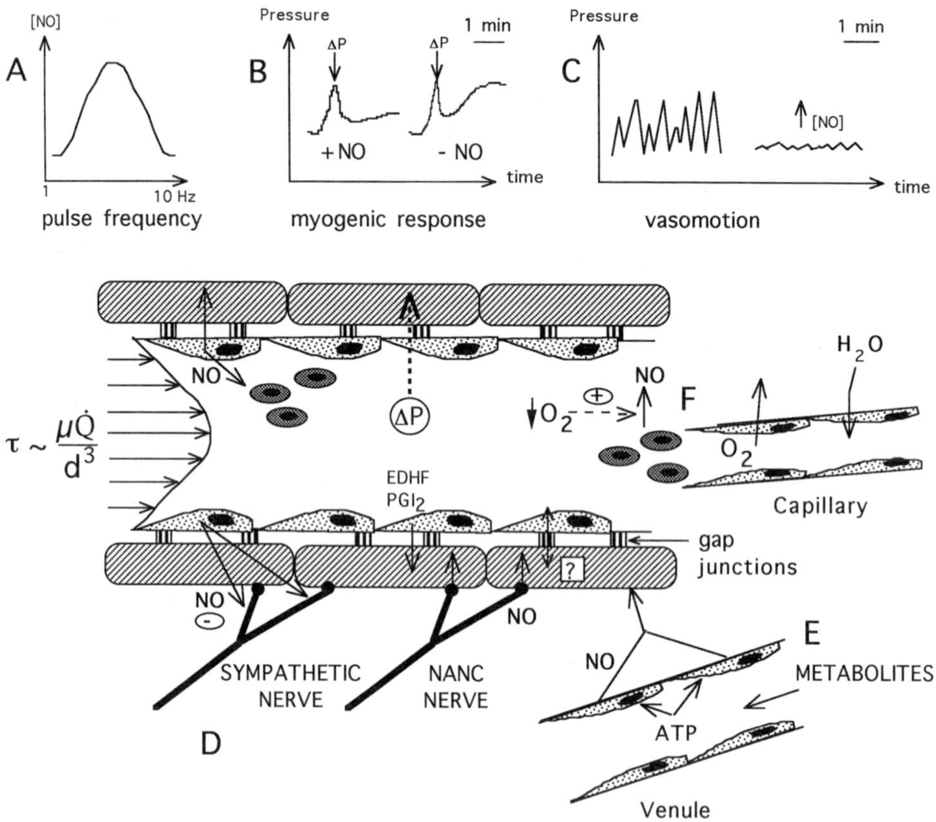

Figure 1. Mechanisms involved in the regulation of arterial tone by NO. The principal haemodynamic stimulus to the endothelium is time-averaged shear stress (τ), which is dependent on perfusate viscosity (μ), flow rate ($\dot{Q}$) and vessel diameter (d) in the case of laminar Poiseuille flow. **A.** The endothelium is also sensitive to the pulsatility of flow with NO release maximal at pulse frequencies ~ 5 Hz. **B.** In isolated vessels acute increases in intravascular pressure promote a slowly-developing myogenic constrictor response that is normally attenuated by NO. **C.** Spontaneous oscillations in vascular calibre (vasomotion) are damped by NO. **D.** Endothelium-derived NO inhibits catecholamine release from sympathetic constrictor nerves, and NO release from non-adrenergic non-cholinergic nerves may also contribute to vascular regulation. **E.** Apart from direct transduction of mechanical stimuli by NO, nucleotides and peptides released "upstream" may stimulate NO release from "downstream" venules which subsequently diffuses to adjacent arterioles and thereby mediates cross-talk between the inflow and outflow of blood. **F.** In an environment where O_2 levels are high, NO is bound and carried by haemoglobin within red cells but can be released distally when O_2 levels fall. The endothelium also influences vascular tone through the release of prostanoids and EDHF, and through direct communication with smooth muscle *via* gap junctions. Integration of complex control mechanisms must ultimately serve to optimize delivery of O_2 and nutrients and modulate tissue pressure and filtration.

NO_2^-/NO_3^-) is also susceptible to the effects of pressure, exhibiting a progressive but reversible ~ 6-fold decrease over the range 0–160 mmHg.[27] By contrast, NO release from cells grown on microcarrier beads appears insensitive to changes in ambient pressure.[28]

NO may also participate in the regulation of flow on subacute time scales that reflect complex iterative adjustments in microvascular flow and pressure and/or time delays in feedback loops that involve neurogenic and metabolic mechanisms of vascular regulation. Perfusion of a segment of rabbit central ear artery isolated between two ligatures thus increases to ~ 50% of its preocclusion control level over 90 minutes as a consequence of NO-dependent diversion of flow through pre-existing collateral pathways.[29] NO-dependent mechanisms similarly contribute to the opening of collateral circulations following occlusion of the canine femoral artery *in vivo*.[30] Sympathetically-mediated constriction will tend to increase endothelial shear stress, but the resulting enhancement of NO synthesis may exert an inhibitory prejunctional effect on catecholamine release and diminish the original constrictor stimulus (Figure 1).[7] Conversely, NO released by circulating agonists will amplify their dilator action by attenuating sympathetic tone. NO may also participate in a countercurrent mechanism that couples the local inflow and outflow of blood, as NO released from venular endothelium can dilate adjacent arterioles.[31] Release of endothelium-dependent agonists such as adenosine 5'-diphosphate (ADP), adenosine 5'-triphosphate (ATP) and substance P during ischaemia/hypoxia[32] could thereby help to match perfusion and metabolism through negative feedback (Figure 1).

Long Term Regulation

Expression of the constitutive endothelial NO synthase (eNOS) increases within hours of exposure of endothelial cells to fluid shear stress, thus explaining the enhanced NO-dependent responses to flow and agonists that follow brief (~ 1 week) exercise training in dogs.[33,34] This upregulation may extend to all beds in which exercise affects flow, as a 3-fold increase in eNOS expression is found in coronary microvessels and epicardial arteries (see also Chapter 13).[33] The endothelial cell is nevertheless sensitive to the precise nature of the mechanical forces imposed upon it, as steady and periodic laminar shear stresses increase eNOS expression and NO synthesis, whereas turbulent flow characterised by high-frequency fluctuations in amplitude and direction is without effect.[35] Cyclic strain at levels comparable to those generated by the pulsatility of blood flow *in vivo* also upregulates eNOS in the absence of flow,[36] so that longitudinal shear is not the only mechanical force regulating ambient NO production. Complex patterns of response to changes in haemodynamic force are therefore likely to contribute to the adaptation of vascular calibre that follows long-term increases or decreases in flow over days or weeks. This is a strictly endothelium-dependent phenomenon that ultimately returns shear stress to control levels,[37,38] although the associated changes in wall structure and composition are multifactorial in aetiology as the endothelium modulates cell proliferation and extracellular matrix production through both NO-dependent and NO-independent mechanisms.[38–40] Creation of an A-V fistula in the rabbit carotid circulation thus leads to increases in carotid artery diameter and remodelling of the media that normalize wall shear stress, but these adaptive changes are only partially attenuated by blockade of eNOS.[38]

As discussed in detail in other chapters of this volume, endothelium-dependent relaxation is impaired in many disease states. In atherosclerotic vessels, dysfunction initially occurs at arterial bifurcations and in the proximal aorta, but extends to the microcirculation and cannot therefore be attributed simply to intimal thickening.[41–43] In hypercholesterolaemic rabbits, flow-induced NO release is attenuated before the response to agonists.[44] Endothelial dysfunction is readily reversed by reductions in plasma cholesterol[45] and exogenous L-arginine[46] (see also Chapters 11, 18 and 19). Cause and effect may be closely interlinked, however, as atheromatous plaques develop preferentially at sites of low shear stress and flow separation, and high levels of flow prevent or reverse smooth muscle proliferation and neointimal thickening.[47,48] This suggests a relationship with the ambient level of NO synthesis, although the cellular mechanisms involved are undoubtedly complex and reflect the balance between NO and other mediators. Expression of the smooth muscle mitogen endothelin-1, for example, is downregulated by shear stress and its release directly inhibited by NO.[35,49] Flow-dependent and agonist-induced dilatation are both impaired in the spontaneously hypertensive rat (SHR), although paradoxically, NO release may still be normal in cascade bioassay (see Chapter 21).[50,51] This may reflect complex interactions with the superoxide anion ($O_2^{\cdot -}$). Hypertension induced by aortic banding causes a ~ 2 fold increase in endothelial $O_2^{\cdot -}$ production after 2 weeks followed by a ~ 4 fold increase in eNOS expression after 6 weeks.[52] Endothelial dysfunction is evident, however, only at 6 weeks and associated with the appearance of nitrotyrosine proteins which are likely to reflect excess formation of the toxic peroxynitrite ion by the interaction of NO and $O_2^{\cdot -}$.[52] Superoxide radicals may similarly impair endothelium-dependent responses in hypercholesterolaemia.[53] In high output cardiac failure induced by creation of an aorto-caval fistula, NO activity is enhanced proximal to the fistula, suggesting that flow-related regulation of eNOS dominates over other mechanisms since endothelium-dependent relaxation is generally attenuated in low output states.[54]

Role of Ca^{2+} in the Nitric Oxide Response to Flow

Close temporal relationships between NO release and the Ca^{2+} transients that are evoked by shear stress, and observations that agonist-induced NO synthesis is decreased in the absence of extracellular Ca^{2+}, are likely to reflect allosteric

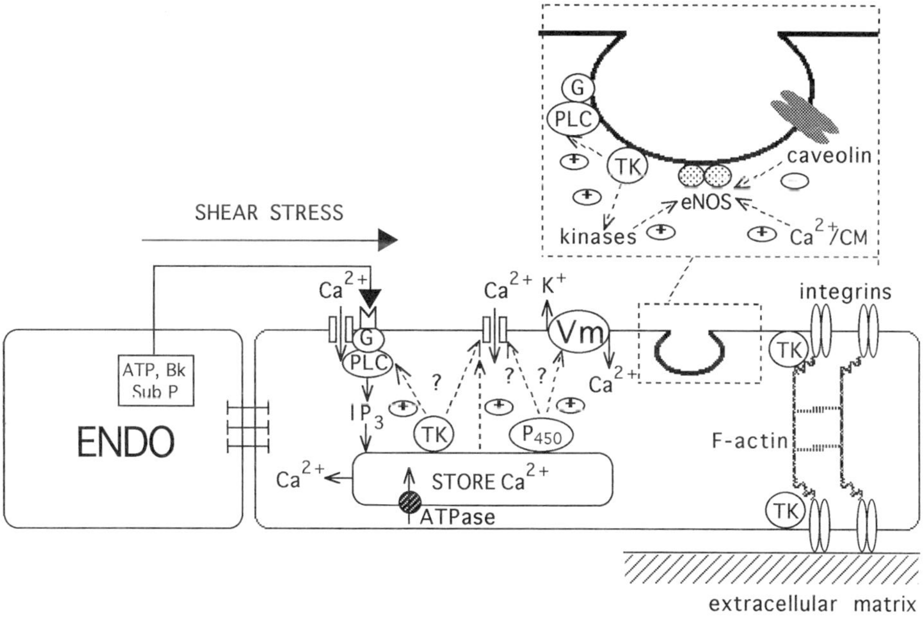

Figure 2. Cellular mechanisms involved in the activation of eNOS by shear forces which are sensed by the endothelial glycocalyx, integrins, G proteins and the cytoskeleton. Shear forces cause release of endothelium-dependent agonists such as ATP, bradykinin and substance P from stores which may then activate adjacent surface receptors. G protein/phospholipase C interactions that may involve tyrosine kinase (TK)-dependent phosphorylation result in IP_3-evoked Ca^{2+} release from internal stores. Release of Ca^{2+} from stores promotes influx of Ca^{2+} *via* a depletion-operated channel through mechanisms that may involve TK-dependent phosphorylation events and arachidonic acid metabolites of cytochrome P_{450} monooxygenase. Shear stress also stimulates a hyperpolarizing K^+ current, and the associated change in membrane potential (V_m) promotes Ca^{2+} entry because endothelial cells do not possess voltage-operated Ca^{2+} channels. Elevations in $[Ca^{2+}]_i$ activate the constitutive Ca^{2+}/calmodulin(CM)-dependent endothelial NO synthase (eNOS) which is preferentially located in membrane caveolae where it is negatively regulated by caveolin. TK-dependent phosphorylation also initiates a phosphorylation cascade that may modulate eNOS activity in a Ca^{2+}-independent fashion. The eNOS-caveolin complex can translocate within the cell resulting in changes in enzyme activity.

regulation of eNOS by the Ca^{2+}/calmodulin complex (Figure 2).[55,56] In cultured endothelial cells shear stress over the range 0.2–10 dyne/cm^2 elevates $[Ca^{2+}]_i$ within 1 to 3 sec of the initiation of laminar flow and this is followed by NO release after 2 to 3 sec.[55] Both responses attain a plateau after ~ 1 min before declining, but in the continued presence of shear stress further elevations in Ca^{2+} and bursts of NO release appear after a latency of ~ 15 min. The oscillatory nature of these responses may result from direct negative feedback by NO on Ca^{2+} influx, as preincubation with an exogenous source of NO abolishes the response to shear but not agonists.[55] *In situ*, $[Ca^{2+}]_i$ transients evoked by flow in arteriolar endothelium may be monophasic, biphasic or oscillatory in different subpopulations of cells.[57]

High resolution imaging has also shown that the response of a given cell may itself be non-uniform as flow evokes $[Ca^{2+}]_i$ transients in distinct nuclear and sub-plasmalemmal pools, which peak after 20–40 secs and whose magnitude and speed of onset are directly related to the shear stress applied.[58,59] Some reports suggest preferential elevations in $[Ca^{2+}]_i$ at the downstream edge of the endothelial cell,[57,58] whereas others demonstrate a repetitive Ca^{2+} wave that originates at the upstream edge of the cell and propagates downstream.[60] Theoretical computations based on the surface geometry of endothelial monolayers suggest large cell-to-cell variations in stress fields, even when all cells are aligned with flow, and an uneven distribution on the surface of individual cells due to their non-uniform thickness.[61] This could explain why the number of cells responding to flow increases with the magnitude of the applied shear stress in a graded fashion.[57,59,60,62] Repeat exposure to flow does not always activate the same subpopulation of cells, also suggesting the existence of a refractory period.[57,62]

Sources of Activating Ca^{2+}

The initial rise in endothelial $[Ca^{2+}]_i$ evoked by shear stress and agonists involves release from inositol 1,4,5-trisphosphate (IP_3)-sensitive endoplasmic reticulum (ER) stores (Figure 2). Stimulation of endothelial cells with bradykinin results in transient (~ 2 min) tyrosine phosphorylation of the PLC-γ_1 isoform of phospholipase C, but activation of this enzyme and an associated increase in IP_3 formation are short-lived because there is also Ca^{2+}-dependent activation of protein tyrosine phosphatases secondary to Ca^{2+} influx and/or release from stores.[63] Tyrosine phosphorylation of PLC-γ_1 by members of the Src kinase family may also be an early event in the response to mechanical forces, as shear stress and strain rapidly activate c-Src (within < 1 min) through mechanisms that are proximal to activation of protein kinase C (PKC).[64,65] Synthesis of IP_3 in response to steady shear stress generally exhibits an early peak at 15–30 secs, although there may also be a delayed rise at ~ 5 mins.[66,67] IP_3 formation is sensitive to oscillating mechanical forces, such that increases and decreases in the frequency of cyclical stretch both stimulate a peak at ~ 10 seconds.[68,69]

Within minutes of the application of shear stress or agonists, endothelial $[Ca^{2+}]_i$ declines towards baseline, generally reaching a plateau that is maintained by influx of extracellular Ca^{2+}. There is accumulating evidence that this influx pathway is regulated by internal ER Ca^{2+} stores. Compounds which empty such stores by blocking Ca^{2+} uptake, such as cycloplazonic acid and thapsigargin, thus stimulate endothelial NO production by promoting a depletion-activated Ca^{2+} influx.[70–72] $[Ca^{2+}]_i$ and NO production subsequently return to basal levels within an hour, but ACh and acute changes in flow are then unable to stimulate further NO release, confirming the role of stores in regulating the response to agonists and shear stress.[72] The depletion-activated pathway is cation-specific with a preference for divalent ions such as Ca^{2+}, Mn^{2+}, and Ba^{2+}, and is thought to involve a transduction cascade of tyrosine and serine-threonine kinases such that Ca^{2+} influx is attenuated by tyrosine kinase inhibitors and prolonged by inhibitors of protein tyrosine phosphorylation.[63,73,74] Depletion of IP_3-sensitive stores by agonists or ATPase inhibitors also stimulates an endothelial microsomal P_{450} monooxygenase to synthesize arachidonic acid metabolites (epoxyeicosatrienoic acids, EETs) that activate Ca^{2+} influx and promote NO synthesis.[75] It is therefore possible that a family of depletion-activated channels, regulated by different pathways, contributes to Ca^{2+} influx in endothelial cells (Figure 2).

Agonists such as ACh, bradykinin and substance P evoke endothelial hyperpolarization, and since the endothelium is devoid of voltage-operated Ca^{2+} channels and therefore electrically non-excitable, this increases the electrochemical gradient for Ca^{2+} entry (Figure 2).[77] Hyperpolarization to substance P follows the opening of Ca^{2+}-activated K^+ channels (K_{Ca}) induced by Ca^{2+} release from internal stores and Ca^{2+} influx.[73] Cytochrome P_{450} monooxygenase products such as EETs may also promote activation of K_{Ca} channels so that pharmacological upregulation of this microsomal enzyme amplifies the hyperpolarization and cation entry induced by ER ATPase inhibitors.[76] Endothelial cells may possess a spectrum of high, medium and low conductance K_{Ca} channels and ATP-sensitive K^+ channels (K_{ATP}).[9,78] Multiple K_{Ca} subtypes appear to participate in the NO release stimulated by the frequency-related component of pulsatile flow and time-averaged shear stress, whereas K_{ATP} channel blockade with glibenclamide attenuates NO release stimulated by time-averaged but not oscillatory shear forces.[9,79,80] Shear stresses equivalent to the normal *in vivo* level (~ 10 dyne/cm^2) appear necessary to increase the open state probability of endothelial K_{ATP} channels, as glibenclamide does not modulate NO release from cells exposed to a shear stress averaging 0.3 dyne/cm^2, which produces only ~ 10% of the maximum achievable hyperpolarization evoked by flow.[22,79] Shear stress also activates an inwardly-rectifying K^+ channel that mediates hyperpolarization and is fully activated at levels within the *in vivo* range, desensitizes slowly and recovers rapidly on cessation of flow.[22] However, single channels enclosed by a micropipette within a cell-attached patch respond to flow after a delay of several minutes so that gating does not require direct exposure to shear stress and may be a secondary phenomenon.[81] The reported species/vessel differences in the effects of K^+ channel antagonists on agonist-induced NO release may reflect the duration of the hyperpolarization induced under different experimental conditions, which may vary from seconds to periods of many minutes.[9,82,83] Indeed, in cultured endothelial cells K^+ channel activity and membrane potential may not always be coupled to NO release stimulated by flow or agonists.[84,85]

Evidence that endothelial Ca^{2+} stores are compartmentalized is provided by experiments with ryanodine, an alkaloid that locks a specific Ca^{2+}-activated release channel in an open sub-conductance state.[72] Ryanodine elevates endothelial $[Ca^{2+}]_i$ but is completely without effect on shear stress- or ACh-induced NO release and does not activate the endothelial P_{450} monooxygenase despite causing store depletion, thus contrasting markedly with the effects of ATPase inhibitors that deplete a wider spectrum of store subtypes.[72,76] Anti-ryanodine receptor antibody binding sites surround the endothelial nucleus and are distributed in a non-homogeneous fashion throughout the cytosol, thus confirming that endothelial Ca^{2+} stores are both functionally and anatomically heterogeneous.[86]

Ca^{2+}-independent Mechanisms of eNOS Activation

There is accumulating evidence that eNOS can also be regulated by Ca^{2+}/calmodulin-independent mechanisms, so that under certain experimental conditions NO synthesis may occur in the absence of extracellular Ca^{2+} or sustained elevations in $[Ca^{2+}]_i$. Mixed Ca^{2+}-dependence and -independence of NO production during the early phase of exposure of human umbilical venous endothelial cells (HUVECs) to prolonged laminar shear stress, followed by an apparently Ca^{2+}-independent delayed component have thus been reported.[87,88] $[Ca^{2+}]_i$-independent NO synthesis after exposure of bovine aortic endothelial cells (BAECs) grown on microcarrier beads to nonlaminar shear stress has also been described.[85]

Phosphorylation at specific sites may constitute a discrete signalling mechanism that differentiates activation of eNOS by shear stress from activation by agonists or the Ca^{2+} ionophore A23187.[89] Indeed, eNOS possesses multiple consensus sequences for PKA, PKC and serine/threonine kinases (possibly including the mitogen-activated protein [MAP] kinase family) that may regulate the activity of the enzyme in a kinase-specific manner (Figure 2).[90] After 1 min exposure to a shear stress of 25 dyne/cm^2 there is a ~ 200% increase in eNOS phosphorylation, whereas there is no change after administration of A23187 at the same time point.[89] NO production may also increase without change in $[Ca^{2+}]_i$ if endothelial cells under chronic low shear stress (5 dyne/cm^2) are exposed to an acute increase (to 25 dyne/cm^2) and then correlates closely with the phosphorylation status of eNOS.[89] Although this NO production is attenuated by tyrosine kinase inhibitors, eNOS is phosphorylated on serine rather than tyrosine residues, suggesting that proximal tyrosine kinase events are transduced to downstream serine-threonine regulatory phosphorylations by a protein kinase cascade.[89] Indeed, the ERK 1 and 2 isoforms of the MAP kinase family are transiently activated by physiological levels of shear stress, with phosphorylation reaching a peak at 5–10 minutes and returning to baseline by 30–60 minutes.[90] Protein tyrosine phosphatase (PTPase) inhibitors can induce sustained NO release in the absence of extracellular Ca^{2+} or following depletion of Ca^{2+} stores, again reflecting Ca^{2+}-independent activation of eNOS.[63]

It has been suggested that shear-dependent activation of the ERK 1 and 2 isoforms of MAP kinase involves a Ca^{2+}-independent isoform of PKC.[89] Cyclic strain stimulates PKC activity, with early translocation to the plasmalemma at 10 sec being followed by sustained elevations in membrane PKC activity starting at 100 sec.[91] This is associated with a biphasic rise in diacylglycerol (DAG) formation secondary to transient breakdown of phosphatidylinositol by PLC and sustained phosphatidylcholine hydrolysis by phospholipase D.[91] It nevertheless remains controversial whether eNOS activation by acute changes in shear stress involves PKC-dependent phosphorylation, as NO release is not attenuated by selective inhibition of PKC.[74]

Mechanotransduction of the Flow Signal

G Proteins

Activation of endothelial PLC by agonists involves G proteins coupled to the NO synthetic pathway that include a pertussis toxin (PTX)-sensitive G_i protein linked to serotonergic, α_2-adrenergic and muscarinic receptors, and a PTX-insensitive G_q protein linked to bradykinin and ATP receptors (Figure 2).[92] G proteins also participate in the earliest transduction events associated with flow, as immunoprecipitation techniques demonstrate activation of heterotrimeric G-protein $G_{\alpha q}/G_{\alpha 11}$ and $G_{\alpha i3}/G_{\alpha o}$ subunits within 1 second of exposure to shear stress,[93] and G_i and G_q subtypes have both been reported to contribute to the NO response to flow.[94] Comparison of step, ramp and impulse changes in flow on NO production suggest the existence of a rapid G protein-dependent response that is sensitive to the rate of change of shear stress and a G protein-independent response to steady shear forces.[95] The general G protein inhibitor $GDP_{\beta}S$ thus blocks NO synthesis in the first hour after exposure of cultured HUVECs to flow, but significant attenuation is no longer apparent after 1 hour.[94] It remains to be determined whether direct linkage of G_i proteins to K^+ and Ca^{2+} channels[61,96] contributes to NO production.

Glycocalyx, Integrins and Cytoskeleton

Surface forces acting on the endothelium are sensed by the glycocalyx and integrins, which are heterodimeric transmembrane glycoproteins, and are ultimately transmitted *via* the cytoskeleton to focal adhesion complexes where integrins bind to the extracellular matrix on the abluminal side of the cell (Figure 2).[61,96] The cytoskeleton is supported by actin in its polymeric or filamentous (F-actin) form, which is organised into microfilament bundles that confer tension to the cell, and a lattice of microtubules formed by polymerization of tubulin, which in conjunction with intermediate filaments, provides cell rigidity.[96] Many secondary interactions are possible as, *inter alia*, cytoskeletal components are linked to stretch-activated channels, G proteins and IP_3-sensitive stores.[96] At the site of endothelial anchorage, signalling proteins such as FAK and paxillin present in focal adhesion complexes are activated by tyrosine-phosphorylation following application of cyclic strain, although this may be relatively slow and related to cytoskeletal rearrangement.[97] Activation of FAK by mechanical forces and c-Src, which is rapidly activated by shear stress, may nevertheless converge in a pathway that leads to activation of MAP kinases.[65,97]

Membrane sensors and the cytoskeleton may act as a single integrated dynamic unit that participates in the mechanotransduction pathway leading to activation of eNOS. Cleavage of sialic acid residues from the glycocalyx with neuraminidase and "stiffening" of the cell membrane with cholesteryl hemisuccinate thus both selectively impair shear-dependent NO responses,[87,98] and the involvement of focal adhesion complexes is reflected by observations that shear stress-induced dilatation is attenuated by a blocking antibody targetted at the integrin β_3 chain and by a peptide that specifically inhibits integrin binding to matrix proteins containing an Arg-Gly-Asp sequence.[99] This peptide also inhibits tyrosine kinase-mediated protein phosphorylation in response to shear stress but does not attenuate the endothelium-dependent response to substance P.[99] In rabbit aorta, endothelial NO release evoked by acute changes in shear stress is attenuated by: cytochalasin B, an F-actin depolymerizing agent; phalloidin, an F-actin stabilizing agent; and colchicine, an inhibitor of tubulin dimerization — whereas none of these agents affects the response to ACh.[100] Cytochalasin D also prevents the rise in $[Ca^{2+}]_i$ evoked by mechanical perturbation of the endothelial cell membrane.[101] Since disruption and stabilization of microfilaments both attenuate the NO response to shear, sensing of mechanical forces may involve dynamic turnover of cytoskeletal elements rather than transmission of force *via* a rigid connection.[100] In mice genetically lacking vimentin, a major intermediate filament cytoskeletal protein, NO dependent flow-related dilatation is impaired in mesenteric resistance arteries.[102] In endothelial cells cultured in the absence of flow, however, neither disruption nor stabilization of actin microfilaments affect the NO response to shear stress or bradykinin, and disruption of microtubules enhances the response to shear.[98] This could reflect differences in cytoskeletal structure, as endothelium cultured in the absence of flow lacks stress fibres formed from actin microfilaments and an organised lattice of microtubules and intermediate filaments, and focal adhesion complexes align in the direction of flow only in response to longitudinal shear forces.[61,96] A 42-kDa G protein present in vesicles prepared from endothelial membranes is activated by shear stress and could therefore represent a primary receptor associated with the lipid bilayer that does not require coupling to cytoskeletal elements.[93] By contrast, a 31-kDa G protein activated by shear forces in intact cells is not activated in such vesicular preparations and may, therefore, require coupling to intracellular components.[93]

Caveolae

Caveolae are invaginations of the plasma membrane that serve as specialized signal transduction domains for the compartmentalization of receptors, G protein subunits and enzymes, including eNOS and non-receptor tyrosine kinases, and could therefore facilitate tightly-regulated interactions of eNOS with other signalling systems that enhance the response to mechanical forces (Figure 2). Uniquely amongst the NOS isoforms, eNOS possesses a substrate site for acyl transferase at its amino terminus which permits sequential N-myristoylation and palmitoylation and subcellular localization within caveolae and in the region of the Golgi apparatus.[103,104] Caveolin, an integral membrane protein which is a key structural component of caveolae, allosterically regulates the activity of eNOS (but not the iNOS and nNOS isoforms) by interacting directly with its oxygenase domain and forming a heteromeric complex that is quantitively dissociated by the Ca^{2+}-calmodulin complex, leading to enzyme activation.[105–108] Dynamic cycling of caveolin and eNOS between caveolae and the Golgi apparatus also regulates enzyme activity.[103,104,108] Exposure of endothelial cells to bradykinin or tyrosine phosphatase inhibitors causes eNOS and caveolin to co-translocate to the cytoskeleton, suggesting a role in subcellular targeting to a specific microenvironment.[63,105,109] Recent evidence suggests that tyrosine kinase-dependent phosphorylation of a protein within the caveolin-eNOS-cytoskeletal complex results in enzyme inhibition.[109] High resolution Ca^{2+} imaging has shown that there is a restricted sub-plasmalemmal space in endothelial cells, as in vascular smooth muscle, so that imaging with Fura 2 may fail to reveal changes in bulk cytosolic $[Ca^{2+}]_i$ when there may be substantial elevations beneath the plasmalemma.[110] Calmodulin is also preferentially localized in caveolae,[106] suggesting that such localized increases in $[Ca^{2+}]_i$ could play an important role in modulating the NO response to shear stress. Indeed, vectorial release of Ca^{2+} from the peripheral ER occurs towards the membrane and is regulated by extrusion *via* the Na^+-Ca^{2+} exchanger, leading to substantial gradients in $[Ca^{2+}]_i$ within the cell.[110]

eNOS Transcription

The mechanisms involved in the upregulation of eNOS transcription by shear stress are Ca^{2+}-dependent[111] and attenuated by K^+ channel inhibition,[34] but are poorly understood. eNOS contains a shear stress responsive element (SSRE) that is present in other endothelial genes such as platelet-derived growth factor β chain and tissue type plasminogen activator, in which the SSRE is thought to regulate expression by mechanical forces.[112] However, G proteins, PLC, PKC and tyrosine kinase-dependent mechanisms that affect transcription of such SSRE-containing genes in response to shear do not appear to participate in the case of eNOS.[34,111] Indeed, there is evidence that activation of PKC downregulates eNOS expression.[113]

Boundary Layer Mass Transport

Movement of agonists across the hydrodynamic boundary layer generated by fluid flow involves both diffusion and convection, and the interplay between these transport processes may affect receptor activation by agents such as ATP and bradykinin which are degraded by ectoenzymes on the endothelial surface. Dull *et al.*[114] were thus able to demonstrate differences in the effects of ATP and its non-hydrolysable analogue $ADP_{\beta}S$ in the presence of flow. Stepwise increases in shear stress produced rapid elevations in endothelial $[Ca^{2+}]_i$ that decayed over minutes, but additional increases in flow evoked further Ca^{2+} transients only in the presence of ATP. Furthermore, abrupt cessation of flow resulted in an immediate drop in $[Ca^{2+}]_i$ in the presence of ATP, whereas there was little change in the recovery rate during an $ADP_{\beta}S$-associated $[Ca^{2+}]_i$ response. These findings are consistent with rapid equilibration of boundary layer and bulk concentrations of $ADP_{\beta}S$, but no such equilibrium for ATP because of dynamic clearance by the 5'-nucleotidase.[114]

Boundary layer effects may also in part determine the consequences of the release of vasoactive agonists from the endothelium itself (Figure 2). Endothelial cells synthesize ACh and ATP and contain heterogeneous cytoplasmic stores of agonists capable of stimulating NO synthesis, including bradykinin, vasopressin, 5-HT and substance P.[32] Flow stimulates release of ATP, and less consistently ACh and substance P, from isolated endothelial cells; substance P, vasopressin and ATP can be detected in the effluent from the rat hindlimb on increasing flow.[32,115] In rat lung dilatation resulting from flow-dependent release of ATP can be blocked by P_2 purinoceptor antagonism, which thereby elevates perfusion pressure[116] and, as discussed in Chapter 11, bradykinin B_2 antagonists promote constriction and reduce flow in the human coronary circulation, confirming the physiological importance of autocrine mechanisms.

In laminar flow the thickness of a developing mass transport boundary layer (the so-called Leveque régime) varies as $\delta \sim (D/\dot{\gamma})^{1/3}$ where $\dot{\gamma}$ is the velocity gradient at the wall (i.e. the shear rate) and D is the diffusivity of the agonist.[117] The rate of accumulation ($\dot{m}$) of an agonist supplied by diffusion across this boundary layer varies as $\dot{m} \sim (D/\delta)\,\Delta C$, where ΔC is the concentration difference between the bulk flow and the cell surface. According to the Stokes-Einstein equation, D is inversely proportional to fluid viscosity, μ, so that $\dot{m} \sim \mu^{-2/3}\ \dot{\gamma}^{1/3}\ \Delta C$.[117] Theoretically, increases in flow should therefore increase agonist accumulation, whereas increases in viscosity under conditions of constant shear rate should result in a decrease.[117] Ando and coworkers found, however, that the $[Ca^{2+}]_i$ transients evoked by flow in the presence of ATP were amplified rather than diminished by increased perfusate viscosity, and that their magnitude correlated closely with absolute shear stress ($\tau = \mu \times \dot{\gamma}$), but not μ or $\dot{\gamma}$ when these parameters were manipulated independently.[117] The interplay between shear stress and receptor-mediated activation events may therefore be more complex than suggested by simple hydrodynamic theory. Indeed, shear stress applied in the presence of ATP causes a larger proportion of confluent cultured cells to exhibit Ca^{2+} transients than in its absence,[62] suggesting the existence of a threshold in individual cells which is modulated by the presence of agonists. Diffusive mass transport does not influence NO release evoked by ACh, as concentration-relaxation curves to this agonist in cascade bioassay are unaffected by increasing perfusate viscosity with dextran either in terms of the potency of ACh or the maximum relaxations observed.[72]

Prostaglandins

In canine epicardial arteries flow-dependent dilatation is insensitive to cyclooxygenase inhibition,[18] but in some vascular beds there is also a contribution from endothelium-derived prostanoids.[12] Shear-related prostacyclin (PGI_2) release parallels that of NO, in that it is linked to a PTX-sensitive G_i-protein, and is enhanced by flow pulsatility and modulated by the profile of the flow wave.[6,118,119] Interactions between the synthetic pathways of the two mediators (see also Chapters 13 and 19) are evident in that exogenous donors of NO increase PGI_2 release from arachidonic acid-stimulated endothelial cells *via* a cGMP-independent mechanism,[120] whereas activation of cyclooxygenase functionally attenuates NO activity by generating $O_2^{\bullet-}$ which interacts to form peroxynitrite.[121] In cultured endothelium, shear stresses of 25 dyne/cm^2 amplify release of both NO and PGI_2 ~ 10 fold above basal levels after 1 hour on a cumulative basis, with PGI_2 synthesis being attenuated by ~ 50% following inhibition of eNOS.[122] Theoretically, NO could activate the constitutive and inducible forms of cyclooxygenase (COX-1 and COX-2) which contain haem at their active sites, but millimolar NO concentrations do not activate purified COX-1 enzyme.[123] Alternatively, peroxynitrite may activate COX-1 and COX-2 by serving as a substrate for the peroxidase activity of these enzymes.[124] The role of free radical reactions in the coupling of shear-induced NO and PGI_2 synthesis thus remains to be fully evaluated, as acute changes in flow stimulate production of the $O_2^{\bullet-}$ anion by the endothelium.[125] Furthermore, expression of manganese superoxide dismutase and cyclooxygenase are upregulated in parallel with eNOS on exposure to steady laminar but not turbulent flow.[126]

Endothelium-dependent Hyperpolarization

Agonists that stimulate NO synthesis may also evoke an endothelium-dependent smooth muscle hyperpolarization that is not attributable to NO or prostanoids and contributes to relaxation.[127] Evidence for the existence of a freely diffusible endothelium-derived hyperpolarizing factor (EDHF) comes from cascade bioassay experiments in which endothelial cells stimulated by bradykinin evoke transient hyperpolarization of downstream vascular myocytes under conditions of combined eNOS and cyclooxygenase blockade.[128] Increases in transmural pressure in a donor vessel enhance the duration but not the amplitude of this hyperpolarizing response, suggesting that mechanical forces also modulate EDHF release.[128] Many studies confirm the involvement of K^+ channel activation but demonstrate considerable regional and species heterogeneity, as the participating subtypes identified include K_{Ca}, K_{ATP} and K_V channels.[127] This high degree of heterogeneity could reflect the existence of more than one EDHF, and a number of candidate mediators have been proposed, including cytochrome P_{450} monooxygenase-derived metabolites such as EETs[128] and anandamide, an endogenous cannabinoid.[129] Indeed, NO can itself activate K_{Ca} channels *via* both cGMP-dependent and direct cGMP-independent mechanisms.[130,131] In guineapig, rat, rabbit and sheep arteries, hyperpolarization to ACh and substance P consists of a transient attributable to an EDHF followed by a sustained component mediated by NO and prostanoids that is enhanced under conditions of stretch.[132] Such variability may reflect experimental conditions, as authentic NO hyperpolarizes arteries whose resting potential is normally ca. – 60 mV (e.g. guineapig mesenteric and coronary), but not those where the potential is ca. –75 mV (e.g. guineapig uterine) unless they are initially depolarized.[133]

Gap Junctional Communication

Direct intercellular endothelial-smooth muscle communication also contributes to the regulation of vascular tone, as NO- and prostanoid-independent relaxations to agonists such as ACh and ATP are depressed by a synthetic inhibitory peptide possessing specific sequence homology with the second extracellular loop of connexin 43, a key component of gap junctions in vascular tissues.[134,135] Although preferential passage of certain dye tracers from endothelium to smooth muscle suggests polarity in the direction of information transfer,[136] activation of small arteries by phenylephrine or high extracellular $[K^+]$ is accompanied by the transmission of a signal from smooth muscle to the endothelium which elevates endothelial $[Ca^{2+}]_i$ and promotes NO release,[137] and the two cell types may be in electrical continuity.[138] The extent to which the signals that allow bidirectional endothelial-smooth muscle gap junctional communication are primarily electrical in nature or carried by Ca^{2+} ions, IP_3 or some other low molecular weight mediator is presently unknown. It also remains to be determined if this pathway contributes to the response to shear forces, a possibility suggested by reports that flow-dependent dilatation in feline femoral and rat basilar arteries is both NO- and prostanoid-independent.[139,140]

Nitric Oxide and Integrated Vascular Control

Efficiency and Optimization

If the vascular tree consisted simply of a system of rigid tubes, left ventricular pressure would need to rise to physiologically unsustainable levels in order to achieve the ~ 5-fold increase in cardiac output that occurs during maximal exercise. In practice, however, mean arterial pressure increases by only 10–20% as a consequence of active responses to changes in flow. Thus, the pressure drop across an isolated segment of rat saphenous artery remains virtually unchanged over a 5-fold range of flow rates if its endothelium is intact, and it has been estimated that ~ 50% of mean aortic pressure would be lost in conducting and distributive arteries alone in the absence of flow-dependent dilatation.[19,20] In the isolated rabbit ear, total hydraulic conductance (G) and the calculated conductance of individual arterial branch generations with diameters ranging from 70–800 μm are almost linearly related to input flow ($\dot{Q}$) at different levels of constrictor tone as a direct consequence of NO activity.[10,11] This implies that the pressure gradient driving flow (i.e. $\dot{Q}/G$) asymptotically approaches a constant value at high flow rates. Stabilization of pressure gradients by the endothelium will contribute to hydraulic "efficiency", particularly in the microcirculation where flow is viscous rather than inertial. Energy losses are directly proportional to the product of pressure gradient and flow (i.e. $\dot{Q}^2/G$) and will therefore be proportional to $\dot{Q}^2$ in the case of a rigid tube (i.e. constant G) but only to $\dot{Q}$ if there is flow-dependent dilatation.[10]

NO may also contribute to geometrical "optimality" in the patterns of arterial branching. Murray[141] hypothesized that a cubic power law relationship between local flow rate and vessel diameter would allow intravascular volume and power losses to be minimized simultaneously, thereby contributing to cardiovascular efficiency by limiting the "metabolic cost" of maintaining a high blood volume and enhancing tissue perfusion for a given level of cardiac work. This "law" requires that the diameter exponent at a bifurcation, x, defined by the equation $d_0^x = d_1^x + d_2^x$ where d_0 is the parent, and d_1 and d_2 the daughter artery diameters, takes a value of exactly 3. In the isolated rabbit ear, the exponent x remains close to the theoretical optimum over a ~ 10-fold range of flow rates at successive arterial bifurcations when there is normal NO activity.[142] High levels of constriction cause deviation from the optimum, but x may then be restored towards 3 by endothelium-dependent agonists such as ACh.[143] Vascular design also conforms to the Murray optimum under resting and vasodilated conditions in the rat cremaster microcirculation *in vivo*[144] and in the human retina when adaptive mechanisms may maintain optimality following chronic reductions in flow caused by head trauma.[145] Mandelbrot[146] has also hypothesized that a diameter exponent of 3 will allow self-similar dichotomously branching architectures to fill the 3-dimensional space available to them as fully as possible, and in the context of the vascular system this would provide maximum surface area for exchange at capillary level. As NO may influence vascular remodelling independently of changes in arterial pressure,[147] the endothelium could thus in theory act as a flow sensor that maintains architectural optimality during acute and long-term changes in flow and during vascular growth.

Stability

If the resistance of proximal feed arteries remains fixed and the supply pressure remains constant, the increase in local perfusion that accompanies the release of vasodilator metabolites from active cells will cause distal pressures to fall as a direct consequence of Poiseuille's Law. Impaired perfusion of adjacent tissue elements (i.e. a microvascular "steal" phenomenon) will, however, be offset by the sequential recruitment of flow-dependent dilatation in successive generations of upstream distributive arteries as this will serve to restore local pressures towards normal. It is also known that localized dilator responses (e.g. induced by agents such as ACh) can propagate electrotonically along the arterial wall with a velocity and amplitude space constant similar to that of conduction in other tissues.[150,151] Such conducted responses will, in theory, interact synergistically with flow-dependent dilatation to minimize local perfusion heterogeneity, by integrating the intensity and nature of stimuli arriving from downstream sites. Dilatation of a daughter vessel at a branch point will increase flow in the parent vessel, thereby decreasing local pressure and flow through the non-dilated daughter branch, but steal effects will be offset by propagation of the dilator response into the non-dilated daughter branch.[150] Pharmacological disruption of the endothelium thus unmasks perfusion inequalities in the rabbit hindlimb and results in the widening of the difference between the lowest and highest tissue O_2 tensions found in different microregions and an overall reduction in O_2 uptake by ~ 40%.[148] Microangiographic techniques also suggest that NO synthesis helps to maintain spatially homogeneous patterns of flow distribution by preserving similar relative resistances in different vascular segments of the rabbit ear under altered conditions of flow and constriction.[10] Much remains to be learnt about the interaction between flow-dependent and conducted control mechanisms, as localized topical application of L-arginine or ACh to distal microvessels can enhance more proximal upstream conducted responses through an NO-dependent mechanism whose fundamental nature remains unknown.[151]

Increases in local transmural pressure lead to constriction (and decreases in pressure to dilatation) through the ability of the smooth muscle cell to sense circumferential wall stress actively.[16,149] This myogenic response can spread electronically to adjacent segments, and there may ultimately be coordinated changes in intravascular pressure over many generations of the same vascular network and iterative local adjustments in resistance.[149] Theoretically, the myogenic response can be regarded as an unstable positive feedback loop in which constriction promotes progressively greater levels of constriction. Flow-dependent dilatation may confer stability by superimposing an opposing feedback mechanism. Indeed, in isolated arteries myogenic activity is potentiated by endothelial denudation and is normally attenuated by NO (Figure 1).[11,13–15] Both control loops contribute to the phenomenon of autoregulation whereby organ resistance is tightly controlled and flow held constant over a wide range of perfusion pressures. Distal metabolic dilatation will cause a fall in upstream pressure, thereby evoking myogenic dilatation of proximal arterioles. In turn, larger "feed" and small distributive arteries will be recruited through flow-dependent mechanisms. In theory, therefore, NO could enhance autoregulation by amplifying myogenic dilatation at low perfusion pressures. Alternatively, NO could oppose autoregulation by attenuating myogenic constriction at high perfusion pressures. eNOS inhibition unmasks an ability of the isolated rabbit ear to autoregulate[11] and extends the weak autoregulatory range of isolated guineapig or rabbit hearts.[152,153] By contrast, in the coronary bed of the concious dog, where there is high intrinsic autoregulatory gain, eNOS inhibition simply results in an increase in the lowest pressure at which autoregulatory mechanisms are activated.[154] The complex interplay between myogenic, flow-dependent and metabolic control mechanisms is discussed in detail in Chapters 9 and 13.

Flexibility

In vivo, local resistance may fluctuate over time, as small arteries and arterioles often contract and relax spontaneously, thereby contributing to local temporal fluctuations in tissue perfusion, a phenomenon known as vasomotion. While the patterns of fluctuation observed are often highly irregular, there is evidence that they are not random, but generated by deterministic nonlinear control mechanisms that can be classified as "chaotic".[155,156] Indeed, in isolated resistance arteries vasomotion is generated by a relatively small number of dominant control variables (probably 4) and involves nonlinearity in the coupling between distinct intracellular and membrane Ca^{2+} oscillators located at the level of the smooth muscle cell.[156] Nonlinear systems possess greater flexibility that linear systems, as minute perturbations can effect changes in state, thereby permitting escape from patterns of behaviour that have become disadvantageous with minimal energy expenditure. Synchronization of the large number of smooth muscle cells present in the vessel wall ultimately depends on communication *via* gap junctions derived from connexin 43.[134]

Experimental observations indicate that vasomotion is an NO-independent phenomenon, and nonlinear mathematical analysis confirms that NO is not a key control parameter.[155] NO may nevertheless profoundly affect the superficial form and amplitude of vasomotion, which can be completely suppressed by micromolar concentrations of ACh.[155,157] The unpredictability of nonlinear systems to perturbation, rather than fundamental differences in smooth muscle control mechanisms, may also explain why NO can also "paradoxically" promote rhythmic activity in otherwise quiescent arteries.[158] Vasomotion is thought to limit extravascular fluid filtration by reducing hydrostatic pressure during periods of low flow, to enhance lymphatic drainage through a pumping action, and to represent a beneficial homeodynamic response during hypoxia and ischaemia by increasing time-averaged vascular conductance.[157] The physiological role of NO in modulating these phenomena remains to be defined, but present evidence suggests that NO activity may serve to damp local fluctuations in resistance to physiologically acceptable levels, and normally suppresses a low frequency component of blood pressure variability *in vivo*.[159]

Flow-dependent, myogenic and metabolic mechanisms of control must ultimately maintain O_2 delivery to distal tissues (Figure 1). NO-dependent transport phenomena involving blood components may facilitate this aim by extending the otherwise short half-life of NO.[4] NO nitrosates the highly-conserved cysteine 93β of haemoglobin, which acts as an allosterically regulated buffer that exchanges NO with low MW intracellular thiols such as glutathione.[161,162] Haemoglobin exists in R (relaxed, high O_2 affinity) and T (tense, low O_2 affinity) forms, allowing the T form to release O_2. Thiol affinity for NO is high in the R structure and low in the T structure so that NO is released in low O_2 tensions. Normally there is a significant O_2 gradient in precapillary vessels, ~50% of the O_2 being lost "upstream" as a consequence of diffusion to draining venules.[161] This functional shunt may promote an allosteric transition from R to T which liberates NO or NO-glutathione (which is exported from erythrocytes) thereby increasing blood flow. Haemoglobin may thus serve both as a sink and a source for dilator NO depending on the local microenvironment. *S*-nitrosation of albumin at its single free thiol cysteine generates *S*-NO-albumin which serves as a reservoir of "stable" NO. Although this adduct itself exhibits relaxant activity, subsequent transnitrosation of low molecular weight thiols, which occurs rapidly and completely *in vitro*, may also generate less diffusion-limited and therefore more biologically active compounds.[163]

Conclusions

NO released by the vascular endothelium plays a pivotal role in the control of perfusion. At the organ level its contribution to stabilising mechanisms such as autoregulation, maintenance of flow distribution, and limitation of cardiac work relative to perfusion are important considerations for cardiovascular function. Mechanical coupling between shear forces and eNOS may involve the glycocalyx, integrins and the cytoskeleton and a variety of second messenger systems and phosphorylation steps, adding a new dimension of complexity to the mechanisms that mediate the interaction between blood flow and the vessel wall. The mechanisms that regulate expression of eNOS in response to mechanical forces nevertheless remain unknown. Much is therefore yet to be learnt about the dynamic relationships between NO and other control mechanisms, especially since its biological chemistry and physiology are so complex. The influence of shear stress on the endothelial cell is thus likely to remain an area of intense research interest.

Acknowledgements

The author is indebted to Dr D Edwards for helpful comments during preparation of the manuscript and Wendy Simons for secretarial assistance.

References

1. Furchgott RF, Zawadzki JV. The obligatory role of endothelial cells in the relaxation of arterial smooth muscle by acetylcholine. *Nature* 1980; **288**: 373–376.
2. Palmer RMJ, Ashton DS, Moncada S. Vascular endothelial cells synthesize nitric oxide from L-arginine. *Nature* 1988; **333**: 664–666.
3. Ignarro LJ. Endothelium-derived nitric oxide: actions and properties. *FASEB J* 1989; **3**: 31–36.
4. Griffith TM, Edwards DH, Lewis MJ *et al.* The nature of endothelium-derived vascular relaxant factor. *Nature* 1984; **308**: 645–647.
5. Rubanyi GM, Romero JC, Vanhoutte PM. Flow-induced release of endothelium-derived relaxing factor. *Am J Physiol* 1986; **250**: H1145–H1149.
6. Pohl U, Busse R, Kuon E, Bassenge E. Pulsatile perfusion stimulates the release of endothelial autacoids. *J Appl Cardiol* 1986; **1**: 215–235.
7. Tesfamariam B, Cohen RA. Inhibition of adrenergic vasoconstriction by endothelial cell shear stress. *Circ Res* 1988; **63**: 720–725.
8. Hutcheson IR, Griffith TM. Release of endothelium-derived relaxing factor is modulated both by frequency and amplitude of pulsatile flow. *Am J Physiol* 1991; **261**: H257–H262.
9. Hutcheson IR, Griffith TM. Heterogeneous populations of K^+ channels mediate NO release to flow but not agonists in rabbit aorta. *Am J Physiol* 1994; **266**: H590–H596.
10. Griffith TM, Edwards DH, Davies RLl *et al.* NO coordinates the behaviour of vascular resistance vessels. *Nature* 1987; **329**: 442–445.
11. Griffith TM, Edwards DH. Myogenic autoregulation of flow may be inversely related to endothelium-derived relaxing factor activity. *Am J Physiol* 1990; **258**: H1171–H1180.
12. Koller A, Kaley G. Prostaglandins mediate arteriolar dilation to increased blood flow velocity in skeletal muscle microcirculation. *Circ Res* 1990; **67**: 529–534.

13. Kuo L, Chilian WM, Davis MJ. Interaction of pressure- and flow-induced responses in porcine coronary resistance vessels. *Am J Physiol* 1991; **261**: H1706–H1715.
14. Kuo L, Arko F, Chilian WM, Davis MJ. Coronary venular responses to flow and pressure. *Circ Res* 1993; **72**: 607–615.
15. Pohl U, Herlan K, Huang A, Bassenge E. NO-mediated, shear-induced dilation opposes myogenic vasoconstriction in small rabbit arteries. *Am J Physiol* 1991; **261**: H2016–H2023.
16. Davis MJ. Myogenic response gradient in an arteriolar network. *Am J Physiol* 1993; **264**: H2168–H2179.
17. Hilton SM. A peripheral arterial conducting mechanism underlying dilatation of the femoral artery and concerned in functional vasodilatation in skeletal muscle. *J Physiol (Lond)* 1959; **149**: 93–111.
18. Holtz J, Förstermann U, Pohl U *et al.* Flow-dependent, endothelium-mediated dilation of epicardial coronary arteries in conscious dogs: Effects of cyclooxygenase inhibition. *J Cardiovasc Pharmacol* 1984; **6**: 1161–1169.
19. Khayutin VM, Nikolsky VP, Rogoza AN, Lukoshkova EV. The effect of pressure drop stabilization in arteries during variations in blood flow rate. *Acta Physiol Scand* 1993; **148**: 285–294.
20. Khayutin VM, Nikolsky VP, Rogoza AN, Lukoshkova EV. Endothelium determines stabilization of the pressure drop in arteries. *Acta Physiol Scand* 1993; **148**: 295–304.
21. Sinoway LI, Hendrickson C, Davidson WR *et al.* Characteristics of flow-mediated brachial artery vasodilation in human subjects. *Circ Res* 1989; **64**: 32–42.
22. Olesen S-P, Clapham DE, Davies PF. Haemodynamic shear stress activates a K^+ current in vascular endothelial cells. *Nature* 1988; **331**: 168–170.
23. Griffith TM, Hutcheson IR, Randall MD, Edwards DH. Role of flow in endothelium-mediated responses. In: *Resistance Arteries, Structure and Function*, Mulvaney MJ, Aalkjaer C, Heagerty AM, Nyborg NCB, Strandgaard S (Eds). Excerpta Medica, Amsterdam 1991: 204–207.
24. Lamontagne D, Pohl U, Busse R. Mechanical deformation of vessel wall and shear stress determine the basal release of endothelium-derived relaxing factor in the intact rabbit coronary vascular bed. *Circ Res* 1992; **70**: 123–130.
25. Canty JM, Schwartz JS. Nitric oxide mediates flow-dependent epicardial coronary vasodilation to changes in pulse frequency but not mean flow in conscious dogs. *Circulation* 1994; **89**: 375–384.
26. Rubanyi GM. Endothelium-dependent pressure-induced contraction of isolated canine carotid arteries. *Am J Physiol* 1988; **255**: H783–H788.
27. Hishikawa K, Nakaki T, Suzuki H *et al.* Transmural pressure inhibits nitric oxide release from human endothelial cells. *Eur J Pharmacol* 1992; **215**: 329–331.

28. Kelm M, Feelisch M, Deussen A *et al.* Release of endothelium derived nitric oxide in relation to pressure and flow. *Cardiovasc Res* 1991; **25**: 831–836.
29. Randall MD, Griffith TM. NO plays a central role in collateral perfusion following arterial occlusion. *Am J Physiol* 1992; **263**: H752–H760.
30. Unthank JL, Nixon JC, Dalsing MC. Acute compensation to abrupt occlusion of rat femoral artery is prevented by NO synthase inhibitors. *Am J Physiol* 1995; **268**: H2523–H2530.
31. Falcone JC, Bohlen HG. NO from rat intestine and skeletal muscle venules causes dilation of arterioles. *Am J Physiol* 1990; **258**: H1515–H1523.
32. Ralevic V, Lincoln J, Burnstock G. Release of vasoactive substances from endothelial cells. In: *Endothelial regulation of vascular tone*, Ryan US, Rubanyi GM (Eds). Marcel-Dekker, New York 1992: 297–328.
33. Sessa WC, Pritchard K, Seyedi N *et al.* Chronic exercise in dogs increases coronary vascular nitric oxide production and endothelial cell nitric oxide synthase gene expression. *Circ Res* 1994; **74**: 349–353.
34. Uematsu M, Ohara Y, Navas JP *et al.* Regulation of endothelial cell nitric oxide synthase mRNA expression by shear stress. *Am J Physiol* 1995; **269**: C1371–C1378.
35. Noris M, Morigi M, Donadelli R *et al.* Nitric oxide synthesis by cultured endothelial cells is modulated by flow conditions. *Circ Res* 1995; **76**: 536–543.
36. Awolesi MA, Sessa WC, Sumpio BE. Cyclic strain upregulates nitric oxide synthase in cultured bovine aortic endothelial cells. *J Clin Invest* 1995; **96**: 1449–1454.
37. Langille BL, O'Donnell F. Reductions in arterial diameter produced by chronic decreases in blood flow are endothelium-dependent. *Science* 1986; **231**: 405–407.
38. Tronc F, Wassef M, Esposito B *et al.* Role of NO in flow-induced remodeling of the rabbit common carotid artery. *Arterioscler Thromb Vasc Biol* 1996; **16**: 1256–1262.
39. Garg UC, Hassid A. Nitric oxide-generating vasodilators and 8-bromo-cyclic guanosine monophosphate inhibit mitogenesis and proliferation of cultured rat vascular smooth muscle cells. *J Clin Invest* 1989; **83**: 1774–1777.
40. De Mey JGR, Dijkstra EH, Vrijdag MJJF. Endothelium reduces DNA synthesis in isolated arteries. *Am J Physiol* 1991; **260**: H1128–H1134.
41. McLenachan JM, Vita J, Fish RD *et al.* Early evidence of endothelial vasodilator dysfunction at coronary branch points. *Circulation* 1990; **82**: 1169–1173.
42. Kuo L, Davis MJ, Cannon MS, Chilian WM. Pathophysiological consequences of atherosclerosis extend into the coronary microcirculation. *Circ Res* 1992; **70**: 465–476.
43. Randall MD, Smith JA, Griffith TM. Hypercholesterolaemia severely impairs NO-dependent collateral perfusion following acute arterial occlusion in rabbit isolated ear. *Br J Pharmacol* 1993; **109**: 838–844.

44. Hutcheson IR, Smith JA, Griffith TM. Abolition of flow-dependent EDRF release before that evoked by agonists in hypercholesterolaemic rabbits. *Br J Pharmacol* 1994; **113**: 190–194.
45. Leung W-H, Lau C-P, Wong C-K. Beneficial effect of cholesterol-lowering therapy on coronary endothelium-dependent relaxation in hypercholesterolaemic patients. *Lancet* 1993; **341**: 1496–1500.
46. Cooke JP, Andon NA, Girerd XJ *et al.* Arginine restores cholinergic relaxation of hypercholesterolemic thoracic rabbit aorta. *Circulation* 1991; **83**: 1057–1062.
47. Kohler TR, Kirkman TR, Kraiss LW *et al.* Increased flow inhibits neointimal hyperplasia in endothelialized vascular grafts. *Circ Res* 1991; **69**: 1557–1565.
48. Mattsson EJR, Kohler TR, Vergel SM, Clowes AW. Increased blood flow induces regression of intimal hyperplasia. *Arterioscler Thromb Vasc Biol* 1997; **17**: 2245–2249.
49. Boulanger C, Lüscher TF. Release of endothelin from porcine aorta. Inhibition by endothelium-derived nitric oxide. *J Clin Invest* 1990; **85**: 587–590.
50. Koller A, Huang A. Impaired nitric oxide-mediated flow-induced dilation in arterioles of spontaneously hypertensive rats. *Circ Res* 1994; **74**: 416–421.
51. Van de Voorde J, Leusen I. Endothelium-dependent and independent relaxation of aortic rings from hypertensive rats. *Am J Physiol* 1986; **250**: H711–H717.
52. Bouloumié A, Bauersachs J, Linz W *et al.* Endothelial dysfunction coincides with an enhanced nitric oxide synthase expression and superoxide anion production. *Hypertension* 1997; **30**: 934–941.
53. Mügge A, Elwell JH, Peterson TE *et al.* Chronic treatment with polyethyleneglycolated superoxide dismutase partially restores endothelium-dependent vascular relaxations in cholesterol-fed rabbits. *Circ Res* 1991; **69**: 1293–1300.
54. Arnal J-F, Schott C, Stoclet J-C. Vascular relaxation and cyclic guanosine monophosphate in a rat model of high output heart failure. *Cardiovasc Res* 1993; **27**: 1651–1656.
55. Kanai AJ, Strauss HC, Truskey GA *et al.* Shear stress induces ATP-independent transient nitric oxide release from vascular endothelial cells, measured directly with a porphyrinic microsensor. *Circ Res* 1995; **77**: 284–293.
56. Griffith TM, Edwards DH, Newby AC *et al.* Production of endothelium-derived relaxant factor is dependent on oxidative phosphorylation and extracellular calcium. *Cardiovasc Res* 1986; **20**: 7–12.
57. Falcone JC, Kuo L, Meininger GA. Endothelial cell calcium increases during flow-induced dilation in isolated arterioles. *Am J Physiol* 1993; **264**: H653–H659.
58. Geiger RV, Berk BC, Alexander RW, Nerem RM. Flow-induced calcium transients in single endothelial cells: spatial and temporal analysis. *Am J Physiol* 1992; **262**: C1411–C1417.
59. Shen J, Luscinskas W, Connolly A *et al.* Fluid shear stress modulates cytosolic free calcium in vascular endothelial cells. *Am J Physiol* 1992; **262**: C384–C390.

60. Yoshikawa N, Ariyoshi H, Ikeda M *et al.* Shear-stress causes polarized change in cytoplasmic calcium concentration in human umbilical vein endothelial cells (HUVECs). *Cell Calcium* 1997; **22**: 189–194.
61. Davies PF. Flow-mediated endothelial mechanotransduction. *Physiol Rev* 1995; **75**: 519–560.
62. James NL, Harrison DG, Nerem RM. Effects of shear on endothelial cell calcium in the presence and absence of ATP. *FASEB J* 1995; **9**: 968–973.
63. Fleming I, Fisslthaler B, Busse R. Interdependence of calcium signaling and protein tyrosine phosphorylation in human endothelial cells. *J Biol Chem* 1996; **271**: 11009–11015.
64. Liu M, Qin Y, Liu J *et al.* Mechanical strain induces $pp60^{src}$ activation and translocation to cytoskeleton in fetal rat lung cells. *J Biol Chem* 1996; **271**: 7066–7071.
65. Takahashi M, Berk BC. Mitogen-activated protein kinase (ERK1/2) activation by shear stress and adhesion in endothelial cells. Essential role for a herbimycin-sensitive kinase. *J Clin Invest* 1996; **98**: 2623–2631.
66. Prasad ARS, Logan SA, Nerem RM *et al.* Flow-related responses of intracellular inositol phsophate levels in cultured aortic endothelial cells. *Circ Res* 1993; **72**: 827–836.
67. Nollert MU, Eskin SG, McIntire LV. Shear stress increases inositol trisphosphate levels in human endothelial cells. *Biochem Biophys Res Commun* 1990; **170**: 281–287.
68. Rosales OR, Sumpio BE. Changes in cyclic strain increase inositol trisphosphate and diacylglycerol in endothelial cells. *Am J Physiol* 1992; **262**: C956–C962.
69. Brophy CM, Mills I, Rosales O *et al.* Phospholipase C: a putative mechanotransducer for endothelial cell response to acute hemodynamic changes. *Biochem Biophys Res Commun* 1993; **190**: 576–581.
70. Hosoki E, Iijima T. Modulation of cytosolic Ca^{2+} concentration by thapsigargin and cyclopiazonic acid in human aortic endothelial cells. *Eur J Pharmacol* 1995; **288**: 131–137.
71. Moritoki H, Hisayama T, Takeuchi S *et al.* Relaxation of rat thoracic aorta induced by the Ca^{2+}-ATPase inhibitor, cyclopiazonic acid, possibly through nitric oxide formation. *Br J Pharmacol* 1994; **111**: 655–662.
72. Hutcheson IR, Griffith TM. Central role of intracellular calcium stores in acute flow- and agonist-evoked endothelial nitric oxide release. *Br J Pharmacol* 1997; **122**: 117–125.
73. Davis MJ, Sharma NR. Calcium-release-activated calcium influx in endothelium. *J Vasc Res* 1997; **34**: 186–195.
74. Ayajiki K, Kindermann M, Hecker M *et al.* Intracellular pH and tyrosine phosphorylation but not calcium determine shear stress-induced nitric oxide production in native endothelial cells. *Circ Res* 1996; **78**: 750–758.

75. Graier WF, Simecek S, Sturek M. Cytochrome P450 mono-oxygenase-regulated signalling of Ca^{2+} entry in human and bovine endothelial cells. *J Physiol* (*Lond*) 1995; **482**: 259–274.
76. Hoebel BG, Kostner GM, Graier WF. Activation of microsomal cytochrome P450 mono-oxygenase by Ca^{2+} store depletion and its contribution to Ca^{2+} entry in porcine aortic endothelial cells. *Br J Pharmacol* 1997; **121**: 1579–1588.
77. Lückhoff A, Busse R. Calcium influx into endothelial cells and formation of endothelium-derived relaxing factor is controlled by the membrane potential. *Pflügers Arch* 1990; **416**: 305–311.
78. Janigro D, West GA, Gordon EL, Winn HR. ATP-sensitive K^+ channels in rat aorta and brain microvascular endothelial cells. *Am J Physiol* 1993; **265**: C812–C821.
79. Cooke JP, Rossitch E, Andon NA *et al.* Flow activates an endothelial potassium channel to release an endogenous vasodilator. *J Clin Invest* 1991; **88**: 1663–1671.
80. Ohno M, Gibbons GH, Dzau VJ, Cooke JP. Shear stress elevates endothelial cGMP. Role of a potassium channel and G protein coupling. *Circulation* 1993; **88**: 193–197.
81. Jacobs ER, Cheliakine C, Gebremedhin D *et al.* Shear activated channels in cell-attached patches of cultured bovine aortic endothelial cells. *Pflügers Arch* 1995; **431**: 129–131.
82. Busse R, Fichtner H, Lückhoff A, Kohlhardt M. Hyperpolarization and increased free calcium in acetylcholine-stimulated endothelial cells. *Am J Physiol* 1988; **255**: H965–H969.
83. Demirel E, Rusko J, Laskey RE *et al.* TEA inhibits ACh-induced EDRF release: endothelial Ca^{2+}-dependent K^+ channels contribute to vascular tone. *Am J Physiol* 1994; **267**: H1135–H1141.
84. Gooch KJ, Frangos JA. Flow- and bradykinin-induced nitric oxide production by endothelial cells is independent of membrane potential. *Am J Physiol* 1996; **270**: C546–C551.
85. O'Neill WC. Flow-mediated NO release from endothelial cells is independent of K^+ channel activation or intracellular Ca^{2+}. *Am J Physiol* 1995; **269**: C863–C869.
86. Lesh RE, Marks AR, Somlyo AV *et al.* Anti-ryanodine receptor antibody binding sites in vascular and endocardial endothelium. *Circ Res* 1993; **72**: 481–488.
87. Hecker M, Mülsch A, Bassenge E, Busse R. Vasoconstriction and increased flow: two principal mechanisms of shear stress-dependent endothelial autacoid release. *Am J Physiol* 1993; **265**: H828–H833.
88. Kuchan MJ, Frangos JA. Role of calcium and calmodulin in flow-induced nitric oxide production in endothelial cells. *Am J Physiol* 1994; **266**: C628–C636.

89. Corson MA, James NL, Latta SE *et al.* Phosphorylation of endothelial nitric oxide synthase in response to fluid shear stress. *Circ Res* 1996; **79**: 984–991.
90. Tseng H, Peterson TE, Berk BC. Fluid shear stress stimulates mitogen-activated protein kinase in endothelial cells. *Circ Res* 1995; **77**: 869–878.
91. Evans L, Frenkel L, Brophy CM *et al.* Activation of diacylglycerol in cultured and endothelial cells exposed to cyclic strain. *Am J Physiol* 1997; **272**: C650–C656.
92. Boulanger CM, Vanhoutte PM. G proteins and endothelium-dependent relaxations. *J Vasc Res* 1997; **34**: 175–185.
93. Gudi SRP, Clark CB, Frangos JA. Fluid flow rapidly activates G proteins in human endothelial cells. Involvement of G proteins in mechanochemical signal transduction. *Circ Res* 1996; **79**: 834–839.
94. Kuchan MJ, Jo H, Frangos JA. Role of G proteins in shear stress-mediated nitric oxide production by endothelial cells. *Am J Physiol* 1994; **267**: C753–C758.
95. Frangos JA, Huang TY, Clark CB. Steady shear and step changes in shear stimulate endothelium *via* independent mechanisms - superposition of transient and sustained nitric oxide production. *Biochem Biophys Res Commun* 1996; **224**: 660–665.
96. Banes AJ, Tsuzaki M, Yamamoto J *et al.* Mechanoreception at the cellular level: the detection, interpretation, and diversity of responses to mechanical signals. *Biochem Cell Biol* 1995; **73**: 349–365.
97. Yano Y, Geibel J, Sumpio BE. Tyrosine phosphorylation of $pp125^{FAK}$ and paxillin in aortic endothelial cells induced by mechanical strain. *Am J Physiol* 1996; **271**: C635–C649.
98. Knudsen, HL, Frangos JA. Role of cytoskeleton in shear stress-induced endothelial nitric oxide production. *Am J Physiol* 1997; **273**: H347–H355.
99. Muller JM, Chilian WM, Davis MJ. Integrin signaling transduces shear stress-dependent vasodilation of coronary arterioles. *Circ Res* 1997; **80**: 320–326.
100. Hutcheson IR, Griffith TM. Mechanotransduction through the endothelial cytoskeleton: mediation of flow- but not agonist-induced EDRF release. *Br J Pharmacol* 1996; **118**: 720–726.
101. Diamond SL, Sachs F, Sigurdson WJ. Mechanically induced calcium mobilization in cultured endothelial cells is dependent on actin and phospholipase. *Arterioscler Thromb* 1994; **14**: 2000–2006.
102. Henrion D, Terzi F, Matrougui K *et al.* Impaired flow-induced dilation in mesenteric resistance arteries from mice lacking vimentin. *J Clin Invest* 1997; **100**: 2909–2914.
103. Liu J, García-Cardeña G, Sessa WC. Palmitoylation of endothelial nitric oxide synthase is necessary for optimal stimulated release of nitric oxide: implications for caveolae localization. *Biochemistry* 1996; **35**: 13277–13281.

104. Robinson LJ, Michel T. Mutagenesis of palmitoylation sites in endothelial nitric oxide synthase identifies a novel motif for dual acylation and subcellular targeting. *Proc Natl Acad Sci USA* 1995; **92**: 11776–11780.
105. Venema VJ, Zou R, Ju H *et al.* Caveolin-1 detergent solubility and association with endothelial nitric oxide synthase is modulated by tyrosine phosphorylation. *Biochem Biophys Res Commun* 1997; **236**: 155–161.
106. Michel JB, Feron O, Sacks D, Michel T. Reciprocal regulation of endothelial nitric-oxide synthase by Ca^{2+}-calmodulin and caveolin. *J Biol Chem* 1997; **272**: 15583–15586.
107. Ju H, Zou R, Venema VJ, Venema RC. Direct interaction of endothelial nitric-oxide synthase and caveolin-1 inhibits synthase activity. *J Biol Chem* 1997; **272**: 18522–18525.
108. Sessa WC, García-Cardeña G, Liu J *et al.* The Golgi association of endothelial nitric oxide synthase is necessary for the efficient synthesis of nitric oxide. *J Biol Chem* 1995; **270**: 17641–17644.
109. Venema VJ, Marrero MB, Venema RC. Bradykinin-stimulated protein tyrosine phosphorylation promotes endothelial nitric oxide synthase translocation to the cytoskeleton. *Biochem Biophys Res Commun* 1996; **226**: 703–710.
110. Graier WF, Paltauf-Doburzynska J, Hill BJF *et al.* Submaximal stimulation of porcine endothelial cells causes focal Ca^{2+} elevation beneath the cell membrane. *J Physiol (Lond)* 1998; **506**: 109–125.
111. Xiao Z, Zhang Z, Diamond SL. Shear stress induction of the endothelial nitric oxide synthase gene is calcium-dependent but not calcium-activated. *J Cell Physiol* 1997; **171**: 205–211.
112. Marsden PA, Heng HHQ, Scherer SW *et al.* Structure and chromosomal localization of the human constitutive endothelial nitric oxide synthase gene. *J Biol Chem* 1993; **268**: 17478–17488.
113. Ohara Y, Sayegh HS, Yamin JJ, Harrison DG. Regulation of endothelial constitutive nitric oxide synthase by protein kinase C. *Hypertension* 1995; **25**: 415–420.
114. Dull RO, Tarbell JM, Davies PF. Mechanisms of flow-mediated signal transduction in endothelial cells: kinetics of ATP surface concentrations. *J Vasc Res* 1992; **29**: 410–419.
115. Milner P, Kirkpatrick KA, Ralevic V *et al.* Endothelial cells cultured from human umbilical vein release ATP, substance P and acetylcholine in response to increased flow. *Proc Royal Soc Lond (B)* 1990; **241**: 245–248.
116. Hassessian H, Bodin P, Burnstock G. Blockade by glibenclamide of the flow-evoked endothelial release of ATP that contributes to vasodilatation in the pulmonary vascular bed of the rat. *Br J Pharmacol* 1993; **109**: 466–472.
117. Ando J, Ohtsuka A, Korenaga R *et al.* Wall shear stress rather than shear rate regulates cytoplasmic Ca^{++} responses to flow in vascular endothelial cells. *Biochem Biophys Res Commun* 1993; **190**: 716–723.

118. Berthiaume F, Frangos JA. Flow-induced prostacyclin production is mediated by a pertussis toxin-sensitive G protein. *FEBS Lett* 1992; **308**: 277–279.
119. Onohara T, Okadome K, Yamamura S *et al.* Simulated blood flow and the effects on prostacyclin production in the dog femoral artery. *Circ Res* 1991; **68**: 1095–1099.
120. Salvemini D, Currie MG, Mollace V. Nitric oxide-mediated cyclooxygenase activation. A key event in the antiplatelet effects of nitrovasodilators. *J Clin Invest* 1996; **97**: 2562–2568.
121. Cohen RA. Dysfunction of vascular endothelium in diabetes mellitus. *Circulation* 1993; **87**: V-67–V-76.
122. Wang W, Diamond SL. Does elevated nitric oxide production enhance the release of prostacyclin from shear stressed aortic endothelial cells? *Biochem Biophys Res Commun* 1997; **233**: 748–751.
123. Tsai A-I, Wei C, Kulmacz R. Interaction between nitric oxide and prostaglandin H synthase. *Arch Biochem Biophys* 1994; **313**: 367–372.
124. Landino LM, Crews BC, Timmons MD *et al.* Peroxynitrite, the coupling product of nitric oxide and superoxide, activates prostaglandin biosynthesis. *Proc Natl Acad Sci USA* 1996; **93**: 15069–15074.
125. Laurindo FRM, Pedro M de A, Barbeiro HV *et al.* Vascular free radical release. *Ex vivo* and *in vivo* evidence for a flow-dependent endothelial mechanism. *Circ Res* 1994; **74**: 700–709.
126. Topper JN, Cai J, Falb D, Gimbrone MA Jr. Identification of vascular endothelial genes differentially responsive to fluid mechanical stimuli: cyclooxygenase-2, manganese superoxide dismutase, and endothelial cell nitric oxide synthase are selectively up-regulated by steady laminar shear stress. *Proc Natl Acad Sci USA* 1996; **93**: 10417–10422.
127. Mombouli J-V, Vanhoutte PM. Endothelium-derived hyperpolarizing factor(s): updating the unknown. *TIPS* 1997; **18**: 252–256.
128. Popp R, Bauersachs J, Hecker M *et al.* A transferable, β-naphthoflavone-inducible, hyperpolarizing factor is synthesized by native and cultured porcine coronary endothelial cells. *J Physiol (Lond)* 1996; **497**: 699–709.
129. Randall MD, Alexander SP, Bennett T *et al.* An endogenous cannabinoid as an endothelium-derived vasorelaxant. *Biochem Biophys Res Commun* 1996; **229**: 114–120.
130. Bolotina VM, Najibi S, Palacino JJ *et al.* Nitric oxide directly activates calcium-dependent potassium channels in vascular smooth muscle. *Nature* 1994; **368**: 850–853.
131. Robertson BE, Schubert R, Hescheler J, Nelson MT. cGMP-dependent protein kinase activates Ca-activated K channels in cerebral artery smooth muscle cells. *Am J Physiol* 1993; **265**: C299–C303.
132. Parkington HC, Tare M, Tonta MA, Coleman HA. Stretch revealed three components in the hyperpolarization of guinea-pig coronary artery in response to acetylcholine. *J Physiol (Lond)* 1993; **465**: 459–476.

133. Tare M, Parkington HC, Coleman HA *et al.* Hyperpolarization and relaxation of arterial smooth muscle caused by nitric oxide derived from the endothelium. *Nature* 1990; **346**: 69–71.
134. Chaytor AT, Evans WH, Griffith TM. Central role of heterocellular gap junctional communication in endothelium-dependent relaxations of rabbit arteries. *J Physiol (Lond)* 1998; **508**: 561–573.
135. Chaytor AT, Evans WH, Griffith TM. Peptides homologous to extracellular loop motifs of connexin 43 reversibly abolish rhythmic contractile activity in rabbit arteries. *J Physiol (Lond)* 1997; **503**: 99–110.
136. Little TL, Xia J, Duling BR. Dye tracers define differential endothelial and smooth muscle coupling patterns within the arteriolar wall. *Circ Res* 1995; **76**: 498–504.
137. Dora KA, Doyle MP, Duling BR. Elevation of intracellular calcium in smooth muscle causes endothelial cell generation of NO in arterioles. *Proc Natl Acad Sci USA* 1997; **94**: 6529–6534.
138. von der Weid PY, Bény J-L. Simultaneous oscillations in the membrane potential of pig coronary artery endothelial and smooth muscle cells. *J Physiol (Lond)* 1993; **471**: 13–24.
139. Melkumyants AM, Balashov SA, Klimachev AN *et al.* Nitric oxide does not mediate flow induced endothelium dependent arterial dilatation in the cat. *Cardiovasc Res* 1992; **26**: 256–260.
140. Fujii K, Heistad DD, Faraci FM. Flow-mediated dilatation of the basilar artery *in vivo. Circ Res* 1991; **69**: 697–705.
141. Murray CD. The physiological principle of minimum work applied to the angle of branching of arteries. *J Gen Physiol* 1926; **9**: 835–841.
142. Griffith TM, Edwards DH. Basal NO activity helps to keep the geometrical configuration of arterial bifurcations close to the Murray optimum. *J Theor Biol* 1990; **146**: 545–573.
143. Griffith TM, Edwards DH, Randall MD. Blood flow and optimal vascular topography: role of the endothelium. *Basic Res Cardiol* 1991; **86 (Suppl 2)**: 89–96.
144. Mayrovitz HN, Roy J. Microvascular blood flow: evidence indicating a cubic dependence on arteriolar diameter. *Am J Physiol* 1983; **245**: H1031–H1038.
145. Rossitti S, Frisen L. Remodelling of the retinal arterioles in descending optic atrophy follows the principle of minimum work. *Acta Physiol Scand* 1994; **152**: 333–340.
146. Mandelbrot BB. *The Fractal Geometry of Nature.* Freeman, New York 1982: 156–165.
147. Numaguchi K, Egashira K, Takemoto M *et al.* Chronic inhibition of NO synthesis causes coronary microvascular remodelling in rats. *Hypertension* 1995; **26**: 957–962.

148. Pohl U, Lamontagne D. Impaired tissue perfusion after inhibition of endothelium-derived nitric oxide. *Basic Res Cardiol* 1991; **86 (Suppl 2)**: 97–105.
149. Rivers RJ. Remote effects of pressure changes in arterioles. *Am J Physiol* 1995; **268**: H1379–H1382.
150. Segal SS, Damon DN, Duling BR. Propagation of vasomotor responses coordinates arteriolar resistances. *Am J Physiol* 1989; **256**: H822–H837.
151. Frame MDS, Sarelius IH. L-arginine-induced conducted signals alter upstream arteriolar responsibity to L-arginine. *Circ Res* 1995; **77**: 695–701.
152. Ueeda M, Silvia SK, Olsson RA. Nitric oxide modulates coronary autoregulation in the guinea pig. *Circ Res* 1992; **70**: 1296–1303.
153. Pohl U, Lamontagne D, Bassenge E, Busse R. Attenuation of coronary autoregulation in the isolated rabbit heart by endothelium derived nitric oxide. *Cardiovasc Res* 1994; **28**: 414–419.
154. Smith TP, Canty JM. Modulation of coronary autoregulatory responses by nitric oxide. *Circ Res* 1993; **73**: 232–240.
155. Griffith TM, Edwards DH. NO suppresses chaotic pressure oscillations in an isolated resistance artery without influencing their intrinsic complexity. *Am J Physiol* 1994; **266**: H1786–H1800.
156. Griffith TM, Edwards DH. Fractal analysis of the role of smooth muscle Ca^{2+}-fluxes in the genesis of chaotic arterial pressure oscillations. *Am J Physiol* 1994; **266**: H1801–H1811.
157. Griffith TM. Temporal chaos in the microcirculation. *Cardiovasc Res* 1996; **31**: 342–358.
158. Jackson WF, Mülsch A, Busse R. Rhythmic smooth muscle activity in hamster aortas is mediated by continuous release of NO from the endothelium. *Am J Physiol* 1991; **260**: H248–H253.
159. Persson PB, Baumann JE, Ehmke H *et al.* Phasic and 24-h blood pressure control by endothelium-derived relaxing factor in conscious dogs. *Am J Physiol* 1992; **262**: H1395–H1400.
161. Jia L, Bonaventura C, Bonaventura J, Stamler JS. *S*-nitrosohaemoglobin: a dynamic activity of blood involved in vascular control. *Nature* 1996; **380**: 221–226.
162. Stamler JS, Jia L, Eu JP *et al.* Blood flow regulation by *S*-nitrosohemoglobin in the physiological oxygen gradient. *Science* 1997; **276**: 2034–2037.
163. Scharfstein JS, Keaney JF Jr, Slivka A *et al.* *In vivo* transfer of nitric oxide between a plasma protein-bound reservoir and low molecular weight thiols. *J Clin Invest* 1994; **94**: 1432–1439.

Chapter 3

Chemistry and Molecular Biology of Nitric Oxide Synthesis

Adrian J Hobbs[1] and Louis J Ignarro[2]

Introduction

Extensive investigation in the 1970s revealed that the smooth muscle relaxant actions of nitrovasodilator drugs, such as glyceryl trinitrate, were due to their ability to raise intracellular guanosine 3',5'-cyclic monophosphate (cGMP) levels. Subsequently, such agents were shown to be able to release NO, which in turn stimulated soluble guanylate cyclase (sGC), the enzyme catalyzing the formation of cGMP. Despite the ability of certain bacteria to synthesize this noxious, highly-reactive gas, mammalian cells were not thought capable of forming NO, and the possibility that NO could represent an endogenous ligand for sGC was ignored. However, identification of EDRF, first described by Furchgott and Zawadzki as NO in 1980,[1] dramatically altered this perception. Indeed, not only did this observation demonstrate that mammalian vascular endothelial cells could produce and release NO to modulate the tone of the underlying smooth muscle, but it also illustrated the profound physiological importance of the NO-cGMP signal transduction system.

The considerable advances in the understanding of the role of NO as an inter- and intra-cellular messenger molecule have further elevated the significance of cGMP as a second messenger. Indeed, the spectrum of biological effects mediated by NO and cGMP has grown exponentially, to include such diverse functions as anticoagulation, non-adrenergic/non-cholinergic (NANC) neurotransmission, phototransduction, vascular and non-vascular smooth muscle relaxation and the cytotoxic and cytostatic actions of immune cells.

This chapter will attempt to provide a brief overview of the physiological synthesis and actions of NO at a chemical level, with particular emphasis on the endogenous formation of *S*-nitrosothiols and their putative biological roles. Additionally, the recent developments in the field following utilization of molecular biology to elucidate the genetic regulation of NO synthase (NOS) will be discussed. For a more comprehensive description of this signalling pathway, the interested reader is referred to several review articles.[2–5]

Nitric Oxide Synthase: Chromosomal Location and Molecular Cloning of Isoforms

Considerable information concerning the physiological and pathophysiological activity and regulation of NOS, and hence NO itself, has been derived from the molecular characterization of the various isoforms of the synthetic enzyme. It is this aspect, therefore, that should be considered first to set the background for the biological and chemical characterization of this unique messenger molecule.

Based upon several criteria including cellular location, regulation of activity and substrate/inhibitor profiles, the NOS enzyme can be divided into three distinct isoforms: firstly, a constitutive form, whose activity is regulated by Ca^{2+} and calmodulin and which is found in neural tissue, both centrally and peripherally (nNOS); a second Ca^{2+}/calmodulin-requiring constitutive enzyme is present in vascular endothelial cells (eNOS); thirdly, a Ca^{2+}-independent isoform isolated from macrophages, vascular smooth muscle cells and hepatocytes following induction with specific cytokines (iNOS). Constitutive isoforms of NOS are all homodimers consisting of subunits ranging from 135–200 kDa.[6–10] Likewise, macrophage iNOS also exists as a homodimer, each subunit approximately 130 kDa.[11–13]

The vast majority of information gleaned about NOS isoforms has only been possible due to the over-expression of enzyme protein in eukaryotic and prokaryotic cell systems. This has provided a plentiful supply of active enzyme with which to conduct structural and kinetic analysis. At first, the individual isozymes were obtained from fresh tissue or cultured cell lines; nNOS was extracted from rat[14] or pig[6] cerebellum, iNOS from mouse peritoneal macrophages (RAW 264.7 cells) stimulated with cytokine cocktails[11] and eNOS from freshly isolated bovine aortic endothelial cells.[7] In recent years, NOS protein overexpression has been achieved by transfection of mammalian embryonic kidney 293 cells,[15] COS-7 kidney fibroblasts[16], and insect (*Spodoptera frugiperda*) cells.[17] Human NOS isozymes have also been cloned and expressed in baculovirus-directed insect cell expression systems.[17,18] More recently, overexpression has been conducted in prokaryotic cells (*E. coli*) which facilitate the production of larger quantities of protein due to their rapid replication.[19] Apparently, the cell types used manufacture sufficient quantities of cofactors to support NOS dimerization and activity, although in many instances, the cells must be supplemented with haemin to provide sufficient haem levels for incorporation into the enzyme. Once expressed, the protein is isolated using 2',5'-ADP sepharose chromatography and eluted with nicotinamide adenine dinucleotide phosphate (NADPH) or 2'-AMP.[20] Further purification by anion-exchange chromatography can be conducted to separate dissociated subunits, if present.[21,22] Identification and purification of individual isoforms has been aided by the generation of monoclonal antibodies.[23,24]

Following the purification of NOS protein, peptide sequences from tryptic digests permitted the definitive molecular cloning of the complementary DNA (cDNA) for each isoform. Initially, two tryptic peptide sequences from purified nNOS were used to design degenerate primers in a polymerase chain reaction (PCR) with a rat brain cDNA library, which amplified a 600bp fragment. In turn, this fragment was used to obtain the full coding from the same library.[25] Subsequently, the cDNAs for eNOS and iNOS have been obtained by probing endothelial cells and activated macrophages, respectively, with the sequence derived from the rat brain library.[26] Although the L-arginine-NO pathway has been identified in many species, including fish, birds, plants and bacteria, NOS has been best studied in mammals.[27] Here, three distinct genes encode each isoform of NOS (iterations of the sequence appear throughout mammalian genomes, yet only one appears to be functional; *personal communication*, Professor IG Charles) and consist of either 26 exons (iNOS and eNOS) or 29 exons (nNOS).[28–30] The isoforms share 50–60% homology and display many similarities to the cytochrome P450 family of enzymes; indeed, it is this analogy that has provided many clues to the structural composition/arrangement and catalytic process mediated by NOS.

Sequence analysis of the predicted primary structure of NOS isoforms has distinguished important functional motifs. All NOS proteins possess a bi-domain structure, and dimerization to homodimers is required for enzymatic activity. The C-terminal half of the NOS protein bears remarkable resemblance to only one other mammalian protein, cytochrome P450 reductase (CPR), and appears to possess the same cofactor binding sites; consequently, this is often referred to as the reductase domain. At the extreme C-terminus is an NADPH binding region, which is conserved in all NOS isoforms and aligns perfectly with CPR.[31] The binding site comprises a glycine-rich region followed by 30 amino acids and a cysteine residue (which is known to bind NADPH due to the latter's ability to protect the Cys residue from sulphydryl oxidizing agents). A highly conserved lysine is also thought to contribute to NADPH binding by interacting with the ribose 2'-phosphate group. The NADPH binding site is followed, in turn, by consensus sequences for flavin mononucleotide (FMN) and flavin adenine dinucleotide (FAD), which comprise approximately 200 and 60 amino acids respectively.[25] NOS, unlike CPR, is a self-sufficient enzyme and the oxygenation of substrate, L-arginine, occurs at a haem-site in the N-terminal portion (oxygenase domain) of the protein. Stoichiometric amounts of haem are present in NOS and are required for catalytic activity.[32,33] Resonance Raman spectroscopy has demonstrated the haem coordination to be pentavalent with a thiolate (fifth) axial ligand.[34] Haem coordination is thought to be provided by Cys-415 (in nNOS) based upon homology to cytochromes P450 and haem incorporation following site-directed mutagenesis.[35] Indeed, this Cys residue is conserved in all NOS isoforms across differing species and corresponds to Cys-200 in human iNOS and Cys-184 in human eNOS. Exons encoding the haem (thiolate) binding sequences are also highly conserved; each encodes a 54 amino acid sequence corresponding to exon 6 (nNOS and iNOS) or exon 4 (eNOS). As expected, close to the haem (catalytic) site is a substrate (L-arginine) binding site, providing further evidence that L-arginine oxidation occurs at the haem iron.[32] Separation of the two domains (*via* trypsin cleavage at a specific alanine residue at the C-terminal end of the oxygenase domain)[36] enabled L-arginine and pterin binding sites to be located to the oxygenase domain.

Bridging the reductase and oxygenase domains is a calmodulin binding site which appears to act as a switch to regulate electron flow between the two regions.[37] Putative calmodulin binding domains have been identified based upon canonical sequences of basic and lipophilic residues. These span a 20–25 amino acid length arising from two exons in each of the mammalian NOS genes.[38] Interestingly, the calmodulin binding domain of eNOS contains a 4 amino acid deletion at its N-terminal aspect, which may confer functional differences in calmodulin binding affinity, although no confirmation of this has been provided to date. NOS appears to assume an α-helical conformation as a consequence of calmodulin binding, analogous to the calmodulin-occupied domain of myosin light-chain kinase. Finally, eNOS possesses consensus sequences for myristoylation/palmitoylation at its N-terminus, which confers its particulate nature (as opposed to the cytosolic location of iNOS). Post-translational modification at these sites is necessary for its membrane association.[39]

Regulation of Nitric Oxide Synthase Expression

Although much has been deduced about the molecular composition of NOS, rather less is known about its regulation of expression. This is of extreme importance in the case of iNOS which, unlike eNOS and nNOS, is not expressed constitutively but induced by the influence of inflammatory mediators and bacterial metabolites. Since iNOS possesses tightly bound calmodulin, it is independent of free Ca^{2+} concentrations and, therefore, once expressed continues to synthesize NO indefinitely. Hence, iNOS activity is regulated by protein expression rather than functional modulation.

Promoter Sequences

The 5'-flanking region of human iNOS possesses approximately 66% homology to its murine counterpart[40] and both contain conserved consensus sequences for nuclear factor kappa B (NF-κB), interferon (IFN)-γ response elements (γ-IRE; thrice repeated) and a tumour necrosis factor (TNF) response element. An A-activator-binding site also exists in the promoter region, which is thought to convey liver-specific gene expression.[41] The murine iNOS promoter (like the human equivalent) possesses an NF-κB binding site at position –76 to –85 which has been shown to bind proteins of the Rel/NF-κB family in response to lipopolysaccharide (LPS).[42] Further upstream, (position –913– to –923) is a γ-IRE which has been shown to bind interferon regulatory factor (IRF)-1 upon stimulation of mouse peritoneal macrophages (RAW 264.7 cells) with IFN-γ.[43] The latter element appears to function as an enhancer and is not able independently to initiate gene expression.[44] However, this promoter sequence arrangement can be reconciled with the inability of IFN-γ alone to induce iNOS expression in RAW 264.7 cells, in contrast to its marked synergism with LPS. In accord with this observation, it is thought that the DNA binding protein NF-κB may impose the greatest degree of transcriptional control on iNOS expression, at least in murine systems; this is discussed in greater detail below. Despite these similarities, the transcriptional control of murine and human iNOS expression is markedly different. In the murine system, it appears that a proximal 1.6 kb 5'-flanking region contains the necessary promoter sequences to induce full gene expression. In the human, however, the corresponding sequence produces little or no gene expression, despite possessing transcription factor consensus sequences. Indeed, if a 16 kb fragment upstream from the coding sequence is cloned, linked to a luciferase reporter gene and transfected into human cell lines, it is still insufficient to promote full gene expression (as measured by luciferase activity).[45] Thus, there is a distinct difference in the requirements for iNOS induction between mouse and human; this may well reflect evolutionary changes in the manner in which NO is utilized to fight pathogenic insults in man. The differences may also explain the reported difficulty in inducing iNOS expression in many human cell lines.[46]

Nuclear Factor Kappa B

NF-κB is a higher eukaryotic transcription factor that plays a pivotal role in the expressional regulation of many genes that encode pro-inflammatory proteins.[47,48] NF-κB is a heterodimeric protein composed most frequently of the DNA-binding proteins p50 and p65 (or Rel-A), and resides in the cytoplasm as a latent form stabilized by an inhibitory protein termed IκB.[48–50] As a consequence of cellular activation by a plethora of pathophysiological mediators, including cytokines and bacterial metabolites, the IκB protein undergoes phosphorylation (by an unidentified protein kinase), which tags the IκB rendering it a high-affinity substrate for a ubiquitin-conjugating enzyme system. In turn, ubiquitination identifies IκB as a substrate for the proteasome, a Ca^{2+}-independent and non-lysosomal multicatalytic proteinase complex. Thus, phosphorylation-regulated ubiquitination results in a rapid degradation of IκB, leading to liberation of the NF-κB heterodimer.[51,52] This separation reveals nuclear localization and DNA-binding sequences which permit the NF-κB to translocate into the nucleus and bind to appropriate regulatory elements in promoter regions on target genes, resulting in gene expression and subsequent protein synthesis.[50] This process is transient and is terminated through a delayed induction of IκB expression by NF-κB itself.[53] The majority of stimuli that activate NF-κB are associated with oxidative stress, as manifested by increased production of reactive oxygen species (ROS), including superoxide ($O_2^{\cdot -}$), hydrogen peroxide (H_2O_2) and hydroxy radical (·OH).[54] Experimental evidence suggests that there is a convergence of distinct stimuli to generate a common ROS, possibly H_2O_2, that serves as an intracellular second messenger in mediating NF-κB activation.[55] It is unknown, however, how these two pathways are coupled (Figure 1).

Since NO was known for many years to play an important role in host defence,[56] it was not surprising when agents known to interfere with NF-κB activity were shown to modulate the induction of iNOS. Firstly, many antioxidants, including pyrrolidinedithiocarbamate (PDTC) and diethyldithiocarbamate (DETC), were demonstrated to inhibit iNOS expression in cultured cells,[57,58] in addition to non-

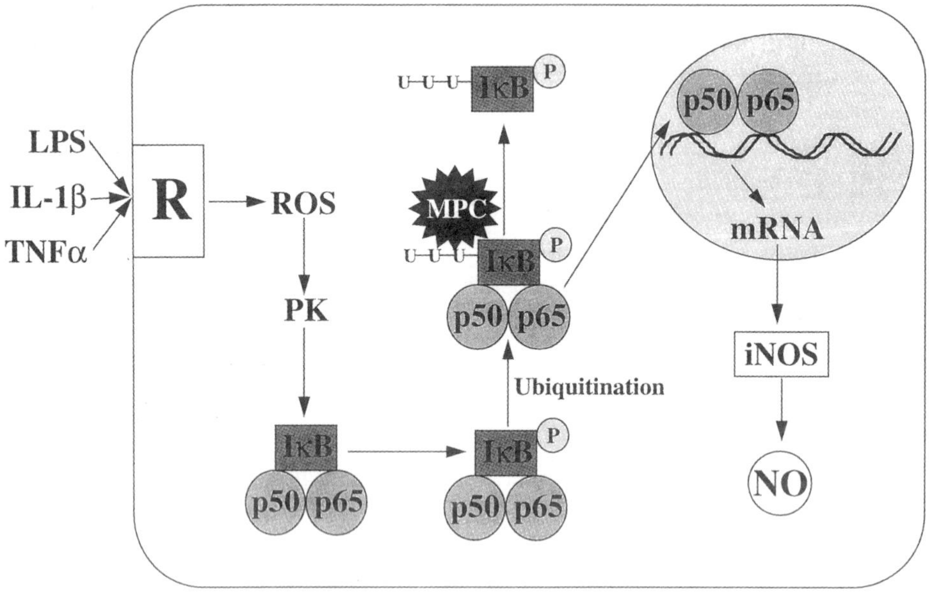

Figure 1. Schematic representation depicting the transcriptional regulation of iNOS by NF-κB. Cytokine activation (e.g. LPS, lipopolysaccharide; TNFα, tumour necrosis factor-α; IL-1β, interleukin-1β) results in the intracellular generation of reactive oxygen species (ROS). The ROS activate an identified protein kinase (PK) which phosphorylates (P) IκB, an inhibitory subunit bound to the NF-κB p50/p65 heterodimer. Phosphorylation of IκB renders it susceptible to a ubiquitin conjugating system. In turn, poly-ubiquitination (U-U-U) results in the degradation of the IκB by the 26S proteasome, a multicatalytic proteinase complex (MPC). The released NF-κB dimer is able to translocate to the nucleus and elicit gene expression.

selective protease inhibitors such as Nα-p-tosyl-L-phenylalanine (TPCK) and Nα-p-tosyl-L-lysine-chloromethylketone (TLCK) and more selective calpain (α-N-acetyl-Leu-Leu-norleucinal) and proteasome inhibitors (Z-IE(O-t-Bu)A-leucinal). Each is believed to prevent the degradation of IκB by the proteasome and thereby prevent liberation of the active NF-κB complex.[59,60] Several other distinct classes of chemical agent have been demonstrated to prevent expression of iNOS *via* inhibition of the NF-κB transduction system. Glucocorticoids such as dexamethasone interfere with iNOS expression in many cell types by reducing the DNA binding capacity of NF-κB.[61–64] Glucocorticoids have also been shown to induce the expression of the IκB protein, thereby preventing the liberation of free NF-κB heterodimer.[65] Non-steroidal anti-inflammatory drugs, including aspirin and sodium salicylate, also appear to down-regulate iNOS expression by interfering with NF-κB activity.[66,67]

Recently, it has become apparent that NO itself possesses the ability to modulate NF-κB activity. This not only has important implications for the feedback regulation of iNOS expression, but since NF-κB is known transcriptionally to regulate the expression of numerous pro-inflammatory proteins (see above), NO may act as a widespread modulator of immune function and host defence. We have recently demonstrated that NO can both enhance and inhibit NF-κB activity and iNOS expression in response to LPS in cultured mouse peritoneal macrophages (RAW 264.7 cells). At lower concentrations, the presence of NO markedly potentiates subsequent NO production (as determined by measurement of stable oxidation products NO_2^-, NO_3^-), whereas at higher concentrations of NO there is a pronounced inhibition. When examined at a more molecular level, utilizing electrophoretic mobility shift assays and Northern blot analysis to assess NF-κB and iNOS mRNA levels respectively, the same biphasic activity profile of NO is observed (Figure 2). Thus, NO can indirectly modulate iNOS expression (and hence its own synthesis) by interfering with the NF-κB transduction pathway. The mechanisms responsible for this biphasic effect are not clear. Certainly, if ROS are responsible for activation of NF-κB, then it could be envisaged that NO (since it is a free radical) may be able to react with and inactivate such species to reduce the degree of activation. Furthermore, it has been demonstrated that NO is able to induce the expression of IκB and inhibit the DNA-binding capacity of NF-κB.[68,69] However, the mechanism(s) underlying NO-induced enhancement of NF-κB activity are rather more obscure. One possibility is that NO may inhibit certain anti-oxidant enzymes and consequently increase the levels of ROS in the local environment, which in turn would elicit a greater degree of NF-κB activation. One putative target in this regard is catalase, which rapidly degrades H_2O_2 to O_2 and H_2O, thus preventing cellular H_2O_2 concentrations from reaching levels where significant Fenton-type chemistry would lead to cytotoxicity; a reduction in catalase activity by NO binding to its ferric catalytic site would raise H_2O_2 levels and possibly stimulate NF-κB activity. Although these hypotheses are somewhat speculative, the potential for NO to play a pivotal role in regulating immune function has been demonstrated and remains to be explored.

The Regulation of High-output Nitric Oxide Production

One essential co-factor for the synthesis of NO by NOS is tetrahydrobiopterin (see above) and it is guanosine triphosphate-cyclohydrolase I (GTP-CHI) that converts GTP to dihydroneopterin, which is the rate-limiting synthetic enzyme. In many cell types, it has been demonstrated that expression of active iNOS protein only occurs if GTP-CHI is co-induced, presumably to provide the necessary tetrahydrobiopterin (BH_4) for NOS catalytic activity.[70,71] 2,4-diamino-6-hydroxypyrimidine (DAHP), a selective inhibitor of GTP-CHI, completely suppresses the elevated levels of BH_4 in response to cytokines in vascular smooth muscle cells. This suggests it is *de novo* synthesis of BH_4 which supports the high-output production of NO by iNOS, rather than the pterin 'salvage' pathway which

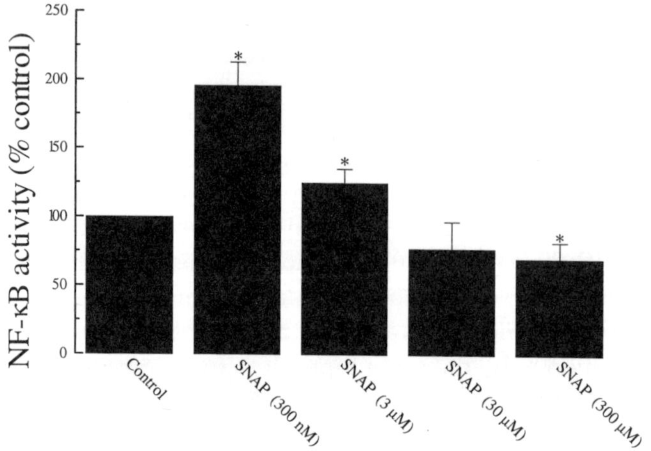

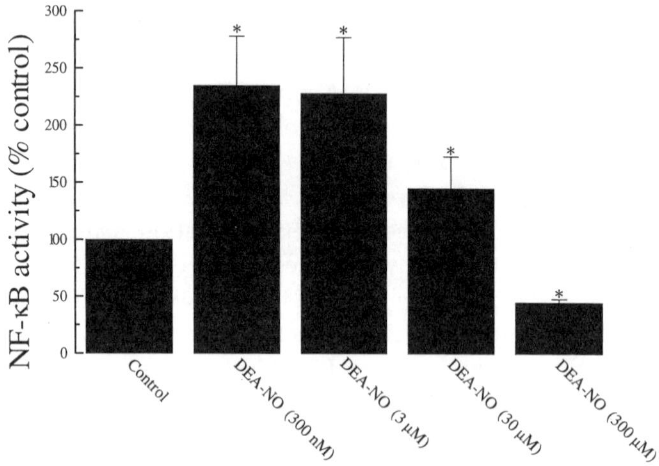

Figure 2. The effect of *S*-nitroso-N-acetylpenicillamine (SNAP, 300 nM–300 μM; upper panel) and sodium (Z)-1-(N,N-diethylamino)diazen-1-ium-1,2-diolate (DEA-NO, 300 nM–300 μM; lower panel) on NF-κB activity in cultured murine peritoneal macrophages (RAW 264.7 cells). Cells (10^6) were incubated with the appropriate test agent for 5 minutes prior to activation with LPS (1 μg/ml). Following a further 90 minute incubation period, cells were harvested, the nuclear fraction isolated, and NF-κB activity measured by electrophoretic mobility shift assay using a ^{32}P-labelled (double-stranded) oligonucleotide probe corresponding to the NF-κB consensus sequence (-GGGGACTTTCC-). n = 6–8 separate experiments; *$P < 0.05$ compared to control (LPS alone; Student's *t*-test).

converts sepiapterin to BH_4.[72] Such observations may have important implications for the treatment of conditions such as septic shock, in which inappropriate iNOS activity results in excessive NO production and profound hypotension; in this case, inhibitors of BH_4 synthesis (e.g. DAHP, methotrexate and N-acetylserotonin) may be able to reduce NO production by limiting the supply of BH_4 to the iNOS enzyme.

Nitric Oxide Synthase: Mechanism of Catalysis

In mammalian cells, NO is generated *via* a five electron oxidation of a terminal guanidinium nitrogen on the amino acid L-arginine. The reaction is both oxygen- and nicotinamide adenine dinucleotide phosphate (NADPH)-dependent and yields the co-product L-citrulline in addition to NO, in a 1:1 stoichiometry.[73,74] NOS is stereospecific for the L-isomer, since D-arginine is not a substrate. With little biochemical precedence, this multi-step, odd-electron oxidation is facilitated by a single enzyme protein, NOS.

Despite extensive research, the precise mechanism by which NOS catalyzes the oxidation of L-arginine to NO remains unclear. Indeed, there is still debate as to whether NO is the actual end-product of this reaction. However, due to the aforementioned similarities between the two enzymes, it appears that many aspects of NOS biochemistry relate directly to cytochrome P-450-facilitated transformations. An important manifestation of this analogy is the requirement of particular cofactors for NOS activity. As with cytochrome P-450,[75] NADPH acts as the source of electrons for oxygen activation and substrate oxidation (as such, NADPH is strictly a co-substrate). It is also believed that the flavins FAD and FMN play an analogous role in shuttling electrons from NADPH to the iron haem. Moreover, the haem moiety of NOS also resembles cytochrome P-450, in that it is linked to the enzyme protein *via* an axial thiolate ligation (from a cysteine residue). Due to this similarity, and the observation that CO and other haem-binding agents inhibit NOS activity, it is believed that the haem component of NOS represents the catalytic centre, responsible for binding and reducing molecular oxygen and subsequent oxidation of substrate.[31,32,76–78] In contrast to cytochrome P-450, NOS also requires BH_4 for maximal activity.[77,78] A final cofactor and important regulator of NOS activity is calmodulin. All isoforms of NOS bind calmodulin, but with differing affinities.[9,79] Purified nNOS does not contain bound calmodulin, but in the presence of Ca^{2+}, association occurs with high affinity. In contrast, iNOS is isolated, containing calmodulin so tightly bound it is regarded as a subunit rather than a cofactor. The role of calmodulin in nNOS, and presumably its function in iNOS, is to transfer electrons between the flavins and haem moiety and hence couple the reductase and oxygenase domains.[80]

The cofactor requirements of NOS are not simply important in aiding catalytic activity, but also appear obligatory in permitting the dimerization of NOS monomers to form active proteins. Interestingly, in purified enzyme preparations (ostensibly with the inducible isoform) the inactive monomeric protein retains its ability to bind calmodulin, FAD and FMN; these monomers, however, lack haem and BH_4. The active, dimeric proteins possess all the cofactors described above, and significantly, dimerization of the monomeric proteins is promoted by the presence of haem, BH_4 and L-arginine.[21] Therefore, it appears as if the latter three cofactors also have a role in maintaining the integrity of the active protein.

The mechanism of NOS-catalyzed oxidation of L-arginine to NO proceeds in at least two distinct steps. The initial reaction involves N-hydroxylation of the guanidinium nitrogen to form N-hydroxy-L-arginine, which is the only identifiable intermediate.[12,81–83] The reaction utilizes one equivalent of NADPH and O_2 to conduct a simple two electron oxidation of nitrogen, and mimics a classical P-450-like hydroxylation, common in the biotransformation of many nitrogen-containing xenobiotics.[84] However, subsequent steps in the conversion of N-hydroxy-L-arginine to NO and L-citrulline remain unclear. Importantly, if NO is the product, NOS must facilitate an odd-electron oxidation, as indicated by the requirement for 0.5 equivalents of NADPH for this process.[12] Such a transformation is therefore difficult to reconcile with cytochrome P-450 ($2e^-$) chemistry. Significantly, oxidation of comparative N-hydroxy-guanidines has shown that nitroxyl (HNO), and not NO, is the preferred nitrogen-oxide product.[85,86] Indeed, this is not surprising since HNO would be the expected two electron oxidation product from N-hydroxy-guanidines. Therefore, it is conceivable that NOS generates HNO from L-arginine, in a four electron process (entirely consistent with cytochrome P-450 enzymology), and that a subsequent $1e^-$ oxidation of this product yields NO.[86,87] Recent studies have supported this potential mechanism of catalysis. First, HNO has been shown to possess biological activity indistinguishable from NO, which seems attributable to the rapid conversion of HNO to NO, by a variety of physiologically relevant oxidants including superoxide dismutase (SOD), oxygen and haemoproteins.[85,86] Second, and in an analogous fashion, SOD has been demonstrated to enhance directly the formation of free NO from L-arginine by NOS.[88] The mechanism responsible for this potentiation was shown to be independent of superoxide anion dismutation and NOS activation. Thus, SOD appears to accelerate the conversion of an intermediate in the L-arginine:NO pathway, such as HNO, to NO, by a $1e^-$ oxidation similar to that described above.

In a similar manner to cytochrome P-450, NOS also appears able to uncouple from its substrate, L-arginine, and generate superoxide anion and hydrogen peroxide *via* the NADPH-dependent reduction of molecular oxygen.[83,89,90] However, the production of ROS appears inversely related to substrate concentration, being virtually absent in the presence of high L-arginine levels. Consequently, under normal physiological conditions, there may be little or no $O_2^{\cdot-}/H_2O_2$ generation by NOS.

An important mechanism controlling NOS activity is a negative feedback mechanism mediated by NO itself. Recent studies have shown that NO generated by nNOS[91] and iNOS[92] is capable of inhibiting subsequent enzymatic activity. Presumably, this inhibition occurs *via* NO ligation to the haem moiety of NOS.

Moreover, BH_4 is capable of preventing and reversing this feedback
and, although the explanation for this is unclear, it may represent a f
for BH_4 as a cofactor.

Inhibition of NOS may prove therapeutically beneficial in treating conditions involving excessive NO production, such as septic shock. Due to the divergent actions of NO, however, efficacious agents will only result from inhibition of selective isoforms of NOS. N^G-substituted analogues of L-arginine (i.e. N^G-methyl-L-arginine and N^G-nitro-L-arginine) have been widely used as tools to probe the physiological importance of NO, due to their competitive, reversible inhibition of NOS.[94,95] Unfortunately, such compounds do not discriminate between NOS isoforms. Of greater interest are the more recent observations that 7-nitroindazole selectively inhibits nNOS associated with nociception,[96] and the inhibitory actions of aminoguanidine may be restricted to iNOS *in vivo*.[97] Indeed, a more detailed understanding of NO synthesis should provide a greater opportunity for the development of selective inhibitors, particularly if differences in cofactor utilization, electron transfer and gene expression can be exploited.

Chemical and Biological Fate of Nitric Oxide

To understand better the biological fate of NO, one must appreciate the complex series of interactions NO can undergo in simple chemical terms before attempting to extrapolate to a biological milieu. Although the biological fate of NO is discussed elsewhere in this volume (Chapter 6), the present and following sections will attempt to explain briefly many of the biologically relevant reactions undertaken by NO and to discuss the importance of *S*-nitrosothiol formation in more detail.

NO is a paramagnetic molecule possessing a single unpaired electron (11 valence electrons). NO is radical in nature, but since its unpaired electron lies in a π (antibonding) orbital, NO is relatively stable; in contrast to many endogenously-formed radicals, particular oxygen-derived (e.g. $O_2^{\bullet -}$, $\bullet$OH), NO does not react with many biological molecules. Nevertheless, the lifetime of NO *in vivo* is relatively short (in the order of a few seconds[98]). Thus, only the faster, direct actions of NO are likely to prove physiologically important. In general, these are encompassed by two chemical interactions: firstly, the reaction of NO with a second paramagnetic species (primarily ROS such as O_2 and $O_2^{\bullet -}$), which results in the formation of a stabilized diamagnetic species; secondly, complexation of NO to a metal, in which the unpaired electron is delocalized to the d orbital. The processes involved in such interactions are discussed briefly below, with emphasis on the formation of *S*-nitrosothiols.

Probably the most important endogenous reaction participated in by NO is the combination with molecular oxygen. In water, NO reacts with molecular oxygen yielding N_2O_4 (Eq. 1a), which dissolutes to produce nitrite (NO_2^-)

and nitrate (NO_3^-) (Eq. 1b). However, the sole identifiable product of the reaction of NO with O_2 in water is nitrite (Eq. 2b),[99] suggesting the existence of an N_2O_3 intermediate (Eq. 2a). As in the gas phase, this process is first order in O_2 and second order in NO. This stoichiometry has very important consequences for the biological half-life of NO because the rate of combination with O_2 is therefore dependent on the square of the NO concentration. Hence, at high NO levels in an aerobic environment, nitrogen dioxide ($\cdot NO_2$) is rapidly formed (see also Chapters 6 and 8); this process is fundamental to the development of air pollution from nitrogen oxides, principally derived from the combustion engine. However, in biological systems, where the concentration of NO is relatively low, its autoxidation is much more languid, and therefore its window of opportunity to exert a biological effect is considerably longer than often perceived.

$$2NO + O_2 \rightarrow N_2O_4 \qquad (1a)$$

$$N_2O_4 + H_2O \rightarrow NO_2^- + NO_3^- + 2H^+ \qquad (1b)$$

$$N_2O_4 + 2NO \rightarrow 2N_2O_3 \qquad (2a)$$

$$N_2O_3 + H_2O \rightarrow 2NO_2^- + 2H^+ \qquad (2b)$$

$$2NO_2^- + Hb(Fe^{2+})O_2 \rightarrow 2NO_3^- + Hb(Fe^{3+}) \qquad (3)$$

$$NO + Hb(Fe^{2+})O_2 \rightarrow NO_3^- + Hb(Fe^{3+}) \qquad (4)$$

The stable oxidation products of NO, inorganic nitrite and nitrate are good parameters with which to assess the production of NO which, due to its radical nature, tends to be short-lived in biological systems. *In vivo*, the majority of nitrite formed (as in Eqs. 1 and 2) reacts rapidly with oxyhaemoglobin to yield nitrate (Eq. 3). Furthermore, NO itself will react with oxyhaemoglobin to give nitrate (Eq. 4). Consequently, in measurements of urine or plasma, the product of NO oxidative metabolism is almost exclusively nitrate.

The biological oxidation of NO results in the formation of N_2O_4 and N_2O_3 intermediates (Eqs. 1 and 2) which can be regarded as donors of the nitrosonium ion NO^+ (Eqs. 5 and 6). The chemistry relating to the transfer of the nitrosonium group has been well studied and is characterized by the nitrosation of a variety of nucleophiles; of particular physiological importance is the reaction of NO^+ with nitrogen and sulphur contained within the amine (Eq. 7) and thiol (Eq. 8) groups of various proteins.

$$N_2O_3 \rightarrow ON^{\delta+} + {}^{\delta-}NO_2 \qquad (5)$$

$$N_2O_4 \rightarrow ON^{\delta+} + {}^{\delta-}NO_3 \qquad (6)$$

$$R_2NH + [NO^+] \rightarrow R_2NNO + H^+ \qquad (7)$$

$$RSH + [NO^+] \rightarrow RSNO + H^+ \qquad (8)$$

In the case of nitrogen, a nitrosamine (R_2NNO) is formed (Eq. 7). Such compounds, particularly those generated by the reaction of NO^+ with secondary amines, form stable structures which are potent mutagens and carcinogens.[100] In the case of sulphur, an *S*-nitrosothiol (RSNO) is the product (Eq. 8). These compounds have taken on profound physiological importance since it has been postulated that these molecules may act as carriers for NO *in vivo*, protecting its biological activity and facilitating delivering to an appropriate target site. This concept will be discussed in detail below.

Since NO is a radical, it reacts at near diffusion-controlled rates with other radicals. In a biological environment this is exemplified by the interaction of NO with superoxide ($O_2^{\cdot -}$), yielding initially peroxynitrite ($ONOO^-$; Eq. 9). Indeed, this reaction proceeds 5 times more rapidly than the disproportionation of superoxide by superoxide dismutase (SOD). The combination of NO and $O_2^{\cdot -}$ to give $ONOO^-$ has received considerable attention over the past few years, since it has been suggested that this may represent a deleterious mechanism resulting in tissue damage.[101] Initially, it was believed that peroxynitrite protonated at physiological pH to give the conjugate acid, peroxynitrous acid ($pK_a = 6.8$ at 25°C), which gave rise to the cytotoxic species $\cdot OH$ and $\cdot NO_2$ (Eq. 10). However, peroxynitrite itself can oxidize substrates, by either one or two electrons, and has a distinctly different reactivity profile from either $\cdot OH$ or $\cdot NO_2$. Furthermore, the toxicity attributed to peroxynitrite formation is considerably less severe than that produced by Fenton-type chemistry, suggesting that $\cdot OH$ and $\cdot NO_2$ may not be responsible. Recently, it has become apparent that CO_2 may react with a peroxynitrite-derived species to elicit cytotoxic effects, although this process remains to be elucidated.[102] An extremely important consideration, however, questions the deleterious nature of peroxynitrite. Peroxynitrous acid rapidly isomerizes to form biologically inert NO_3^- (< 1s; Eq. 10), which suggests that $ONOO^-$ formation may simply represent an inactivation pathway for NO and/or superoxide.

$$NO + \cdot O_2^- \rightarrow ONOO^- \tag{9}$$

$$ONOO^- + H^+ \rightarrow ONOOH \rightarrow [\cdot OH + \cdot NO_2] \rightarrow NO_3^- + H^+ \tag{10}$$

NO also reacts with many biologically important metal ions; this is particularly true of transition metals such as iron, whose d-orbitals are of similar energy and may thereby delocalize the unpaired electron of NO. The result is the formation of metal-nitrosyl complexes, which have been studied for many years. Because of the structural similarities between NO and dioxygen, the former has been used as a tool for probing the interaction of O_2 with metals.[103] This is exemplified by numerous metalloproteins which bind dioxygen as an integral part of their function and which also avidly bind NO (e.g. cytochrome P-450, haemoglobin, cytochrome oxidase, NOS). Indeed, it is the interaction of NO with such metalloproteins that is responsible for NO-induced modulation of the activity of many of these enzymes.

For instance, NO can complex with cytochrome oxidase to inhibit mitochondrial respiration,[104] and exert a negative feedback influence upon its own synthesis by binding the iron-haem of NOS.[91,92] Probably the most important NO-metal interaction in physiology is the binding of NO to the iron-haem of sGC, which results in the catalytic conversion of GTP to cGMP and which mediates by far the vast majority of biological effects of NO.[105]

Physiological Generation and Relevance of *S*-nitrosothiols

S-nitrosothiols (or thionitrites) are readily formed from thiols and an electrophilic nitrosating agent, X-NO, which is capable of delivering an NO^+ (or equivalent) species (Eqs. 5, 6 and 8). It is possible to generate aqueous solutions of all *S*-nitrosothiols by treatment of a thiol solution with sodium nitrite under acid conditions, and such solutions are often sufficiently stable to enable experimentation to be conducted after the appropriate adjustment of the pH, particularly if light is excluded (*S*-nitrosothiols are relatively stable at low pH but the S-N bond is photolabile). The stable *S*-nitrosothiols are red or green solids, essentially representing secondary (*S*-nitrosoglutathione, GSNO) and tertiary (e.g. *S*-nitroso-N-acetylpenicillamine, SNAP) structures. It has been known for many years that *S*-nitrosothiols (RSNO) decompose in aqueous solution to yield the corresponding disulphide and NO (Eq. 11). This reaction has been of particular importance in the past few years as an experimental tool, and an alternative NO source to many organic nitrates which suffer from the development of tolerance.

$$2RSNO \rightarrow RSSR + 2NO \qquad (11)$$

S-nitrosothiols can also partake in transnitrosation reactions in which the nitroso group is transferred to a second molecule. The NO group exchange between an *S*-nitrosothiol and a thiol has been demonstrated (Eq. 12), although in many instances the *S*-nitrosothiols decompose to disulphides and release their NO moiety[106] (Eq. 11). It is thought that transnitrosation proceeds *via* an attack of the thiolate anion, since the pH profile favours this mechanism.[107] This reaction occurs considerably faster than the decomposition of *S*-nitrosothiols and proceeds quite readily at physiological pH,[107] which may have important consequences for the biological activity of this family of compounds.

$$RSNO + R'S^-(H^+) \leftrightarrow RS^-(H^+) + R'SNO \qquad (12)$$

The presence of Cu^{2+} also has a profound effect on the rate of decomposition (and consequently NO release) from *S*-nitrosothiols in aqueous solution.[108] The Cu^{2+} is catalytic, and concentrations as low as 1 μM can facilitate the release of NO. Thus, the Cu^{2+} levels found in distilled water are often sufficient to bring about this decomposition, which may explain the discrepancies reported in the half-lives of *S*-nitrosothiols in solution.[109] Essentially, the process can be halted by the addition of EDTA, which complexes the Cu^{2+}. Hence, *S*-nitrosothiol solutions are fairly stable at room temperature, if kept in the dark and the concentration of free Cu^{2+} is low. On the basis of extensive kinetic studies,[110,111] it is believed the following process is responsible for the catalytic effect of Cu^{2+} on *S*-nitrosothiol decomposition (Eqs. 13 and 14).

$$RS^- + Cu^{2+} \rightarrow RS^\cdot + Cu^+ \quad (13)$$

$$RSNO + Cu^+ \rightarrow RS^- + Cu^{2+} + NO \quad (14)$$

$$RSNO + RS^\cdot \rightarrow RSSR + NO \quad (15)$$

Since the identification of EDRF as NO, there has been considerable interest in the possible biological role of *S*-nitrosothiols. This interest originated from the possibility that EDRF itself may be an *S*-nitrosothiol rather than free NO; if NO were 'protected' within an *S*-nitrosothiol (or similar NO-containing/releasing compound) then its biological activity would be preserved and its physiological half-life prolonged. There were several observations supporting this idea which described discrepancies between the biological activity of EDRF and NO.[112,113] Certain investigators suggested that the biological activity of EDRF actually more closely resembled an *S*-nitrosothiol than NO.[114] Furthermore, there was considerable controversy over the identity of the non-adrenergic/non-cholinergic (nitrergic) transmitter released by certain parasympathetic nerves supplying the respiratory, gastrointestinal and urogenital tracts; it was suggested that free NO would represent an extremely inefficient and unreliable neurotransmitter due to its labile nature, and that the nerves more likely released an NO-containing or releasing compound (such as an *S*-nitrosothiol) which would preserve the activity of NO within the synaptic cleft.[115,116] As a whole, these observations provided evidence, albeit indirect, for putative physiological roles for *S*-nitrosothiols.

Probably the first demonstration of the physiological importance of *S*-nitrosothiols was in the late 1970s and early 1980s, during investigation into the mechanism of activation of sGC. Initially, Craven & DeRubertis[117] demonstrated that crude preparations of hepatic sGC required the addition of reduced haemoproteins in order to be responsive to NO and other nitroso compounds. Subsequently, it was shown that haem and haemoglobin were able to inhibit enzyme activation by nitroso compounds, an effect that could be reversed by the addition of dithiothreitol.[118] This phenomenon was explained by the finding that dithiothreitol and other thiols (or free sulphydryls) could react chemically with the nitroso compounds to form the corresponding *S*-nitrosothiol.[119,120] In turn, such *S*-nitrosothiols were found to be more resistant to the inhibitory actions of haemoglobin with respect to enzyme activation, providing an explanation for the enhancing action of these compounds on nitroso-stimulation of enzyme activity. More extensive studies demonstrated that nitroglycerin was only able to activate sGC in the presence of cysteine,[120] and sodium nitrite (which only activated sGC at high concentrations (> 10 mM)) potently stimulated enzyme activity in the presence of a thiol moiety. In combination, these observations led to the hypothesis that NO is responsible for the vascular relaxant effects of organic nitrate and nitrite esters (and nitroprusside) *via* the formation of an *S*-nitrosothiol intermediate.[121] Further work examined whether integral thiol groups possessed by sGC played a role in enzyme activation. Partially purified enzyme preparations were found to be rapidly (but reversibly) inactivated in a 100% oxygen atmosphere, whereas protein maintained under nitrogen displayed no such loss of activity.[122] This oxygen-dependent inactivation could be reversed by the addition of dithiothreitol. Further, various thiol oxidising and alkylating agents (including disulphides themselves) were able to mimic the inhibition brought about by oxygen exposure. The discovery that the putative catalytic site of sGC possessed two juxtaposed -SH (thiol) groups,[123] and that enzyme-bound thiols were involved in catalytic activity,[124,125] confirmed that *S*-nitrosothiol intermediates, probably interacting with the -SH groups at the enzyme's catalytic site, were indeed necessary for stimulation of activity.

The physiological relevance of *S*-nitrosothiols was re-affirmed by the more recent observations that many sulphur-containing proteins form adducts with NO *in vivo*.[126] The majority of NO circulates in plasma as an adduct with albumin, *S*-nitrosoalbumin.[127] *S*-nitrosoalbumin possesses EDRF-like properties in that it can elicit vasodilatation and decrease platelet reactivity.[128] Other NO-adducts with biologically important thiols have also been reported. *S*-nitrosoglutathione, which can be found in up to mM concentrations in human bronchial fluid,[129] may be important in protecting the biological activity of NO in the oxygen-rich environment of the lung. *S*-nitrosocysteine has been shown to be a product of endothelial cells,[114] and may represent a second `EDRF'. Finally, *S*-nitroso-tissue plasminogen activator has been shown to enhance the catalytic efficiency of plasminogen activation in the presence of fibrin.[130] Thus, *S*-nitrosylation of thiol-containing proteins appears to represent a form of post-translational modification which regulates function. Further examples exist. Glyceraldehyde-3-phosphate dehydrogenase undergoes ADP-ribosylation which inhibits enzyme function, and this process seems to be regulated by the formation of a transient *S*-nitrosothiol intermediate.[131,132] *S*-nitrosylation of the N-methyl-D-aspartate receptor appears to be neuroprotective.[133]

Recently, a novel role for *S*-nitrosylation has been proposed.[134] Haemoglobin is a tetramer comprising two α and two β subunits. In human haemoglobin, each subunit possesses a haem moiety, and additionally, the β subunits contain highly conserved (in all mammals and birds) -SH groups, provided by cysteine residues (Cysβ93).[135] *In vivo*, it is believed that haemoglobin severely restricts the sphere of action of NO[136] by a direct scavenging activity. However, it appears as if this interaction of NO with haemoglobin has a mutual effect on function. NO will react with oxyhaemoglobin to give nitrate and methaemoglobin

(Eq. 4). Yet, the NO contained by *S*-nitrosothiols does not undergo the same decomposition; rather, it participates in a transnitrosation reaction in which an NO^+ equivalent is transferred from low molecular weight *S*-nitrosothiols to the Cys residues in haemoglobin to give *S*-nitrosohaemoglobin. Interestingly, in the R-state (oxyhaemoglobin), this transnitrosation reaction is considerably more rapid than with deoxyhaemoglobin (T-state). Furthermore, the reaction can be prevented by the prior treatment of haemoglobin with N-ethylmaleimide, an agent which oxidizes thiol-containing residues. Transnitrosation from low molecular weight thiols to haemoglobin also occurs through erythrocyte membranes, and intra-erythrocytic *S*-nitrosohaemoglobin can be detected (by photolysis-chemiluminescence) in arterial and venous blood, the levels being some ten times higher in the arterial circulation.

From these observations, is has been concluded that *S*-nitrosohaemoglobin is probably formed during passage of erythrocytes through the lung, where the oxygen tension is high and therefore the presence of nitrosating species such as N_2O_3 and N_2O_4 is enhanced (see Eqs. 5–8). Since transnitrosation between *S*-nitrosohaemoglobin and glutathione (which is present in millimolar concentrations in erythrocytes) occurs readily, both intra- and inter-cellularly, it is assumed that the NO-related activity (i.e. vascular smooth muscle relaxation) is conveyed from the erythrocyte to the surrounding tissue by transnitrosation of low molecular weight thiols (principally GSNO), which then release their NO spontaneously or *via* reaction with other thiols (Eqs. 5–8). This idea is supported by the fact that the metal centre of haemoglobin does not scavenge the NO-like activity associated with transnitrosation of its thiol group, suggesting that it is not free NO that is released. Interestingly, the rate of release of NO from *S*-nitrosohaemoglobin is enhanced in the T-state (deoxy), which suggests that O_2 binding modulates the NO-carrying capacity of the molecule.

Such findings present an interesting hypothesis. The arterial-venous difference suggests that *S*-nitrosohaemoglobin acts as an NO donor in the systemic circulation, a facet enhanced at low oxygen tensions. In the microcirculation, where blood pressure is controlled and oxygen delivery accomplished, NO release from *S*-nitrosohaemoglobin may have a profound influence on vascular tone and subsequent delivery of oxygen. In highly metabolic tissues (e.g. skeletal muscle during exercise) the pO_2 is low and oxygen demand high. Release of O_2 from haemoglobin is favoured under these conditions and the deoxyhaemoglobin formed would be more susceptible to release its thiol-bound NO. In turn, the NO would cause vasodilatation, thereby increasing perfusion and augment subsequent oxygen delivery to the tissues. Moreover, NO may also serve to promote efficient utilization of this oxygen, since it can regulate mitochondrial respiration.[137] The existence of a nitrosated haemoglobin may also explain, at least in part, the hypertensive effect of cell-free haemoglobin, often administered to patients for cardiovascular resuscitation. This novel hypothesis is extremely intriguing and provides evidence:

(a) that haemoglobin has a dual function in carrying oxygen and NO; and (b) that nitrosation of proteins can significantly alter function.

Many aspects of the above scenario need clarification, however. For instance, it is not known what mechanisms are responsible for the formation of *S*-nitrosohaemoglobin in the pulmonary circulation. Also, the transnitrosation reactions involved in the transfer of NO from *S*-nitrosohaemoglobin to low-molecular weight thiols, and the delivery of those molecules to their site of action (i.e. the sGC in the vascular smooth muscle), are slow processes and may not significantly increase the local concentration of NO, since endothelial cells are known to increase their own NO output in response to shear stress and metabolic products such as lactate.[138] Nevertheless, the above hypothesis draws attention to the likely physiological significance of *S*-nitrosothiols, which should become more apparent with continued research in the next few years.

Biological Activity of N-hydroxy-L-arginine

It is worth mentioning at this juncture, albeit briefly, the possible biological significance of N-hydroxy-L-arginine (NOHA) formation. Initial studies of crude NOS homogenates followed by purified enzyme preparations suggested that NOHA was the only identifiable intermediate in the catalytic conversion of L-arginine to L-citrulline plus NO; further, the vast majority of NOHA was thought to stay tightly bound to NOS during catalysis.[139,140] Recently it has become clear that levels of free NOHA can build substantially, particularly during high-output NO production by iNOS. For instance, we have found that stimulation of rat aortic endothelial cells in culture with LPS + IFN-γ + TNF-α + interleukin (IL)-1β results in the 24 h accumulation of approximately 500 μM NO (measured as NO_2^- + NO_3^-) and L-citrulline, but additionally 170 μM NOHA.[141] Thus, it appears as if NOHA may serve as a biological effector molecule in its own right and should not be simply regarded as an inactive intermediate. If this is the case, what would its biological action be? One possible function is to reduce the activity of arginase, of which NOHA is a relatively potent inhibitor (K_i 10–35 μM).[142] In most cells, which do not possess a full urea cycle, arginase is responsible for the conversion of arginine to ornithine, which is the precursor for polyamine synthesis and therefore involved in cell growth and proliferation. Thus, NOHA may represent an endogenous inhibitor of cell proliferation and may turn out to be at least as important as NO in exerting a cytostatic influence under certain physiological and pathophysiological conditions. The ability of NOHA to inhibit arginase activity may also be important when one considers the role that arginase itself plays in regulating NOS activity by competing for the substrate, L-arginine.[143] By reducing arginase activity, NOHA may 'shunt' the utilization of L-arginine in the favour of NOS, thereby ensuring the availability of sufficient substrate to support high-output NO production.

Summary and Conclusions

The ability of NO to utilize the sGC-cGMP system for signal transduction has extended the role of this pathway to both intra- and inter-cellular communication and regulation. As a consequence, the perceived importance of this signal transduction pathway has grown exponentially over the past decade. The multidisciplinary approach outlined in this article has combined molecular biology, biochemistry, physiology and chemistry to probe the genetic regulation and biological actions of this unique, ubiquitous messenger molecule. Further elucidation of the processes underlying NOS expression, NO synthesis and the physiological actions attributed to NO will be required to identify targets for selective therapeutic intervention.

Despite these tremendous advances, many questions remain unanswered. Particularly, what mechanisms regulate human NOS expression and why are these so different from our murine counterparts? Why does the human genome possess multiple copies of the iNOS gene, and can these be linked to racial differences in NOS activity (affecting, for example, immunity and cardiovascular homeostasis)? Does NOS have a multi-faceted role in biology, not only to synthesize NO, but also N-hydroxy-L-arginine, $O_2^{\bullet-}$ and HNO? Despite these uncertainties, it is clear that due to the ubiquitous nature of NO, many pathological states may be attributed to its inappropriate functioning. Indeed, conditions such as hypertension (Chapter 21), septic shock (Chapter 22) and impotence have already been linked to aberrations in the synthesis and release of NO. It is of fundamental importance therefore, that answers to the above questions and many others are understood in their entirety, so that agents can be developed to modulate the action of NO artificially, and be able to correct any pathophysiological dysfunction.

References

1. Furchgott RF, Zawadzki JV. The obligatory role of endothelial cells in the relaxation of arterial smooth muscle by acetylcholine. *Nature* 1980; **288**: 373–376.
2. Ignarro LJ. Haem-dependent activation of cytosolic guanylate cyclase by nitric oxide: a widespread signal transduction mechanism. *Biochem Soc Trans* 1992; **20**: 465–469.
3. Schmidt HHHW, Lohmann SM, Walter U. The nitric oxide and cGMP signal transduction system: regulation and mechanism of action. *Biochim Biophys Acta* 1993; **1178**: 153–175.
4. Mayer B. Nitric oxide/cyclic GMP-mediated signal transduction. *Ann New York Acad Sci* 1994; **733**: 357–364.
5. Murad F. Regulation of cytosolic guanylyl cyclase by nitric oxide: the NO-cyclic GMP signal transduction system. *Adv Pharmacol* 1994; **26**: 19–33.

6. Mayer B, John M, Bohme E. Purification of a Ca^{2+}/calmodulin-dependent nitric oxide synthase from porcine cerebellum. Cofactor-role of tetrahydrobiopterin. *FEBS Lett* 1990; **277**: 215–219.
7. Pollock JS, Förstermann U, Mitchell JA *et al.* Purification and characterization of particulate endothelium-derived relaxing factor synthase from cultured and native bovine aortic endothelial cells. *Proc Natl Acad Sci USA* 1991; **88**: 10480–10484.
8. Schmidt HHHW, Murad F. Purification and characterization of a human NO symthase. *Biochem Biophys Res Commun* 1991; **181**: 1372–1377.
9. Schmidt HHHW, Pollock JS, Nakane M *et al.* Purification of a soluble isoform of sGC-activating factor synthase. *Proc Natl Acad Sci USA* 1991; **88**: 365–369.
10. Hiki K, Hattori R, Kawai C, Yui Y. Purification of insoluble nitric oxide synthase from rat cerebellum. *J Biochem* 1992; **111**: 556–558.
11. Hevel JM, White KA, Marletta MA. Purification of the inducible murine macrophage nitric oxide synthase. Identification as a flavoprotein. *J Biol Chem* 1991; **266**: 22789–22791.
12. Stuehr DJ, Cho HJ, Kwon NS *et al.* Purification and characterization of the cytokine-induced macrophage nitric oxide synthase; an FAD- and FMN-containing flavoprotein. *Proc Natl Acad Sci USA* 1991; **86**: 7773–7777.
13. Yui Y, Hattori R, Kosuga K *et al.* Purification of nitric oxide synthase from rat macrophages. *J Biol Chem* 1991; **266**: 12544–12547.
14. Bredt DS, Snyder SH. Isolation of nitric oxide synthetase, a calmodulin-requiring enzyme. *Nature* 1990; **87**: 682–685.
15. McMillan K, Bredt DS, Hirsch DJ *et al.* Cloned, expressed rat cerebellar nitric oxide synthase contains stoichiometric amounts of heme, which binds carbon monoxide. *Proc Natl Acad Sci USA* 1992; **89**: 11141–11145.
16. Nishida K, Harrison DG, Navas JP. Molecular cloning and characterization of the constituive bovine aortic endothelial cell nitric oxide synthase. *J Clin Invest* 1992; **90**: 2092–2096.
17. Charles IG, Chubb A, Gill R *et al.* Cloning and expression of a rat neuronal nitric oxide synthase coding sequence in a baculovirus/insect cell system. *Biochem Biophys Res Commun* 1993; **196**: 1481–1489.
18. Moss DW, Wei X, Liew FY *et al.* Enzymatic characterisation of recombinant murine inducible nitric oxide synthase. *Eur J Pharmacol* 1995; **289**: 41–48.
19. Fossetta JD, Niu XD, Lunn CA *et al.* Expression of human inducible nitric oxide synthase in Escherichia coli. *FEBS Lett* 1996; **379**: 135–138.
20. Mayer B, Klatt P, List BM *et al.* Large-scale purification of rat brain nitric oxide synthase from a baculovirus overexpression system. *Meths Enzymol* 1996; **268**: 420–427.
21. Hevel JM, Marletta MA. Macrophage nitric oxide synthase: relationship between enzyme-bound tetrahydrobiopterin and synthase activity. *Biochemistry* 1992; **31**: 7160–7165.

22. Baek KJ, Thiel BA, Lucas F, Stuehr DJ. Macrophage nitric oxide synthase subunits. *J Biol Chem* 1993; **268**: 21120–21129.
23. Bredt DS, Hwang PM, Snyder SH. Localisation of nitric oxide synthase indicating a neural role for nitric oxide. *Nature* 1991; **347**: 768–770.
24. Dinerman JL, Dawson TM, Schell MJ *et al.* Endothelial nitric oxide synthase localised to hippocampal pyramidal cells: implications for synaptic plasticity. *Proc Natl Acad Sci USA* 1990; **91**: 4214–4218.
25. Bredt DS, Glatt CE, Hwang PM *et al.* Cloned and expressed nitric oxide synthase structurally resembles cytochrome P450 reductase. *Nature* 1991; **351**: 714–718.
26. Nathan C, Xie QW. Regulation of biosynthesis of nitric oxide. *J Biol Chem* 1994; **269**: 13725–13728.
27. Nathan C. Nitric oxide as a secretory product of mammalian cells. *FASEB J* 1992; **6**: 3051–3064.
28. Geller DA, Lowenstein CJ, Shapiro RA *et al.* Molecular cloning and expression of inducible nitric oxide synthase from human hepatocytes. *Proc Natl Acad Sci USA* 1993; **90**: 3491–3495.
29. Miyahara K, Kawamoto T, Sase K *et al.* Cloning and structural characterization of the human endothelial nitric-oxide-synthase gene. *Eur J Biochem* 1994; **223**: 719–726.
30. Hall AV, Antoniou H, Wang Y *et al.* Structural organization of the human neuronal nitric oxide synthase gene (NOS1). *J Biol Chem* 1994; **269**: 33082–33090.
31. Djordjevic S, Roberts DL, Wang M *et al.* Crystallization and preliminary X-ray studies of NADPH-cytochrome P450 reductase. *Proc Natl Acad Sci USA* 1995; **92**: 3214–3218.
32. Stuehr DJ, Ikeda-Saito M. Spectral characterization of brain and macrophage nitric oxide synthases. Cytochrome P450-like hemoproteins that contain a flavin semiquinone radical. *J Biol Chem* 1992; **267**: 20547–20550.
33. White KA, Marletta MA. Nitric oxide synthase is a cytochrome P450-type hemoprotein. *Biochemistry* 1992; **31**: 6627–6631.
34. Wang J, Stuehr DJ, Ikeda-Saito M, Rousseau DL. Heme coordination and structure of the catalytic site in nitric oxide synthase. *J Biol Chem* 1993; **268**: 22255–22258.
35. McMillan K, Masters BS. Prokaryotic expression of the heme- and flavin-binding domains of rat neuronal nitric oxide synthase as distinct polypeptides: identification of the heme-binding proximal thiolate ligand as cysteine-415. *Biochemistry* 1995; **34**: 3686–3693.
36. Sheta EA, McMillan K, Masters BS. Evidence for a bidomain structure of constitutive cerebellar nitric oxide synthase. *J Biol Chem* 1994; **269**: 15147–15153.

37. Stuehr DJ, Abu-Soud HM, Rousseau DL. Control of electron transfer in neuronal nitric oxide synthase by calmodulin, substrate analogs, and nitric oxide. *Adv Pharmacol* 1995; **34**: 207–213.
38. Lowe PN, Smith D, Stammers DK *et al.* Identification of the domains of neuronal nitric oxide synthase by limited proteolysis. *Biochem J* 1996; **314**: 55–62.
39. Busconi L, Michel T. Recombinant endothelial nitric oxide synthase: post-translational modifications in a baculovirus expression system. *Mol Pharmacol* 1995; **47**: 655–659.
40. Chartrain NA, Geller DA, Koty PP *et al.* Molecular cloning, structure, and chromosomal localization of the human inducible nitric oxide synthase gene. *J Biol Chem* 1994; **269**: 6765–6772.
41. Kaling M, Kugler W, Ross K *et al.* Liver-specific gene expression: A-activator-binding site, a promoter module present in vitellogenin and acute phase genes. *Mol Cell Biol* 1991; **11**: 93–101.
42. Xie QW, Kashiwabara Y, Nathan C. Role of transcription factor NF-κB/Rel in induction of nitric oxide synthase. *J Biol Chem* 1994; **269**: 4705–4708.
43. Martin E, Nathan C, Xie QW. Role of interferon regulatory factor 1 in induction of nitric oxide synthase. *J Exp Med* 1994; **180**: 977–984.
44. Lowstein CJ, Alley EW, Raval P *et al.* Macrophage nitric oxide synthase gene: two upstream regions mediate induction by interferon gamma and lipopolysaccharide. *Proc Natl Acad Sci USA* 1993; **90**: 9730–9734.
45. De Vera ME, Shapiro RA, Nussler AK *et al.* Transcriptional regulation of human inducible nitric oxide synthase (NOS2) gene by cytokines: initial analysis of the human NOS2 promoter. *Proc Natl Acad Sci USA* 1995; **93**: 1054–1059.
46. Weinberg JB, Misukonis MA, Shami PJ *et al.* Human mononuclear phagocytic inducible nitric oxide synthase (iNOS): analysis of iNOS mRNA, iNOS protein, biopterin, and nitric oxide production by blood monocytes and peritoneal macrophages. *Blood* 1995; **86**: 1184–1195.
47. Lenardo MJ, Baltimore D. NF-κB: a pleitropic mediator of inducible and tissue-specific gene control. *Cell* 1989; **58**: 227–229.
48. Baeuerle PA. The inducible transcription factor NF-κB: regulation by distinct protein subunits. *Biochim Biophys Acta* 1991; **1072**: 63–80.
49. Baeuerle PA, Baltimore D. IκB: a specific inhibitor of the NF-κB transcription factor. *Science* 1988; **242**: 540–546.
50. Urban MB, Schreck R, Baeuerle PA. NF-κB contacts DNA by a heterodimer of the p50 and p65 subunit. *EMBO J* 1991; **10**: 1817–1825.
51. Palombella VJ, Rando OJ, Goldberg AL, Maniatis T. The ubiquitin-proteasome pathway is required for processing the NF-κB1 precursor protein and the activation of NF-κB. *Cell* 1994; **78**: 773–785.
52. Traenckner EB, Pahl HL, Henkel T *et al.* Phosphorylation of human IκB-α on serines 32 and 36 controls IκB-α proteolysis and NF-κB activation in response to diverse stimuli. *EMBO J* 1995; **14**: 2876–2883.

53. Henkel T, Machleidt I, Alkalay I *et al.* Rapid proteolysis of IκB-α in response to phorbol ester, cytokines and lipopolysaccharide is a necessary step in the activation of NF-κB. *Nature* 1993; **365**: 182–185.
54. Schreck R, Albermann K, Baeuerle PA. NF-κB: an oxidative stress-response transcription factor of eukaryotic cells. *Free Radic Res Commun* 1992; **17**: 221–237.
55. Schreck R, Reiber P, Baeuerle PA. Reactive oxygen intermediates as apparently widely used messengers in the activation of NF-κB transcription factor and HIV-1. *EMBO J* 1991; **10**: 2247–2258.
56. Hibbs JB, Taintor RR, Vavrin Z, Rachlin EM. Nitric oxide: a cytotoxic activated macrophage effector molecule. *Biochem Biophys Res Commun* 1988; **157**: 87–94.
57. Sherman MP, Aeberhard EE, Wong VZ *et al.* Pyrrolidine dithiocarbamate inhibits induction of nitric oxide synthase activity in rat alveolar macrophages. *Biochem Biophys Res Commun* 1993; **191**: 1301–1308.
58. Mulsch B, Schray-Utz B, Mordvinteev PI *et al.* Diethyldithiocarbamate inhibits induction of macrophage NO synthase. *FEBS Lett* 1993; **321**: 215–218.
59. Griscavage JM, Wilk S, Ignarro LJ. Serine and cysteine proteinase inhibitors prevent nitric oxide production by activated macrophages by interfering with transcription of the inducible NO synthase gene. *Biochem Biophys Res Commun* 1995; **215**: 721–729.
60. Griscavage JM, Wilk S, Ignarro LJ. Inhibitors of the proteasome pathway interfere with induction of nitric oxide synthase in macrophages by blocking activation of nuclear factor κB. *Proc Natl Acad Sci USA* 1996; **93**: 3308–3312.
61. Radomski MW, Palmer RMJ, Moncada S. Glucocorticoids inhibit the expression of an inducible, but not the constitutive, nitric oxide synthase in vascular endothelial cells. *Proc Natl Acad Sci USA* 1990; **87**: 10043–10047.
62. Di Rosa M, Radomski MW, Carnuccio R, Moncada S. Glucocorticoids inhibit the induction of nitric oxide synthase in macrophages. *Biochem Biophys Res Commun* 1990; **171**: 1246–1252.
63. Ray A, Prefontaine KE. Physical association and functional antagonism between the p65 subunit of transcription factor NF-κB and the glucocorticoid receptor. *Proc Natl Acad Sci USA* 1994; **91**: 752–756.
64. Kleinert H, Euchenhofer C, Ihrig-Biedert I, Förstermann U. Glucocorticoids inhibit the induction of nitric oxide synthase II by down-regulating cytokine-induced activity of transcription factor nuclear factor-κB. *Mol Pharmacol* 1996; **49**: 15–21.
65. Auphan N, DiDonato JA, Rosette C *et al.* Immunosuppression by glucocorticoids: inhibition of NF-κB activity through induction of IκB synthesis. *Science* 1995; **270**: 286–290.
66. Aeberhard EE, Henderson SA, Arabolos NS *et al.* Nonsteroidal anti-inflammatory drugs inhibit expression of the inducible nitric oxide synthase gene. *Biochem Biophys Res Commun* 1995; **208**: 1053–1059.

67. Kopp E, Ghosh S. Inhibition of NF-κB by sodium salicylate and aspirin. *Science* 1994; **265**: 956–959.
68. Peng HB, Libby P, Liao JK. Induction and stabilization of IκB-α by nitric oxide mediates inhibition of NF-κB. *J Biol Chem* 1995; **270**: 14214–14219.
69. Matthews JR, Botting CH, Panico M *et al.* Inhibition of NF-κB DNA binding by nitric oxide. *Nucleic Acids Res* 1996; **24**: 2236–2242.
70. Di Silvio M, Geller DA, Gross SS *et al.* Inducible nitric oxide synthase activity in hepatocytes is dependent on the coinduction of tetrahydrobiopterin synthesis. *Adv Exp Med Biol* 1993; **338**: 305–308.
71. Gross SS, Levi R, Madera A *et al.* Tetrahydrobiopterin synthesis is induced by LPS in vascular smooth muscle and is rate-limiting for nitric oxide production. *Adv Exp Med Biol* 1993; **338**: 295–300.
72. Gross SS, Levi R. Tetrahydrobiopterin synthesis. An absolute requirement for cytokine-induced nitric oxide synthase generation by vascular smooth muscle. *J Biol Chem* 1992; **267**: 25722–25729.
73. Tayeh MA, Marletta MA. Macrophage oxidation of L-arginine to nitric oxide, nitrite, and nitrate: tetrahydrobiopterin is required as a cofactor. *J Biol Chem* 1989; **264**: 19654–19658.
74. Bush PA, Gonzalez NE, Griscavage JM, Ignarro LJ. Nitric oxide synthase from cerebellum catalyzes the formation of equimolar quantities of nitric oxide and citrulline from L-arginine. *Biochem Biophys Res Commun* 1992; **185**: 960–966.
75. Wattanabe Y, Groves JT. Molecular mechanisms of oxygen activation by P450. In: *The Enzymes, 3rd ed, Vol 20,* Sigman DS (Ed). Academic Press, San Diego 1992: 405–452.
76. Klatt P, Schmidt K, Mayer B. Brain nitric oxide synthase is a hemoprotein. *Biochem J* 1992; **288**: 15–17.
77. Kwon NS, Nathan CF, Stuehr DS. Reduced biopterin as a cofactor in the generation of nitrogen oxides by murine macrophages. *J Biol Chem* 1989; **265**: 20496–20501.
78. Mayer B, John M, Heinzel B *et al.* Brain nitric oxide synthase is a biopterin- and flavin-containing multifunctional oxidoreductase. *FEBS Lett* 1991; **288**: 187–191.
79. Cho HJ, Xie Q-W, Calatcay J *et al.* Calmodulin is a subunit of nitric oixde synthase from macrophages. *J Exp Med* 1992; **176**: 599–604.
80. Abu-Soud HM, Stuehr DJ. Nitric oxide synthases reveal a role for calmodulin in controlling electron transfer. *Proc Natl Acad Sci USA* 1993; **90**: 10769–10772.
81. Wallace GC, Fukuto JM. Synthesis and bioactivity of N^{ω}-hydroxyarginine: a possible intermediate in the biosynthesis of nitric oxide from arginine. *J Med Chem* 1991; **34**: 1746–1748.

82. Pufahl RA, Nanjappan PG, Woodard RW, Marletta MA. Mechanistic probes of N-hydroxylation of L-arginine by inducible nitric oxide synthase from murine macrophages. *Biochemistry* 1992; **31**: 6822–6828.
83. Klatt P, Schmidt K, Uray G, Mayer B. Multiple catalytic functions of brain nitric oxide synthase. *J Biol Chem* 1993; **268**: 14781–14787.
84. Cho AK, Lindeke B. *Biotransformation of Organic Nitrogen Compounds*. Karger, Basel 1988.
85. Fukuto JM, Wallace GC, Hszeih R, Chaudhuri G. Chemical oxidation of N-hydroxyguanidine compounds. *Biochem Pharmacol* 1992; **43**: 607–613.
86. Fukuto JM, Stuehr DJ, Feldman PL *et al.* Peracid oxidation of an N-hydroxyguanidine compound: a chemical model for the oxidation of N-hydroxy-L-arginine by nitric oxide synthase. *J Med Chem* 1993; **36**: 2666–2670.
87. Stuehr DJ, Griffiths OW. Mammalian nitric oxide synthases. In: *Advances in Enzymology*, Meister A (Ed). Wiley, New York 1992: 287–346.
88. Hobbs AJ, Fukuto JM, Ignarro LJ. Formation of free nitric oxide from L-arginine by nitric oxide synthase: direct enhancement of generation by superoxide dismutase. *Proc Natl Acad Sci USA* 1994; **91**: 10992–10996.
89. Heinzel B, John M, Klatt P *et al.* Ca^{2+}/calmodulin-dependent formation of hydrogen peroxide by brain nitric oxide synthase. *Biochem J* 1992; **281**: 627–630.
90. Pou S, Pou WS, Bredt DS *et al.* Generation of superoxide by purified brain nitric oxide synthase. *J Biol Chem* 1992; **267**: 24173–24176.
91. Rogers NE, Ignarro LJ. Constitutive nitric oxide synthase from cerebellum is reversibly inhibited by nitric oxide formed from L-arginine. *Biochem Biophys Res Commun* 1993; **189**: 242–249.
92. Griscavage JM, Rogers NE, Sherman MP, Ignarro LJ. Inducible nitric oxide synthase from a rat alveolar macrophage cell line is inhibited by nitric oxide. *J Immunol* 1993; **151**: 6329–6337.
93. Griscavage JM, Fukuto JM, Komori Y, Ignarro LJ. Nitric oxide inhibits neuronal nitric oxide synthase by interacting with the heme prosthetic group: role of tetrahydrobiopterin in modulating the inhibitory action of nitric oxide. *J Biol Chem* 1994; **269**: 21644–21649.
94. Moncada S, Palmer RMJ, Higgs AE. Nitric oxide: physiology, pathophysiology and pharmacology. *Pharmacol Rev* 1991; **43**: 109–141.
95. Feldman PL, Griffiths OW, Stuehr DJ. The surprising life of nitric oxide. *Chem Eng News* 1993 (Dec. 20): 26–38.
96. Moore PK, Wallace P, Gaffen Z *et al.* Characterization of the novel nitric oxide synthase inhibitor 7-nitro indazole and related indazoles: antinociceptive and cardiovascular effects. *Br J Pharmacol* 1993; **110**: 219–224.
97. Joly GA, Ayres M, Chelly F, Kilbourn RG. Effects of N^G-methyl-L-arginine, N^G-nitro-L-arginine, and aminoguanidine on constitutive and inducible nitric

oxide synthase in rat aorta. *Biochem Biophys Res Commun* 1994; **199**: 147–154.

98. Ignarro LJ. Signal transduction mechanisms involving nitric oxide. *Biochem Pharmacol* 1991; **41**: 485–490.
99. Wink DA, Darbyshire JF, Nims RW *et al.* Reactions of the bioregulatory agent nitric oxide in oxygenated aqueous media: determination of the kinetics for oxidation and nitrosation by intermediates generated in the NO/O_2 reaction. *Chem Res Toxicol* 1993; **6**: 23–27.
100. Miwa M, Stuehr DJ, Marletta MA *et al.* Nitrosation of amines by stimulated macrophages. *Carcinogenesis* 1987; **8**: 955–958.
101. Radi R, Beckman JS, Bush KM, Freeman BA. Peroxynitrite-induced membrane lipid peroxidation: the cytotoxic potential of superoxide and nitric oxide. *Arch Biochem Biophys* 1991; **288**: 481–487.
102. Uppu RM, Squadrito GL, Pryor WA. Acceleration of peroxynitrite oxidations by carbon dioxide. *Arch Biochem Biophys* 1996; **327**: 335–343.
103. Radi R. Reactions of nitric oxide with metalloproteins. *Chem Res Toxicol* 1996; **9**: 828–835.
104. Torres J, Darley-Usmar V, Wilson MT. Inhibition of cytochrome c oxidase in turnover by nitric oxide: mechanism and implications for control of respiration. *Biochem J* 1995; **312**: 169–175.
105. Hobbs AJ, Ignarro LJ. Nitric oxide-cyclic GMP signal transduction. *Meth Enzymol* 1996; **269**: 134–149.
106. Park JW. Reaction of *S*-nitrosoglutathione with sulphydryl groups in proteins. *Biochem Biophys Res Commun* 1988; **152**: 916–920.
107. Barnett DA, McAninly J, Williams DLH. Transnitrosation between nitrosothiols and thiols. *J Chem Soc, Perkins Trans* 1994; **2**: 1131–1133.
108. Askew SC, Barnett DJ, McAninly DJ, Williams DLH. Catalysis by Cu^{2+} of nitric-oxide release from *S*-nitrosothiols (RSNO). *J Chem Soc, Perkins Trans* 1995; **2**: 741–745.
109. Kowaluk EA, Fung B. Spontaneous liberation of nitric oxide cannot account for *in vitro* vascular relaxation by *S*-nitrosothiols. *J Pharmacol Exp Ther* 1990; **255**: 1256–1264.
110. Dicks AP, Swift HR, Williams DLH *et al.* Identification of Cu^{+} as the effective reagent in nitric-oxide formation from *S*-nitrosothiols (RSNO). *J Chem Soc, Perkins Trans* 1996; **2**: 481–487.
111. Williams DLH. *S*-nitrosothiols and role of metal ions in decomposition to nitric oxide. *Meth Enzymol* 1996; **268**: 299–308.
112. Shikano K, Ohlstein EH, Berkowitz BA. Differential selectivity of endothelium-derived relaxing factor and nitric oxide in smooth muscle. *Br J Pharmacol* 1987; **92**: 483–485.
113. Kelm M, Schrader J. Control of coronary vascular tone by nitric oxide. *Circ Res* 1990; **66**: 1561–1567.

114. Myers PR, Minor RL, Guerra R *et al.* Vasorelaxant properties of endothelium-derived relaxing factor more closely resemble *S*-nitrosocysteine than nitric oxide. *Nature* 1990; **345**: 161–163.
115. Hobbs AJ, Tucker JF, Gibson A. Differentiation by hydroquinone of relaxations induced by endogenous and exogenous nitrates in non-vascular smooth muscle: role of superoxide anions. *Br J Pharmacol* 1991; **104**: 645–650.
116. Gibson A, Babbedge R, Brave SR *et al.* An investigation of some *S*-nitrosothiols, and of hydroxy-arginine, on the mouse anococcygeus. *Br J Pharmacol* 1992; **107**: 715–721.
117. Craven PA, DeRubertis FR. Restoration of the responsiveness of purified guanylate cyclase to nitrosoguanidine, nitric oxide, and related activators by heme and hemeproteins. Evidence for the involvement of the paramagnetic nitrosyl-heme complex in enzyme activation. *J Biol Chem* 1978; **253**: 8433–8443.
118. Ohlstein EH, Barry BK, Gruetter DY, Ignarro LJ. Methemoglobin blockade of coronary arterial sGC activation by nitroso compounds and its reversal by dithiothreitol. *FEBS Lett* 1979; **102**: 316–320.
119. Ignarro LJ, Edwards JC, Gruetter DY *et al.* Possible involvement of *S*-nitrosothiols in the activation of guanylate cyclase by nitroso compounds. *FEBS Lett* 1980; **110**: 275–278.
120. Ignarro LJ, Gruetter CA. Requirement of thiols for activation of coronary arterial guanylate cyclase by glyceryl trinitrate and sodium nitrite: possible involvement of *S*-nitrosothiols. *Biochim Biophys Acta* 1980; **631**: 221–231.
121. Ignarro LJ, Lippton H, Edwards JC *et al.* Mechanism of vascular smooth muscle relaxation by organic nitrates, nitrites, nitroprusside and nitric oxide. Evidence for the involvement of *S*-nitrosothiols as active intermediates. *J Pharmacol Exp Ther* 1981; **218**: 739–749.
122. Ignarro LJ, Kadowitz PJ, Baricos WH. Evidence that regulation of hepatic guanylate cyclase activity involves interactions between catalytic site -SH groups and both substrate and activator. *Arch Biochem Biophys* 1981; **208**: 75–86.
123. Craven PA, DeRubertis FR. Effects of thiol inhibitors on hepatic guanylate cyclase activity. *Biochim Biophys Acta* 1978; **524**: 231–244.
124. DeRubertis FR, Craven PA. Activation of hepatic guanylate cyclase by N-methyl-N′-nitro-N-nitrosoguanidine. Effects of thiols, N-ethylmaleimide, and divalent cations. *J Biol Chem* 1977; **252**: 5804–5814.
125. Braughler JM, Mittal CK, Murad F. Effects of thiols, sugars, and proteins on nitric oxide activation of guanylate cyclase. *J Biol Chem* 1979; **254**: 12450–12454.
126. Stamler JS, Simon DI, Osborne JA *et al.* *S*-nitrosylation of products with nitric oxide: synthesis and characterisation of biologically active compounds. *Proc Natl Acad Sci USA* 1992; **89**: 444–448.

127. Stamler JS, Jaraki O, Osborne JA *et al.* Nitric oxide circulates in mammalian plasma primarily as an *S*-nitroso adduct of serum albumin. *Proc Natl Acad Sci USA* 1992; **89**: 7674–7677.
128. Keaney JF, Stamler JS, Simon DI *et al.* *S*-nitroso-bovine serum albumin: an NO adduct with EDRF-like properties. *J Clin Invest* 1993; **91**: 1582–1589.
129. Gaston B, Reilly J, Drazen JN *et al.* Endogenous nitrogen oxides and bronchodilator *S*-nitrosothiols in human airways. *Proc Natl Acad Sci USA* 1993; **90**: 10957–10961.
130. Stamler JS, Simon DI, Jaraki O *et al.* *S*-nitrosylation of tissue-type plasminogen activator confers vasodilatory and anti-platelet properties on the enzyme. *Proc Natl Acad Sci USA* 1992; **89**: 8087–8091.
131. Molina Y, Vedia L, McDonald B *et al.* Nitric oxide-induced *S*-nitrosylation of glyceraldehyde-3-phosphate dehydrogenase inhibits enzyme activity and increases endogenous ADP-ribosylation. *J Biol Chem* 1993; **267**: 24929–24932.
132. Mohr S, Stamler JS, Brune B. Mechanism of covalent modification of glyceraldehyde-3-phosphate dehydrogenase at its active site thiol by nitric oxide, peroxynitrite, and related nitrosating agents. *FEBS Lett* 1994; **348**: 223–227.
133. Lipton SA, Choi Y-B, Lei SZ. A redox-based mechanism for the neuroprotective and neurodestructive effects of nitric oxide and related nitroso compounds. *Nature* 1993; **364**: 626–632.
134. Jia L, Bonaventura C, Bonaventura J, Stamler JS. *S*-nitrosohaemoglobin: a dynamic activity of blood involved in vascular control. *Nature* 1996; **380**: 221--226.
135. Olson JS. Stopped-flow, rapid mixing measurements of ligand binding to hemoglobin and red cells. *Meth Enzymol* 1981; **76**: 631–651.
136. Lancaster JR. Simulation of the diffusion and reaction of endogenously produced nitric oxide. *Proc Natl Acad Sci USA* 1994; **91**: 8137–8141.
137. Shen W, Hintze TH, Wolin MS. Nitric oxide. An important signaling mechanism between vascular endothelium and parenchymal cells in the regulation of oxygen consumption. *Circulation* 1995; **92**: 3505–3512.
138. Fleming I, Busse R. Control and consequences of endothelial nitric oxide formation. *Adv Pharmacol* 1995; **34**: 187–206.
139. Stuehr DJ, Kwon NS, Nathan CF *et al.* N^{ω}-hydroxy-L-arginine is an intermediate in the biosynthesis of nitric oxide from L-arginine. *J Biol Chem* 1991; **266**: 6259–6263.
140. Wallace GC, Gulati P, Fukuto JM. N^{ω}-hydroxy-L-arginine: a novel arginine analog capable of causing vasorelaxation in bovine intrapulmonary artery. *Biochem Biophys Res Commun* 1991; **176**: 528–534.
141. Buga GM, Singh R, Pervin S. Arginase activity in endothelial cells: inhibition by N^{G}-hydroxy-L-arginine during high-output NO production. *Am J Physiol* 1996; **271**: H1988–H1998.

142. Boucher JL, Custot J, Vadon S *et al.* N^{ω}-hydroxy-L-arginine, an intermediate in the L-arginine to nitric oxide pathway, is a strong inhibitor of liver and macrophage arginase. *Biochem Biophys Res Commun* 1994; **203**: 1614–1621.
143. Daghigh F, Fukuto JM, Ash DE. Inhibition of rat liver arginase by an intermediate in NO biosynthesis, N^{G}-hydroxy-L-arginine: implications for the regulation of nitric oxide biosynthesis by arginase. *Biochem Biophys Res Commun* 1994; **202**: 174–180.

Chapter 4

The Action of Nitrovasodilator Drugs

Christopher M Rembold

Introduction

Nitrovasodilators are defined as agents which contain nitrogen oxides and induce vascular smooth muscle relaxation. The prototypical nitrovasodilator is nitroglycerin, which has long been employed as a treatment for angina pectoris. As detailed in other chapters of this book, some nitrovasodilators, e.g. nitroprusside, appear to elaborate NO spontaneously.[1] Other nitrovasodilators, e.g. nitroglycerin, require enzymatic action to produce NO.[2,3] This chapter will concentrate on the mechanisms whereby nitrovasodilators and NO induce smooth muscle relaxation.

The specific mechanisms whereby nitrovasodilators and NO induce relaxation are controversial. NO diffuses into smooth muscle cells, where it activates the soluble form of guanylate cyclase (sGC).[4] NO also may activate other enzymes that may be involved in relaxation.[5] Activated sGC produces 3',5'-cyclic guanosine monophosphate (cGMP) which activates cGMP-dependent protein kinase (PKG),[6] and may also activate cAMP-dependent protein kinase (PKA; this requires larger increases in [cGMP] for activation).[7,8] Relaxation is hypothesized to result from specific protein phosphorylation by these enzymes. Unfortunately, few target phosphoproteins have been identified.

Rather than discussing the controversies regarding the target(s) for NO and the phosphoproteins involved in relaxation, this review will concentrate on the end mechanisms resulting from exposure of smooth muscle to nitrovasodilators, discussing: (1) the proposed agonist-activated contractile mechanisms; (2) the proposed nitrovasodilator-activated relaxant mechanisms; and (3) experiments evaluating the relative importance of these mechanisms.

Contractile Mechanisms in Vascular Smooth Muscle

There are at least four mechanisms responsible for arterial smooth muscle contraction (Figure 1);[9] the first three mechanisms cause contraction by increasing myoplasmic $[Ca^{2+}]$ ($[Ca^{2+}]_i$).

(a) High extracellular $[K^+]$ ($[K^+]_o$) and contractile agonists depolarize vascular smooth muscle which increases Ca^{2+} influx through L-type Ca^{2+} channels.

(b) Contractile agonists induce formation of 1,4,5-inositol trisphosphate (IP_3) which releases Ca^{2+} from the sarcoplasmic reticulum.

(c) Contractile agonists also increase Ca^{2+} influx by directly opening Ca^{2+}-permeable channels: specifically, agonists can open both non-selective cation channels and L-channels without a change in membrane potential.

The increased $[Ca^{2+}]_i$ resulting from these three mechanisms binds to and activates calmodulin, which activates myosin light chain kinase (MLCK). This enzyme phosphorylates the 20 kDa myosin light chain (MLC) on Ser19, a reaction that increases the actin-activated myosin ATPase activity, thereby inducing contraction.

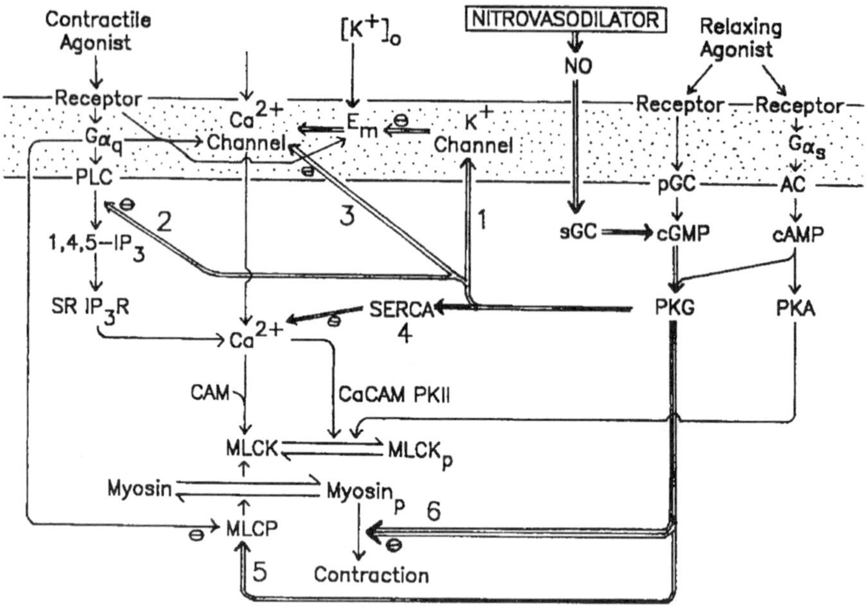

Figure 1. Schema of proposed mechanisms for nitrovasodilator-induced relaxation of arterial smooth muscle. The mechanisms hypothesized to be responsible for nitrovasodilator induced relaxation are shown in double lines and the numbers refer to the mechanisms described below and in the text. Nitrovasodilators produce nitric oxide (NO) which activates soluble guanylate cyclase (sGC), which produces cGMP, and thereby activates cGMP-dependent protein kinase (PKG). Activated PKG then is proposed to activate six mechanisms: (1) a K^+ channel, inducing hyperpolarization and decreased Ca^{2+} influx; (2) inhibition of IP_3 generation and Ca^{2+} release from the sarcoplasmic reticulum; (3) direct inhibition of Ca^{2+} influx through L channels; (4) enhancement of Ca^{2+} sequestration by SERCA or efflux by plasma membrane Ca^{2+} pumps/Na^+/Ca^{2+} exchange (*latter not shown*); (5) decreases in the Ca^{2+} sensitivity of phosphorylation, probably by stimulation of MLCP; and (6) uncoupling force from MLC phosphorylation. This scheme is by nature incomplete. For example, it is possible that some actions of nitrovasodilators do not involve cGMP and/or PKG. Also shown: (i) proposed contractile mechanisms; (ii) proposed mechanisms for cAMP-dependent relaxation *via* cAMP-dependent protein kinase (PKA). Abbreviations: $[K^+]_o$, extracellular $[K^+]$; $G\alpha_q$ and $G\alpha_s$, G proteins; E_m, membrane potential; PLC, phospholipase C; 1,4,5-IP_3, 1,4,5 inositol trisphosphate; cGMP, guanosine 3',5'-cyclic monophosphate; cAMP, adenosine 3',5'-cyclic monophosphate; SR IP_3R, sarcoplasmic reticulum IP_3 receptor; SERCA, sarcoplasmic endoplasmic reticulum calcium ATPase; CAM, calmodulin; CaCAM PKII, Ca^{2+}-calmodulin protein kinase II; MLCK, myosin light chain kinase; $Myosin_p$, phosphorylated 20 kDa light chain of myosin; MLCP, myosin light chain phosphatase; pGC, particulate guanylate cyclase; sGC, soluble guanylate cyclase; AC, adenylate cyclase.

(d) Contractile agonists also induce contraction by increasing MLC phosphorylation, independent of increases in $[Ca^{2+}]_i$ (i.e. increases in the $[Ca^{2+}]_i$ sensitivity of MLC phosphorylation). The increased MLC phosphorylation appears to be caused by G protein-mediated inhibition of MLC phosphatase.

Mechanisms of Nitrovasodilator-induced Smooth Muscle Relaxation

There are at least six specific mechanisms hypothesized to account for nitrovasodilator-mediated relaxation of smooth muscle (Figure 1). Four of these mechanisms are the reverse of the above contractile mechanisms and two are unique to relaxation. The first four mechanisms involve nitrovasodilator-induced alteration in $[Ca^{2+}]_i$ and the latter two involve changes in the cellular response to changes in $[Ca^{2+}]_i$.

It has long been known that nitrovasodilators and NO decrease $[Ca^{2+}]_i$ in isolated smooth muscle cells[10–12] and intact tissues.[13] Some of this reduction in $[Ca^{2+}]_i$ appears to be caused by a decrease in Ca^{2+} influx, because nitroglycerin is known to reduce Mn^{2+} influx in histamine-stimulated intact swine carotid artery.[14] This result suggests that nitrovasodilators can reduce $[Ca^{2+}]_i$ at least partially either by repolarization or by a direct effect on Ca^{2+} channels.

Repolarization

NO induces repolarization in some,[15] but not all, smooth muscles.[16,17] In some tissues, it appears that NO-dependent repolarization is not necessary for NO-dependent relaxation.[18] It has been proposed that NO, through increases in [cGMP] and activation of PKG, increases the open probability of K^+ channels, repolarizing the cell which decreases Ca^{2+} influx through L-type channels.[19–22] A recent study in pituitary cells suggests that PKG phosphorylates and activates a protein phosphatase 2A that dephosphorylates and activates a K^+ channel, inducing repolarization.[23] It is possible that a similar mechanism exists in smooth muscle. It has also been proposed that NO can directly activate K^+ channels.[22]

Inhibition of Ca^{2+} Release

Preincubation of rat aorta with nitroprusside attenuated agonist-induced increases in mass of inositol monophosphate (IP_1)[24] and IP_3.[25] Supporting the importance of this mechanism was the finding that preincubation of swine carotid artery with nitroprusside attenuated histamine-induced increases in $[Ca^{2+}]_i$ and the resulting contraction.[26]

Inactivation of Ca^{2+} Channels

Nitroprusside,[27] NO,[28] and 8-bromo-cGMP[29] decreased L-channel conductance in patch-clamped vascular smooth muscle cells. These findings showed that repolarization may not be necessary for decreased Ca^{2+} influx through L-channels.

Enhancement of Ca^{2+} Efflux and/or Sequestration

There are several proposed mechanisms for nitrovasodilator-dependent increases in Ca^{2+} extrusion and sequestration: (a) PKG can phosphorylate phospholamban (see also Chapter 1), a process that activates the sarcoplasmic reticulum Ca^{2+} ATPase by removing the inhibition caused by unphosphorylated phospholamban.[30–32] (b) nitroglycerin appears to activate a plasma membrane Ca^{2+} ATPase in isolated cells.[33,34] (c) 8-bromo-cGMP increased Na^+/Ca^{2+} exchange in isolated rat aortic smooth muscle cells.[35] Supporting these latter two mechanisms was the finding that nitroglycerin enhanced $^{45}Ca^{2+}$ efflux in acetylcholine-stimulated coronary artery cells.[36]

Reduction in the Myoplasmic $[Ca^{2+}]$ Sensitivity of Myosin Light Chain Phosphorylation

Nitroprusside-induced relaxation of intact smooth muscle was first reported to occur without a decrease in $[Ca^{2+}]_i$, i.e. there was a decrease in the $[Ca^{2+}]_i$ sensitivity of force.[37] In α-toxin skinned smooth muscle, cGMP induced a relaxation when $[Ca^{2+}]$ was clamped at a constant level.[38] In the intact swine carotid, nitrovasodilators decreased the $[Ca^{2+}]_i$ sensitivity of force, partially by decreasing the $[Ca^{2+}]_i$ sensitivity of MLC phosphorylation.[39] However, this decrease in the $[Ca^{2+}]_i$ sensitivity of MLC phosphorylation was only seen initially after addition of nitroglycerin. At time points after 5 min, relaxation was primarily caused by an uncoupling of force from MLC phosphorylation[13] (see below). In Triton X-100 skinned guinea-pig mesenteric arterioles, addition of cGMP-preactivated PKG reduced MLC phosphorylation and contractile force proportionally despite a constant $[Ca^{2+}]_i$, suggesting that cGMP primarily decreased the $[Ca^{2+}]_i$ sensitivity of phosphorylation in these skinned tissues.[40] The mechanism for cGMP-dependent decreases in the $[Ca^{2+}]_i$ sensitivity of phosphorylation is unknown. It does not appear to be caused by site A MLCK phosphorylation, because nitroprusside decreased the $[Ca^{2+}]_i$ sensitivity of MLC phosphorylation without increasing site A MLCK phosphorylation in intact swine carotid artery.[39]

Uncoupling of Force from Myosin Light Chain Phosphorylation

In the intact swine carotid, nitrovasodilators also decreased force below that expected from the degree of MLC phosphorylation (Figure 2).[13,41] This has been termed 'uncoupling of force from MLC phosphorylation'. Prior studies revealed a sigmoidal steady-state relation between MLC phosphorylation and stress (force normalized to cross-sectional area).[42–44] Forskolin, which increases [cAMP] rather than [cGMP], relaxed swine carotid artery, but did not alter the relation between

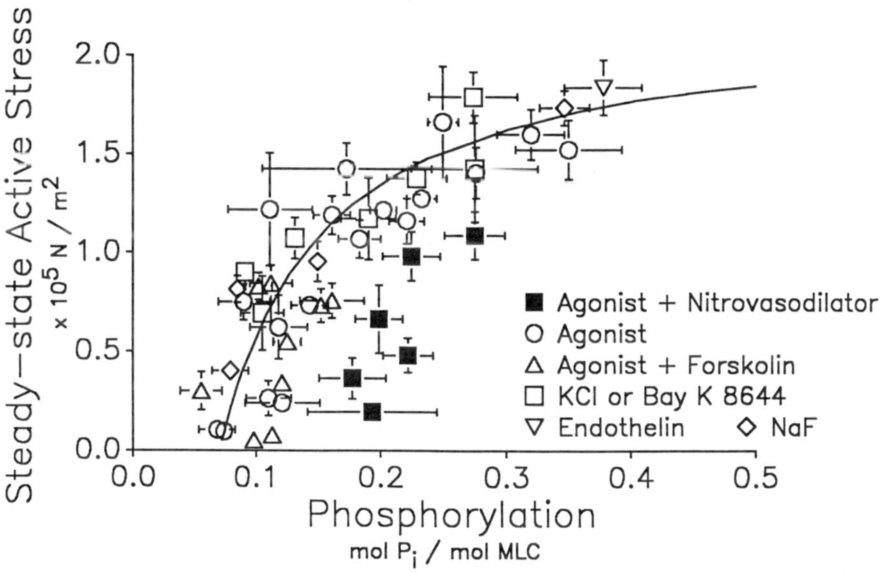

Figure 2. The effect of nitrovasodilators on the steady-state relationship between myosin light chain phosphorylation and contractile stress (force normalized to cross-sectional area). The filled squares represent data from swine carotid arterial tissues contracted with 3 or 10 μM histamine and relaxed with either 10 or 100 μM sodium nitroprusside or 10 or 100 μM nitroglycerin, and frozen 120 minutes later. The open symbols represent data previously obtained during control contractions induced with agonists (phenylephrine or histamine, open circles), high $[K^+]_o$ with or without Bay K 8644 (open squares), endothelin (open inverted triangles), sodium fluoride (open diamonds), or agonists plus forskolin to induce relaxation by increasing cAMP (open triangles) (data from Refs. 44–46).

MLC phosphorylation and force.[45,46] These data suggest that nitrovasodilators are the only physiological stimuli that uncouple force from MLC phosphorylation. Other pharmacological treatments also can uncouple force from MLC phosphorylation in smooth muscle: Ca^{2+} depletion/repletion,[47,48] high extracellular $[Mg^{2+}]$,[49] and okadaic acid.[50] These findings suggest that regulatory systems other than MLC phosphorylation may also modulate contractile force in some smooth muscles under certain conditions. There are a number of other candidate regulatory systems for force in smooth muscle. (1) The thin filament proteins caldesmon and calponin can alter thin filament function.[51–54] (2) Myosin can both bind Ca^{2+} and be phosphorylated on the heavy chain.[51,55] (3) Phorbol diesters, presumably by activation of protein kinase C, can induce contraction by an unknown cascade.[56–58] Finally, (4) some investigators propose non crossbridge-mediated stress in the form of cytoskeletal domain crosslinks.[56,59] It is possible that nitroglycerin uncouples MLC phosphorylation from force by these and/or other mechanisms.

Relative Importance of Mechanisms of Nitrovasodilator-induced Relaxation

The above studies suggest that nitrovasodilators could induce relaxation by one or more of the six proposed mechanisms. Very little has been published evaluating the relative importance of these six mechanisms in an intact smooth muscle. It is possible that some of these mechanisms are only activated in isolated smooth muscle cells and/or with very high concentrations of nitrovasodilators. We evaluated the relative importance of these mechanisms in the rat tail artery.[60] When tissues were stimulated with 1 μM phenylephrine or depolarized with high $[K^+]_o$, addition of nitroglycerin induced relaxation without significant changes in membrane potential or Fura 2 estimated $[Ca^{2+}]_i$. These data suggest that the latter two mechanisms — i.e. #5 (decreasing the Ca^{2+} sensitivity of MLC phosphorylation), and #6 (uncoupling force from MLC phosphorylation) — were the only mechanisms activated by nitroglycerin when rat tail artery was stimulated with 1 μM phenylephrine or depolarized with high $[K^+]_o$. Since studies in the swine carotid primarily showed that uncoupling force from MLC phosphorylation was more important than reduction in the Ca^{2+} sensitivity of MLC phosphorylation,[13] it is reasonable to conclude that uncoupling force from MLC phosphorylation is an important mechanism in these two arterial preparations.

When tissues were stimulated with submaximal concentrations of phenylephrine (0.1–0.3 μM), only part of the relaxation was caused by a decrease in the $[Ca^{2+}]_i$ sensitivity of MLC phosphorylation or uncoupling of force from MLC phosphorylation.[60] The other part of the nitroglycerin-induced relaxation was associated with a partial repolarization and reductions in Fura 2 estimated $[Ca^{2+}]_i$. The decline in Fura 2 estimated $[Ca^{2+}]_i$ was that predicted by the depolarization, i.e. there was no nitroglycerin-induced alteration in the relation between membrane potential (E_m) and Fura 2 estimated $[Ca^{2+}]_i$.[60] This result suggests that repolarization (mechanism #1) was responsible for nitroglycerin-induced reduction in Ca^{2+} influx[14] and $[Ca^{2+}]_i$. This result also suggests that mechanisms that alter $[Ca^{2+}]_i$ without altering E_m — i.e. #3 (direct inhibition of Ca^{2+} channels), and #4 (enhancement of Ca^{2+} efflux/sequestration) - were not activated by nitroglycerin in these tissues. Both of these mechanisms would reduce $[Ca^{2+}]_i$ below that expected given the E_m. Our results differ from a number of prior studies showing that increases in [cGMP] decreased L-channel conductance in patch-clamped cells,[27,29] increased Ca^{2+} efflux,[33–36] and increased Ca^{2+} sequestration.[30–32] These studies were performed in either cultured or patch-clamped cells and typically involved high [nitroglycerin] or large increases in [cGMP]. Phenotypic modulation from culturing and dialysis of intracellular constituents from patch clamping could alter the results. Additionally, our results were performed with relatively low concentrations of nitroglycerin. It is possible that other mechanisms are invoked by supraphysiological [nitroglycerin] and huge increases in [cGMP]. If that were the case, these other mechanisms are less physiologically relevant because relaxation would already

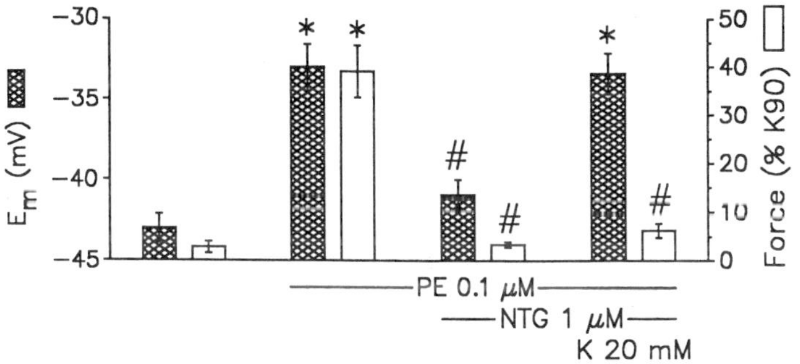

Figure 3. E_m (mV, left-hand bar) and force (as a percent of the response after 5 min to 90 mM $[K^+]_o$, right-hand bar) in rat tail artery tissues treated with cumulative additions of 0.1 μM phenylephrine (PE), 1.0 μM nitroglycerin (NTG), and 20 mM $[K^+]_o$ (all at 5 min intervals, n = 7). *, significantly different from unstimulated; #, significantly different from phenylephrine 1.0 μM alone, by the Newman-Keuls test (data replotted from Ref. 60).

have occurred from decreases in $[Ca^{2+}]$ sensitivity or repolarization prior to the invocation of these other mechanisms.

We examined the relative importance of nitroglycerin-induced repolarization in 0.1 μM phenylephrine-induced contractions by reversing the nitroglycerin-induced repolarization with high $[K^+]_o$ (Figure 3).[60] Stimulation with 0.1 μM phenylephrine induced a modest depolarization and contraction. Addition of 1 μM nitroglycerin repolarized and relaxed the tissues. Further addition of 20 mM $[K^+]_o$ depolarized the tissues back to the level similar to that observed with 0.1 μM phenylephrine alone. Addition of 20 mM $[K^+]_o$ increased force by 104 ± 38% from the value observed with phenylephrine and nitroglycerin. However, this force was only 16% of the value observed with 0.1 μM phenylephrine alone, despite similar E_m values. These data suggest that nitroglycerin relaxed the phenylephrine-stimulated tail artery primarily by uncoupling force from MLC phosphorylation and/or decreasing the $[Ca^{2+}]_i$ sensitivity of MLC phosphorylation.

The only mechanism so far not discussed is #2 (inhibition of Ca^{2+} release). Evaluation of Ca^{2+} release requires different experimental conditions, i.e. removal of extracellular Ca^{2+}. Pretreatment with nitroglycerin in the absence of extracellular Ca^{2+} attenuated 100 μM phenylephrine-induced Ca^{2+} release.[60] This effect was confined to the first 10 seconds after phenylephrine stimulation, suggesting that this mechanism primarily altered only the initial responses. Prior studies also revealed that release of intracellular Ca^{2+} only occurs with relatively high concentrations of contractile agonists in the rat tail and other arteries.[9,61] These data suggest that release of intracellular Ca^{2+} and its inhibition by nitroglycerin may have less importance than the other mechanisms in sustained smooth muscle contraction and relaxation.

Comparison with Endothelium-dependent Relaxation

This chapter has concentrated on the action of exogenous nitrovasodilators. The endothelium also produces NO, as detailed elsewhere in this book. It should be noted that endothelial-dependent relaxation is more complex than that observed with exogenous nitrovasodilators because the endothelium releases at least two relaxing autacoids: NO and endothelium-derived hyperpolarizing factor (EDHF). There are conflicting reports on the roles of these two agents.[18,62,63] For example, ADP[64] and acetylcholine[65,66] induce relaxation *via* the endothelial release of NO and/or EDHF.[67,68] In some tissues, ADP-induced relaxation is not associated with hyperpolarization[64] or with changes in [cGMP].[67] The endothelium-dependent relaxation induced by acetylcholine[69] or ATP (*via* P_2 receptor stimulation)[70] can occur without inducing hyperpolarization. Therefore, extrapolation of data from nitrovasodilators to mediators secreted by the endothelium may be subject to errors.

Conclusion

There are at least six mechanisms proposed to be involved in nitrovasodilator-induced vascular smooth muscle relaxation. In intact arteries, the most important appears to be uncoupling of force from MLC phosphorylation. The mechanism for this effect is unknown. Nitrovasodilators also appear to induce relaxation by inducing repolarization, inhibiting Ca^{2+} release, and decreasing the $[Ca^{2+}]_i$ sensitivity of MLC phosphorylation. There is no evidence in intact arteries that nitrovasodilators directly inhibit Ca^{2+} channels or alter Ca^{2+} efflux or sequestration.

Acknowledgements

The author would like to thank Nancy McDaniel, Dee Van Riper, Xiao-Liang Chen, and Barbara Weaver for their help in this research. C. M. Rembold is a Lucille P. Markey Scholar and grants from the Lucille P. Markey Charitable trust, the National Institutes of Health, the Virginia affiliate of the American Heart Association, and the Jeffress trust supported the work.

References

1. Kowaluk EA, Seth P, Fung H-L. Metabolic activation of sodium nitroprusside to nitric oxide in vascular smooth muscle. *J Pharmacol Exp Ther* 1992; **262**: 916–922.
2. Seth P, Fung H-L. Biochemical characterization of a membrane-bound enzyme responsible for generating nitric oxide from nitroglycerin in vascular smooth muscle cells. *Biochem Pharmacol* 1993; **46**: 1481–1486.

3. Salvemini D, Pistelli A, Vane J. Conversion of glyceryl trinitrate to nitric oxide in tolerant and non-tolerant smooth muscle and endothelial cells. *Br J Pharmacol* 1993; **108**: 162–169.
4. Bredt DS, Snyder SH. Nitric oxide: a physiologic messenger molecule. *Annu Rev Biochem* 1994; **63**: 175–195.
5. McDonald LJ, Moss J. Nitric oxide and NAD-dependent protein modification. *Mol Cell Biochem* 1994; **138**: 201–206.
6. Lincoln TM, Cornwell TL, Komalavilas P, Boerth N. Cyclic GMP-dependent protein kinase in nitric oxide signaling. *Methods Enzymol* 1996; **269**: 149–166.
7. Lincoln TM, Cornwell TL, Taylor AE. cGMP-dependent protein kinase mediates the reduction of Ca^{2+} by cAMP in vascular smooth muscle cells. *Am J Physiol* 1990; **258**: C399–C407.
8. Francis SH, Noblett BD, Todd BW *et al.* Relaxation of vascular and tracheal smooth muscle by cyclic nucleotide analogs that preferentially activate purified cGMP-dependent protein kinase. *Mol Pharmacol* 1988; **34**: 506–517.
9. Chen X-L, Rembold CM. Phenylephrine contracts rat tail artery by one electromechanical and three pharmacomechanical mechanisms. *Am J Physiol* 1995; **268**: H74–H81.
10. Word RA, Casey ML, Kamm KE, Stull JT. Effects of cGMP on $[Ca^{2+}]_i$, myosin light chain phosphorylation, and contraction in human myometrium. *Am J Physiol* 1991; **260**: C861–C867.
11. Rashatwar SS, Cornwell TL, Lincoln TM. Effects of 8-bromo-cGMP on Ca^{2+} levels in vascular smooth muscle cells: possible regulation of Ca^{2+}-ATPase by cGMP-dependent protein kinase. *Proc Natl Acad Sci USA* 1987; **84**: 5685–2689.
12. Kobayashi S, Kanaide H, Nakamura M. Cytosolic-free calcium transients in cultured vascular smooth muscle cells: microfluorometric measurements. *Science* 1985; **229**: 553–555.
13. McDaniel NL, Chen X-L, Singer HA *et al.* Nitrovasodilators relax arterial smooth muscle by decreasing $[Ca^{2+}]_i$, $[Ca^{2+}]_i$ sensitivity, and uncoupling stress from myosin phosphorylation. *Am J Physiol* 1992; **263**: C461–C467.
14. Chen X-L, Rembold CM. Cyclic nucleotide dependent regulation of Mn^{2+} influx, $[Ca^{2+}]_i$, and arterial smooth muscle relaxation. *Am J Physiol* 1992; **263**: C468–C473.
15. Tare M, Parkington HC, Coleman HA *et al.* Hyperpolarization and relaxation of arterial smooth muscle caused by nitric oxide derived from the endothelium. *Nature* 1990; **346**: 69–71.
16. Komori K, Lorenz RR, Vanhoutte PM. Nitric oxide, ACh, and electrical and mechanical properties of canine arterial smooth muscle. *Am J Physiol* 1988; **255**: H207–H212.

17. Brayden JE. Membrane hyperpolarization is a mechanism of endothelium-dependent cerebral vasodilation. *Am J Physiol* 1990; **259**: H668–H673.
18. Plane F, Garland CJ. Differential effects of acetylcholine, nitric oxide and levcromakalim on smooth muscle membrane potential and tone in the rabbit basilar artery. *Br J Pharmacol* 1993; **110**: 651–656.
19. Williams DL, Jr., Katz GM, Roy-Contancin L, Reuben JP. Guanosine 5'-monophosphate modulates gating of high-conductance Ca^{2+}-activated K^+ channels in vascular smooth muscle cells. *Proc Natl Acad Sci USA* 1988; **85**: 9360–9364.
20. Thornbury KD, Ward SM, Dalziel HH *et al.* Nitric oxide and nitrosocysteine mimic nonadrenergic, noncholinergic hyperpolarization in canine proximal colon. *Am J Physiol* 1991; **261**: G553–G557.
21. Robertson BE, Schubert R, Hescheler J, Nelson MT. cGMP-dependent protein kinase activates Ca-activated K channels in cerebral artery smooth muscle cells. *Am J Physiol* 1993; **265**: C299–C303.
22. Koh SD, Campbell JD, Carl A, Sanders KM. Nitric oxide activates multiple potassium channels in canine colonic smooth muscle. *J Physiol (Lond)* 1995; **489**: 735–743.
23. White RE, Lee AB, Shcherbatko AD *et al.* Potassium channel stimulation by natriuretic peptides through cGMP-dependent phosphorylation. *Nature* 1993; **361**: 263–266.
24. Rapoport RM. Cyclic guanosine monophosphate inhibition of contraction may be mediated through inhibition of phosphatidylinositol hydrolysis in rat aorta. *Circ Res* 1986; **58**: 407–410.
25. Langlands JM, Diamond J. The effect of phenylephrine on inositol 1,4,5-trisphosphate levels in vascular smooth muscle measured using a protein binding assay system. *Biochem Biophys Res Commun* 1990; **173**: 1258–1265.
26. Shishibori T, Yamashita K, Bandoh J *et al.* Presence of Ca^{2+}-sensitive and -insensitive SM22α isoproteins in bovine aorta. *Biochem Biophys Res Commun* 1996; **229**: 225–230.
27. Clapp LH, Gurney AM. Modulation of calcium movements by nitroprusside in isolated vascular smooth muscle cells. *Pflügers Arch* 1991; **418**: 462–470.
28. Blatter LA, Wier WG. Nitric oxide decreases $[Ca^{2+}]_i$ in vascular smooth muscle by inhibition of the calcium current. *Cell Calcium* 1994; **15**: 122–131.
29. Ishikawa T, Hume JR, Keef KD. Regulation of Ca^{2+} channels by cAMP and cGMP in vascular smooth muscle cells. *Circ Res* 1993; **73**: 1128–1137.
30. Lincoln TM, Cornwell TL. Towards an understanding of the mechanism of action of cyclic AMP and cyclic GMP in smooth muscle relaxation. *Blood Vessels* 1991; **28**: 129–137.
31. Twort CH, Van Breemen C. Cyclic guanosine monophosphate-enhanced sequestration of Ca^{2+} by sarcoplasmic reticulum in vascular smooth muscle. *Circ Res* 1988; **62**: 961–964.

32. Kimura M, Kimura I, Kobayashi S. Relationship between cyclic AMP-dependent protein kinase activation and Ca uptake increase of sarcoplasmic reticulum fraction of hog biliary muscles relaxed by cholecystokinin-C-terminal peptides. *Biochem Pharmacol* 1982; **31**: 3077–3083.
33. Furakawa K, Tawada Y, Shigekawa M. Regulation of the plasma membrane Ca^{2+} pump by cyclic nucleotides in cultured vascular smooth muscle cells. *J Biol Chem* 1988; **263**: 8058–8065.
34. Popescu LM, Foril CP, Hinescu M *et al.* Nitrogylcerin stimulates the sarcolemmal Ca^{++}-extrusion ATPase of coronary smooth muscle cells. *Biochem Pharmacol* 1985; **34**: 1857–1860.
35. Furukawa K, Ohshima N, Tawada-Iwata Y, Shigekawa M. Cyclic GMP stimulates Na^+/Ca^{2+} exchange in vascular smooth muscle cells in primary culture. *J Biol Chem* 1991; **266**: 12337–12341.
36. Itoh T, Kuriyama H, Ueno H. Mechanisms of the nitroglycerine-induced vasodilation in vascular smooth muscles of the rabbit and pig. *J Physiol* (*Lond*) 1983; **343**: 233–252.
37. Karaki H, Sato K, Ozaki H, Murakami K. Effects of sodium nitroprusside on cytosolic calcium level in vascular smooth muscle. *Eur J Pharmacol* 1988; **156**: 259–266.
38. Nishimura J, Van Breemen C. Direct regulation of smooth muscle contractile elements by second messengers. *Biochem Biophys Res Commun* 1989; **163**: 929–935.
39. Van Riper DA, McDaniel NL, Rembold CM. Myosin light chain kinase phosphorylation in nitrovasodilator induced swine carotid artery relaxation. *Biochim Biophys Acta Mol Cell Res* 1997; **1355**: 323–330.
40. Pfitzer G, Boels PJ. Differential skinning of smooth muscle: a new approach to excitation-contraction coupling. *Blood Vessels* 1991; **28**: 262–267.
41. Bárány M, Bárány K. Dissociation of relaxation and myosin light chain dephosphorylation in porcine uterine muscle. *Arch Biochem Biophys* 1993; **305**: 202–204.
42. Ratz PH, Hai C-M, Murphy RA. Dependence of stress on cross-bridge phosphorylation in vascular smooth muscle. *Am J Physiol* 1989; **256**: C96–C100.
43. Rembold CM, Murphy RA. Myoplasmic $[Ca^{2+}]$ determines myosin phosphorylation in agonist-stimulated swine arterial smooth muscle. *Circ Res* 1988; **63**: 593–603.
44. Rembold CM. Modulation of the $[Ca^{2+}]$-sensitivity of myosin phosphorylation in intact swine arterial smooth muscle. *J Physiol* (*Lond*) 1990; **429**: 77–94.
45. McDaniel NL, Rembold CM, Richard HL, Murphy RA. cAMP relaxes arterial smooth muscle predominantly by decreasing cell $[Ca^{2+}]$. *J Physiol* (*Lond*) 1991; **439**: 147–160.

46. McDaniel NL, Rembold CM, Murphy RA. Cyclic nucleotide dependent relaxation in vascular smooth muscle. *Can J Physiol Pharmacol* 1994; **72**: 1380–1385.
47. Gerthoffer WT. Dissociation of myosin phosphorylation and active tension during muscarinic stimulation of tracheal smooth muscle. *J Pharmacol Exp Ther* 1987; **240**: 8–15.
48. Aburto TK, Lajoie C, Morgan KG. Mechanisms of signal transduction during α_2-adrenergic receptor-mediated contraction of vascular smooth muscle. *Circ Res* 1993; **72**: 778–785.
49. D'Angelo EKG, Singer HA, Rembold CM. Magnesium relaxes arterial smooth muscle by decreasing intracellular [Ca^{2+}] without changing intracellular [Mg^{2+}]. *J Clin Invest* 1992; **89**: 1988–1994.
50. Tansey MG, Hori M, Karaki H *et al.* Okadaic acid uncouples myosin light chain phosphorylation and tension in smooth muscle. *FEBS Lett* 1990; **270**: 219–221.
51. Kamm KE, Stull JT. Regulation of smooth muscle contractile elements by second messengers. *Annu Rev Physiol* 1989; **51**: 299–313.
52. Ngai PK, Walsh MP. Inhibition of smooth muscle actin-activated myosin Mg^{2+}-ATPase activity by caldesmon. *J Biol Chem* 1984; **259**: 13656–13659.
53. Sobue K, Muramoto Y, Inui M *et al.* Control of actin-myosin interaction of gizzard smooth muscle by calmodulin- and caldesmon-linked flip-flop mechanism. *Biomed Res* 1982; **2**: 188–196.
54. Tsunekawa S, Takahashi K, Abe M *et al.* Calpain proteolysis of free and bound forms of calponin, a troponin T-like protein in smooth muscle. *FEBS Lett* 1989; **250**: 493–496.
55. Ito M, Hartshorne DJ. Phosphorylation of myosin as a regulatory mechanism in smooth muscle. *Prog Clin Biol Res* 1990; **327**: 57–72.
56. Rasmussen H, Takuwa Y, Park S. Protein kinase C in the regulation of smooth muscle contraction. *FASEB J* 1987; **1**: 177–185.
57. Rembold CM, Murphy RA. [Ca^{2+}]-dependent myosin phosphorylation in phorbol diester stimulated smooth muscle contraction. *Am J Physiol* 1988; **255**: C719–C723.
58. Jiang MJ, Morgan KG. Intracellular calcium levels in phorbol ester-induced contractions of vascular muscle. *Am J Physiol* 1987; **253**: H1365–H1371.
59. Small JV, Furst DO, De Mey JGR. Localization of filamin in smooth muscle. *J Cell Biol* 1986; **102**: 210–220.
60. Chen XL, Rembold CM. Nitroglycerin relaxes rat tail artery primarily by lowering Ca^{2+} sensitivity and partially by repolarization. *Am J Physiol* 1996; **271**: H962–H968.
61. Rembold CM, Murphy RA. Histamine concentration and Ca^{2+} mobilization in arterial smooth muscle. *Am J Physiol* 1989; **257**: C122–C128.

62. Parsons SJW, Hill A, Waldron GJ *et al.* The relative importance of nitric oxide and nitric oxide-independent mechanisms in acetylcholine-evoked dilatation of the rat mesenteric bed. *Br J Pharmacol* 1994; **113**: 1275–1280.
63. Waldron GJ, Garland CJ. Contribution of both nitric oxide and a change in membrane potential to acetylcholine-induced relaxation in the rat small mesenteric artery. *Br J Pharmacol* 1994; **112**: 831–836.
64. Brayden JE. Hyperpolarization and relaxation of resistance arteries in response to adenosine diphosphate. Distribution and mechanism of action. *Circ Res* 1991; **69**: 1415–1420.
65. Feletou M, Vanhoutte PM. Endothelium-dependent hyperpolarization of canine coronary smooth muscle. *Br J Pharmacol* 1988; **93**: 515–524.
66. Keef KD, Bowen SM. Effect of ACh on electrical and mechanical activity in guinea pig coronary arteries. *Am J Physiol* 1989; **257**: H1096–H1103.
67. Brayden JE, Wellman GC. Endothelium-dependent dilation of feline cerebral arteries: role of membrane potential and cyclic nucleotides. *J Cereb Blood Flow Metab* 1989; **9**: 256–263.
68. Nagao T, Vanhoutte PM. Characterization of endothelium-dependent relaxations resistant to nitro-L-arginine in the porcine coronary artery. *Br J Pharmacol* 1992; **107**: 1102–1107.
69. McPherson GA, Angus JA. Evidence that acetylcholine-mediated hyperpolarization of the rat small mesenteric artery does not involve the K^+ channel opened by cromakalim. *Br J Pharmacol* 1991; **103**: 1184–1190.
70. Keef KD, Pasco JS, Eckman DM. Purinergic relaxation and hyperpolarization in guinea pig and rabbit coronary artery: role of the endothelium. *J Pharmacol Exp Ther* 1992; **260**: 592–600.

Chapter 5

Nitric Oxide Synthase Biology: Insights Gained from 'Knockout' Mice

Andreas Papapetropoulos, Venetia Zachariou
and William C Sessa

Introduction

The pivotal discoveries that L-arginine is the precursor for endothelium-derived relaxing factor in the vasculature and for the cytotoxic activity and production of nitrogen oxides in activated macrophages has led to the identification of the pathways involved in NO synthesis.[1-5] Endothelial[6-8] and neuronal[9-12] isoforms of the NO synthase enzyme (eNOS and nNOS, respectively) are expressed under baseline conditions, and are therefore termed 'constitutive'. This distinguishes them from an inducible iNOS isoform which is expressed following stimulation of most types of nucleated cells with cytokines such as tumour necrosis factor (TNF)-α and interleukin (IL)-1.[13-16] The genes for these isoforms are located on different chromosomes,[17-21] and cDNA cloning and expression of recombinant NOS has shown that all three isoforms contain N-terminal oxygenase and C-terminal reductase domains which are separated by a calmodulin binding site.[6-9,13-16] The topography of NOS suggests that NADPH binds and transfers electrons to flavin 'electron docking stations', which subsequently transfer electrons to haem iron to activate O_2 for the hydroxylation of L-arginine. A second redox cycle then generates NO and citrulline from hydroxy-L-arginine.[1,22] The Ca^{2+}/calmodulin complex binds reversibly to eNOS and nNOS and thereby leads to enzyme activation.[1] By contrast, calmodulin is normally tightly bound to iNOS so that NO formation by this isoform is dependent on the supply of L-arginine rather than $[Ca^{2+}]$.[23] In the absence of L-arginine, all NOS isoforms are capable of generating superoxide anions through the reduction of O_2,[24] thereby potentially resulting in the formation of radical species that lead to the oxidization of proteins and/or lipids (see Chapter 18). Several excellent reviews discuss the historical perspectives of NO research, the pharmacology of NO and the biochemistry of the NOS isoforms.[1-5,11,22,25-27] This chapter will focus on novel aspects of the molecular regulation of the NO synthases, by examining recent advances in the area of NOS isoform 'knockout' animals. Mice with targeted disruption of the NOS genes (−/− and +/−) can provide powerful tools to examine more directly the involvement of specific NOS isoforms in different pathophysiological processes and to elucidate other control mechanisms that may compensate for the lack of a specific NOS isoform.

Control of Nitric Oxide Synthase Activity by Dimerization and Subcellular Localization

All active NOS proteins exist in the cell as homodimers, with dimerization being critically dependent on haem and stabilized by L-arginine and tetrahydrobiopterin (BH_4).[22,26,28-32] Recent experiments using site-directed mutagenesis and yeast two-hybrid screening have identified key residues important for dimerization and BH_4 binding.[29,31] Mutation of a cysteine residue critical for haem coordination to alanine thus abolishes eNOS dimer formation.[29,33] Incubation of eNOS purified from insect cells with BH_4 and L-arginine increases the fraction of eNOS present in its dimeric form, thus suggesting that L-arginine and BH_4 cooperate to facilitate dimerization.[29] Manipulation of the cellular mechanisms that control assembly of the NOS dimer may ultimately allow the development of selective inhibitors of the different NOS isoforms. Indeed, a newly-described protein inhibitor of nNOS (named PIN) has been shown to promote the de-dimerization of nNOS.[34]

Specific subcellular targeting of NOS may place the enzyme in a microenvironment that permits efficient coupling of NOS to extracellular stimuli, or near to substrates that are critical for intracellular signalling by NO.[35–44] nNOS contains a PDZ (PSD-95, discs large, ZO-1) motif based on the residues GLGF found at positions 27–30 of its amino terminus, which is responsible for localizing nNOS to the sarcolemma in skeletal muscle and to synaptic junctions in brain.[37] This involves a protein-protein interaction between the PDZ motif of nNOS with similar motifs found in α_1-syntrophin, postsynaptic density-95 or postsynaptic density-93 proteins in skeletal muscle myocytes and neurones.[37,38] NOS inhibitors increase electrically-driven muscle contraction to a modest extent, suggesting that localization of nNOS to the sarcolemma is functionally important.[10] However, it remains unclear whether localization primarily affects the intrinsic activity of the enzyme or its ability to be coupled to stimulus-dependent excitation-contraction.

In freshly isolated bovine endothelial cells, cultured bovine and human endothelial cells, and COS cells transfected with eNOS cDNA, $> 95\%$ of the NOS activity is found in the particulate fraction.[6,45,46] Incubation of bovine endothelial cells with labelled myristic acid followed by immunoprecipitation of eNOS, has revealed the incorporation of myristate into eNOS.[40] N-myristoylation is a rare, co-translational modification that is important for protein localization, and permits enzymes to be orientated near their endogenous substrates and to initiate or respond to compartmentalized cellular signals. Site directed mutagenesis of glycine-2 to alanine converts the membrane associated eNOS to a cytosolic form and reduces its ability to generate NO in intact cells, even though the soluble eNOS mutant behaves in an identical fashion to wild-type eNOS with respect to cofactor requirements, affinity for L-arginine, and inhibition by guanidino-substituted arginine analogues.[43,44] These data support the notion that the compartmentalization of eNOS in discrete regions of the endothelial cell is required for optimal activation. eNOS also undergoes palmitoylation at cysteine residues 15 and 26. Mutation of the palmitoylation sites does not block N-myristoylation or influence its membrane asscociation or kinetics.[41] However, NO release from human embryonic kidney cells stably transfected with palmitoylation mutant forms of eNOS (single or double mutants) produce less NO in response to ionomycin stimulation.[41]

It is now known that myristoylation is necessary for palmitoylation and that both N-myristoylation and palmitoylation are necessary for Golgi targeting of the enzyme. Recently, this has been shown unequivocably using eNOS coupled to green flourescent protein (GFP) to track movement of the enzyme within living cells. Indeed, the first 35 amino acids, in addition to N-myristoylation and palmitoylation, are required for Golgi targeting of GFP under these conditions,[47] with palmitoylation serving as a molecular 'zip code' to direct eNOS to specialized regions of the plasma membrane called caveolae.[41] Caveolae ('small caves') were identified over 40 years ago as non-clathrin coated invaginations of the plasma membrane in endothelial cells, epithelial cells, smooth muscle cells and fibroblasts. Recent evidence suggests that they act as signal transduction centres in which there is clustering of important signalling molecules such as G-protein α subunits, Src family members and the platelet-derived growth factor receptor.[48,49] Recent observations suggest that eNOS interacts with caveolin-1 directly and that this interaction can negatively regulate eNOS.[50,51] These data raise several important questions. Does eNOS in caveolae equate to eNOS bound to caveolin? If all NOS is bound to caveolin, then how does one get NO release? Are there other proteins that dissociate NOS and caveolin-1? These questions are further complicated by the presence of both caveolin and NOS in the Golgi and plasma membrane and the movement of eNOS in and out of the plasma membrane. Identification of the proteins that influence eNOS trafficking and activation and real time imaging of eNOS-GFP constructs in endothelial cells will undoubtedly be useful in elucidating the subcellular mechanisms that contribute to the regulation of this important enzyme.

Subcellular localization of iNOS is less well studied and has been found to depend on the species and cell types studied. iNOS isolated from transformed macrophages is primarily soluble,[52] whereas iNOS from primary isolates of mouse macrophages is 40% membrane-associated[53] and in human neutrophils is almost all (> 90%) membrane-associated.[54] Membrane-associated iNOS is localized in the perinuclear region and in post-Golgi vesicles in primary isolates from both mice and humans. Since iNOS is not fatty acylated *via* N-myristoylation, palmitoylation or prenylation, other (as yet unidentified) elements must determine the membrane association of this NOS isoform.

Elucidation of the Functions of Nitric Oxide Utilizing `Knockout' Mice

eNOS Knockout Mice

Two groups have independently generated mice with targeted disruption of the eNOS locus, as demonstrated by Western blot analysis[55] and immunocytochemistry.[56] Such mice are indistinguishable in appearance, behaviour and organ histology from wild-type animals, with the exception that the phenotype manifested by homozygous eNOS mutant mice is ~ 8% smaller in body weight than wild-type littermates.[56] This is consistent with reports that NO synthase inhibitors reduce food intake.[57] As discussed elsewhere in this volume, endothelium-dependent relaxations to agonists such as acetylcholine (ACh), bradykinin and substance P are inhibited by selective NOS inhibitors to a variable degree, depending on the vascular bed and species studied. Analogously, endothelium-dependent relaxations of preconstricted aortic rings to ACh are completely abolished in eNOS knockout mice, whereas responses to sodium nitroprusside, an exogenous source of NO, are preserved.[55] Additionally, Huang *et al.*[55] reported a mean arterial blood pressure (MABP) of 97 ± 8 mmHg for +/+ *vs* 117 ± 10 mmHg for −/− in awake mice, whereas Shesely *et al.*[56] reported values of 122 ± 2 and 140 ± 2 mmHg for +/+ *vs* −/−, respectively. Similar differences were observed in anaesthetized mice. Mice with one copy of the eNOS gene had mild, but not statistically significant, elevations in MABP (about 4 mmHg) compared to wild-type animals.[56] An intriguing observation was made after systemic administration of the NOS inhibitor N^G-nitro-L-arginine (L-NA) which caused a paradoxical fall in blood pressure in urethane-anaesthetized −/− mice, while producing the expected rise in +/+ animals.[55] This suggests that such pharmacological inhibitors either have targets other than eNOS or that non-endothelial NOS isoforms participate in the regulation of blood pressure. Plasma levels of renin are increased approximately 2-fold in both male and female mice that lack eNOS.[56] However, it remains uncertain how this is linked to loss of eNOS activity as kidney renin mRNA is slightly decreased,[56] and there is conflicting evidence for the role of NO in modulating renin release.[58–60]

Initial observations suggested that there were no differences in heart rate between +/+ and −/− mice measured under urethane anaesthesia.[55] In conscious homozygote mutant mice, however, heart rate is significantly lower than in wild type animals (670 ± 11 *vs* 709 ± 5 beats/min).[56] The reduced heart rate in −/− mice may either be due to a compensatory resetting of the baroreceptor reflex in response to chronic increase in blood pressure, or may arise from a direct effect of the lack of NO on heart rate. In spite of the hypertension and decreased heart rate observed in eNOS knockout mice, heart/body weight ratios do not reveal histological evidence for adaptive cellular hypertrophy.[55,56]

As discussed in Chapter 22, NOS inhibitors may be protective, without effect, or even detrimental for the survival of animals in septic shock. The contribution of eNOS to the lethal response following endotoxin administration has also been examined in eNOS knockout mice. +/+, +/− and −/− animals showed similar susceptibility to lipopolysaccharide (LPS) injection with 20%, 30% and 20% surviving after 96 hours, suggesting that NO produced from the endothelial isoform of NOS does not affect the outcome of septic shock in this model.[56] Studies with mice lacking the eNOS gene have also helped clarify the role of NO in cerebral ischaemia (see also Chapter 10). eNOS mutant mice thus display enlarged infarcts after middle cerebral (MCA) occlusion.[61]

Many other aspects of eNOS biology remain to be investigated using mice with targetted disruption of the eNOS locus. Pharmacological experiments demonstrating the role of endothelium-derived NO in inhibiting platelet aggregation, leucocyte adhesion and smooth muscle proliferation have yet to be repeated in mice lacking the eNOS protein. Another, relatively poorly investigated, autocrine action of endothelial NO is its involvement in angiogenesis. NO has recently been shown to stimulate angiogenesis both *in vivo* and *in vitro*[62,63] and may contribute to the angiogenic properties of some growth factors.[64] Although eNOS knockout mice are viable and apparently have a grossly normal vasculature, more detailed examination of their vascular tree may reveal specific abnormalities. Indeed, recent evidence indicates that mice lacking eNOS exhibit abnormal vascular remodelling in response to chronic decreases in blood flow.[65]

nNOS Knockout Mice

Expression of nNOS in the nervous system is localized primarily to discrete populations of neurones in the cerebellum, olfactory bulb, hippocampus, cortex, striatum, basal forebrain and brain stem.[66] In the peripheral nervous system, NO is present in non-adrenergic, non-cholinergic neurones and mediates relaxation of smooth muscle, a function important for gut peristalsis and penile erection.[25,67–69] Mice with targeted disruption of the nNOS locus (–/–) have been generated by homologous recombination techniques.[70] Such animals lack NADPH-diaphorase staining, nNOS mRNA and protein expression in neuronal tissues, and exhibit a pronounced loss in NOS catalytic activity in the nervous tissues examined. However, residual NOS activity (0.2–7%) can be detected in homozygous mutant mice in brain structures such as the hippocampus, cortex and caudate-putamen and this has been attributed to novel neuronal NOS isoforms which are smaller in size (125 and 136 kDa *vs* 155 kDa) than the prototype nNOS.[37] Huang *et al.*[70] have shown that neurones that normally express nNOS do not degenerate in nNOS null mutant mice, which are viable, fertile and possess unperturbed brain anatomy. Moreover, NOS-expressing peripheral tissues in nNOS knockout mice have a histologically normal appearance, and mean arterial blood pressure does not differ from littermate controls. The most notable anatomical difference between nNOS homozygous mutants and their wild-type siblings is a grossly enlarged stomach with hypertrophy in the pyloric sphincter and the circular muscle layer, a phenotype that resembles the human disorder of infantile pyloric stenosis.[71]

Mice deficient in nNOS have been used to address the role of NO in ischaemic cerebral tissue injury (see also Chapter 10).[72] Following MCA occlusion, infarct volumes and associated neurological deficit were reduced in nNOS –/– mice compared to littermate siblings at 24 and 72 hours. Furthermore, administration of N^G-nitro-L-arginine methyl ester (L-NAME) to nNOS –/– mice, to inhibit eNOS, increased tissue damage. These observations, taken together with data from the experiments with eNOS –/– mice discussed above, suggest that neuronal NO can exacerbate acute ischaemic injury, whereas vascular NO protects after MCA occlusion. The role of NO in ischaemic damage and glutamate neurotoxicity has also been investigated *in vitro*.[73] Neurotoxicity to N-methyl-D-aspartate (NMDA) was attenuated in cortical cultures from nNOS −/− mice compared to wild-type animals, while pathways stimulated by NO that result in neurotoxicity are still intact in knockout mice, as assessed by the effects of NO donors. In addition, cortical cultures from nNOS −/− mice were markedly protected from neurotoxicity after 60 minutes of combined oxygen-glucose deprivation compared with neurones from wild-type animals.[73]

nNOS and eNOS Double Knockout Mice

Pharmacological experiments with NOS inhibitors have yielded conflicting results concerning the role of NO as a retrograde messenger during long term potentiation (LTP).[25,74–79] Moreover, LTP is normal in both eNOS and nNOS knockout mice.[80] Since immunohistochemical data suggest that eNOS and nNOS are both expressed in neurones in the hippocampus,[79,81] doubly mutant mice have been used to address the role of NO in LTP. These animals again reveal no gross anatomical differences from wild-type controls.[80] NOS activity in the hippocampus of nNOS −/− mice was 1.7 ± 0.2% of control, whereas there was a slight elevation (131 ± 27% of control) in eNOS −/− mice, suggesting that there may be a compensatory increase in nNOS in eNOS −/− animals. Doubly mutant mice have reduced NOS activity compared to nNOS −/− animals. Interestingly, doubly mutant mice showed reduced LTP in stratum radiatum, but not in stratum oriens. Long-term depression was also normal in nNOS and eNOS mutant animals. Taken together, these data suggest that both eNOS and nNOS-derived NO is involved in LTP in stratum radiatum and that each isoform can compensate for loss of the other. In addition, there is a NOS-dependent component of LTP in the stratum radiatum, whereas LTP in the stratum oriens is largely NO-independent.

iNOS Knockout Mice

Unlike nNOS and eNOS, iNOS is normally not expressed *in vivo* under physiological conditions.[26,27,82] Data from mice with a null mutation of the transforming growth factor (TGF)-β1 gene suggest that iNOS expression *in vivo* is under the negative control of this cytokine, since TGF-β1 −/− mice exhibit high serum NO_2^-/NO_3^- levels, and iNOS expression both at the mRNA and protein level has been demonstrated in the heart and kidney of these animals.[83] Mice with targeted disruption of the iNOS locus (+/− and −/−) are viable and indistinguishable from wild-type animals in terms of gross appearance, reproduction, histology of all major organs and blood tests (glucose, alkaline phosphatase and cholesterol among others).[84–86] iNOS gene knockout was confirmed by Northern blot analyses of organs from animals treated with endotoxin (LPS); iNOS mRNA was induced in the hearts, kidneys, livers and spleens of wild-type mice injected with LPS, but not in iNOS null mutants. The specificity of the iNOS gene disruption was confirmed using eNOS as control; eNOS expression in the hearts of iNOS −/− mice was not different from wild-type animals.[85]

Pharmacological experiments have implicated NO in local inflammatory responses, oedema formation and host-defence mechanisms.[1,3,27] To test the role of iNOS in host defence directly, MacMicking *et al.*[85] and Wei *et al.*[86] injected −/− mice with *Listeria monocytogenes* and *Leishmania major*, respectively. Wild-type and heterozygous mutant mice achieved spontaneous healing after injection with 10^6 *L. major* promastigotes, whereas mutant mice were susceptible to infection and developed visceral disease. Similarly, iNOS mutant mice succumbed to inocula of Listeria that were 10-fold lower than the ones required to kill wild-type mice and exhibited 100-fold greater bacterial burdens in the liver and spleen. In addition, activated macrophages from iNOS −/− animals failed to restrict replication of lymphoma cells *in vitro*.[85] Taken together, these data suggest that isoform selective antagonism may increase susceptibility to infection and dampen immunosurveillance mechanisms.

As discussed in detail in Chapter 22, iNOS is generally thought to contribute to the hypotension and lethality associated with septic shock.[87] Initial experiments in anaesthetized mice demonstrated that LPS (1 mg/kg) induced a time-dependent decrease in blood pressure that reached 64% 3½ hours after the LPS injection.[85] This dose of LPS was lethal for all the wild-type animals within 5 hours of injection. In contrast, iNOS +/− and −/− mice showed only a 30% and 15% decrease in blood pressure, respectively, and no lethality to the same dose of LPS. The small but significant reduction in blood pressure observed in the iNOS mutant mice after LPS administration suggests that mediators other than NO contribute to the hypotensive effects of LPS. It should be noted that serum NO_2^-/NO_3^- levels increased after injection of wild-type mice with LPS, whereas no measurable changes in serum NO_2^-/NO_3^- were observed in iNOS −/− animals. However, changes in plasma levels of TNF-α, IL-1β and IL-6, as well as serum levels of lactate dehydrogenase, aspartate and alanine aminotransferase, were comparable in both genotypes. When conscious mice were injected with 5 or 15 mg/kg LPS, iNOS knockout mice were not protected against LPS lethality; however, a higher dose of LPS (30 mg/kg) was not as lethal in −/− mice, suggesting that iNOS deficiency confers partial resistance to LPS under some conditions.[85] Indeed, Wei *et al.* reported that iNOS null mutant mice were completely resistant to LPS (10 mg/kg)-induced lethality.[86] By contrast, Laubach *et al.*[84] found that iNOS +/− and −/− mice exhibited similar survival rates after intraperitoneal injection of LPS; in a different model of septic shock in which conscious animals were injected with heat-inactivated *P. acnes* followed by low amounts of LPS after 6 days, lack of iNOS in null mutant mice did not confer resistance to LPS-induced death.[85] There is therefore accumulating evidence that effective therapeutic strategies designed to reduce the mortality of septic shock in the clinical setting will be more complex than merely administering a selective inhibitor of iNOS.

Conclusions and Future Directions

Understanding the biochemical, molecular and cellular details of how mammalian cells produce NO will undoubtedly lead to innovative therapeutic strategies to augment or inhibit NO production. Recent advances in enzymology and cell biology are opening new arenas for understanding the molecular requirements that control NOS assembly and activity. Finally, with the advent of NOS isoform knockout mice, investigators will be able to assess definitively the role of NO in biology and disease and to ascertain what control mechanisms engage after the loss of a given NOS isoform.

References

1. Moncada S, Palmer RMJ, Higgs EA. Nitric oxide: physiology, pathophysiology, and pharmacology. *Pharmacol Rev* 1991; **43**: 109–142.
2. Sessa WC. The nitric oxide synthase family of proteins. *J Vasc Res* 1994; **31**: 131–143.
3. Lowenstein CJ, Dinerman JL, Snyder SH. Nitric oxide: a physiologic messenger. *Ann Intern Med* 1994; **120**: 227–237.
4. Lowenstein CJ, Snyder SH. Nitric oxide, a novel biologic messenger. *Cell* 1992; **70**: 705–707.
5. Förstermann U, Closs EI, Pollock JS *et al.* Nitric oxide synthase isozymes. Characterization, purification, molecular cloning, and functions. *Hypertension* 1994; **23**: 1121–1131.
6. Sessa WC, Harrison JK, Barber CM *et al.* Molecular cloning and expression of a cDNA encoding endothelial cell nitric oxide synthase. *J Biol Chem* 1992; **267**: 15274–15276.
7. Janssens SP, Shimouchi A, Quertermous T *et al.* Cloning and expression of a cDNA encoding human endothelium-derived relaxing factor/nitric oxide synthase. *J Biol Chem* 1992; **267**: 14519–14522.
8. Lamas S, Marsden PA, Li GK *et al.* Endothelial nitric oxide synthase: molecular cloning and characterization of a distinct constitutive enzyme isoform. *Proc Natl Acad Sci USA* 1992; **89**: 6348–6352.
9. Bredt DS, Hwang PM, Glatt CE *et al.* Cloned and expressed nitric oxide synthase structurally resembles cytochrome P-450 reductase. *Nature* 1991; **351**: 714–718.
10. Kobzik L, Reid MB, Bredt DS, Stamler JS. Nitric oxide in skeletal muscle. *Nature* 1994; **372**: 546–548.

11. Förstermann U, Kleinert H, Gath I *et al.* Expression and expressional control of nitric oxide synthases in various cell types. *Adv Pharmacol* 1995; **34**: 171–186.
12. Wang Y, Marsden PA. Nitric oxide synthases: gene structure and regulation. *Adv Pharmacol* 1995; **34**: 71–90.
13. Evans T, Carpenter A, Cohen J. Purification of a distinctive form of endotoxin-induced nitric oxide synthase from rat liver. *Proc Natl Acad Sci USA* 1992; **89**: 5361–5365.
14. Lowenstein CJ, Glatt CS, Bredt DS, Snyder SH. Cloned and expressed macrophage nitric oxide synthase contrasts with the brain enzyme. *Proc Natl Acad Sci USA* 1992; **89**: 6711–6715.
15. Lyons CR, Orloff GJ, Cunningham JM. Molecular cloning and functional expression of an inducible nitric oxide synthase from a murine macrophage cell line. *J Biol Chem* 1992; **267**: 6370–6374.
16. Xie QW, Cho HJ, Calaycay J *et al.* Cloning and characterization of inducible nitric oxide synthase from mouse macrophages. *Science* 1992; **256**: 225–228.
17. Chartrain NA, Geller DA, Koty PP *et al.* Molecular cloning, structure, and chromosomal localization of the human inducible nitric oxide synthase gene. *J Biol Chem* 1994; **269**: 6765–6772.
18. Hall AV, Antoniou H, Wang Y *et al.* Structural organization of the human neuronal nitric oxide synthase gene (NOS1). *J Biol Chem* 1994; **269**: 33082–33090.
19. Marsden PA, Schappert KT, Chen HS *et al.* Molecular cloning and characterization of human endothelial nitric oxide synthase. *FEBS Lett* 1992; **307**: 287–293.
20. Marsden PA, Heng HH, Scherer SW *et al.* Structure and chromosomal localization of the human constitutive endothelial nitric oxide synthase gene. *J Biol Chem* 1993; **268**: 17478–17488.
21. Xu W, Charles IG, Moncada S *et al.* Mapping of the genes encoding human inducible and endothelial nitric oxide synthase (NOS2 and NOS3) to the pericentric region of chromosome 17 and to chromosome 7, respectively. *Genomics* 1994; **21**: 419–422.
22. Griffith OW, Stuehr DJ. Nitric oxide synthases: properties and catalytic mechanism. *Annu Rev Physiol* 1995; **57**: 707–736.
23. Cho HJ, Xie QW, Calaycay J *et al.* Calmodulin is a subunit of nitric oxide synthase from macrophages. *J Exp Med* 1992; **176**: 599–604.
24. Pou S, Pou WS, Bredt DS *et al.* Generation of superoxide by purified brain nitric oxide synthase. *J Biol Chem* 1992; **267**: 24173–24176.
25. Bredt DS, Snyder SH. Nitric oxide, a novel neuronal messenger. *Neuron* 1992; **8**: 3–11.
26. Nathan C, Xie QW. Nitric oxide synthases: roles, tolls, and controls. *Cell* 1994; **78**: 915–918.

27. Nathan C. Nitric oxide as a secretory product of mammalian cells. *FASEB J* 1992; **6**: 3051–3064.
28. Klatt P, Pfeiffer S, List BM *et al.* Characterization of heme-deficient neuronal nitric-oxide synthase reveals a role for heme in subunit dimerization and binding of the amino acid substrate and tetrahydrobiopterin. *J Biol Chem* 1996; **271**: 7336–7342.
29. Venema RC, Ju H, Zou R *et al.* Subunit interactions of endothelial nitric-oxide synthase. Comparisons to the neuronal and inducible nitric-oxide synthase isoforms. *J Biol Chem* 1997; **272**: 1276–1282.
30. Tzeng E, Billiar TR, Robbins PD *et al.* Expression of human inducible nitric oxide synthase in a tetrahydrobiopterin (H_4B)-deficient cell line: H_4B promotes assembly of enzyme subunits into an active dimer. *Proc Natl Acad Sci USA* 1995; **92**: 11771–11775.
31. Cho HJ, Martin E, Xie QW *et al.* Inducible nitric oxide synthase: identification of amino acid residues essential for dimerization and binding of tetrahydrobiopterin. *Proc Natl Acad Sci USA* 1995; **92**: 11514–11518.
32. Klatt P, Schmidt K, Lehner D *et al.* Structural analysis of porcine brain nitric oxide synthase reveals a role for tetrahydrobiopterin and L-arginine in the formation of an SDS-resistant dimer. *EMBO J* 1995; **14**: 3687–3695.
33. Chen PF, Tsai AL, Wu KK. Cysteine 184 of endothelial nitric oxide synthase is involved in heme coordination and catalytic activity. *J Biol Chem* 1994; **269**: 25062–25066.
34. Jaffrey SR, Snyder SH. PIN: an associated protein inhibitor of neuronal nitric oxide synthase. *Science* 1996; **274**: 774–777.
35. Busconi L, Michel T. Endothelial nitric oxide synthase. N-terminal myristoylation determines subcellular localization. *J Biol Chem* 1993; **268**: 8410–8413.
36. Busconi L, Michel T. Endothelial nitric oxide synthase membrane targeting. Evidence against involvement of a specific myristate receptor. *J Biol Chem* 1994; **269**: 25016–25020.
37. Brenman JE, Chao DS, Gee SH *et al.* Interaction of nitric oxide synthase with the postsynaptic density protein PSD-95 and α1-syntrophin mediated by PDZ domains. *Cell* 1996; **84**: 757–767.
38. Brenman JE, Chao DS, Xia H *et al.* Nitric oxide synthase complexed with dystrophin and absent from skeletal muscle sarcolemma in Duchenne muscular dystrophy. *Cell* 1995; **82**: 743–752.
39. García-Cardeña G, Oh P, Liu J *et al.* Targeting of nitric oxide synthase to endothelial cell caveolae *via* palmitoylation: implications for nitric oxide signaling. *Proc Natl Acad Sci USA* 1996; **93**: 6448–6453.
40. Liu J, Sessa WC. Identification of covalently bound amino-terminal myristic acid in endothelial nitric oxide synthase. *J Biol Chem* 1994; **269**: 11691–11694.

41. Liu J, García-Cardeña G, Sessa WC. Palmitoylation of endothelial nitric oxide synthase is necessary for optimal stimulated release of nitric oxide: implications for caveolae localization. *Biochemistry* 1996; **35**: 13277–13281.
42. Robinson LJ, Michel T. Mutagenesis of palmitoylation sites in endothelial nitric oxide synthase identifies a novel motif for dual acylation and subcellular targeting. *Proc Natl Acad Sci USA* 1995; **92**: 11776–11780.
43. Sessa WC, Barber CM, Lynch KR. Mutation of N-myristoylation site converts endothelial cell nitric oxide synthase from a membrane to a cytosolic protein. *Circ Res* 1993; **72**: 921–924.
44. Sessa WC, García-Cardeña G, Liu J *et al.* The Golgi association of endothelial nitric oxide synthase is necessary for the efficient synthesis of nitric oxide. *J Biol Chem* 1995; **270**: 17641–17644.
45. Pollock JS, Förstermann U, Mitchell A *et al.* Purification and characterization of particulate endothelium-derived relaxing factor synthase from cultured and native bovine aortic endothelial cells. *Proc Natl Acad Sci USA* 1991; **88**: 10480–10484.
46. Förstermann U, Pollock JS, Schmidt HH *et al.* Calmodulin-dependent endothelium-derived relaxing factor/nitric oxide synthase activity is present in the particulate and cytosolic fractions of bovine aortic endothelial cells. *Proc Natl Acad Sci USA* 1991; **88**: 1788–1792.
47. Liu J, Hughes TE, Sessa WC. The first 35 amino acids and fatty acid acylation sites determine the molecular targeting of endothelial nitric oxide synthase into the Golgi region of cells: a green fluorescent protein study. *J Cell Biol* 1997; **137**: 1525–1535.
48. Li S, Okamoto T, Chun M *et al.* Evidence for a regulated interaction between heterotrimeric G proteins and caveolin. *J Biol Chem* 1995; **270**: 15693–15701.
49. Liu P, Ying Y, Ko Y-G, Anderson RGW. Localization of platelet-derived growth factor-stimulated phosphorylation cascade in caveolae. *J Biol Chem* 1996; **271**: 10299–10303.
50. García-Cardeña G, Martasek P, Masters BS *et al.* Dissecting the interaction between nitric oxide synthase (NOS) and caveolin. Functional significance of the NOS caveolin binding domain *in vivo*. *J Biol Chem* 1997; **272**: 25437–25440.
51. García-Cardeña G, Fan R, Stern DF *et al.* Endothelial nitric oxide synthase is regulated by tyrosine phosphorylation and interacts with caveolin-1. *J Biol Chem* 1996; **271**: 27237–27240.
52. Stuehr DJ, Cho HJ, Kwon NS *et al.* Purification and characterization of the cytokine-induced macrophage nitric oxide synthase: an FAD- and FMN-containing flavoprotein. *Proc Natl Acad Sci USA* 1991; **88**: 7773–7777.
53. Vodovotz Y, Russel D, Xie Q-W *et al.* Vesicle membrane association of nitric oxide synthase in primary mouse macrophages. *J Immunol* 1995; **154**: 2914–2925.

54. Wheeler MA, Smith SD, García-Cardeña G *et al.* Bacterial infection induces nitric oxide synthase in human neutrophils. *J Clin Invest* 1997; **99**: 110–116.
55. Huang PL, Huang Z, Mashimo H *et al.* Hypertension in mice lacking the gene for endothelial nitric oxide synthase. *Nature* 1995; **377**: 239–242.
56. Shesely EG, Maeda N, Kim HS *et al.* Elevated blood pressures in mice lacking endothelial nitric oxide synthase. *Proc Natl Acad Sci USA* 1996; **93**: 13176–13181.
57. Morley JE, Kumar VB, Mattammal MB *et al.* Inhibition of feeding by a nitric oxide synthase inhibitor: effects of aging. *Eur J Pharmacol* 1996; **311**: 15–19.
58. Kurtz A, Kaissling B, Busse R, Baier W. Endothelial cells modulate renin secretion from isolated mouse juxtaglomerular cells. *J Clin Invest* 1991; **88**: 1147–1154.
59. Salazar FJ, Pinilla JM, Lopez F *et al.* Renal effects of prolonged synthesis inhibition of endothelium-derived nitric oxide. *Hypertension* 1992; **20**: 113–117.
60. Sigmon DH, Carretero OA, Beierwaltes WH. Endothelium-derived relaxing factor regulates renin release *in vivo*. *Am J Physiol* 1992; **263**: F256–F261.
61. Huang Z, Huang PL, Ma J *et al.* Enlarged infarcts in endothelial nitric oxide synthase knockout mice are attenuated by nitro-L-arginine. *J Cereb Blood Flow Metab* 1996; **16**: 981–987.
62. Papapetropoulos A, Desai KM, Rudic RD *et al.* Nitric oxide synthase inhibitors attenuate transforming growth factor β_1-stimulated capillary organization *in vitro*. *Am J Pathol* 1997; **150**: 1835–1844.
63. Ziche M, Morbidelli L, Masini E *et al.* Nitric oxide mediates angiogenesis *in vivo* and endothelial cell growth and migration *in vitro* promoted by substance P. *J Clin Invest* 1994; **94**: 2036–2044.
64. Papapetropoulos A, García-Cardeña G, Madri JA, Sessa WC. Nitric oxide production contributes to the angiogenic properties of vascular endothelial growth factor in human endothelial cells. *J Clin Invest* 1997; **100**: 3131–3139.
65. Rudic RD, Shesely EG, Maeda N *et al.* Direct evidence for the importance of endothelium-derived nitric oxide in vascular remodelling. *J Clin Invest* 1998; **101**: 731–736.
66. Bredt DS, Hwang PM, Snyder SH. Localization of nitric oxide synthase indicating a neural role for nitric oxide. *Nature* 1990; **347**: 768–770.
67. Burnett AL, Lowenstein CJ, Bredt DS *et al.* Nitric oxide: a physiologic mediator of penile erection. *Science* 1992; **257**: 401–403.
68. Desai KM, Sessa WC, Vane JR. Involvement of nitric oxide in the reflex relaxation of the stomach to accommodate food or fluid. *Nature* 1991; **351**: 477–479.
69. Rajfer J, Aronson WJ, Bush PA *et al.* Nitric oxide as a mediator of relaxation of the corpus cavernosum in response to nonadrenergic, noncholinergic neurotransmission. *N Engl J Med* 1992; **326**: 90–94.
70. Huang PL, Dawson TM, Bredt DS *et al.* Targeted disruption of the neuronal nitric oxide synthase gene. *Cell* 1993; **75**: 1273–1286.

71. Vanderwinden JM, Mailleux P, Schiffmann SN *et al.* Nitric oxide synthase activity in infantile hypertrophic pyloric stenosis. *N Engl J Med* 1992; **327**: 511–515.
72. Huang Z, Huang PL, Panahian N *et al.* Effects of cerebral ischemia in mice deficient in neuronal nitric oxide synthase. *Science* 1994; **265**: 1883–1885.
73. Dawson VL, Kizushi VM, Huang PL *et al.* Resistance to neurotoxicity in cortical cultures from neuronal nitric oxide synthase-deficient mice. *J Neurosci* 1996; **16**: 2479–2487.
74. Chetkovich DM, Klann E, Sweatt JD. Nitric oxide synthase-independent long-term potentiation in area CA1 of hippocampus. *Neuroreport* 1993; **4**: 919–922.
75. Cummings JA, Nicola SM, Malenka RC. Induction in the rat hippocampus of long-term potentiation (LTP) and long-term depression (LTD) in the presence of a nitric oxide synthase inhibitor. *Neurosci Lett* 1994; **176**: 110–114.
76. Doyle C, Holscher C, Rowan MJ, Anwyl R. The selective neuronal NO synthase inhibitor 7-nitro-indazole blocks both long-term potentiation and depotentiation of field EPSPs in rat hippocampal CA1 *in vivo*. *J Neurosci* 1996; **16**: 418–424.
77. Gribkoff VK, Lum-Ragan JT. Evidence for nitric oxide synthase inhibitor-sensitive and insensitive hippocampal synaptic potentiation. *J Neurophysiol* 1992; **68**: 639–642.
78. Haley JE, Malen PL, Chapman PF. Nitric oxide synthase inhibitors block long-term potentiation induced by weak but not strong tetanic stimulation at physiological brain temperatures in rat hippocampal slices. *Neurosci Lett* 1993; **160**: 85–88.
79. O'Dell TJ, Huang PL, Dawson TM *et al.* Endothelial NOS and the blockade of LTP by NOS inhibitors in mice lacking neuronal NOS. *Science* 1994; **265**: 542–546.
80. Son H, Hawkins RD, Martin K *et al.* Long-term potentiation is reduced in mice that are doubly mutant in endothelial and neuronal nitric oxide synthase. *Cell* 1996; **87**: 1015–1023.
81. Wendland B, Schweizer FE, Ryan TA *et al.* Existence of nitric oxide synthase in rat hippocampal pyramidal cells. *Proc Natl Acad Sci USA* 1994; **91**: 2151–2155.
82. Moncada S, Higgs EA. Molecular mechanisms and therapeutic strategies related to nitric oxide. *FASEB J* 1995; **9**: 1319–1330.
83. Vodovotz Y, Geiser AG, Chesler L *et al.* Spontaneously increased production of nitric oxide and aberrant expression of the inducible nitric oxide synthase *in vivo* in the transforming growth factor β1 null mouse. *J Exp Med* 1996; **183**: 2337–2342.
84. Laubach VE, Shesely EG, Smithies O, Sherman PA. Mice lacking inducible nitric oxide synthase are not resistant to lipopolysaccharide-induced death. *Proc Natl Acad Sci USA* 1995; **92**: 10688–10692.

85. MacMicking JD, Nathan C, Hom G *et al.* Altered responses to bacterial infection and endotoxic shock in mice lacking inducible nitric oxide synthase. *Cell* 1995; **81**: 641–650.
86. Wei XQ, Charles IG, Smith A *et al.* Altered immune responses in mice lacking inducible nitric oxide synthase. *Nature* 1995; **375**: 408–411.
87. Kilbourn RG, Szabó C, Traber DL. Beneficial versus detrimental effects of nitric oxide synthase inhibitors in circulatory shock: lessons learned from experimental and clinical studies. *Shock* 1997; **7**: 235–246.

Chapter 6

The Normal Metabolic Fate of Nitric Oxide

Åke Wennmalm and Lennart Jungersten

Introduction

Proper knowledge about the normal metabolic fate of NO is essential from several points of view. Differences in metabolism between various body fluids and tissues may exist, and can be assumed to have biological significance. Variations in the fraction of NO being transformed into active and inactive metabolites may occur in parallel to pathophysiological states. Insight into the metabolic pathways, and their relative importance qualitatively and quantitatively, may also yield an ability to assess the endogenous formation of NO in normal man under various conditions, and in patients with conditions affecting NO formation.

Metabolism may be considered to involve both formation and breakdown or inactivation. Other chapters in this book deal with the chemistry and molecular biology of NO synthesis (Chapter 3), with the NO synthases (NOS) (Chapters 5 and 7), and with measurement of NO and NOS activity (Chapter 8). This review will therefore mainly deal with the inactivation and catabolic transformation of NO, and with the excretion of NO and its inactivation products. Furthermore, focus will be on inactivation and catabolic transformation in normal man. It has become increasingly evident during the last few years that the inactivation of NO may play a crucial role in several disease states, particularly in the cardiovascular system. In fact, premature inactivation of NO may be as important in the development of endothelial dysfunction as insufficient new synthesis. However, quantitative data on such aspects of NO inactivation are hitherto lacking and will have to await further clinical investigations.

Data on tissue differences in NO metabolism are slowly emerging. Unless otherwise stated, statements and conclusions in this chapter will refer to events in the immediate vicinity of the cardiovascular system, particularly close to the vascular endothelium. Little is known about species differences in NO metabolism; nothing has been reported on such differences that are of biological importance. For the purposes of this review, therefore, metabolism of NO will be considered to be — at least qualitatively — similar in all mammals.

Regulation of NO formation is complex and incompletely known. Substrate availability does not seem to be a rate-limiting factor under normal conditions (but may well be so in some disease states). Rather, the formation rate seems to be regulated at the level of the enzyme converting L-arginine to NO, i.e. NOS. Both expression and activity of this enzyme are probably regulated *via* transcriptional and/or translational procedures, *via* surface receptors, or *via* availability of cofactors. Most authors seem to support the idea that there is an ongoing 'basal' rate of NO synthesis in tissues, and that stimulation of NO formation *via* receptors or other regulatory mechanisms operate on top of this 'basal' formation. Whether or not this division into basal and stimulated formation is biologically correct is not yet known. In addition, sufficient data are not available to evaluate possible differences in breakdown or inactivation of NO formed as a result of ongoing 'basal' formation, versus formation as a result of enhanced synthesis due to one or several of these factors. Therefore, breakdown or inactivation of NO will be considered as one single issue in this chapter, i.e. irrespectively of the conditions under which the NO is formed.

Physiological Aspects

Preliminary data from our laboratory indicate that there is no diurnal variation and no seasonal variation in the levels of nitrate in plasma (unpublished data). This suggests that the endogenous formation of NO is stable during day and night, and also during the year. In females, the level of nitrate in plasma is somewhat lower than in males,[1] and the nitrate level decreases with age.[2] In healthy women, both variations[1] and absence of variations[3] in plasma nitrate levels during the menstrual cycle have been reported. These contradictory results may possibly be explained by methodological discrepancies.

Changes in plasma nitrate levels, suggestive of variations in NO formation, have been demonstrated during some defined physiological conditions. For instance, during acute physical exercise there is an enhanced formation of NO.[4,5] The origin of the enhanced NO formation is not known, but has been proposed to be based on shear stress-induced activity of NOS in the vascular system.[6] In both well-trained and untrained subjects, plasma nitrate is increased by about 15–20% above the basal (i.e. resting) level after a period of physical exercise.[5] In well-trained subjects, the basal level of nitrate in plasma is higher than in untrained controls, indicating a positive correlation between physical fitness and NO formation.[5] Such a relation may be based on a similar mechanism as observed in animals, i.e. that chronic exercise upregulates the expression of NOS.[6] If this mechanism is operating also in humans, a larger amount of enzyme in fit subjects can perhaps explain an enhanced release of NO also at rest, as reflected by higher basal levels of nitrate in these subjects. The observed effects of physical activity and physical fitness on NO formation, and hence also on plasma nitrate levels, must probably be taken into account when evaluating studies on the endogenous formation of NO in humans.

Protein Binding of Nitric Oxide in Blood

Although protein binding of NO may not appear as an ordinary inactivation mechanism or part of a metabolic degradation of this agent, such binding may nevertheless play a role in the fate of NO originating from the vascular endothelium. A few remarks on protein binding therefore seem warranted.

After the early observation by Needleman *et al.*[7] that thiol availability is critical in the development of the vasodilator effect of the organic nitrate class of drugs, several authors assessed whether endogenous NO may also require some type of adduct formation in order to be operative as a vasodilator. In some support of this idea, it was observed that the reduced thiol N-acetylcysteine was able to potentiate platelet inhibition by NO, and that this effect was associated with increased intracellular guanosine 3',5'-cyclic monophosphate (cGMP).[8] Another study reported that EDRF was much more likely to be a nitrosylated compound, such as a nitrosothiol, than authentic NO.[9] The rapid formation under physiological conditions of bioactive nitrosoproteins also raised the possibility that they played a more general role in regulation of cellular processes.[10]

Since these studies clearly indicated that NO not only preserved, but even enhanced and prolonged its biological activity by forming *S*-nitrosothiols, another hypothesis was presented. According to this hypothesis, NO formed in the vascular endothelium might bind to reduced thiols in the plasma proteins, thereby stabilizing and extending its biological activity. A consequence of such an adduct formation of locally formed NO with plasma proteins would be that NO is not a local, but rather a circulating vasodilator hormone. In this context, it is of interest that nitrosylation of glutathione,[11] albumin,[12] and tissue-type plasminogen activator[12,13] transfers vasodilator and anti-platelet properties to these compounds.

Studies more directly aimed at elucidating the role of nitrosylated plasma proteins have also been reported. Stamler and co-workers analysed the total amount of *S*-nitrosothiols in human plasma with a method combining UV photolytic cleavage of the *S*-NO bond before reaction with ozone for chemiluminescence detection.[14] They reported that human plasma contained about 7 μmol/L *S*-nitrosothiols, of which 96% were *S*-nitrosoproteins; 82% of the latter were accounted for by *S*-nitroso-serum albumin. In the same study, it was reported that administration of the NOS inhibitor N^G-monomethyl-L-arginine (L-NMMA) in rabbits decreased *S*-nitrosothiols in plasma by about 40% and in parallel increased mean arterial blood pressure by about 22%.[14] In another study, the same group reported that synthesized *S*-nitroso-bovine serum albumin, when infused in dogs, displayed distinct vasodilator properties.[15] Thus, although being less potent than *i.v.* nitroglycerine on a molar basis when given as a bolus, *S*-nitroso-bovine serum albumin was found equipotent with nitroglycerine when given as an *i.v.* infusion. The authors suggested that endogenously formed *S*-nitroso-serum albumin might serve as a reservoir with which plasma levels of highly reactive, short-lived free NO could be regulated for the maintenance of vascular tone.[14]

More recently, nitrosylation of haemoglobin has also been proposed to be physiologically significant. According to this proposal, haemoglobin would be *S*-nitrosylated in the lungs when the erythrocytes are oxygenated,[16] and the NO group would be released during arterial-venous transit. In this dynamic cycle, the vasoactivity of the *S*-nitrosylated haemoglobin would be supported by release of *S*-nitrosothiols from the red blood cells. The physiological consequences of this cycle would be that haemoglobin is involved in the control of blood pressure as well as in the dissociation of oxygen in the peripheral tissues.[16]

Degradation of Nitric Oxide in Aqueous Solution

The metabolism of NO in living tissues appears to depend on the composition of the fluid into which the compound is released. The biological half-life of NO is usually reported to be 5 seconds or less.[17] Observations on such short half-lives are usually based on experiments in the cardiovascular system. NO released from cells outside the cardiovascular system will reach other cells, interstitial fluid, or non-haemoglobin-containing compartments such as e.g. the subarachnoid space, the peritoneal cavity, or joints. In aqueous, haemoglobin-free solution, NO has been reported to have a half-life of at least 8 minutes, which is considerably longer than the reported half-life *in vivo*. Obviously, the metabolic degradation of NO differs between haemoglobin-containing and haemoglobin-free media. These metabolic pathways will therefore be discussed separately below, to be followed by an overview of the integrated metabolism of NO *in vivo*.

In air, NO is almost immediately oxidized to NO_2, which in turn is dimerized to N_2O_4. If the latter dimer is introduced into water, both nitrite (NO_2^-) and nitrate (NO_3^-) are formed. It was earlier assumed that the same would occur if NO was directly released into an aqueous solution, i.e. that NO_2^- and NO_3^- would be formed in roughly equal amounts. This does not, however, seem to be the case. It rather appears that NO reacts with water and dissolved oxygen to form mainly or exclusively nitrite ions and protons.[18–20] Whether nitrite is the only reaction product or, alternatively, minute amounts of nitrate may be formed as well, is still an open question. Ignarro and co-workers reported that NO in aqueous solution containing oxygen was oxidized primarily to nitrite with little or no formation of nitrate,[18] and Lewis & Deen reported that (within experimental uncertainty of a few percent) the rate of nitrite formation in watery solution equalled the rate of reaction of NO.[19] However, Hogg and co-workers found that the ratio between nitrite and nitrate also varied as a function of the rate of generation of NO, and that NO in air-saturated solutions generated substantial amounts of nitrate if released at a rate less than 10^{-17} M/s.[20] For practical purposes, however, NO inactivation in watery solution may be regarded to be exclusively *via* formation of nitrite.

The single metabolic degradation to nitrite does not seem to be valid if cells are present in the solution to which NO is added. Activated macrophages in suspension generated both nitrite and nitrate; their formation appeared at approximately equal and constant rates.[21] Kinetic analysis of the reaction revealed that only half of the nitrite formed could be accounted for by the reaction of NO with molecular oxygen. To test the possibility that the excess formation of nitrite was due to reactions involving superoxide ion and/or peroxynitrite, the authors added superoxide dismutase to the cell suspension under study.[21] Superoxide dismutase markedly reduced the ratio of nitrate to nitrite, consistent with the hypothesis that nitrate in the medium primarily resulted from a reaction between NO and superoxide ion.[21] Extrapolation to *in vivo* conditions would imply that degradation of NO in haemoglobin-free fluids is dependent on the degree of oxidative stress in the system. If little or no oxidative stress is prevailing, only — or at least mainly — nitrite is formed. With increasing oxidative stress — i.e. in the presence of superoxide anion or other reactive products — the formation of nitrate will gradually increase. A tissue subjected to severe ischaemia might be assumed to suffer from a combination of hypoperfusion/hypoxia and increased formation of superoxide anion. Under such circumstances, nitrate formation should be the main metabolic route for endogenous NO, even if no haemoglobin were present.

Addition of oxyhaemoglobin to an aqueous solution changes the pattern of reaction products from NO dramatically. In a study in our laboratory, NO incubated with arterialized blood *in vitro* was almost quantitatively converted to nitrate in less than 2 minutes, even at room temperature.[22] Venous, i.e. less oxygenated, blood was also able to convert NO to nitrate, although at a slower rate. A similar dependence on oxyhaemoglobin was observed by Ignarro *et al.*,[18] who reported that oxymyoglobin — but not methaemoglobin (metHb) — also fulfilled the requirements to enable the conversion of NO to nitrate.

Metabolism of Nitric Oxide in Haemoglobin- or Myoglobin-containing Solution

Long before NO was known as an endogenously formed biological mediator, metabolic balance studies proposed that nitrate ion was synthesized in the body. In a long-term balance study of this type, it was observed that subjects on a constant low ingestion of nitrate (< 180 μmol/day) excreted an average of four times as much nitrate in the urine as the amount ingested.[23] The endogenous synthesis of nitrate appeared independent of the amount ingested.[23] In an attempt to identify the source of the endogenous nitrate, a long-term infusion of labelled ammonia was given to rats; conversion of NH_3 to NO_3^- could, however, only explain about half of the endogenous formation estimated.[24] The source of the remaining nitrate formed was unidentified.

The remarkable affinity of NO for haemoglobin was initially noted more than a hundred years ago.[25] In a somewhat more recent study, Gibson & Roughton[26] found that NO has about 1500 times greater affinity for haemoglobin than has carbon monoxide. These and other authors also found that the nitrosohaemoglobin (HbNO) formed was a highly stable molecule.[25,27] Despite this, Yoshida and co-workers,[28] who made one of the very first studies on the metabolic fate of inhaled NO, were unable to demonstrate any accumulation of HbNO in the blood of rats directly injected *i.v.* with ^{15}N-labelled NO. Instead, they observed markedly elevated amounts of ^{15}N-labelled nitrate in serum and urine from the rats. Their conclusions were that inhaled NO primarily reacted with haemoglobin and was changed to nitrite and nitrate, and that these metabolites were transferred to the

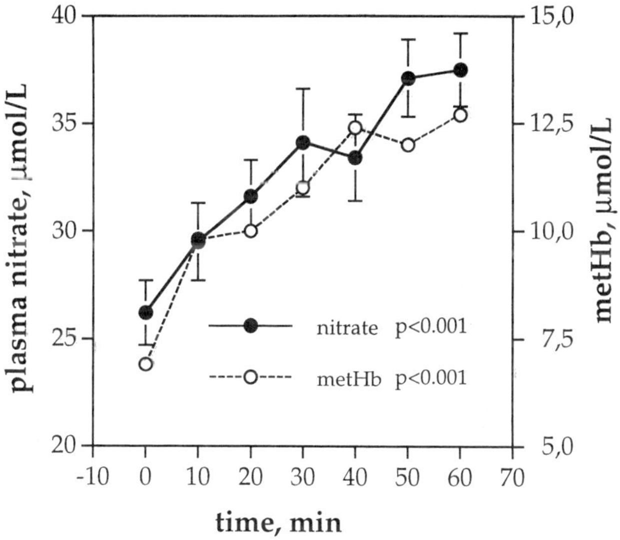

Figure 1. Plasma levels of nitrate and whole blood concentrations of methaemoglobin (metHb) during inhalation of 25 ppm NO. Inhalation started at time 0 min and lasted for 60 min. Data are presented as mean ± SEM of eight experiments. *P* values (by Analysis of Variance) refer to changes in the levels of plasma nitrate and whole blood metHb with time.

serum, from whence they could either react with tissues or be excreted in urine in the form of nitrate.[28]

The results of Yoshida and co-workers proved to be remarkably correct. Except for their proposal that HbNO was formed as a first step in the conversion of NO to nitrate — NO reacts in fact with oxygenated haemoglobin (see below) — these authors pinpointed all the essential features in the inactivation and excretion of NO under physiological conditions. In a later study, the same authors clarified further the details of NO metabolism in rats and mice.[29] In that study, they also reported that a certain amount of the nitrate formed in the blood is discharged into the oral cavity through the salivary glands and transformed to nitrite. Part of this nitrite is converted to nitrogen gas in the stomach. Nitrate in the intestine is partly reduced to ammonia through nitrite, reabsorbed into the body, and converted to urea.[29]

Direct studies of NO metabolism in humans have been performed in our laboratory.[30] We studied the degradation of NO in healthy subjects inhaling 25 ppm NO for 60 minutes and in patients with severe heart failure inhaling 20, 40, and 80 ppm NO in consecutive 10-minute periods (Figure 1). During inhalation of NO, the plasma levels of nitrate increased progressively, both in the healthy subjects (from 26 to 38 μmol/L) and in the patients (from 72 to 90 μmol/L). MetHb also increased in the healthy subjects (from 7 to 13 μmol/L), as well as in the patients (from 19 to 42 μmol/L). No change in HbNO was detected, either in

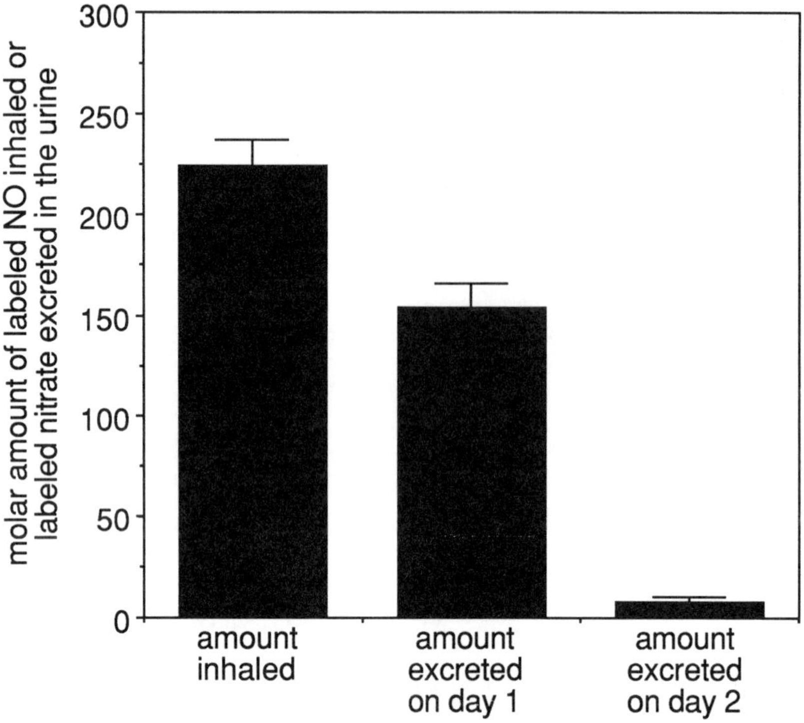

Figure 2. Retained total amount of ^{15}N-labelled NO (^{15}NO) during inhalation, and amount of $^{15}NO_3^-$ excreted in the urine on days 1 and 2 after the inhalation of ^{15}NO. Data are presented as mean ± s.e.m. of 8 experiments.

the healthy subjects or in the patients. It was concluded: (i) that uptake into the red blood cells with subsequent conversion to nitrate and metHb is a major metabolic pathway for endogenously formed NO; and (ii) that the occurrence of HbNO *in vivo* probably indicates liberation of NO into partly deoxygenated blood.[30]

In a later series, we assessed quantitatively the importance of the NO/nitrate pathway in man.[31] Healthy humans inhaled 25 ppm ^{15}NO for 1 hour, and the plasma and urine levels of ^{15}N-labelled nitrate were followed (Figure 2). About 90% of the ^{15}NO inhaled was retained in the lungs. More than 70% of the retained ^{15}NO was recovered as ^{15}N-labelled nitrate in the urine.[31] The fate of the remaining NO could not be identified.

The metabolism of inhaled NO may not necessarily be an adequate model for the degradation of endogenous NO formed in the endothelium of the systemic vasculature. Inhaled NO is, after retention in the lungs, almost immediately exposed to oxygenated haemoglobin in the pulmonary circulation. In contrast, NO formed in the systemic vasculature may be exposed either to oxygenated haemoglobin (after diffusion into the vascular lumen), or to media without haemoglobin (after

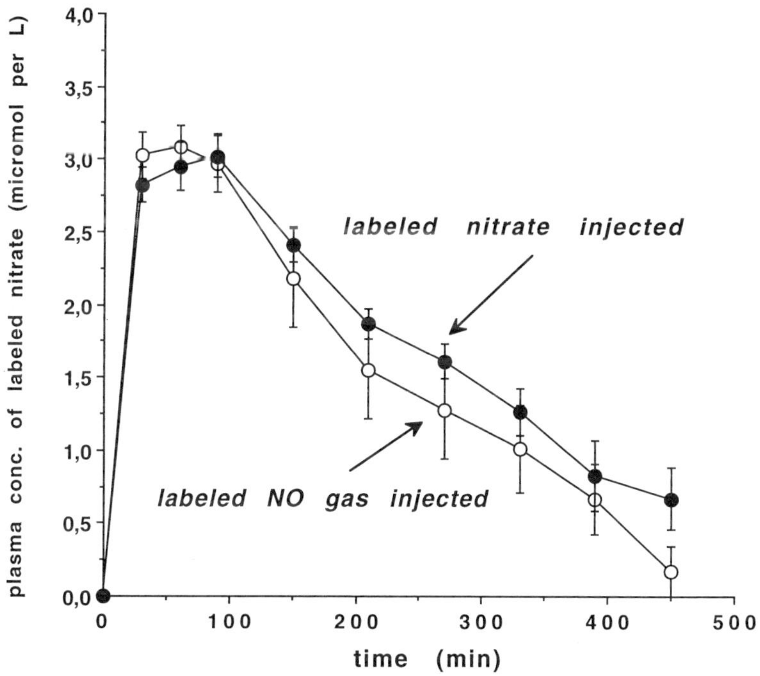

Figure 3. Plasma concentrations of $^{15}NO_3^-$ following subcutaneous injection of 0.4 μmol ^{15}NO gas or $K^{15}NO_3$ solution. (From Ref. 32, with permission from the publishers.)

diffusion to the deeper layers of the vascular wall). Certainly, myoglobin-containing cells, such as vascular smooth muscle cells, are present in the vascular wall. Although firm data to support this are not available, these cells may be assumed to convert NO to nitrate. However, some NO may be degraded in the interstitial fluid in the vessel wall, or in cells not containing myoglobin. That fraction of the NO released is probably converted to nitrite or nitrate, depending on the oxidative stress prevailing in the tissue (cf. above). Hence, NO diffusing abluminally may be metabolized differently from NO reaching the blood, and the pattern of reaction products may depend on the physiological conditions in the tissue.

To assess the fate of NO not immediately exposed to haemoglobin- or myoglobin-containing fluids, we injected ^{15}NO subcutaneously in rats and followed the appearance of $^{15}NO_3^-$ in their plasma (Figure 3).[32] Control rats were injected with the same molar mass of authentic $^{15}NO_3^-$ at the same site. The area under the plasma decay curve for rats injected with ^{15}NO amounted to 89% of the corresponding area for animals given $^{15}NO_3^-$. This demonstrates that ^{15}NO, when given extravascularly, is mainly transformed into $^{15}NO_3^-$. Hence, nitrate also appears to be a major stable metabolite of endogenous NO that does not primarily diffuse into the vascular lumen following formation.[32]

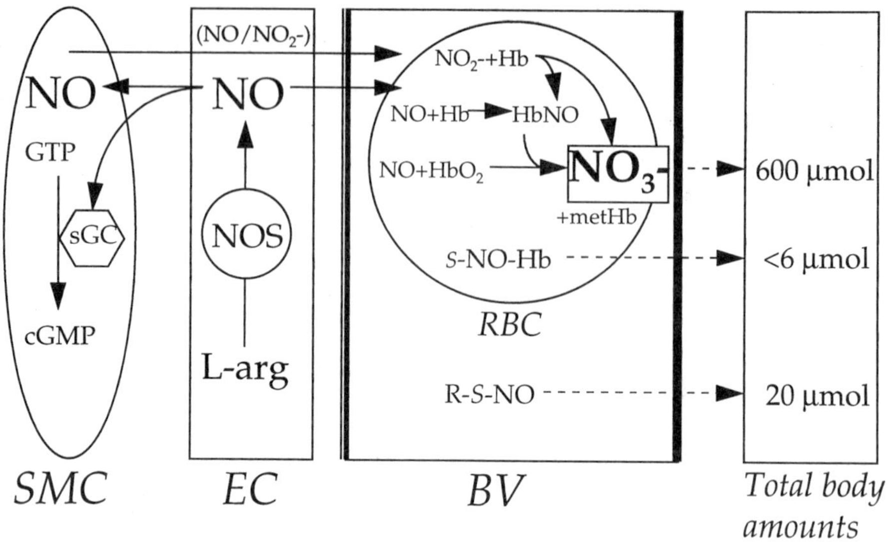

Figure 4. Diagrammatic presentation of the proposed metabolism of NO *in vivo*. NO is generated from L-arginine (L-arg) in endothelial cells (EC), and diffuses to the lumen of the blood vessels (BV), from which it is taken up in the red blood cells (RBC), or abluminally to the smooth muscle cells (SMC). In the RBC, NO reacts with oxygenated haemoglobin (HbO_2) to form NO_3^- and methaemoglobin (metHb), or with haemoglobin (Hb) to form nitroso-haemoglobin (HbNO). HbNO may be further degraded to metHb and NO_3^- when the oxygen pressure in the blood is increasing during passage through the lungs. If NO is converted to nitrite (NO_2^-) before being taken up in the RBC, the same products are formed, i.e. HbNO, metHb and NO_3^-. Alternatively to being converted to nitrate in the red blood cells, NO may be bound to plasma proteins (R-*S*-NO) or to the globin part of the haemoglobin (*S*-NO-Hb). NO_3^- formed in the red blood cells diffuses to the plasma from which it is cleared *via* the kidneys.

Summary of Nitric Oxide Metabolism *In Vivo*

NO endogenously formed in the cardiovascular system may be released luminally or abluminally. NO immediately reaching the blood is taken up in the red blood cells to be quantitatively transformed to nitrate, by reaction with oxygenated haemoglobin (see Figure 4). Some NO will meet non-oxygenated haemoglobin in the blood; such NO will nitrosylate the Fe^{2+} to a fairly stable HbNO adduct. Upon passage through the lungs, when the partial pressure for oxygen increases, the O_2/NO partial pressure balance in the blood will probably force the HbNO complex to dissociate. The released NO will then find another oxy-Hb molecule and be converted to nitrate. Alternatively, it will nitrosylate *S*-groups in the haemoglobin molecule. This assumption is based on the fact that the level of *S*-NO groups is higher in haemoglobin from arterial compared to venous blood.[16] As a side product of the reaction between oxy-Hb and NO, metHb will be formed in parallel to the appearance of nitrate. This metHb will be reduced back to normal Hb by the house-keeping reductase present in the erythrocytes. Transformation to nitrate covers 70–90% of NO formed in the body. The normal plasma level of nitrate in man is about 30 μM.[33]

As mentioned above, some NO in the red blood cells will be transferred to *S*-groups in the protein part of the haemoglobin molecule instead of being attached to the haem group. This transfer seems to take place in the lungs.[16] *S*-nitrosylated NO has been suggested to play a role in the control of blood pressure and the delivery of oxygen to tissues. *S*-nitrosylation of haemoglobin is, however, probably of minor importance in the degradation of NO. The levels of *S*-nitrosylated Hb in arterial and venous blood have been reported to be about 300 and 30 μmol/L, respectively.[16]

A moderate fraction of NO seems to nitrosylate other protein fractions in the plasma. Plasma has been reported to contain about 7 μmol/L *S*-nitrosothiols, of which the major part is *S*-nitroso-serum albumin. *S*-nitrosylated plasma proteins may be interesting from a physiological point of view, but are probably unimportant in the degradation of NO in the blood.

NO diffusing abluminally may also be converted to nitrate. This may occur following reaction with myoglobin[18] in the vascular smooth muscle cells, or after reaction with superoxide anion.[19] The latter is obviously an event restricted to tissues with increased oxidative stress, e.g. following exposure to hypoxia/ischaemia or in states of inflammation. Nitrate formed in the tissues in this way may readily diffuse into the systemic circulation. Under non-oxidative conditions, NO in tissues appears to be converted to nitrite, as is NO taken up in tissue cells not containing haemoglobin or myoglobin. Nitrite thus formed can diffuse freely into the systemic circulation to be taken up in the red blood cells, where it is converted to nitrate.[30]

Little is known about the fate of NO formed in the central nervous system or in peripheral nitrergic nerves. In cerebrospinal fluid, the nitrite/nitrate ratio is considerably higher than that in plasma (concentrations of nitrite and nitrate about 3 and 10 μmol/L, respectively), indicating that degradation of NO in this medium follows the principles discussed above for inactivation in aqueous solution.[34] Also, the fate of NO formed in challenged immunocompetent cells is largely unknown. It has been demonstrated that plasma nitrate is markedly increased in subjects suffering from a gastrointestinal infection, probably indicating conversion of NO by (inflammatory) cells in the gut.[35] Judging from these data, NO released from tissues outside the vascular system would be metabolized in the same way as endothelial NO. This issue does require additional study, however.

Excretion of Nitric Oxide and Its Metabolites

The levels of nitrate in plasma and urine display large inter- and intra-individual variations. The nitrate level in plasma is normally about 30 μmol/L, but may be up to 1000 μmol/L under certain pathological conditions. The urinary excretion of nitrate may range in parallel, from 500 to 5000 μmol/24 h.[5,23,36] These large variations in nitrate plasma level and excretion can be explained by: (i) differences in the intake of nitrate/nitrite-containing food (see below), (ii) variations in the endogenous formation of NO (see above), and (iii) variations in the elimination/excretion of nitrate or NO.

Nitrate is mainly eliminated *via* the kidneys. The excretion of nitrate is slow, i.e. the half-life in plasma is about 8 hours and the clearance is about 20–30 ml/min.[30,33,37] Since nitrate is mainly excreted in the urine, its level in plasma/urine is directly influenced by the renal function. Thus, an increase in the basal plasma level of nitrate may not necessarily indicate that the endogenous formation of NO is increased.

Nitrate is distributed within a volume corresponding to about 30% of body weight.[33] The distribution volume includes not only the extracellular volume (ECV) but probably also the volume of the erythrocytes. This assumption is based on the fact that NO is metabolized to nitrate in the red cell fraction.[22] On the basis of knowledge about the plasma concentration under basal conditions (~ 30 μmol/L) and the distribution volume for nitrate (~ 20 L),[33] the total body amount of nitrate may be estimated to about 600 μmol.

Metabolic studies on nitrate balance have shown that 40–60% of ingested ^{15}N-labelled nitrate is recovered in the urine.[23,36] This degree of recovery is similar to that obtained in inhalation studies using ^{15}N-labelled NO. About 70% of the ^{15}N-labelled NO was excreted as ^{15}N-labelled nitrate.[31] The fate of the remaining nitrate and NO (about 30%) is not clarified, but a metabolic route resulting in the formation of small amounts of urea has been observed.[29] Yet, there appear to exist other pathways *via* which nitrate is further metabolized in the body.

The formation of nitrosothiols is a result of the reaction between NO and specific thiols in proteins (cf. above). The level of such nitroso-compounds in plasma is low, and consequently such NO adducts represent a relatively small fraction of all NO formed, released and metabolized in the body. The metabolic fate of the nitrosothiols is not known. Part of them may be transported to the lungs and exhaled as NO, or alternatively, converted to and excreted as nitrate.

A higher NO content in exhaled than in inhaled air has been demonstrated, indicating that there is a respiratory net loss of NO. In exhaled air, NO is present at a concentration of about 30 ppb.[38] Under resting conditions, this concentration of exhaled NO corresponds to a nitrate elimination of approximately 9 μmol/24 h. Thus, only a minor part of total body formation of NO is eliminated by exhalation.

Confounding Factors

Direct measurement of NO in tissue fluids is, as discussed earlier in this chapter, not a feasible means to assess its formation rate. Instead, stable NO metabolites or adducts have to be measured. The possibility of confounding factors must then be carefully reviewed. In particular, analysis of plasma or urine nitrate concentrations requires careful precautions. A few words on the possible confounding factors for nitrate in plasma or urine therefore seem warranted.

The most obvious confounding factor to deal with is the nitrate present in different kinds of foodstuff, e.g. vegetables, ham and sausages.[36] After intake of nitrate-rich food, the level of nitrate in plasma is rapidly elevated.[33] Accordingly, a period of nitrate/nitrite restriction is an absolute requirement if nitrate in plasma is to be used as an index of the endogenous NO formation. On the basis of the half-life for nitrate in plasma (about 8 hours),[33] and assuming that about 98% ($6 \times t_{½}$) of exogenous nitrate should be cleared before sampling, dietary restriction has to be maintained for at least a period of 48 hours.

In city air, the content of NO_x (the term 'NO_x' is often applied to indicate the sum of NO and NO_2 in air) is about 50 μg/m^3 (1–2 nmol/L). If this amount were taken up in the lungs and converted into nitrate in the blood, it would contribute no more than about 10 μmol to the 24-hour excretion of nitrate under resting conditions. Tobacco smoke may contain NO in a concentration as high as 1000 ppm/40 mL puff[39] and, theoretically, smoking might acutely influence the nitrate level substantially. However, studies on the acute effects of smoking on nitrate levels in humans have yielded negative results.[33] When calculating the amount of nitrate from NO in three cigarettes (30 puffs) and taking into account the distribution volume for nitrate, smoking three cigarettes would not contribute to more than 2–3 μmol/L. Thus, the contribution of nitrate from inhaled NO in normal ambient air and tobacco smoke seems negligible.

Summary

The normal metabolic fate of NO may be summarized as follows:

- In tissues, i.e. outside the blood vessels, NO is metabolized to nitrite or nitrate, depending on the prevailing degree of oxidative stress in the tissue.
- In the vascular lumen, NO is either bound to plasma proteins or retained in the red blood cells.
- In the red blood cells, NO is either bound to haemoglobin or degraded to nitrate by oxyhaemoglobin.
- Conversion to nitrate covers 75–90% of all NO formed in the body.
- Nitrate is distributed in the extracellular fluid and in the red blood cells.
- Nitrate is eliminated from the plasma *via* the kidneys.
- The normal plasma concentration of nitrate is about 30 μmol/L, and the 24-hour formation of nitrate is 0.5–1 μmol.
- Plasma nitrate concentration or the urinary excretion of nitrate may be used to reflect the endogenous formation of NO, provided that alimentary intake of nitrite/nitrate is previously restricted for 48 hours.

References

1. Jilma B, Kastner J, Mensik C *et al.* Sex differences in concentrations of exhaled nitric oxide and plasma nitrate. *Life Sci* 1996; **58**: 469–476.
2. Reckelhoff JF, Kellum JA, Blanchard EJ *et al.* Changes in nitric oxide precursor, L-arginine, and metabolites, nitrate and nitrite, with aging. *Life Sci* 1994; **55**: 1895–1902.
3. Rosselli M, Imthurm B, Macas E *et al.* Circulating nitrite/nitrate levels increase with follicular development: indirect evidence for estradiol mediated NO release. *Biochem Biophys Res Commun* 1994; **202**: 1543–1552.
4. Bode-Böger SM, Böger RH, Shröder EP, Frölich JC. Exercise increases systemic nitric oxide production in men. *J Cardiovasc Risk* 1994; **1**: 173–178.
5. Jungersten L, Ambring A, Wall B, Wennmalm Å. Both physical fitness and acute exercise regulate nitric oxide formation in healthy humans. *J Appl Physiol* 1997; **82**: 760–764.
6. Sessa WC, Pritchard K, Seyedi N *et al.* Chronic exercise in dogs increases coronary vascular nitric oxide production and endothelial cell nitric oxide synthase gene expression. *Circ Res* 1993; **73**: 829–838.
7. Needleman P, Jakshik B, Johnson ME. Sulfhydryl requirement for relaxation of vascular smooth muscle. *J Pharmacol Exp Ther* 1973; **187**: 324–331.
8. Stamler J, Mendelsohn ME, Amarante P *et al.* N-Acetylcysteine potentiates platelet inhibition by endothelium-dependent relaxing factor. *Circ Res* 1989; **65**: 789–795.
9. Myers PR, Minor Jr RL, Guerra Jr R *et al.* Vasorelaxant properties of the endothelium-dependent relaxing factor more closely resembles *S*-nitroso-cysteine than nitric oxide. *Nature* 1990; **345**: 161–163.
10. Stamler JS, Simon DI, Osborne JA *et al.* *S*-Nitrosylation of proteins with nitric oxide: synthesis and characterization of biologically active compounds. *Proc Natl Acad Sci USA* 1992; **89**: 444–448.
11. Radomski MS, Rees DD, Dutra A, Moncada S. *S*-Nitroso-glutathione inhibits platelet activation *in vitro* and *in vivo*. *Br J Pharmacol* 1992; **107**: 745–749.
12. Simon DI, Stamler JS, Jaraki O *et al.* Antiplatelet properties of protein *S*-nitrosothiols derived from nitric oxide and endothelium-derived relaxing factor. *Arterioscler Thromb* 1993; **13**: 791–799.
13. Stamler JS, Simon DI, Jaraki O *et al.* *S*-nitrosylation of tissue-type plasminogen activator confers vasodilatory and antiplatelet properties on the enzyme. *Proc Natl Acad Sci USA* 1992; **89**: 8087–8091.
14. Stamler JS, Jaraki O, Osborne J *et al.* Nitric oxide circulates in mammalian plasma primarily as an *S*-nitroso adduct of serum albumin. *Proc Natl Acad Sci USA* 1992; **89**: 7674–7677.
15. Keaney JF Jr, Simon DI, Stamler JS *et al.* NO forms an adduct with serum albumin that has endothelium-derived relaxing factor-like properties. *J Clin Invest* 1993; **91**: 1582–1589.

16. Jia L, Bonaventura C, Bonaventura J, Stamler JS. *S*-nitrosohaemoglobin: a dynamic activity of blood involved in vascular control. *Nature* 1996; **380**: 221–226.
17. Moncada S, Palmer RMJ, Higgs EA. Nitric oxide: physiology, pathophysiology, and pharmacology. *Pharmacol Rev* 1991; **43**: 109–142.
18. Ignarro LJ, Fukuto JM, Griscavage JM *et al.* Oxidation of nitric oxide in aqueous solution to nitrite but not nitrate: comparison with enzymatically formed nitric oxide from L-arginine. *Proc Natl Acad Sci USA* 1993; **90**: 8103–8107.
19. Lewis RS, Deen WM. Kinetics of the reaction of nitric oxide with oxygen in aqueous solution. *Chem Res Toxicol* 1994; **7**: 568–574.
20. Hogg N, Singh RJ, Joseph J *et al.* Reactions of nitric oxide with nitronyl nitroxides and oxygen: prediction of nitrite and nitrate formation by kinetic simulation. *Free Radic Res* 1995; **22**: 47–56.
21. Lewis RS, Tamir S, Tannenbaum SR, Deen WM. Kinetic analysis of the fate of nitric oxide synthesized by macrophages *in vitro*. *J Biol Chem* 1995; **270**: 29350–29355.
22. Wennmalm Å, Benthin G, Petersson AS. Dependence of the metabolism of nitric oxide (NO) in healthy human blood on the oxygenation of its red cell haemoglobin. *Br J Pharmacol* 1992; **106**: 507–508.
23. Green LC, Ruiz de Luzuriaga K, Wagner DA *et al.* Nitrate biosynthesis in man. *Proc Natl Acad Sci USA* 1981; **78**: 7764–7768.
24. Wagner DA, Moldawer LL, Pomposelli JJ *et al.* Nitrate biosynthesis in the rat. Precursor-product relationships with respect to ammonia. *Biochem J* 1985; **232**: 547–551.
25. Hermann L. Ueber die Wirkungen des Stickoxydulgases auf das Blut. *Arch Anat Physiol Lpz* 1865; 469–481.
26. Gibson QH, Roughton FJ. Further studies on the kinetics and equilibria of the reaction of nitric oxide with haemoproteins. *Proc R Soc Lond B Biol Sci* 1965; **163**: 197–205.
27. Moore EG, Gibson QH. Cooperativity in the dissociation of nitric oxide from hemoglobin. *J Biol Chem* 1976; **251**: 2788–2794.
28. Yoshida K, Kasama K, Kitabatake M *et al.* Metabolic fate of nitric oxide. *Int Arch Occup Environ Health* 1980; **46**: 71–77.
29. Yoshida K, Kasama K. Biotransformation of nitric oxide. *Environ Health Perspect* 1987; **73**: 201–205.
30. Wennmalm Å, Benthin G, Edlund A *et al.* Metabolism and excretion of nitric oxide in humans. An experimental and clinical study. *Circ Res* 1993; **73**: 1121–1127.
31. Westfelt UN, Benthin G, Lundin S *et al.* Conversion of inhaled nitric oxide to nitrate in man. *Br J Pharmacol* 1995; **114**: 1621–1624.
32. Benthin G, Björkhem I, Breuer O *et al.* Transformation of subcutaneous NO into nitrate in the rat. *Biochem J* 1997; **323**: 853–858.

33. Jungersten L, Edlund A, Petersson AS, Wennmalm Å. Plasma nitrate as an index of nitric oxide formation in man: analyses of kinetics and confounding factors. *Clin Physiol* 1996; **16**: 369–379.
34. Samuelsson H, Wolff T, Wennmalm Å, Hedner T. Nitric oxide in central pain transmission as evaluated by CSF nitrite concentrations in pain patients. Submitted for publication.
35. Jungersten L, Edlund A, Hafström LO *et al.* Plasma nitrate as an index of immune system activation in animals and in man. *J Clin Lab Immunol* 1993; **40**: 1–4.
36. Mitchell HH, Shonle HA, Grindley HS. The origin of nitrates in the urine. *J Biol Chem* 1916; **24**: 461–490.
37. Cortas NK, Wakid NW. Pharmacokinetic aspects of inorganic nitrates ingestion in man. *Pharmacol Toxicol* 1991; **68**: 192–195.
38. Gustafsson LE, Leone AM, Persson MG *et al.* Endogenous nitric oxide is present in the exhaled air of rabbits, guinea pigs and humans. *Biochem Biophys Res Commun* 1991; **181**: 852–857.
39. Norman V, Keith CH. Nitrogen oxides in tobacco smoke. *Nature* 1965; **205**: 915–916.

Chapter 7

Inhibitors of Nitric Oxide Biosynthesis

Garry J Southan, Csaba Szabó and Christoph Thiemermann

Introduction

In the present chapter, the various approaches aimed at inhibition of NO synthesis are reviewed, with particular emphasis on the issue of selectivity with respect to the three known isoforms of NO synthase (NOS: the constitutive endothelial isoform, eNOS; the constitutive neuronal isoform, nNOS; and the inducible isoform, iNOS). The most common approach is related to inhibitors that interfere with binding of L-arginine at the active site. Subsequent sections discuss the alternative approaches, aimed at limiting NO biosynthesis by interfering with cofactor availability or binding, L-arginine uptake or recycling and other possible strategies. Throughout this chapter, the characteristics and properties of the various inhibitors are linked to their reported *in vivo* pharmacological actions (under physiological and pathophysiological conditions). Finally, a separate section is devoted to endogenous inhibitors of NO biosynthesis and their potential relevance to various pathophysiological conditions.

Biosynthesis of Nitric Oxide

The synthesis of NO by NOS involves the oxidation of one of the guanidino nitrogens of L-arginine (Figure 1, R = H). L-arginine is recognized and bound by a portion of NOS so that the guanidino group is adjacent to the catalytic site, which consists of an iron-containing haem group, iron protoporphyrin IX. The iron is coordinated to the tetradentate porphyrin in a planar arrangement and to the thiol of a cysteine residue. The sixth coordination site is available for the binding and activation of molecular oxygen for subsequent reaction with substrate.[1] The ferric iron of the haem prosthetic group is reduced by transfer of an electron from nicotinamide adenine dinucleotide phosphate (NADPH) and binds and activates molecular oxygen, which is then cleaved, with one oxygen atom released as water and the other incorporated onto one of the terminal guanidino nitrogens of arginine, to yield N^G-hydroxy-L-arginine (Figure 1, R = –OH). Subsequent activation of a further molecule of oxygen facilitates the further oxidation of N^G-hydroxy-L-arginine to NO and citrulline (Figure 2, S = O).[1,2]

NOS subunits are divided into a reductase domain, binding flavin-adenine dinucleotide (FAD) and flavin mononucleotide (FMN) and NADPH (even in the monomer), and an oxygenase domain which contains the binding sites for substrate, haem and tetrahydrobiopterin (BH_4). Only in the dimeric form, however, can iNOS and nNOS bind BH_4 and haem. Consequently, nNOS and iNOS are active only as homodimers. Since the oxygenase domains of the monomers overlap in the dimer [3] and both haems must be present for catalysis,[4] there is the possibility that both haems are in the same catalytic site. A further cofactor, calmodulin, appears to be essential for NOS activity and is thought to control the electron transfer from the flavins to the haem. The binding of calmodulin to native (inactive)

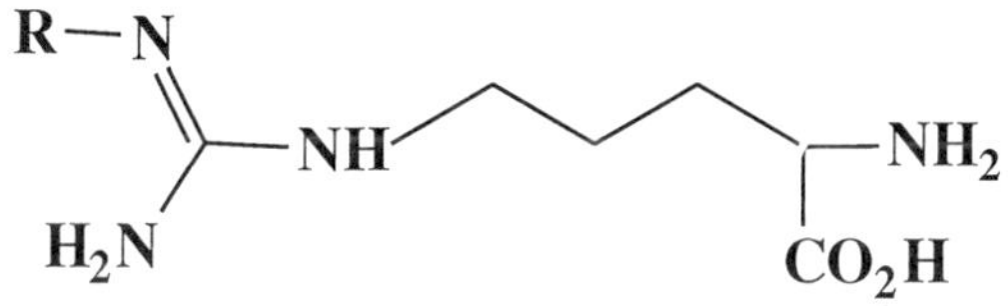

Figure 1

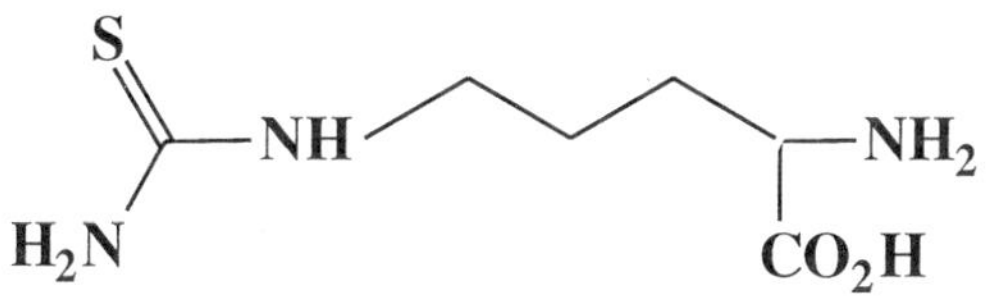

Figure 2

constitutive isoforms of NOS is promoted by calcium ions. Consequently, the activities of nNOS and eNOS are regulated by the concentrations of intracellular calcium. In contrast, calmodulin is permanently bound to iNOS so that the activity of iNOS is independent of intracellular calcium.

Active Site Inhibitors of Nitric Oxide Synthase

Inhibitors in this category are described in detail below, and listed for convenience in Table 1.

L-arginine Analogues

Most studies that have used inhibition of NOS to determine the involvement of NO in a physiological or pathophysiological process have employed analogues of L-arginine which, by virtue of their structural similarity to L-arginine, bind at the substrate binding site of NOS. Even though many derivatives have been prepared, most reports relied upon the same few examples. Indeed, inhibition of a given biological process by the prototypic NOS inhibitor, N^G-monomethyl-L-arginine (L-NMA or L-NMMA; Figure 1, R = $-CH_3$), is generally accepted as strong evidence for the involvement of NO. L-NMA shows similar potency against all isoforms of NOS and is readily taken up by cells. The D-isomer is inactive and often used as a negative control. Other N^{ω}-alkyl-L-arginines have been prepared and tested for NOS inhibitory activity. Some L-arginine analogue NOS inhibitors occur endogenously (see later section).

Table 1. The principal active site inhibitors of nitric oxide biosynthesis.

Inhibitor	Comments	Refs.
Analogues of L-arginine		
N^G-(Mono)methyl-L-arginine (L-NMA, L-NMMA)	Prototypical NOS inhibitor with little discrimination between isoforms	141
N^G-Nitro-L-arginine (L-NA)	Potent inhibitor of constitutive isoforms (nNOS and eNOS)	141
N^G-Nitro-L-arginine methyl ester (L-NAME)	Weak NOS inhibitor *per se*, but convenient prodrug for L-NA	141
N^6-(1-iminoethyl)-L-lysine (L-NIL)	Marked selectivity for iNOS *in vitro* and *in vivo*	27
N^5-(1-iminoethyl)-L-ornithine (L-NIO)	Potent, but little isoform selectivity	27
L-Thiocitrulline	Potent inhibitor of all NOS isoforms	29, 141
S-Methyl-L-thiocitrulline	Shows a preference for nNOS amongst human isoforms	30, 141
S-Ethyl-L-thiocitrulline		30
Canavanine	*In vivo* selectivity for iNOS; beneficial in shock	34–38
Isothioureas		
S-Methylisothiourea (*S*-Methyl-ITU)	Shows selectivity for iNOS; beneficial effects in models of septic shock	39, 48–50
S-Ethyl-ITU	Extremely potent, but not isoform selective	40, 41
S-Isopropyl-ITU	Extremely potent, some preference for human iNOS; benefits in early hemorrhagic shock	40, 53
2-Aminothiazoline	Potent, not isoform selective	40, 41
2-Amino-5,6-dihydro-6-methyl-4H-1,3-thiazine (AMT)	Potent and selective inhibitor of iNOS	42
S-(2-aminoethyl)-ITU (AET)	Precursor of MEG and 2-aminothiazoline, selective for iNOS	14, 41, 71
S-(3-aminopropyl)-ITU	Non-isoform selective NOS inhibitor, by virtue of rearranging to mercaptopropyl-guanidine (see below)	41, 71
S-ethyl-N-[4-(trifluoromethyl)phenyl]-ITU	30-fold selectivity for human nNOS *vs.* eNOS	45
S,S'-[1,3-phenylene-bis(1,2-ethanediyl)]bis-ITU (1,3-PB-ITU)	Potent and selective inhibitor of iNOS. Limited uptake by cells. Some toxicity *in vivo*	40

Table 1. (*Continued*)

Inhibitor	Comments	Refs.
Guanidines and Amidines		
Aminoguanidine	Selective for iNOS. Minimal effects on normal BP. Inexpensive. Some non-specific effect	39, 60, 142
Mercaptoethyl-guanidine (MEG)	Selective iNOS inhibitor. Additional effects include peroxynitrite scavenging. Exerts anti-shock and anti-inflammatory actions	67, 71, 75, 76
Mercaptopropyl-guanidine	Non-selective NOS inhibitor and peroxynitrite scavenger	67, 71
Guanidinoethyldisulphide (GED)	Effective iNOS inhibitor *in vivo* and *in vitro*	71, 74
Iminopiperidine	Potent NOS inhibitor with preference for iNOS	78, 79
4-Methyl-2-aminopyridine	Modest *in vivo* preference for iNOS	80, 81
1H-Pyrazole-1-carboxamidine (PCA)	Effective NOS inhibitor. 4-methyl derivative has 10-fold selectivity for iNOS	83
N-(3-(aminomethyl)benzyl)-acetamidine (1400 W)	iNOS specific, irreversible inhibitor, with considerable selectivity maintained *in vivo*	13
Imidazoles and Indazoles		
7-Nitroindazole (7-NI)	Non-selective *in vitro*, but nNOS selective *in vivo*. Exerts minimal pressor effects *in vivo*	86, 89
3-bromo-7-NI	More potent against (rat) nNOS than 7-NI	89
1-(2-trifluoromethyl phenyl)imidazole (TRIM)	Weaker against eNOS. Antinociceptive *in vivo*	87

Studies using L-NMA and related inhibitors demonstrated the role of NO in the regulation of a whole array of physiological processes. A basal production of NO keeps the vasculature in a continuous state of vasodilatation, *via* a guanosine 3',5'-cyclic monophosphate (cGMP)-dependent process. Consequently, inhibition of eNOS has potent vasoconstrictor and pressor effects. These effects have become generally accepted as indicators to assess the degree of the eNOS inhibition elicited by a NOS inhibitor *in vivo* (see also below). Moreover, a basal production of NO from eNOS inhibits adhesion of platelets and neutrophils to the intima of blood vessels. The NOS isoforms present in the central nervous system (eNOS and nNOS) play important roles in neurotransmission, memory generation and regulation of cerebral blood flow. For more detailed discussions on the physiological roles of NO, the reader is referred to other chapters of this book.

N^G-nitro-L-arginine (L-NA; Figure 1, R = $-NO_2$) has been used as an inhibitor of NOS for about 10 years and shows a general preference for the constitutive NOS isoforms when compared to L-NMA. L-NA is a neutral amino acid at physiological pH and so enters cells by diffusion. It has a limited solubility in physiological solution (~ 5 mM), so that the methyl ester (L-NAME) is often used as it is conveniently more soluble. However, esterification of the carboxyl group can markedly reduce the inhibitory potency. L-NAME is over an order of magnitude less potent towards iNOS than is L-NA,[5] but is readily hydrolyzed to L-NA *in vivo*.[6] Using esters of NOS inhibitors as prodrugs may offer the opportunity for selective targeting of NOS in various tissues (with specific esterase activities).

The mechanisms of inhibition of NOS activity by L-arginine derivatives vary, but all involve occupation of the substrate binding site. Some (cyclopropyl-L-arginine and L-NA) are competitive inhibitors of NOS as their effects can be reversed by L-arginine. However, in contrast to its reversible inhibition of iNOS, L-NA has been reported to cause irreversible inactivation of nNOS.[7] Under certain conditions, L-NA behaves as an irreversible inhibitor due to conformational change within the L-arginine binding site, resulting in a very slow dissociation. The inhibitory effects of L-NMA can initially be largely reversed by L-arginine, but prolonged exposure of iNOS to L-NMA results in a time-dependent irreversible inactivation of the enzyme, which is preceded by an NADPH-dependent hydroxylation of the inhibitor to N-hydroxy-N-methyl-L-arginine. [8] As with other mechanistic inhibitors, for instance aminoguanidine, enzyme turnover is required for permanent inactivation. Such irreversible inhibition can be associated with alterations of inhibitor binding to reduce the rate of dissociation of the inhibitor/NOS complex or chemical modification of the inhibitor and/or enzyme. This is probably the case for the irreversible component of the inhibition of NOS by N^G-allyl-L-arginine, where the allyl part of the inactivator is bound to the haem and both oxygen and NADPH are required for inactivation.[9] Such differences in the binding characteristics of NOS isoforms offer a potential rational basis for drug development.

In addition to the investigations into the role of eNOS- and nNOS-derived NO, L-NMA and L-NAME were widely used to elucidate the pathophysiological importance of iNOS-derived NO. Beneficial effects of NOS inhibition have been shown in various forms of shock, inflammation and central nervous system disorders,[10] despite the fact that the above-mentioned L-arginine based inhibitors lack substantial isoform selectivity (L-NMA) or are even selective towards the constitutive, rather than the inducible isoform of NOS (L-NAME).[11,12] In circulatory shock and in various forms of inflammation, NO derived from iNOS has been shown to have pathophysiological actions, whereas, in many cases, eNOS has beneficial and protective roles. Therefore, inhibition of eNOS may be detrimental and mask the beneficial effects of iNOS inhibition, as in endotoxin-induced intestinal injury,[13] liver dysfunction[14,15] and cardiovascular decompensation.[16] Consequently, inhibitors that can selectively target one isoform may realize the full potential of NOS inhibition.[17]

L-arginine based inhibitors can behave differently *in vivo* and *in vitro* due to a variety of effects other than their inhibition of NOS. Some of these effects are related to the arginine/NO system and some are unrelated to NO. For example, L-NMA affects transport of arginine into cells[18] and can also act as a substrate for NO production by NOS in the rat aorta and rabbit pulmonary artery.[19,20]

L-NA has been shown to inhibit the activities of other enzymes besides NOS. Inhibition of arginase activity is perhaps not surprising (although L-NMA and L-NAME do not inhibit arginase), as arginase catalyzes the conversion of L-arginine to L-ornithine, but L-NA also attenuates brain catalase activity *in vitro*.[21] L-NAME inhibits the activity of P450 apoenzymes which, at high doses of L-NAME, may have consequences for the metabolism of other drugs.[22] L-NAME presumably has the potential to elicit the same effects as L-NA as it is hydrolyzed to L-NA *in vivo*, although its disposition/distribution will be different.[23] Hydrolysis to L-NA *in vivo* is rapid,[24] but partial hydrolysis *in vitro* may lead to possible complications. Similar considerations apply when using other esters and amides of arginine analogues, especially when the ester has additional properties to the parent compound. For example, L-NAME and esters of L-arginine, but not other NOS inhibitors tested (such as L-NMA), have been reported to act as muscarinic receptor antagonists.[25] Among other arginine analogues, L-aminoarginine (Figure 1, R = $-NH_2$) has been reported to cause epileptiform seizures in dogs,[26] an effect which is probably unrelated to blockade of NO in the central nervous system, since similar actions have not been reported with L-NMA or L-NAME.

L-NA and L-NMA, as well as many other NOS inhibitors, contain the guanidine group as found in L-arginine, though variously substituted. Replacement of one of the guanidine nitrogens has led to the development of novel NOS inhibitors. N^6-(1-iminoethyl)-L-lysine (L-NIL; Figure 3, n = 4), N^7-(1-iminoethyl)-L-homolysine (Figure 3, L-NIH, n = 5) and N^5-(1-iminoethyl)-L-ornithine (L-NIO; Figure 3, n = 3) all contain the acetamidine group, but while L-NIO and L-NIH show no better selectivity than L-NMA or L-NA, L-NIL exhibits a 30-fold selectivity for iNOS over nNOS.[27] L-NIL showed potent anti-inflammatory potency in a rat model of adjuvant arthritis,[27] without raising blood pressure, suggesting lack of inhibition of the eNOS at the therapeutically useful dose.

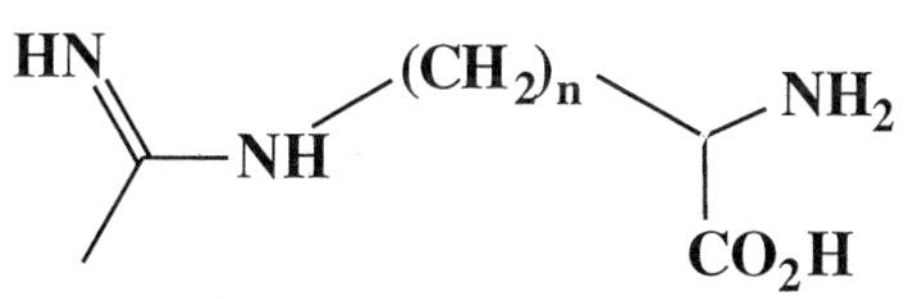

Figure 3

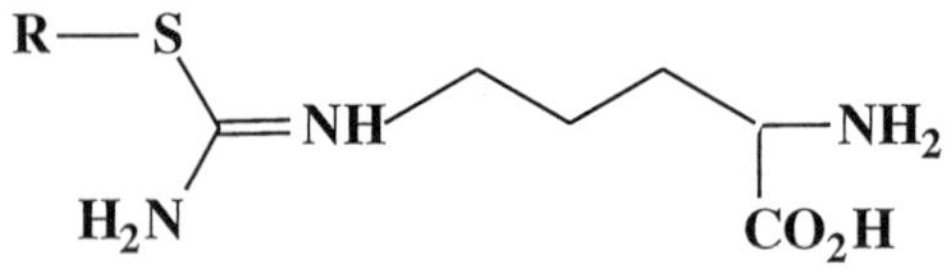

Figure 4

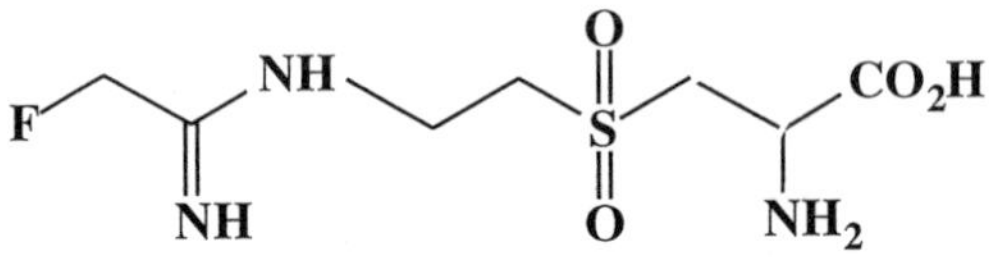

Figure 5

Recently, several classes of sulphur-containing drugs have been reported to be effective against NOS. Some of them are amino acid derivatives of L-citrulline.[28] L-thiocitrulline (Figure 2) is an extremely potent inhibitor of all NOS isoforms and is a potent pressor agent in normal and endotoxaemic rats.[29] It is competitive with L-arginine and binds to the haem moiety of NOS, probably as a sixth (axial) ligand. Homo-L-thiocitrulline is a weaker inhibitor that does not appear to interact with the haem in the same way.[29] *S*-Methyl and *S*-ethyl thiocitrulline (Figure 4, $R = CH_3$ and $-CH_2CH_3$ respectively), which can also be considered as isothioureas (see below), are also potent, reversible inhibitors and are competitive with L-arginine. Unlike L-thiocitrulline, they do not appear to interact with the haem iron.[28] Although *S*-methyl- and *S*-ethyl-thiocitrulline are potent pressor agents in normal rats,[28] they show a preference for human nNOS over eNOS and iNOS.[30] However, their therapeutic usefulness in various neurodegenerative diseases that are associated with over-activity of nNOS, may be limited due to their poor uptake into cells.[30]

While modifications of the guanidine group have resulted in some novel inhibitors with relatively high potency, alteration of the amino acid function is not so well tolerated. Esters of the carboxylic acid reduce potency (e.g. esters of L-NA) and removal of the α-amino or the α-carboxyl groups from arginine generally reduce potency.[31] Replacement of the carboxylic acid function with phosphinic or phosphorous acid groups produces compounds with only a weak inhibitory effect.[32]

Various alterations of the 3-carbon chain that links the guanidine group to the amino acid group of L-arginine or L-arginine based inhibitors have been made with variable results. Insertion of a heteroatom into the chain has, in at least one case, produced compounds of promising selectivity: 2-amino-6-(1-imino-2-fluoroethyl)-4,4-dioxo-4-thiahexanoic acid (Figure 5) and the

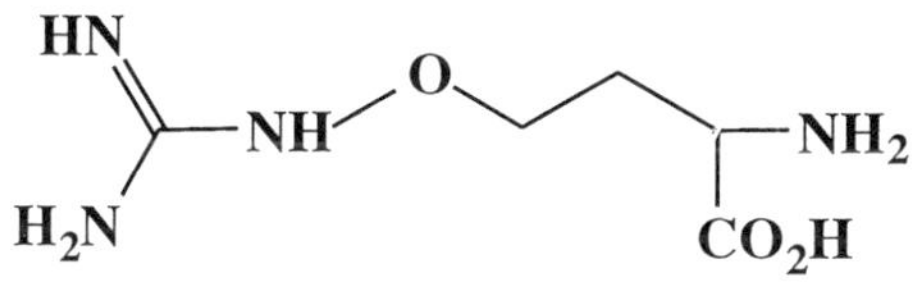

Figure 6

non-fluorinated derivative (Figure 5, F = H) showed > 10-fold selectivity for purified human iNOS over eNOS, but much greater apparent selectivity (> 100-fold) when their effects against iNOS (induced in endothelium-denuded rings by treatment with lipopolysaccharide, LPS) and eNOS (in normal intact rings) were compared in rat aortic rings.[33] Similarly weak effects against eNOS were reflected in the lack of effect on blood pressure of normal mice by the non-fluorinated reagent, although the effect of the more potent fluorinated analogue was not reported. However, these compounds are also potent inhibitors of purified human nNOS.[33]

Another analogue of arginine bearing a heteroatom in the α-aminopentanoic acid side chain is canavanine (Figure 6), a constituent of many plants. In this compound, the C^{δ}-of arginine (adjacent to the guanidine group) is replaced by an oxygen atom to yield, in effect, an O-substituted hydroxyguanidine. Such compounds can be effective inhibitors of NOS (see below), as is the case for L-canavanine. Early work showed canavanine to be effective at inhibiting NO production by immunostimulated macrophages and neutrophils,[34] but without effect on nonadrenergic/noncholinergic stimulation (an nNOS-dependent phenomenon) of rat anococcygeus muscle.[35] L-canavanine augments the contractile responses to phenylephrine of endothelium-denuded aorta from endotoxin treated rats, while having no effect on endothelium-dependent relaxations elicited by acetylcholine.[36] This selectivity for iNOS is also apparent *in vivo*: in rats, L-canavanine attenuates endotoxin-induced hypotension and vascular hyporeactivity to noradrenaline at doses that do not raise mean arterial pressure in normal rats.[37] Furthermore, in models of rodent endotoxaemia, L-canavanine has been shown to reduce endotoxin-induced changes to liver and kidneys,[38] reduce the increase in plasma nitrate levels and significantly improve survival.[37]

Non-amino Acid-based Inhibitors

Isothioureas

As discussed above, the commonly used amino acid-based NOS inhibitors have offered only limited selectivity towards the NOS isoforms. The promise that selective inhibition of iNOS or nNOS activity would offer significant benefits over

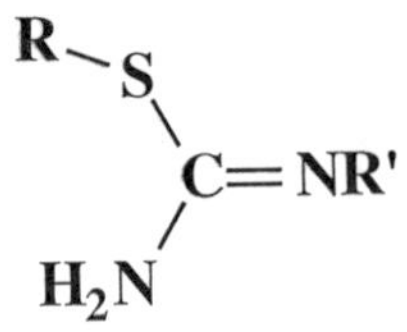

Figure 7

indiscriminate inhibition of NOS in inflammatory diseases and various forms of shock (iNOS), as well as degenerative neuronal diseases (nNOS), led to intensive efforts towards the development of non-amino acid compounds with selective effects on these isoforms.

S-Substituted isothioureas (ITUs; Figure 7) have been described as very potent inhibitors of NOS with variable selectivity towards the NOS isoforms.[39–41] ITUs have been demonstrated to perturb the haem spectra of NOS, suggesting binding close to, or interaction with, the haem centre.[40] *S*-ethyl-ITU (Figure 7, R = $-CH_2CH_3$, R′ = H) and *S*-isopropyl-ITU (Figure 7, R = $-CH(CH_3)_2$, R′ = H) appear to be the most potent members of the group of *S*-alkyl-ITUs. They do not show any preference for a particular isoform in rodent models,[39,41] but isopropyl-ITU shows some selectivity towards iNOS in human systems.[40]

Substitution of either or both of the amidine nitrogens of simple *S*-substituted isothioureas greatly reduces their potency towards iNOS, as exemplified by *S*-methyl-ITUs.[41] Notable exceptions are cyclic derivatives where the sulphur is attached *via* a carbon chain to one of the nitrogens: 2-aminothiazoline[40,41] and methyl-substituted 2-aminothiazoles (Figure 8)[40] show limited isoform selectivity, but variously substituted 2-amino-5,6-dihydro-1,3-thiazines (Figure 9) show selectivity for iNOS.[42] An example of such a compound is 2-amino-5,6-dihydro-6-methyl-4H-1,3-thiazine (AMT, see Table 1).

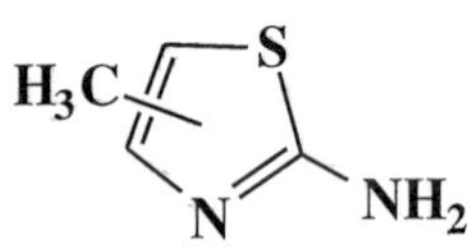

Figure 8

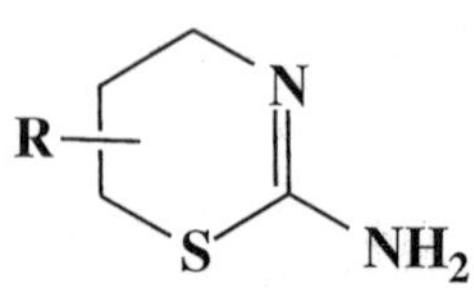

Figure 9

Although a decrease in potency towards nNOS is also seen upon N-substitution of ITUs, it is less pronounced than with iNOS. Indeed, N-phenyl isothioureas (Figure 9, R' = phenyl) demonstrate a preference for human nNOS.[43–45] For example, *S*-ethyl-ITU has a slight preference for iNOS over nNOS[40,44] but *S*-ethyl-N-[4-(trifluoromethyl)phenyl]-ITU (Figure 7, R = $-CH_2CH_3$, R' = phenyl) has a marked selectivity for human nNOS over iNOS.[44,45] N-Amino-*S*-methylisothiourea (Figure 7, R = $-CH_3$, R' = NH_2) also shows a 30-fold selectivity towards nNOS on isolated enzyme, but does not appear to enter cells (Wolff & Southan, unpublished observations).

S-Benzyl-ITU ('U-19451A'; Figure 7, R = benzyl, R' = H) has reduced potency when compared to *S*-methyl-ITU, but has a limited selectivity for nNOS and iNOS over eNOS.[46] A further group of very potent ITUs are of the bis-ITU type, where the sulphur atoms of two isothiourea units are linked by a carbon chain, sometimes containing unsaturated (hetero)cyclic rings. Some of these bis-ITUs, for example *S,S'*-[1,3-phenylene-bis (1,2-ethanediyl)]bis-TU (1,3-PB-ITU, Table 1) and its analogue 1,4-PB-ITU, show marked selectivity for human iNOS when compared with human eNOS and nNOS preparations.[40] Unfortunately, poor cellular uptake and direct cytotoxic effects will probably limit their therapeutic usefulness.[40]

S-Methyl-ITU shows a preference for iNOS over eNOS in rodents.[39] In rodent models of endotoxic shock, *S*-methyl-ITU caused a restoration of the blood pressure and pressor (vasoconstrictor) responsiveness.[39,47] Its pressor effect (similarly to that of aminoguanidine, but unlike L-NMA or L-NA) is larger in endotoxin-treated rats than in control animals.[39] Moreover, *S*-methyl-ITU reduces the endotoxin-induced increase in the plasma markers of hepatic injury, reduces endotoxin-induced pulmonary leak in rats, and improves 24-hour survival in endotoxin-treated mice and in rabbits challenged with live bacteria.[39,48,49] *S*-methyl-ITU was shown to improve left ventricular function after acute myocardial infarction in rabbits, while L-NA failed to do so.[50] It is, perhaps, pertinent that LPS-induced endotoxaemia causes an increase in uptake of ^{11}C-labelled *S*-methyl-ITU by liver and lungs — both organs that express iNOS in response to LPS — while uptake by the brain remained very low, presumably due to the low lipophilicity of *S*-methyl-ITU.[51]

S-Methyl-ITU or other derivatives may cause tachyphylaxis — pressor responses to subsequent injections of the agent or of other ITUs are attenuated in anaesthetized animals. Indeed, this phenomenon has been noted for related pressor agents (isoureas, amidines).[52] Thus, it appears that ITUs, along with other derivatives of amidines (see below), possess additional properties that affect the cardiovascular system.

S-isopropyl-ITU has been shown to provide potent inhibition of eNOS and to improve blood pressure and survival of pigs and rats in the early phase of severe haemorrhagic shock, without altering cardiac output or splanchnic and renal perfusion.[53] In haemorrhagic shock, therefore, inhibition of eNOS may elicit more beneficial effects than deleterious effects, though this is a view not universally accepted. However, other workers have reported improved survival together with reduced gastric injury in rats subjected to acute, irreversible haemorrhagic shock after NOS inhibition with L-NAME.[54] *S*-Ethyl-ITU has been shown to cause a marked and sustained increase in blood pressure in a hyperdynamic model of endotoxin shock in the sheep.[55]

S-methyl-ITU[39] and *S*-(2-aminoethyl)-ITU (AET; Figure 7, R = $-CH_2CH_2NH_2$, R′ = H) are also effective in preventing endotoxin-induced liver failure,[14] although the effects of the latter drug are presumably due to its metabolite, mercaptoethylguanidine (MEG), as discussed below. The same two compounds proved to be more effective than L-NA, L-NMA, aminoguanidine and other inhibitors at attenuating NO biosynthesis from chondrocytes and cartilage explants taken from patients with osteoarthritis (while being the least effective against bovine eNOS activity).[56]

Aminoguanidine and related inhibitors

Most models of the L-arginine binding/catalytic site invoke several regions that are potential targets for drug binding. While some groups are reticent to publicize their findings in detail, there are some generalizations. There is a 'guanidine binding site' that recognizes one of the guanidino nitrogens of arginine and can interact with guanidines or related compounds that contain a similar feature. Hydrogen bonding may well be involved. The haem iron can accommodate compounds that act as a sixth ligand. Nitrogen heterocycles and sulphur-containing compounds may interact in this way. The BH_4 binding site is probably in the near vicinity as several inhibitors interact with the binding sites of both BH_4 and L-arginine. Further areas that can accommodate certain structural features are becoming apparent as the number and variety of reported NOS inhibitors increases.

An increasing number of compounds containing the 'amidine' group are being reported as potent and/or selective inhibitors of the various isoforms of NOS. These generally contain an unsubstituted amidine nitrogen, which suggests that interaction with the guanidine binding site may contribute to binding to NOS. Of the various types, guanidines (Figure 10) were soon investigated due to their similarity to L-arginine (itself a substituted guanidine). While guanidine itself is not an inhibitor, simple substituted guanidines have modest effects on NOS. Methylguanidine (an endogenous substance) and ethylguanidine (Figure 10, R = $-CH_3$ or $-CH_2CH_3$, R′ = H), do inhibit NOS, although they are weaker than L-NMA. 1,1-dimethylguanidine is somewhat more potent against iNOS than methylguanidine.[57] It appears to be equivalent to L-NMA in its effect on eNOS in mesenteric artery preparations and causes rapid increases in blood pressure *in vivo*.[57] Arcaine (N,N′- ethanylbiguanidine), containing two guanidine groups, shows moderate inhibition of NOS.[58] Certain O-substituted hydroxyguanidines, such as O-(t-butyl)hydroxyguanidine (Figure 10, R = $-OC(CH_3)_3$, R′ = H), are as

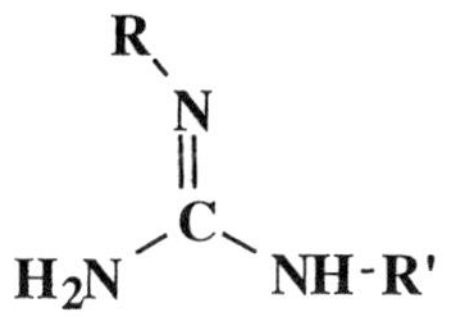

Figure 10

effective as L-NMA against isolated NOS enzymes but, like the guanidines mentioned above, display little selectivity (GJ Southan, unpublished observations). However, the example of L-canavanine exhibits some preference for iNOS *in vivo*, as discussed above.

Perhaps the guanidine that has received most attention is aminoguanidine (Figure 10, R = $-NH_2$, R = H) due to the early recognition of its selectivity towards iNOS.[59] This and its low acute toxicity pointed towards potential therapeutic value. Aminoguanidine is nearly equipotent with L-NMA in its ability to inhibit NO production by immunostimulated macrophages, but is less potent than L-NMA on eNOS and nNOS activities in cell homogenates.[39,60] *In vivo*, it has only limited pressor effects in anaesthetized rats. There are now numerous reports on the beneficial effects of aminoguanidine in various experimental models of inflammation and shock. *In vitro*, the inhibition of the activity of semi-purified enzyme or of NOS activity in cells or tissues homogenates is time dependent.[60] This phenomenon is due to a mechanism-based inhibition whereby the enzyme turnover results in an irreversible inactivation by aminoguanidine. Inactivation of various catalytic units of NOS by aminoguanidine (NADPH oxidase and citrulline formation but not the cytochrome c reductase activity) is dependent on an active NOS and the appropriate NOS cofactors.[61] A similar time-dependent mechanism is also found for alkyl-substituted aminoguanidines (Figure 10, R = $-NH_2$, R′ = alkyl) although they are generally less potent and not as selective. However, 1-amino-2-ethylguanidine (Figure 10, R = $-NH_2$, R′ = $-CH_2CH_3$) has a potency and selectivity for iNOS similar to that of L-NIL. The hydrazine function seems to confer the ability for mechanism-based inhibition: *S*-methyl isothiosemicarbazide (Figure 7, R = $-CH_3$, R′ = $-NH_2$), which can be regarded as N-amino-*S*-methylisothiourea, seems to inactivate nNOS selectively (30-fold selectivity for bovine nNOS compared with eNOS) by a similar mechanism (Wolff & Southan, unpublished observations). This is in contrast to the inhibition by other *S*-alkyl-isothioureas (see above) and guanidines, which can be reversed by dilution or excess L-arginine. Aminoguanidine shows rather limited selectivity towards iNOS in certain experimental systems: for example, when comparing its effects on eNOS and iNOS from intestinal homogenates.[62] In addition, relatively large doses of aminoguanidine are required for substantial inhibition of iNOS *in vivo*.[60,63]

Aminoguanidine has been shown to restore contractile responses in vascular tissues of animals treated with endotoxin *in vivo* and *in vitro* (an iNOS-dependent phenomenon).[59,60] Furthermore, aminoguanidine improves survival rate in mice treated with endotoxin.[60] Other effects of aminoguanidine that are tentatively linked to its inhibitory effect on iNOS include anti-inflammatory actions: it inhibits the progression of experimental autoimmune encephalomyelitis,[63] as well as preventing the endotoxin-induced acute pulmonary extravasation.[48] Aminoguanidine (but not L-NAME) was effective in inhibiting the development of multiple organ failure in a model of endotoxic shock in the rat.[15] The related compound N'-hydroxy-aminoguanidine (Figure 10, R = –OH, R' = $-NH_2$) has been reported to attenuate LPS-induced liver damage in rats.[64] This compound is an inhibitor of iNOS with a potency similar to that of aminoguanidine, but there may well be conversion of the hydroxy derivative to aminoguanidine (or *vice versa*) *in vivo*.

Aminoguanidine has been extensively studied in diabetes, where it appears to exert NO-dependent and NO-independent beneficial effects. Aminoguanidine prevents the interleukin-1 induced decrease in glucose secretion in islets of Langerhans and reduces the glucose-induced vascular dysfunction in rats.[59] Its effects in inhibiting diabetic vascular dysfunction also appear to be NO-mediated.[65]

In addition to inhibiting the catalytic activity of iNOS, high concentrations of aminoguanidine have been reported to inhibit the process of iNOS induction.[66] Moreover, aminoguanidine has numerous NO-independent effects. These include inhibition of polyamine catabolism, histamine metabolism and the activity of catalase and other copper- or iron-containing enzymes.[10, 60] Recently, it has also been discovered that aminoguanidine directly inhibits peroxynitrite-induced oxidative processes *in vitro*.[67] This is a beneficial effect, since peroxynitrite is produced in various forms of reperfusion injury and inflammation and exerts potent cytotoxic effects.[68] However, peroxynitrite is produced from NO by all three isoforms of NOS. Thus, *in vivo*, it is difficult to distinguish between the beneficial effects of aminoguanidine that are due to inhibition of iNOS, scavenging of peroxynitrite and other actions as listed above. In addition to the above mentioned guanidines, several other compounds that have been reported to inhibit NOS also contain the guanidino group. Inhibition of NOS activity by α-guanidinoglutaric acid[58] and by 3-amino-1,2,4-triazole[69] has been reported, but only for one isoform each (nNOS and iNOS respectively).

Guanidines may exhibit pharmacological properties that are independent of their effects on NOS. These include diverse vascular effects including increased blood pressure by the smaller alkyl guanidines (and guanidine) which may, at least partially, be related to the inhibition of NOS, while those with bulky substituents possess a smooth muscle relaxing action.[52] They may also inhibit membrane Na^+,K^+-ATPase.[70]

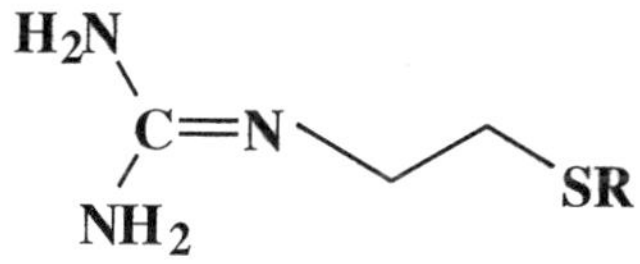

Figure 11

Mercaptoalkyl-guanidines

As early as 1995, we observed that the inhibitory characteristics of aminoethyl-isothiourea do not follow the structure-activity relationship observed for other isothioureas.[41] The reason for this irregularity is that the pharmacological effects of AET (Figure 7, R = $-CH_2CH_2NH_2$, R′ = H) are due to its rearrangement to MEG (Figure 11, R = H).[71] *S*-Aminoethyl- and *S*-aminopropyl-isothioureas, as well as their corresponding seleno derivatives,[72] where the amine is primary or secondary, are capable of intramolecular rearrangement at neutral and basic pH to give the corresponding mercaptoalkyl-guanidine. This occurs readily, especially at higher pH. For longer chain length or a tertiary amino group, the rearrangement is negligible.[71,72]

MEG and other mercaptoalkyl-guanidines are far more effective than L-NMA at inhibiting iNOS in whole cells and do so in an L-arginine-dependent manner. They appear to be much less effective against eNOS both *in vitro* and *in vivo* and cause only slight pressor responses in anaesthetized rats.[71] Mercaptoalkyl-guanidines and related compounds are oxidized to disulphides in aqueous solution, a process that also can occur *in vivo*.[73] The disulphide dimer resulting from the oxidation of MEG is guanidinoethyldisulphide (GED), itself a potent reversible inhibitor of NOS, which is competitive with L-arginine and non-competitive with BH_4.[74] It is interesting that the experimental results on whole cells and tissues suggest a better selectivity (iNOS/eNOS) for GED than would be anticipated from the modest 4-fold selectivity seen with purified enzymes.[74] It is conceivable that a preferential uptake by cells may play a part in the relative effectiveness of some compounds *in vivo*. Substituents on the sulphur, as in *S*-methyl- or *S*-ethyl-mercaptoethyl-guanidines (Figure 11, R = $-CH_3$ or $-CH_2CH_3$), appear to reduce the potency of NOS inhibition,[71] but maintain the selectivity towards iNOS as compared to eNOS (Southan and Szabó, unpublished observations). Such compounds may have the advantage that they do not dimerize.

From the above mentioned members of the mercaptoalkyl-guanidine class, *in vivo* reports are only available with MEG and GED (as well as the reports with aminoethyl-isothiourea, where MEG is the likely biologically active principle). Both MEG and GED have been reported to have potent anti-inflammatory effects. For instance, they improved the cardiovascular parameters and reduced the mortality rate in a murine model of endotoxin shock.[71,74] MEG is also effective in the delayed phase of haemorrhagic shock.[75] Furthermore, MEG has recently been shown to reverse the course of disease development in a collagen-induced model of arthritis in the rat.[76]

While these anti-inflammatory effects of MEG and GED are remarkable, it is not clear what proportion of these effects is, in fact, related to inhibition of iNOS. MEG (largely due to its free thiol group) is a potent scavenger of peroxynitrite,[67] and is a scavenger of oxygen-derived free radicals, which is the probable mode of its protective action in radiation injury.[73] Furthermore, MEG has a modest direct inhibitory effect on the activity of cyclooxygenase.[77]

Amidines

Guanidines and isothioureas have the common structural element, –C (=NH)-NH_2. If this function is attached to another carbon, the compound is generally termed an amidine. Simple alkyl amidines (Figure 12) such as propionamidine (Figure 12, R = –CH_3) and butyramidine (Figure 12, R = –CH_2CH_3), show modest activity towards NOS,[78] while the cyclic amidine, 2-iminopiperidine (Figure 13) is a more potent inhibitor of rat iNOS than is L-NMA.[78,79] They are generally powerful pressor agents in the rat, although some show a preference for iNOS over eNOS when compared to L-NMA. A study of cyclic amidines confirmed the modest preference of iminopiperidine for iNOS and reported similar preferences for some substituted homopiperidines (7-membered cyclic analogues) using the human enzymes.[79] This contrasts with some straight chain amidines which show a general preference for human nNOS over human eNOS, a trend enhanced by the inclusion of a sulphur into the alkyl chain.[79]

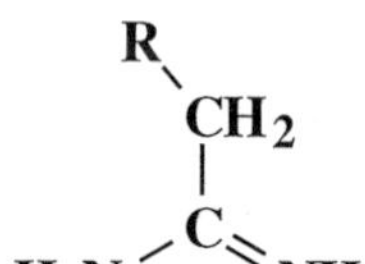

Figure 12

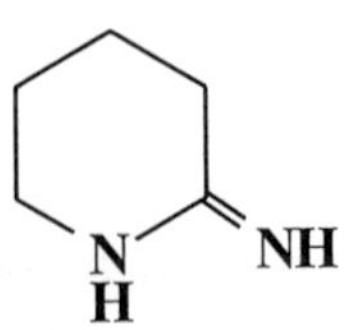

Figure 13

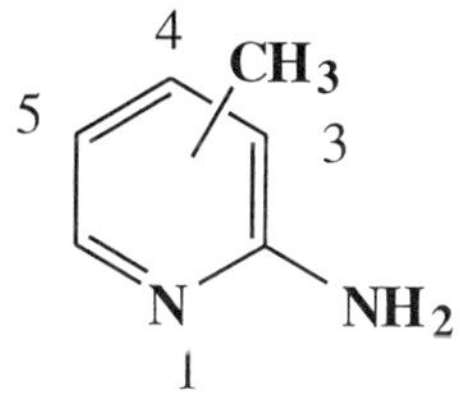

Figure 14

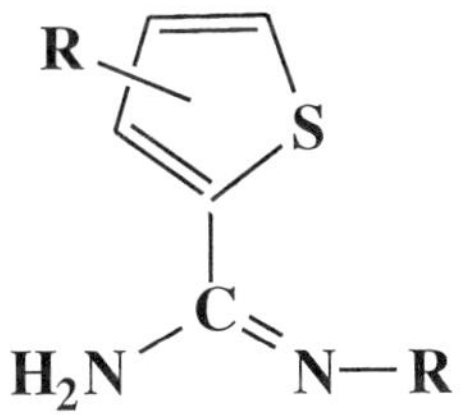

Figure 15

Some methyl-substituted 2-aminopyridines (Figure 14) also show potent inhibition of NOS in rat macrophages and smooth muscle.[80] While 4,6-dimethylaminopyridine demonstrated strong pressor effects,[80] its derivative 4-methyl-2-aminopyridine showed a '6.9 fold *in vivo* selectivity' for iNOS over eNOS,[81] although a similar selectivity was not seen on the human recombinant enzyme.[81] Heterocyclic amidines of the type Het-C(= NH)-NH_2 show little evidence of a preference for a particular isoform, although they can be effective inhibitors.[44] However, substitution of one of the amidine nitrogens with (substituted) benzyl or phenyl groups endows marked selectivity for nNOS, at least on the isolated enzymes, as exemplified by N-substituted thiophene-2-carboxamidines (Figure 15).[82] This preference for nNOS by N-substituted compounds appears to be true for other amidines and isothioureas.

A further group of heterocyclic carboxamidine inhibitors are the 1-pyrazole carboxamidines (Figure 16). The parent, unsubstituted 1H-pyrazole-1-carboxamidine (PCA) is considerably more potent than L-NMA and equally effective on all isoforms.[83] Substitution of the ring with methyl groups reduces the potency, but not uniformly, so that a preference for iNOS over eNOS can be seen, which is approximately 10-fold in the case of 4-methyl-PCA.[83] Although, superficially, the PCA structure incorporates aminoguanidine, the mechanism of inhibition is competitive with L-arginine, unlike aminoguanidine (see above). Another structure that incorporates aminoguanidine is 1,2-diaminobenzimidazole. Although the effects are weak, the 1-(methylamino) derivative is effective against nNOS, but inactive against iNOS and eNOS.[84] Again it is noteworthy that

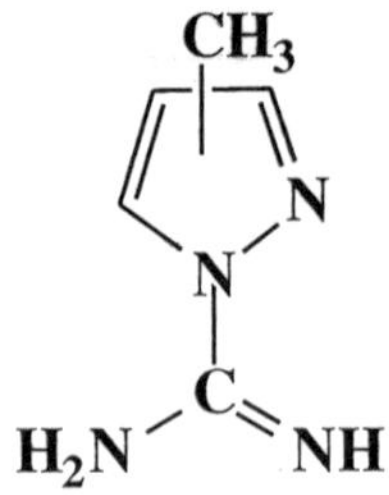

Figure 16

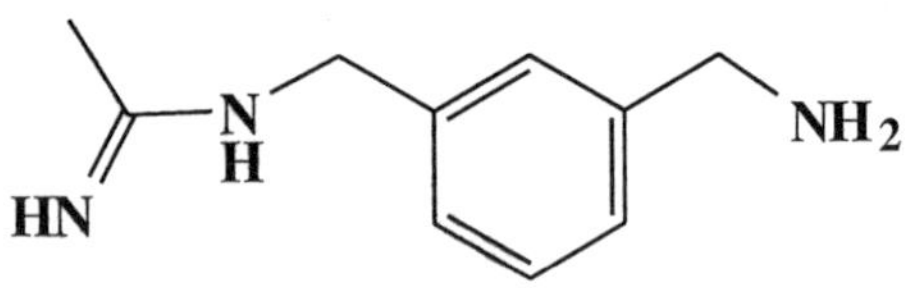

Figure 17

substitution of one of the nitrogens favours inhibition of nNOS (or disfavours iNOS and eNOS). Even N-benzyl-PCA, though a weak inhibitor, favours nNOS.[83] It seems that N-substitution is not well tolerated by iNOS or eNOS. An interesting exception is the N-substituted acetamidine, N-(3-(aminomethyl)benzyl)acetamidine (Figure 17). This compound ('1400 W') acts as an irreversible inhibitor of iNOS with a 5000-fold selectivity over eNOS on isolated human enzyme.[13] A considerable iNOS-selectivity was maintained on vascular ring models *in vitro* and, to a lesser extent, in models of endotoxin-induced vascular leakage *in vivo*. The selectivity for isolated iNOS versus nNOS was 200-fold. Although 1400 W showed a similar toxicity profile to the bis-isothiourea from which it was conceptually derived, these effects were observed at higher doses — e.g. rapid death at 50 mg/kg bolus injection of 1400 W. However, significant biological activity against iNOS-mediated injury was seen at doses 100 times lower.

Another compound with a leaning towards nNOS is pentamidine, an anti-microbial agent, that shows no effect on iNOS from RAW 264.7 macrophages, but inhibits nNOS non-competitively.[85]

Some amidines that are weak inhibitors of NOS are also potent pressor agents or have disproportionate pressor activity when compared to their ability to inhibit NOS (e.g. butyramidine[52,78]), suggesting that these small amidines may exert their pressor actions through more than one mechanism. Long chain amidines (and long chain guanidines), on the other hand, have been observed to exhibit *de*pressor actions. There are no published data on the potential anti-inflammatory effects of amidines.

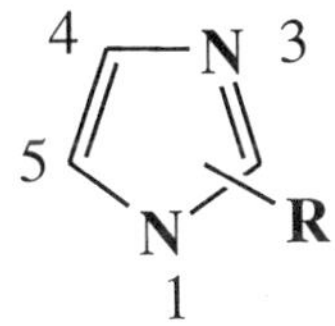

Figure 18

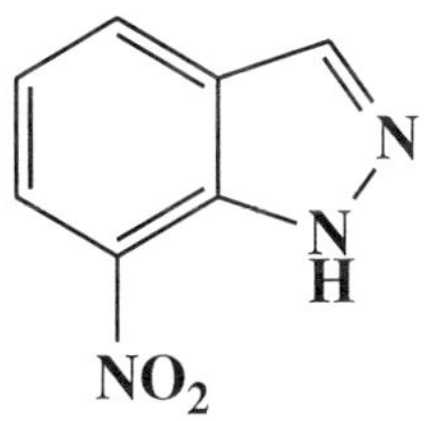

Figure 19

Imidazoles and indazoles

Nitrogen-containing heterocycles are known to inhibit haem-dependent reactions of cytochrome P450s and have been shown to inhibit the activity of NOS as well.

Imidazole (Figure 18) and its phenyl-substituted derivatives have been shown to inhibit NOS activity.[86] These agents do not prevent the haem-independent cytochrome c reductase activity of NOS, which is consistent with their binding as (axial) ligands to the haem in the catalytic site. Of the phenylimidazoles studied, only 1-phenylimidazole (Figure 18, R = phenyl) shows appreciable selectivity, with a k_i for macrophage iNOS fifty times lower than for nNOS or eNOS. Its binding characteristics vary with isoform, being competitive with BH_4 and/or L-arginine, suggesting interaction with binding sites in addition to the haem. The close proximity and isoform specificity of the binding sites of BH_4, guanidine and possibly the haem, as well as other areas, offers a potential for isoform-selective or even specific inhibition. 1-(2-trifluoromethylphenyl)imidazole (TRIM) is a weak inhibitor of eNOS, but considerably more potent against iNOS and nNOS activities, and shows anti-nociceptive effects *in vivo*.[87] In a more recent *in vitro* study, more than 30 cytochrome P450-sensitive substrates and inhibitors were tested for potential inhibitory activity on human iNOS, and only chlorzoxazone, a cytochrome P450 2E1 substrate, showed significant iNOS inhibitory activity.[88]

7-nitroindazole (7-NI; Table 1 and Figure 19) shows potent anti-nociceptive effects, related to inhibition of nNOS, but with minimal pressor effects *in vivo*. This is despite being an extremely potent inhibitor of NOS activity in homogenates of endothelial cells.[89] This would suggest that the basis of selectivity lies in the differential uptake of the inhibitor into endothelial cells compared with neuronal cells. The inhibition of nNOS by 7-NI is competitive with both L-arginine and BH_4, but non-competitive with L-arginine and competitive with BH_4 for iNOS and eNOS.[86] Similar anti-nociceptive effects have been observed with L-NA but not with L-NMA. Recently, 7-NI has been successfully used against nNOS to protect against neurotoxicity elicited by occlusion of the middle cerebral artery[90] in a murine model. 7-NI has been reported to exhibit NO-independent vascular actions, such as an endothelium-independent relaxant effect of various arterial smooth muscles.[91]

Miscellaneous

A variety of other compounds has been reported to inhibit NOS activity, of which porphyrin is the most curious.[92] Protoporphyrin IX inhibits citrulline formation by all three isoforms of NOS in a reversible manner, while zinc protoporphyrin IX shows a time- and concentration-dependent inactivation. Protoporphyrin IX also inhibits nitrite production by cytokine-stimulated murine RAW 264.7 cells, although this effect is accompanied by cellular cytotoxicity. The anti-ulcer drug, Benexate, a relatively large mono-substituted guanidine, was shown to attenuate the activity of purified nNOS in a manner competitive with L-arginine.[93] Other compounds that inhibit NOS include iminobiotin,[94] ebselen and derivatives,[95] tannin,[96] phencyclidine,[97] straight chain alcohols[98] and methylene blue.[99]

Cofactor-based and Alternative Approaches for Inhibition of Nitric Oxide Synthase

Differences in cofactor requirements could, in theory, be exploited selectively to prevent NO production by a particular isoform. Indeed, calmodulin antagonists such as calmidazolium[100] and W-7 can block endothelium-dependent relaxation of rat aorta at doses that have no effect on relaxations due to iNOS. However, due to numerous other calmodulin-dependent biological processes, it is doubtful whether this approach has therapeutic potential. The same is true of targeting the flavins with flavoprotein inhibitors such as diphenyleneiodonium: although it potently and irreversibly inhibits NOS activity, probably by its effects on the FAD binding site,[101] its effects are general to many flavoproteins.

BH_4 is an essential cofactor for activity of all NOS isoforms. Its function is not clear, but may involve allosteric activation, a redox role or stabilization of the active form of NOS.[1] Potential selective modulation of NOS may be achieved *via* agents directed at the biopterin binding site of NOS. 5-aryl (e.g. chlorophenyl)-substituted 2,4-diaminopyrimidines[102] and pterin derivatives[103] have been claimed as inhibitors of NOS, the former as selective for nNOS as measured in brain cytosol and veratrine-stimulated slices of brain. Some compounds, such as 7-NI and GED, associate with both the L-arginine and BH_4 binding sites, although the nature of the inhibition may differ between isoforms (see above). An alternative strategy involving BH_4 relies upon inhibiting the increased synthesis of BH_4 that occurs concomitantly with synthesis of iNOS in response to immuno-stimulation. Inhibition of BH_4 production by 2,4,-diamino-6-hydroxypyrimidine (DAHP), a selective inhibitor of guanosine triphosphate (GTP) cyclohydrolase I, the rate-limiting enzyme in the production of BH_4, inhibits NO production by iNOS.[104] In addition, it has been shown that N-acetylserotonin, phenprocoumon and dicumarol, which inhibit the synthesis of BH_4 by blocking sepiapterin reductase, also reduce cytokine-induced NO production.[104] It appears that such acute treatment may not affect normal BH_4 levels. For example, phenprocoumon and dicumarol, while inhibiting iNOS activity, do not block NO production by eNOS.[105] *In vitro*, DAHP has been shown to reduce NO production in endotoxic shock.[106] Moreover, N-acetylserotonin pretreatment has been shown to reduce endotoxin-induced hypotension in the rat,[107] although this may be due to its ability to inhibit the induction of iNOS.[108] BH_4 is also a cofactor for tyrosine hydroxylase and tryptophan hydroxylase (among other enzymes),[109] but the criteria for BH_4 activation of iNOS may be sufficiently different to allow enzyme-specific inhibition, especially in an acute setting.

The idea of starving NOS of its cofactors can also be applied to the substrate, L-arginine. iNOS utilizes large amounts of L-arginine and this may result in an increased uptake of L-arginine from extracellular sources. The extensive extracellular L-arginine requirement of cells expressing iNOS may offer a target for pharmacological intervention: administration of arginase can profoundly reduce plasma arginine levels and shows beneficial effects in a model of septic shock.[110] Induction of iNOS is also associated with the induction of arginosuccinate synthase, which, by stimulating the 're-cycling' of L-citrulline to L-arginine, may support iNOS with its substrate supplied from endogenous sources.[111] This may provide another target for limiting arginine availability to iNOS. In some cells, the same stimuli that induce iNOS also enhance the membrane transport system for L-arginine.[112] The predominant transport system for arginine is the Y+ transporter for cationic amino acids such as L-lysine and L-ornithine [112] which may compete with arginine and inhibit its uptake. Several arginine analogues, some of them NOS inhibitors (e.g. L-NMA and L-NIO), can also inhibit L-arginine uptake,[18] while themselves being taken up into cells.[113]

Endogenous Inhibitors of Nitric Oxide Synthase Activity

L-NMA, N^G,N^G-dimethyl-arginine (symmetric dimethylarginine or SDMA) and N^G,N'^G-dimethylarginine (asymmetric dimethylarginine or ADMA) were first isolated from the hydrolysate of proteins contained in bovine brain in 1971.[114]

Biological N-methylation of basic amino acid residues in proteins may occur while the polypeptide is synthesized (on the ribosome) as well as on nascent polypeptides in the absence of protein biosynthesis.[115] For instance, the methyl group of methionine is incorporated into L-arginine contained in e.g. myosin of embryonic skeletal muscle cells, resulting in the formation of L-NMA, SDMA or ADMA.[116] In the rabbit, the plasma concentrations of L-NMA, ADMA and SDMA are very low (~ 1 μmol/ml), but all three of these methylated amino acids can be found in extracts of brain, kidney, liver and spleen as well as in the cerebrospinal fluid.[116] In the rat, the highest concentrations of methylarginines are found in the spleen and the brain (cerebellum, olfactory bulb).[117] The cerebellar grey matter, hippocampus and hypothalamus of the bovine brain also contains large amounts of L-NMA and ADMA, while the concentration of SDMA is similar in all regions of the brain. Interestingly, regions of the brain which contained larger amounts of NOS activity also contained higher concentrations of L-NMA or ADMA, suggesting that these endogenous inhibitors of NOS activity may actually regulate NOS activity.[117]

The endogenous mono- or dimethylarginines may be metabolized by N^G,N^G-dimethylarginine dimethylaminohydrolase. For instance, this enzyme catalyses the hydrolytic cleavage of the dimethylamino moiety of SDMA, resulting in the formation of dimethylamine and L-citrulline. This enzyme is highly specific for SDMA and L-NMA, and the K_m values for these amino acids are 0.18 and 0.36 mM, respectively.[118] In addition, dimethylarginine:pyruvate aminotransferase (EC 2.6.1.44) catalyses the transaminations of the guanidino N-methylated L-arginines L-NMA, SDMA and ADMA and the alpha amino group of L-ornithine to pyruvate and glyoxylate.[118] It has been proposed that the metabolism of L-NMA by N^G,N^G-dimethylarginine dimethylaminohydrolase accounts for the shorter duration of the haemodynamic effects of L-NMA (and hence the duration of the inhibition of NOS activity) when compared to those of L-NA in the anaesthetized rat.[19] Indeed, endothelial cells metabolize L-NMA, but not L-NA, to L-citrulline and subsequently to L-arginine[19]. Interestingly, high concentrations of N^G,N^G-dimethylarginine dimethylaminohydrolase are found in the kidney, pancreas, liver, brain and aorta, as well as in peritoneal neutrophils and macrophages of the rat.[119] Most notably, N^G,N^G-dimethylarginine dimethylaminohydrolase (immunoreactivity) has now also been detected in pancreas, kidney and liver of humans.[120]

In healthy humans, SDMA and ADMA are excreted in the urine (presumably upon degradation of tissue proteins) and the normal ratio of SDMA to ADMA is 0.98 (range: 0.71–1.33).[121] ADMA is also found in the plasma of humans and more than 10 mg of ADMA are excreted in the urine over 24 h.[122] In 1992, Vallance and colleagues reported that the elimination of ADMA in patients with end-stage chronic renal failure, who have little or no urine output, is substantially reduced or abolished so that the circulating concentrations of ADMA rise sufficiently to inhibit NOS activity. Thus, these authors proposed that the accumulation of ADMA and the subsequent reduction in NO formation may contribute to the hypertension and immune dysfunction associated with chronic renal failure.[122]

In activated murine macrophages, ADMA is a more potent inhibitor of NOS activity than L-NMA or methylguanidine. Moreover, L-NMA (IC_{50}: ~ 5 mM) and ADMA (IC_{50}: ~ 18 mM) inhibit endothelium-dependent relaxation of the human saphenous vein.[123] In rats, the topical administration of ADMA causes constriction of the basilar artery (*in situ*) and produces a marked inhibition of the vasodilator responses elicited by acetylcholine. In addition, ADMA (1–100 mM) causes a concentration-dependent inhibition of nNOS activity in the rat brain (IC_{50}: ~ 2 mM).[124] Similarly, ADMA reduces the diameter of cerebral arterioles in the rabbit and attenuates the endothelium-dependent vasodilatation caused by acetylcholine in these vessels. Thus, it has been suggested that ADMA may be an important endogenous modulator of cerebral vascular tone under resting conditions and in response to vasoactive stimuli.[124]

Prior to the realization that ADMA is a potent inhibitor of NOS activity, there were several studies (in the 1970s) which document an enhanced formation or urinary excretion of ADMA in patients. For instance, the urinary concentration of ADMA in patients with liver disease, including chronic active hepatitis, is significantly higher than the urinary concentration of SDMA (ratio: 0.79; range: 0.49–1.30).[121] In contrast, the ratio of the urinary concentration of SDMA to ADMA is not significantly different (from the one in healthy volunteers, see above) in patients with multiple sclerosis, cerebrovascular accident, cancer or systemic lupus erythematosus.[121] Interestingly, the urine of patients with Duchenne muscular dystrophy contains large quantities of ADMA and the ratio of ADMA to SDMA is significantly increased when compared to healthy volunteers, patients with other types of neuromuscular disease or patients with muscle atrophy resulting from traumatic paralysis.[125] It should, however, be stressed that it is unclear whether: (a) the enhanced excretion of ADMA in the above disorders is associated with elevated plasma or tissue levels of ADMA, (b) tissue levels of ADMA are sufficient to inhibit NOS activity, and (c) ADMA is in any way related to the pathophysiology of the respective disorders.

There is some evidence that the plasma levels of ADMA increase in animal models of disease. For instance, graded haemorrhage (up to 40% of the calculated blood volume) in pigs leads to a significant increase in the serum levels of ADMA from 3.4 to 5.2 mM, suggesting that the vasoconstriction caused by ADMA may help to maintain blood pressure in conditions of acute hypovolaemia.[126] Interestingly, feeding of rabbits with a diet containing high fat and high cholesterol also results in a significant increase in the serum levels of ADMA. This is not due to an impairment in renal function, as the serum levels of creatinine remained normal.[127] Thus, it is possible that an increase in the tissue or serum levels of ADMA (due to either an enhanced formation or a reduced metabolism) contributes to the 'endothelial dysfunction' (e.g. impairment of responses to endothelium-dependent vasodilators) in animals and man with hypercholesterolaemia. Although the dietary supplementation with L-arginine does not reduce the plasma concentration of ADMA (which was elevated by 2-fold) in cholesterol-fed rabbits, L-arginine restored the urinary nitrite excretion, suggesting that an increase in the ratio of L-arginine to ADMA explains the restoration of NO formation in this model of hypercholesterolaemia.[127] The intra-coronary infusion of L-arginine in patients with coronary heart disease also restores the vasodilator responses to endothelium-dependent vasodilators. Thus, it is possible that this (and other) vasodilator effects of L-arginine in patients e.g. with hypercholesterolaemia are due to the reversal (by L-arginine) of the competitive inhibition of NOS activity by endogenous ADMA.

An impairment of endothelium-dependent vasodilator responses is also a feature of the endothelial dysfunction occurring after endothelial cell injury. For instance, balloon denudation of the rabbit carotid artery leads within 6 weeks to the regeneration of the endothelium with a significant neointima formation as well as a significant attenuation of endothelium-dependent relaxation. The concentration of L-NMA and ADMA (SDMA concentration below detection limit) in these regenerated endothelial cells is significantly higher, while the concentration of L-arginine is significantly lower, than the equivalent values in 'normal' endothelial cells. These findings suggest that the accumulation of endogenous inhibitors of NOS activity, together with a reduction of the intracellular concentration of L-arginine, contributes importantly to the impairment in NO formation by regenerated endothelial cells.[128]

During acute pulmonary allograft rejection, endothelial dysfunction not only occurs in the transplanted lung, but also in arteries of the transplant recipient. For instance, the vasodilator responses elicited by acetylcholine in the brachial artery of transplant recipients (measured by venous occlusion plethysmography) are significantly reduced during acute rejection.[129] However, in these patients the plasma levels of ADMA and L-arginine were not significantly different from those measured in healthy volunteers, suggesting that the endothelial dysfunction associated with acute transplant rejection is not due to inhibition of NOS activity by ADMA.[129] Similarly, there is little evidence that an increase in the plasma/tissue levels of ADMA contributes to the reduction in NOS activity in the placenta or the endothelial dysfunction associated with pre-eclampsia.[130]

Other endogenous guanidines have been reported to inhibit the activity of NOS: methylguanidine and 1,1-dimethylguanidine, as discussed previously, are weak to moderate inhibitors found in urine. Methylguanidine, as with various other guanidines, accumulates in pathophysiological conditions such as renal failure.[123] Though methylguanidine is a putative uraemic toxin and a convulsive and pressor agent at pharmacological doses,[131] its inhibition of NOS is probably of little physiological significance. Possibly of more relevance is the endogenous convulsant, α-guanidinoglutaric acid (GGA), which is as effective an inhibitor of nNOS activity in homogenates of rat brain as L-NMA.[58] In addition, L-arginine increases the threshold of seizures induced by GGA, suggesting that suppression of NOS may be involved in GGA-induced convulsions.[58]

Summary and Future Directions

Over the last decade, inhibitors of NOS activity have been instrumental in characterizing the roles of NO in a wide range of physiological and pathophysiological processes. Selective inhibitors of iNOS and nNOS have also been described and their use has demonstrated that enhanced formation of NO plays a crucial role in a variety of inflammatory and neurodegenerative diseases. Intensive efforts are underway to further develop and test highly potent, selective, and non-toxic inhibitors of iNOS and nNOS.

Although L-NMA is about equipotent for constitutive and inducible NOS activity, this compound has been shown to be effective in restoring the hypotension in animal models of septic shock. Furthermore, in several studies L-NMA has improved survival rate. L-NMA also improves blood pressure in humans with sepsis and reduces the catecholamine requirement in order to maintain blood pressure. A large-scale, phase III study of L-NMA ('546C88'), run by Glaxo-Wellcome, is under way to investigate efficacy in terms of long-term survival.[17] While L-NMA can be viewed as an effective, first-generation drug, this agent is unlikely to be useful in chronic inflammatory disorders, because of its effects on eNOS. In chronic inflammatory disorders, it is more likely that more novel, selective inhibitors of iNOS will be examined.

Regarding the role of endogenous inhibitors, there is good evidence that L-NMA and ADMA are endogenous inhibitors of NOS activity, which may modulate the formation of NO (by all isoforms of NOS) under physiological and pathophysiological conditions. There is support for the concept that an increase in the tissue or plasma levels of ADMA contributes to the impairment of endothelium-dependent vasodilatation in some cardiovascular disorders (e.g. hypercholesterolaemia). The restoration of endothelium-dependent vasodilator responses in hypercholesterolaemic patients by infusion of L-arginine may be (at least in part) due to the competitive reversal of the ADMA-induced inhibition of NOS activity.

Although studies with various NOS inhibitors provided good evidence for the regulation of NO in various cardiovascular functions, more recent studies, using eNOS, nNOS and iNOS knockout animals have provided a somewhat complicated picture. Although it appears that eNOS knockout mice have elevated blood pressure, when compared to wild-type controls,[132] it has been also shown that eNOS-knockout animals still demonstrate what was previously known as an 'endothelium-dependent relaxant response' to acetylcholine.[133,134] The role of nNOS in stroke and neuro-injury appears to be confirmed with nNOS knockouts and with combination studies using NOS inhibitors in eNOS and nNOS knockouts.[135–137] However, the degree of protection against shock and inflammation provided by the lack of iNOS gene is widely variable between different investigators.[138–140] One obvious reservation against the use of knockout animals is that these animals are likely to develop compensatory mechanisms after genetic ablation of a given gene, and therefore the responses seen in the knockout animals may or may not be relevant for the intact animal. There is no doubt that additional studies (pharmacological studies and studies using genetically engineered animals) will be needed to clarify the role of the various NOS isoforms in physiology and pathophysiology.

Acknowledgements

This work was supported, in part, by a grant from the National Institutes of Health (R29GM54773) to C.S. C.T. is a Senior Fellow of the British Heart Foundation.

References

1. Marletta MA. Nitric oxide synthase structure and mechanism. *J Biol Chem* 1993; **268**: 12231–12234.
2. Kerwin JF Jr, Lancaster JR, Feldman PL. Nitric oxide: a new paradigm for second messengers. *J Med Chem* 1995; **38**: 4343–4362.
3. Ghosh DK, Stuehr DJ. Macrophage NO synthase: characterization of isolated oxygenase and reductase domains reveals a head-to-head subunit interaction. *Biochemistry* 1996; **34**: 801–807.
4. Xie QW, Leung M, Fuortes M *et al.* Complementation analysis of mutants of nitric oxide synthase reveals that the active site requires two hemes. *Proc Natl Acad Sci USA* 1996; **93**: 4891–4896.
5. Southan GJ, Gross SS, Vane JR. Esters and amides of N^G-nitro-L-arginine act as prodrugs in their inhibition of inducible nitric oxide syntase. In: *The Biology of Nitric Oxide. 4. Enzymology, Biochemistry and Immunology,* Moncada S, Feelisch M, Busse R, Higgs EA (Eds). Portland Press, London 1995: 4–7.
6. Pfeiffer S, Leopold E, Schmidt K *et al.* Inhibition of nitric oxide synthesis by N^G-nitro-L-arginine methyl ester (L-NAME): requirement for bioactivation to the free acid, N^G-nitro-L-arginine. *Br J Pharmacol* 1996; **118**: 1433–1440.
7. Klatt P, Schmidt K, Brunner F, Mayer B. Inhibitors of brain nitric oxide synthase: binding kinetics, metabolism and enzyme inactivation. *J Biol Chem* 1994; **269**: 1674–1680.
8. Nishimura JS, Martasek P, McMillan K *et al.* Modular structure of neuronal nitric oxide synthase: localization of the arginine binding site and modulation by pterin. *Biochem Biophys Res Commun* 1995; **210**: 288–294.

9. Zhang HQ, Dixon RP, Marletta MA, Silverman RB. Mechanism of inactivation of neuronal nitric oxide synthase by N^{ω}-allyl-L-arginine. *Abstracts of the 213th ACS National Meeting Part 2*, April 13–17, 1997; Medical Chemistry Section: 016.
10. Southan GJ, Szabó C. Selective pharmacological inhibition of distinct nitric oxide synthase isoforms. *Biochem Pharmacol* 1996; **51**: 383–394.
11. Gross SS, Stuehr DJ, Aisaka K *et al.* Macrophage end endothelial cell nitric oxide synthesis: cell-type selective inhibition by N^{G}-aminoarginine, N^{G}-nitroarginine and N^{G}-methylarginine. *Biochem Biophys Res Commun* 1990; **170**: 96–103.
12. Lambert LE, Whitten JP, Baron BM *et al.* Nitric oxide synthesis in the CNS, endothelium and macrophages differs in its sensitivity to inhibition by arginine analogues. *Life Sci* 1991; **48**: 69–75.
13. Garvey EP, Oplinger JA, Furfine ES *et al.* 1400 W is a slow, tight binding, and highly selective inhibitor of inducible nitric-oxide synthase *in vitro* and *in vivo*. *J Biol Chem* 1997; **272**: 4959–4963.
14. Thiemermann C, Ruetten H, Wu CC, Vane JR. The multiple organ dysfunction syndrome caused by endotoxin in the rat: attenuation of liver dysfunction by inhibitors of nitric oxide synthase. *Br J Pharmacol* 1995; **116**: 2845–2851.
15. Wu C-C, Ruetten H, Thiemermann C. Comparison of the effects of aminoguanidine and N^{ω}-nitro-L-arginine methyl ester on the multiple organ dysfunction caused by endotoxaemia in the rat. *Eur J Pharmacol* 1996; **300**: 99–104.
16. Nava E, Palmer RMJ, Moncada S. The role of nitric oxide in endotoxic shock: effects of N^{G}-monomethyl-L-arginine. *J Cardiovasc Pharmacol* 1992; **20 (Suppl 12)**: S132–S134.
17. Kilbourn RG, Szabó C, Traber D. Beneficial versus detrimental effects of nitric oxide synthase inhibitors in circulatory shock: lessons learned from experimental and clinical studies. *Shock* 1997; **7**: 235–246.
18. Bogle RG, Moncada S, Pearson JD, Mann GE. Identification of inhibitors of nitric oxide synthase that do not interact with the endothelial cell L-arginine transporter. *Br J Pharmacol* 1992; **105**: 768–770.
19. Hecker M, Mitchell JA, Harris HJ *et al.* Endothelial cells metabolize N^{G}-monomethyl-L-arginine to L-citrulline and subsequently to L-arginine. *Biochem Biophys Res Commun* 1990; **167**: 1037–1043.
20. Archer SL, Hampl V. N^{G}-monomethyl-L-arginine causes nitric oxide synthesis in isolated arterial rings: trouble in paradise. *Biochem Biophys Res Commun* 1992; **188**: 590–596.
21. Robertson CA, Green BG, Niedzwiecki L *et al.* Effect of nitric oxide synthase inhibitors on rat liver arginase. *Biochem Biophys Res Commun* 1993; **197**: 523–528.
22. Barakat M, El-Kadi AOS, Du Souich P. *In vivo, ex vivo* and *in vitro* effects of L-NAME and L-arginine on the metabolism of theophylline in the rabbit. *Drug Metab Dispos* 1997; **25**: 191–195.

23. Krejcy K, Schwarzacher S, Raberger G. Distribution and metabolism of N^G-nitro-L-arginine and N^G-nitro-L-arginine methylester in canine blood *in vitro*. *Naunyn-Schmiedebergs Arch Pharmacol* 1993; **347**: 342–345.
24. Brouillet E, Roeda D, Valette H *et al*. *In vivo* metabolites of N^ω-nitro-L-arginine methyl ester: methanol and Nw-nitro-L-arginine. *Eur J Pharmacol* 1995; **293**: 487–490.
25. Buxton ILO, Cheek DJ, Eckman D *et al*. N^G-nitro-L-arginine methyl ester and other alkyl esters of arginine are muscarinic receptor antagonists. *Circ Res* 1993; **72**: 387–395.
26. Cobb JP, Natanson C, Hoffman WD *et al*. N^ω-amino-arginine, an inhibitor of nitric oxide synthase, raises vascular resistance but increases mortality rates in awake canines challenged with endotoxin. *J Exp Med* 1992; **176**: 1175–1182.
27. Connor J, Manning P, Settle S *et al*. Suppression of adjuvant-induced arthritis by selective inhibition of inducible nitric oxide synthase. *Eur J Pharmacol* 1995; **273**: 15–24.
28. Narayanan K, Griffith OW. Synthesis of L-thiocitrulline, L-homothiocitrulline, and *S*-methyl-L-thiocitrulline: a new class of potent nitric oxide synthase inhibitors. *J Med Chem* 1994; **37**: 885–887.
29. Frey C, Narayanan K, McMillan K *et al*. L-Thiocitrulline. A stereospecific, heme-binding inhibitor of nitric-oxide synthases. *J Biol Chem* 1994; **269**: 26083–26091.
30. Furfine E, Harmon M, Paith J *et al*. Potent and selective inhibition of human nitric oxide synthases. *J Biol Chem* 1994; **269**: 26677–26683.
31. Yokoi I, Kabuto H, Habu H *et al*. Structure-activity relationships of arginine analogues on nitric oxide synthase activity in the rat brain. *Neuropharmacology* 1994; **33**: 1261–1265.
32. Cowart M, Kowaluk EA, Kohlhaas EL *et al*. Synthesis of phosphorus-containing amino acid analogs as inhibitors of nitric oxide synthase. *Biorg Med Chem Lett* 1996; **6**: 999–1002.
33. Hodson HF, Palmer RMJ, Sawyer DA *et al*. *PCT Patent Appl* WO 95/34534 (1995).
34. Faint RW, Mackie IJ, Machin SJ. Platelet aggregation is inhibited by a nitric oxide-like factor released from human neutrophils *in vivo*. *Br J Haematol* 1991; **77**: 539–545.
35. Gillespie JS, Liu XR, Martin W. The effects of L-arginine and N^G-monomethyl L-arginine on the response of the rat anoccygeus muscle to NANC nerve stimulation. *Br J Pharmacol* 1989; **98**: 1080–1082.
36. Umans JG, Samsel RW. L-Canavanine selectively augments contraction in aortas from endotoxaemic rats. *Eur J Pharmacol* 1992; **210**: 343–346.
37. Liaudet L, Feihl F, Rosselet A *et al*. Beneficial effects of L-canavanine, an selective inhibitor of inducible nitric oxide synthase, during rodent endotoxaemia. *Clin Sci* 1996; **90**: 369–377.

38. Fatehi-Hassanabad Z, Burns H, Aughey EA *et al.* Effects of L-canavanine, an inhibitor of inducible nitric oxide synthase, on endotoxin mediated shock in rats. *Shock* 1996; **6**: 194–200.
39. Szabó, C, Southan GJ, Thiemermann C. Beneficial effects and improved survival in rodent models of septic shock with *S*-methyl-isothiourea sulfate, a novel, potent and selective inhibitor of inducible nitric oxide synthase. *Proc Natl Acad Sci USA* 1994; **91**: 12472–12476.
40. Garvey EP, Oplinger JA, Tanoury GJ *et al.* Potent and selective inhibition of human nitric oxide synthases. *J Biol Chem* 1994; **269**: 26669–26676.
41. Southan GJ, Szabó C, Thiemermann C. Isothioureas: potent inhibitors of nitric oxide synthases with variable isoform selectivity. *Br J Pharmacol* 1995; **114**: 510–516.
42. Nakane M, Klinghofer V, Kuk JE *et al.* Novel potent and selective inhibitors of inducible nitric oxide synthase. *Mol Pharmacol* 1995; **47**: 831–834.
43. Orplinger JA, Shearer BG, Bigham E *et al. PCT Patent Appl* WO 95/09619 (1995).
44. Moore WM, Webber RK, Fok KF *et al.* Inhibitors of human nitric oxide synthase isoforms with the carbamidine moiety as a common structural element. *Bioorg Med Chem* 1996; **4**: 1559–1564 .
45. Shearer BG, Lee S, Oplinger JA *et al.* Substituted N-phenylisothioureas: potent inhibitors of human nitric oxide synthase with neuronal isoform selectivity. *J Med Chem* 1997; **40**: 1901–1905.
46. Stratman NC, Fici GJ, Sethy VH. U-19451A: a selective inducible nitric oxide synthase inhibitor. *Life Sci* 1996; **59**: 945–951.
47. Martinez MC, Muller B, Stoclet JC, Andriantsitohaina R. Alteration by lipopolysaccharide of the relationship between intracellular calcium levels and contraction in rat mesenteric artery. *Br J Pharmacol* 1996; **118**: 1218–1222.
48. Arkovitz M, Garcia VF, Wispe JR, Szabó C. Selective inhibition of the inducible isoform of nitric oxide synthase inhibits endotoxin-induced pulmonary leak. *J Pediatr Surg* 1996; **31**: 1009–1015.
49. Aranow JS, Zhuang J, Wang H *et al.* A selective inhibitor of inducible in nitric oxide synthase prolongs survival in a rat model of bacterial peritonitis: comparison with two nonselective strategies. *Shock* 1996; **5**: 116–121.
50. Wildhirt SM, Suzuki H, Wolf P-W *et al. S*-methylisothiourea inhibits inducible nitric oxide synthase and improves left ventricular performance after acute myocardial infarction. *Biochem Biophys Res Commun* 1996; **227**: 328–333.
51. Zhang J, McCarthy TJ, Moore WM *et al.* Synthesis and evaluation of two positron-labeled nitric oxide synthase inhibitors, *S*-[^{11}C]methylisothiourea and *S*-(2-[^{18}F]fluoroethyl)isothiourea, as potential positron emission tomography tracers. *J Med Chem* 1996; **39**: 5110–5118.
52. Fastier FN. Structure activity relationships of amidine derivatives. *Pharmacol Rev* 1962; **14**: 37–90.

53. Vromen A, Szabó C, Southan G, Salzman AL. Effects of *S*-isopropyl isothiourea, a potent inhibitor of nitric oxide synthase, in severe hemorrhagic shock. *J Appl Physiol* 1996; **81**: 707–715.
54. Zingarelli B, Squadrito F, Altavilla D *et al.* Evidence for a role of nitric oxide in hypovolemic hemorrhagic shock. *J Cardiovasc Pharmacol* 1992; **19**: 982–986.
55. Booke M, Schenarts PJ, Shin C *et al.* A new non-amino acid inhibitor of nitric oxide synthase (3936W920) in hyperdynamic ovine sepsis [Abstract]. *Shock* 1995; **3 (Suppl 1)**: S72.
56. Jang D, Szabó C, Murrell G. Isothioureas are specific and potent inhibitors of inducible nitric oxide synthase in cartilage. *Eur J Pharmacol* 1996; **312**: 341–347.
57. Hasan K, Heesen BJ, Corbett JA *et al.* Inhibition of nitric oxide formation by guanidines. *Eur J Pharmacol* 1993; **249**: 101–106.
58. Yokoi I, Kabuto H, Habu H, Mori A. α-Guanidinoglutaric acid, an endogenous convulsant, as a novel nitric oxide synthase inhibitor. *J Neurochem* 1994; **63**: 1565–1567.
59. Corbett A, Tilton RG, Chang K. Aminoguanidine, a novel inhibitor of nitric oxide formation, prevents diabetic vascular dysfunction. *Diabetes* 1992; **41**: 552–556.
60. Wu CC, Chen SJ, Szabó C *et al.* Aminoguanidine inhibits the delayed circulatory failure in endotoxic shock in the anaesthetized rat. *Br J Pharmacol* 1995; **114**: 1666–1672.
61. Wolff DJ, Lubeskie A. Aminoguanidine is an isoform-selective, mechanism-based inactivator of nitric oxide synthase. *Arch Biochem Biophys* 1995; **316**: 290–301.
62. Laszlo F, Evans SM, Whittle BJR. Aminoguanidine inhibits both constitutive and inducible nitric oxide synthase isoforms in rat intestinal microvasculature *in vivo*. *Eur J Pharmacol* 1994; **272**: 169–175.
63. Cross AH, Misko TP, Lin RF *et al.* Aminoguanidine, an inhibitor of inducible nitric oxide synthase, ameliorates experimental autoimmune encephalomyelitis in SJL mice. *J Clin Invest* 1994; **93**: 2684–2690.
64. Ruetten H, Southan GJ, Abate A, Thiemermann C. Attenuation of endotoxin-induced multiple organ dysfunction by 1-amino-2-hydroxyguanidine, a potent inhibitor of inducible nitric oxide synthase. *Br J Pharmacol* 1996; **118**: 261–270.
65. Tilton RG, Hang K, Hasan KS *et al.* Prevention of diabetic vascular dysfunction by guanidines. Inhibition of nitric oxide synthase versus advanced glycation end-product formation. *Diabetes* 1993; **42**: 221–232.
66. Joshi PC, Grogan JB, Thomae KR. Effect of aminoguanidine on *in vivo* expression of cytokines and inducible nitric oxide synthase in the lungs of endotoxemic rats. *Res Commun Mol Path Pharmacol* 1996; **91**: 339–346.

67. Szabó C, Ferrer-Sueta G, Zingarelli B *et al.* Mercaptoethylguanidine and related guanidine nitric oxide synthase inhibitors react with peroxynitrite and protect against peroxynitrite-induced oxidative damage. *J Biol Chem* 1997; **272**: 9030–9036.
68. Szabó C. The role of peroxynitrite in the pathophysiology of shock, inflammation and ischemia-reperfusion injury. *Shock* 1996; **6**: 79–88.
69. Buchmüller-Rouiller Y, Schneider P, Betz-Corradin S *et al.* 3-amino-1,2,4-triazole inhibits macrophage NO synthase. *Biochem Biophys Res Commun* 1992; **183**: 150–155.
70. David P, Mayan H, Tal DM, Karlish SJD. Guanidium derivatives act as high affinity antagonists of Na^+ ions in occlusion sites of Na^+,K^+-ATPase. *J Biol Chem* 1992; **267**: 1141–1149.
71. Southan GJ, Zingarelli B, O'Connor M *et al.* Spontaneous rearrangement of aminoalkylisothioureas into mercaptoalkylguanidines, a novel class of nitric oxide synthase inhibitors with selectivity towards the inducible isoform. *Br J Pharmacol* 1996; **117**: 619–632.
72. Southan GJ, Salzman AL, Szabó C. Potent inhibition of the inducible isoform of nitric oxide synthase by aminoethylselenourea and related compounds. *Life Sci* 1996; **58**: 1139–1148.
73. Kozak A, Arient M. Metabolism and radioprotective effect of AET and MEG in organism. *Strahlentherapie* 1973; **146**: 367–373.
74. Szabó C, Bryk R, Zingarelli B *et al.* Pharmacological characterization of guanidinoethyldisulphide (GED), a novel inhibitor of nitric oxide synthase with selectivity towards the inducible isoform. *Br J Pharmacol* 1996; **118**: 1659–1668.
75. Zingarelli B, Ischiropoulos H, Salzman AL, Szabó C. Mercaptoethylguanidine, and inhibitor of nitric oxide synthase, and a scavenger of peroxynitrite, exerts beneficial effects in hemorrhagic shock [Abstract]. *FASEB J* 1997; **11**: A68.
76. Brahn E, Banquerigo, ML, Firestein *et al.* Beneficial effects of mercaptoethylguanidine, an inhibitor of nitric oxide synthase and a scavenger of peroxynitrite, in collagen-induced arthritis: inhibition of synovitis and suppression of TNF-α, collagenase and stromeolysin [Abstract]. *FASEB J* 1997; **11**: A530.
77. Zingarelli B, Southan GJ, Gilad E *et al.* Mercaptoalkylguanidines are direct inhibitors of cyclooxygenase activity. *Br J Pharmacol* 1997; **120**: 357–366.
78. Southan GJ, Szabó C, O'Connor MP *et al.* Amidines are potent inhibitors of nitric oxide synthases: preferential inhibition of the inducible isoform. *Eur J Pharmacol* 1995; **291**: 311–318.
79. Moore WM, Webber RK, Fok KF *et al.* 2-Iminopiperidine and other 2-iminoazaheterocycles as potent inhibitors of human nitric oxide synthase isoforms. *J Med Chem* 1996; **39**: 669–672.

80. Southan GJ, Szabó C, Salzman AL. 2-Aminopyridines are competitive inhibitors of nitric oxide synthases with potent pressor actions. *Pharmacol Commun* 1996; **7**: 275–286.
81. Faraci WS, Nagal AA, Verdries KA *et al.* 2-amino-4-methylpyridine as a potent inhibitor of inducible NO synthase activity *in vitro* and *in vivo*. *Br J Pharmacol* 1996; **19**: 1101–1108.
82. Gentile RJ, Murray RJ, MacDonald JE, Shakespeare WC. *PCT Patent Appl* WO 95/05363 (1995).
83. Southan GJ, Gauld D, Lubeskie A *et al.* Inhibition of nitric oxide synthase with pyrazole-1- carboxamidine and related compounds. *Biochem Pharmacol* 1997; **54**: 409–417.
84. Hamley P, Tinker AC. 1,2-diaminobenzimidazoles: selective inhibitors of nitric oxide synthase derived from aminoguanidine. *Bioorg Med Chem Lett* 1995; **5**: 1573–1576.
85. Baer HP, Schmidt K, Mayer B, Kukovetz WR. Pentamidine does not interfere with nitrite formation in acivated RAW 264.7 macrophages but inhibits constitutive brain nitric oxide synthase. *Life Sci* 1995; **57**: 1973–1980.
86. Wolff DJ, Lubeskie A, Umansky C. The inhibition of the constitutive bovine endothelial nitric oxide synthase by imidazole and indazole agents. *Arch Biochem Biophys* 1994; **314**: 360–366.
87. Handy RL, Harb HL, Wallace P *et al.* Inhibition of nitric oxide synthase by 1-(2-trifluoromethylphenyl) imidazole (TRIM) *in vitro*: antinociceptive and cardiovascular effects. *Br J Pharmacol* 1996; **119**: 423–431.
88. Grant SK, Green BG, Wang R *et al.* Characterization of inducible nitric-oxide synthase by cytochrome P-450 substrates and inhibitors. Inhibition by chlorzoxazone. *J Biol Chem* 1997; **272**: 977–983.
89. Babbedge RC, Bland-Ward PA, Hart SL, Moore PK. Inhibition of rat cerebellar nitric oxide synthase by 7-nitroindazole and related substituted indazoles. *Br J Pharmacol* 1993; **110**: 225–228.
90. Yoshida T, Limmroth V, Irikura K, Moskowitz MA. The NOS inhibitor, 7-nitroindazole, decreases focal infarct volume but not the response to topical acetylcholine in pial vessels. *J Cereb Blood Flow Metab* 1994; **14**: 924–929.
91. Medhurst AD, Greenlees C, Parsons AA, Smith SJ. Nitric oxide synthase inhibitors 7- and 6- nitroindazole relax smooth muscle *in vitro*. *Eur J Pharmacol* 1994; **256**: R5–R6.
92. Wolff DJ, Naddelman RA, Lubeskie A, Saks DA. Inhibition of nitric oxide synthase isoforms by porphyrins. *Arch Biochem Biophys* 1996; **333**: 27–34.
93. Arimoto T, Yoshikawa T, Koromi Y, Kumagai Y. Inhibition of constitutive nitric oxide synthase by benexate. *Life Sci* 1996; **59**: 953–959.
94. Sup SJ, Green BG, Grant SK. 2-Iminobiotin is an inhibitor of nitric oxide synthases. *Biochem Biophys Res Commun* 1994; **204**: 962–968.

95. Hatchett RJ, Gryglewski RJ, Mlochowski J *et al.* Carboxyebselen: a potent and selective inhibitor of endothelial nitric oxide synthase. *J Physiol Pharmacol* 1994; **45**: 55–67.
96. Chiesi M, Schwaller R. Inhibition of constitutive endothelial NO-synthase activity by tannin and quercetin. *Biochem Pharmacol* 1995; **49**: 495–501.
97. Osawa Y, Davila J. Phencyclidine, a psychotomimetic agent and drug of abuse, is a suicide inhibitor of brain nitric oxide synthase. *Biochem Biophys Res Commun* 1993; **194**: 1435–1439.
98. Chen Q, LaBella FS. Inhibition of nitric oxide synthase by straight chain and cyclic alcohols. *Eur J Pharmacol* 1997; **321**: 355–360.
99. Mayer B, Brunner F, Schmidt K. Inhibition of nitric oxide synthesis by methylene blue. *Biochem Pharmacol* 1993; **45**: 367–374.
100. Schini VB, Vanhoutte PM. Inhibitors of calmodulin impair the constitutive but not the inducible nitric oxide synthase activity in the rat aorta. *J Pharmacol Exp Ther* 1992; **261**: 553–559.
101. Stuehr DJ, Fasehun OA, Kwon NS *et al.* Inhibition of macrophage and endothelial cell nitric oxide synthase by diphenyleneiodonium and its analogs. *FASEB J* 1991; **5**: 98–103.
102. Bigham EC, Reinhard JF, Moore PK *et al. PCT Patent Appl* WO 94/14780 (1994).
103. Werner ER, Pitters E, Schmidt K *et al.* Identification of the 4-amino analogue of tetrahydrobiopterin as a dihydropteridine reductase inhibitor and a potent antagonist of rat neuronal nitric oxide synthase. *Biochem J* 1996; **320**: 193–196.
104. Gross SS, Levi R. Tetrahydrobiopterin synthesis. An absolute requirement for cytokine-induced nitric oxide generation by vascular smooth muscle. *J Biol Chem* 1992; **267**: 25722–25729.
105. Schoedon G, Blau N, Schneemann M *et al.* Nitric oxide production depends on preceding tetrahydrobiopterin synthesis by endothelial cells: selective suppression of induced nitric oxide production by sepiapterin reductase inhibitors. *Biochem Biophys Res Commun* 1994; **199**: 504–510.
106. Bune AJ, Brand MP, Heales SJ *et al.* Inhibition of tetrahydrobiopterin synthesis reduces *in vivo* nitric oxide production in experimental endotoxic shock. *Biochem Biophys Res Commun* 1996; **220**: 13–19.
107. Klemm P, Ostrowski J, Morath T *et al.* N-acetylserotonin prevents the hypotension induced by bacterial lipopolysaccharide in the rat. *Eur J Pharmacol* 1993; **250**: R9–R10.
108. Klemm P, Hecker M, Stockhausen C *et al.* Inhibition by N-acetyl-5-hydroxytryptamine of nitric oxide synthase expression in cultured cells and in the anaesthetized rat. *Br J Pharmacol* 1995; **115**: 1175–1181.
109. Kapatos G, Hirayama K, Hasegawa H. Tetrahydrobiopterin turnover in cultured rat sympathetic neurons: developmental profile, pharmacologic sensitivity, and relationship to norepinephrine synthesis. *J Neurochem* 1992; **59**: 2048–2055.

110. Griffith OW, Park KH, Aisaka K *et al.* The role of plasma arginine in nitric oxide synthesis: studies with arginase-treated guinea pigs and rats. In: *The Biology of Nitric Oxide. I. Physiological and Clinical Aspects,* Moncada S, Marletta MA, Hibbs JB, Higgs EA (Eds). Portland Press, London 1992: 6–9.
111. Nussler AK, Billiar TR, Liu ZZ, Morris SM. Coinduction of nitric oxide synthase and arginosuccinate synthetase in a murine macrophage cell line. *J Biol Chem* 1994; **269**: 1257–1261.
112. Schott CA, Vetrovsky P, Stoclet JC. Cationic amino acids inhibit the effects of L-arginine in rat aorta exposed to lipopolysaccharide. *Eur J Pharmacol* 1993; **236**: 155–157.
113. Forray MI, Angelo S, Boyd CAR, Deves R. Transport of nitric oxide synthase inhibitors through cationic amino acid carriers in human erythrocytes. *Biochem Pharmacol* 1995; **50**: 1963–1968.
114. Nakajima T, Matsuoka Y, Kakimoto Y. Isolation and identification of N^G-monomethyl, N^G, N^G-dimethyl- and N^G,N'^G-dimethylarginine from the hydrolysate of proteins of bovine brain. *Biochim Biophys Acta* 1971; **230**: 212–222.
115. Morse RK, Vergenes JP, Malloy J, McMagnus IR. Sites of biological methylation of proteins in cultured chick mucle. *Biochemistry* 1975; **14**: 4316–4325.
116. McDermott JR. Studies on the catabolism of N^G-methylarginine, N^G,N^G-dimethylarginine and N^G,N'^G-dimethylarginine in the rabbit. *Biochem J* 1976; **154**: 179–184.
117. Ueno S, Sano A, Kotani K *et al.* Distribution of free methylarginines in rat tissues and in the bovine brain. *J Neurochem* 1992; **59**: 2012–2016.
118. Ogawa T, Kimoto M, Sasoaka K Purification and properties of a new enzyme, N^G,N^G- dimethylarginine dimethylaminohydrolase, from rat kidney. *J Biol Chem* 1989; **264**: 10205–10209.
119. Kimoto M, Tsuji H, Ogawa T, Sasoaka K. Detection of N^G,N^G-dimethylarginine dimethylaminohydrolase in the nitric oxide-generating system of rats using a monoclonal antibody. *Arch Biochem Biophys* 1993; **300**: 657–662.
120. Kimoto M, Whitley GS, Tsujii H, Ogawa TJ. Detection of N^G,N^G-dimethylarginine dimethylaminohydrolase in human tissues using a monoclonal antibody. *Biochemistry* 1995; **117**: 237–238.
121. Carnegie PR, Fellows FC, Symington GR. Urinary excretion of methylarginine in human disease. *Metabolism* 1977; **26**: 531–537.
122. Vallance P, Leone A, Calver A *et al.* Accumulation of an endogenous inhibitor of nitric oxide synthesis in chronic renal failure. *Lancet* 1992; **339**: 572–575.
123. MacAllister RJ, Whitley GS, Vallance P. Effects of guanidino and uremic compounds on nitric oxide pathways. *Kidney Int* 1994; **45**: 737–742.
124. Faraci FM, Brian JE Jr, Heistad DD. Response of cerebral vessels to an endogenous inhibitor of nitric oxide synthase. *Am J Physiol* 1995; **269**: 1522–1527.

125. Inoue R, Miyake M, Kanazawa A *et al.* Decrease of 3-methylhistidine and increase in N^G,N^G-dimethylarginine in the urine of patients with muscular dystrophy. *Metabolism* 1979; **28**: 801–804.
126. Aneman A, Backman V, Snyegg J *et al.* Accumulation of an endogenous inhibitor of nitric oxide synthase during graded hemorrhagic shock. *Circ Shock* 1994; **44**: 111–114.
127. Bode-Böger SM, Böger RH, Kienke S *et al.* Elevated L-arginine/dimethylarginine ratio contributes to enhanced systemic NO production by dietary L-arginine in hypercholesterolemic rabbits. *Biochem Biophys Res Commun* 1996; **219**: 598–603.
128. Azuma H, Sato J, Hamasaki H *et al.* Accumulation of endogenous inhibitors of nitric oxide synthesis and decreased content of L-arginine in regenerated endothelial cells. *Br J Pharmacol* 1995; **115**: 1001–1004.
129. Scherston H, Kirno K, Ekroth R *et al.* Impaired endothelium-mediated vasodilatation in the peripheral vasculature of patients with acute pulmonary allograft rejection. *J Heart Lung Transplant* 1996; **15**: 556–563.
130. Lopez-Jaramillo P, Navaez M, Calle A *et al.* Cyclic guanosine 3',5' monophosphate cencentrations in pre-eclampsia: effects of hydralazine. *Br J Obstet Gynaecol* 1996; **103**: 33–38.
131. Fastier FN, Smirk F. Circulatory properties of iso-thioureas, guanidines, iso-ureas and amidines. *J Pharmacol Exp Ther* 1947; **89**: 256–270.
132. Shesely EG, Maeda N, Kim HS *et al.* Elevated blood pressures in mice lacking endothelial nitric oxide synthase. *Proc Natl Acad Sci USA* 1996; **93**: 13176–13181.
133. Meng W, Ma J, Ayata C *et al.* ACh dilates pial arterioles in endothelial and neuronal NOS knockout mice by NO-dependent mechanisms. *Am J Physiol* 1996; **271**: H1145–H1150.
134. Huang PL, Fishman MC. Genetic analysis of nitric oxide synthase isoforms: targeted mutation in mice. *J Mol Med* 1996; **74**: 415–421.
135. Huang Z, Huang PL, Panahian N *et al.* Effects of cerebral ischemia in mice deficient in neuronal nitric oxide synthase. *Science* 1994; **265**: 1883–1885.
136. Huang Z, Huang PL, Ma J *et al.* Enlarged infarcts in endothelial nitric oxide synthase knockout mice are attenuated by nitro-L-arginine. *J Cereb Blood Flow Metab* 1996; **16**: 981–987.
137. Hara H, Huang PL, Panahian N *et al.* Reduced brain edema and infarction volume in mice lacking the neuronal isoform of nitric oxide synthase after transient MCA occlusion. *J Cereb Blood Flow Metab* 1996; **16**: 605–611.
138. MacMicking JD, Nathan C, Hom G *et al.* Altered responses to bacterial infection and endotoxic shock in mice lacking inducible nitric oxide synthase. *Cell* 1995; **81**: 641–650.
139. Wei XQ, Charles IG, Smith A *et al.* Altered immune responses in mice lacking inducible nitric oxide synthase. *Nature* 1995; **375**: 408–411.

140. Laubach VE, Shesely EG, Smithies O, Sherman PA. Mice lacking inducible nitric oxide synthase are not resistant to lipopolysaccharide-induced death. *Proc Natl Acad Sci USA* 1995; **92**: 10688–10692.
141. Griffith OW, Kilbourn RG. Nitric oxide synthase inhibitors: amino acids. *Methods Enzymol* 1996; **268**: 375–392.
142. Corbett JA, McDaniel ML. Selective inhibition of inducible nitric oxide synthase by aminoguanidine. *Methods Enzymol* 1996; **268**: 398–408.

Chapter 8

Measurement of Nitric Oxide and Nitric Oxide Synthase Activity

Evangelos D Michelakis, A Tuan Dinh-Xuan, Khier Djaballah, Evelyne Souil and Stephen L Archer

Introduction

NO is a free radical and, although not particularly unstable in low concentrations and in the absence of O_2, it is very unstable in biological models, primarily due to the presence of O_2, O_2 radicals and oxidizing agents such as oxyhaemoproteins. NO tends to exist as a gas and is poorly soluble in water. It has a very high partition coefficient of > 20 (molecules per unit volume gas/molecules per unit volume solution) and does not undergo any kind of hydration reaction. This property allows the small molecule to pass freely across cell boundaries.

The fate of NO in the presence of O_2 depends on the phase in which the reaction occurs. In the *gas phase*, NO reacts with O_2 to give the brown gas nitrogen dioxide ($\cdot NO_2$), which in turn can be dimerized to N_2O_4 or combine with NO to form N_2O_3 (Eqs. 1a,b).

$$2\cdot NO + O_2 \rightarrow 2\cdot NO_2 \rightarrow N_2O_4 \quad (1a)$$

$$\cdot NO + \cdot NO_2 \rightarrow N_2O_3 \quad (1b)$$

In the *liquid phase* — in the absence of haemoproteins, i.e. in plasma, ultrafiltrate or water — NO is oxidized to nitrite (NO_2^-) (Eqs. 2a,b; 3a,b).

$$2\cdot NO + O_2 \rightarrow N_2O_4 \quad (2a)$$

$$N_2O_4 + H_2O \rightarrow NO_2^- + NO_3^- + 2H^+ \quad (2b)$$

$$N_2O_4 + 2\cdot NO \rightarrow 2N_2O_3 \quad (3a)$$

$$N_2O_3 + H_2O \rightarrow 2NO_2^- + 2H^+ \quad (3b)$$

In *whole blood*, NO_2^- is rapidly converted to NO_3^- (Eq. 4).

$$2NO_2^- + Hb(Fe^{2+})O_2 \rightarrow 2NO_3^- + Hb(Fe^{3+}) \quad (4)$$

Nitrate is thus the dominant species of NO oxidation products (NO_x) *in vivo*. The conversion of NO into nitrite and nitrate by these reactions varies in each system, so both nitrite and nitrate concentrations should be measured. In biological systems also, NO reacts with the superoxide anion ($O_2^{\cdot -}$) to form peroxynitrite ($ONOO^-$).

Because of its very short half life and its susceptibility to post-synthetic oxidation, measuring NO is a challenge. In this chapter, we shall mainly discuss four methods used to measure NO or NO synthase (NOS): the chemiluminescence assay for NO, the L-citrulline assay to study NOS enzymatic activity, NOS immunohistochemistry, and the NO electrode. We shall emphasize the importance of using several of these methods, when one wants to study complex conditions (like hypoxia) that can affect the NO/NOS pathway at many levels.

Chemiluminescence

One of the fastest, most sensitive and specific assays to measure NO is the chemiluminescence reaction between NO and ozone (O_3). Using the same technique, one can measure NO in the gas phase (for example, in the human breath or in the gas above cultured cells, 'headspace') or NO oxidation products (NO_2^- and NO_3^-) in biological fluids (for example, in vascular rings or perfused organs). Chemiluminescence is quickly superceding the traditional and slower Griess reaction method for determining NO_2^- in biological fluids. In the latter method, after the conversion of NO_3^- to NO_2^- by the enzyme nitrate reductase, the concentration of NO_2^- is determined spectrophotometrically as a coloured azo dye product of the Griess reaction that absorbs visible light at 540 nm.

Chemiluminescence was first used to measure NO (which at that time was considered to be just an atmospheric pollutant) in 1970 by Fontijn[1] and was modified to its current form in 1980 by Zafiriou and McFarland.[2] In 1987, chemiluminescence was used to prove that the ubiquitous endogenous vasodilator EDRF (endothelium-derived relaxing factor) was NO[3,4] and this boosted interest in NO research dramatically. It also highlighted the importance of measuring NO in biological fluids.

The NO chemiluminescence assay is based on the gas-phase interaction between NO and O_3. Some of the NO_2 produced by the reaction of NO with O_3 is in the excited state[5] (Eq. 5).

$$NO + O_3 \rightarrow NO_2^* + O_2 \tag{5}$$

As the unstable electrons of NO_2^* return to their original ground state, they dissipate energy (hv) (Eq. 6).

$$NO_2^* \rightarrow NO_2 + hv \tag{6}$$

This energy can either be used in reactions between NO_2 and other gas molecules, or released as a photon.[5] Light emission is linearly related to the NO content of the sample[2,6] and increases significantly when the pressure in the reaction chamber is reduced below 0.3 atm.[7] This is probably because the concentration of other gas molecules that can quench the emitted energy (by reacting with NO_2) is reduced. Furthermore, the light emission is inversely related to temperature.[7] The reaction of NO with O_3 is fast (rate constant at room temperature is 10^{-7} l mol^{-1} s^{-1}),[5] and this allows for on-line monitoring of NO levels. In contrast to the reaction of O_3 with NO, which has a very low energy of activation (10.5 kJ), other nitrogen oxides such as nitrogen dioxide have much higher energies and react much more slowly.[8] Thus, although NO_2 is present, it will not affect the chemiluminescence reaction between the O_3 and NO. The same is true for other substances (pollutants) like CO_2, C_2H_4, CO, NH_3 or SO_2.[1] The high speed and the specificity make this reaction very attractive for use in an assay to measure NO.

The NO chemiluminescence reaction is always performed in the gas phase. If liquid specimens are measured, the NO is 'stripped' into the gas phase prior to entry into the analyzer. Furthermore, nitrogen oxides must first be reduced to NO before they can be measured. The fact that the assay is exclusively in gas phase further enhances its specificity; to be detected, a substance must both be volatile and react with O_3, emitting light. Following a brief description of the NO analyzer, we shall discuss two of the major applications of this assay, the measurement of nitrogen oxides in the liquid phase and the measurement of NO in the gas phase, as in exhaled NO.

The NO Analyser

Like all luminometers, the basic structure of the chemiluminescence NO analyzer (NOA) is simple: it consists of a reaction chamber, a light detector which collects the emitted light and transforms it to an electrical signal which is fed to a recorder (Figure 1). The reaction chamber, where the chemiluminescence reaction takes place, has two inputs: from one the samples to be analyzed are injected (sample inlet) and from the other the O_3 (which is produced by an internal O_3 generator using electrostatic discharge) is supplied. As O_3 concentration is kept significantly high, it is never the rate-limiting factor in the reaction with NO. The pressure in the chamber is constantly kept low (1–15 mmHg) with a vacuum pump. This serves three purposes: (a) the decrease in pressure results in higher light signal, by reducing the number of gas molecules that absorb some of the energy emitted by NO_2^* as discussed above; (b) the vacuum constantly removes O_2, stabilizing any NO present; and (c) the vacuum facilitates the flow of samples from the inlet port. Gorimar[7] showed that at a stable temperature, the light output of the $NO + O_3$ reaction varies directly as the sample flow rate and inversely as the pressure, and is independent of the chamber volume and O_3 concentration (assuming a stable supply of O_3 in large excess compared with the quantity of NO). The flow rate from the sample inlet into the reaction chamber is controlled by a needle valve; it is adjusted so that authentic NO gives rapidly rising peaks with no slurring of the signal. Flow is also controlled, to some extent, by a reducer frit in the inflow line (see Figure 1), which reduces flow to prevent excessively rapid transit throughout the chamber. This is a key factor and if the reducer is clogged by particulate matter from the air, the response time may become very slow. Conversely, without the reducer, the transit time may be too fast to detect NO, due to the powerful vacuum in the chamber. The effluent from the chamber (O_3, NO, NO_2) is externally vented, but is first passed through a charcoal filter (Figure 1), which eliminates most of the nitrogen dioxide species.

A photomultiplier is the light detector of choice because it is mostly sensitive to low levels of light at the red end of the spectrum and NO_2^* emits weak red and near-infrared light. Although the peak intensity of the emitted light is at

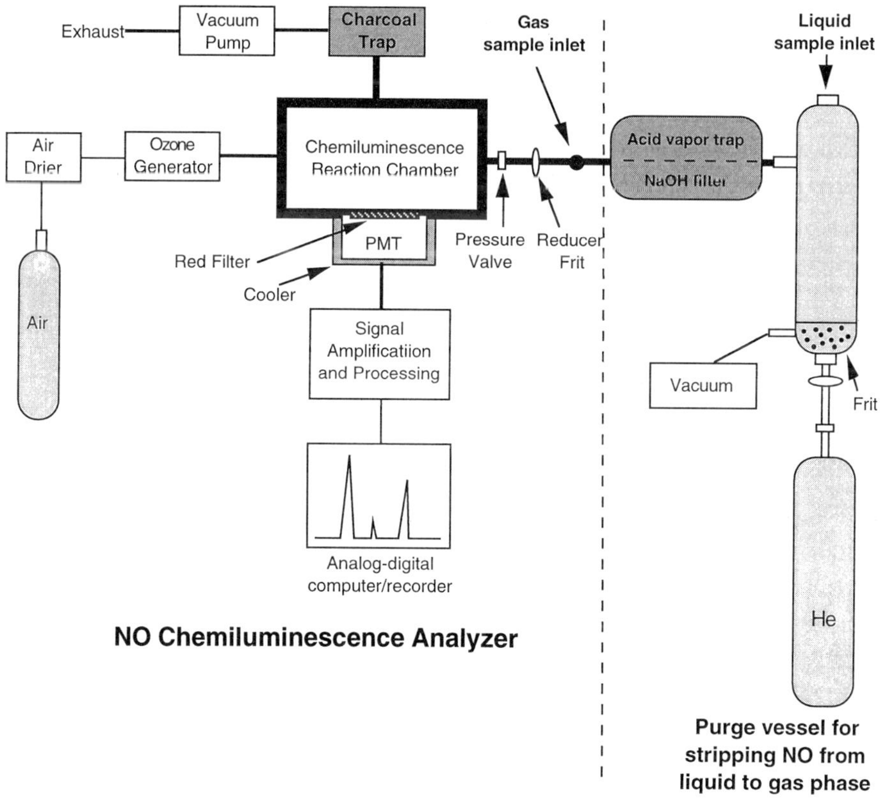

Figure 1. A schematic outline of the NO chemiluminescence analyzer and the purge vessel for stripping NO from liquid to gas phase.

~ 1100 nm (640–3000 nm range), the sensitivity of photomultipliers to detect wavelengths below 900 nm ensures very sensitive measurements throughout the entire range of emissions.[7]

Several factors can increase the background noise (dark current) of the photomultiplier. Exposure to large amounts of NO leads to enormous light emissions and increases dark current, which can take hours to stabilize. For this reason, when large amounts of NO are expected in the sample, it should be diluted. If a light leak occurs, exposing the photomultiplier to ambient light, the recovery period is often more than a day. This may be expected when the photomultiplier tube is changed or when the chamber is opened for service.[9] The photomultiplier might also detect light with wavelengths below 640 nm, like the emission from alkenes and sulphur-containing species or from the poorly understood interaction between O_3 and the reaction chamber wall. For this reason, a red cut-off filter is interposed between the reaction chamber and the photomultiplier.[10] An important determinant of dark current is temperature; for every 10° increase in the temperature, the dark current doubles.[9] Since most internal coolers keep the analyzer's temperature at a fixed level below the ambient temperature, the room temperature should be kept cool and relatively constant. Optimally, the reaction chamber temperature should be kept below 5°C.

Most analyzers give an analogue signal. This can be output to a recorder, an analogue/digital converter, allowing visualization of the signal contours. This is preferable to relying solely on the analyzer's light emitting diode display. A good analogue/digital converter program gives information about the signal which helps to identify artifacts and permits data analysis. The analogue/digital converter, software and computer are usually purchased from third party providers (for example, Mac Lab).

NO Measurement in Gas Samples

NO has recently been shown to be present in human breath.[11–13] The field of exhaled NO research is new and expanding. The following is the method we use in our laboratory to measure NO in human breath.

To calibrate the NOA we connect a source of NO-free air (nitrogen) to the gas inlet portion. Medical air used in many hospitals contains at least 3 ppb (parts per billion) of NO. After equilibration is achieved (5 minutes) the recorded value

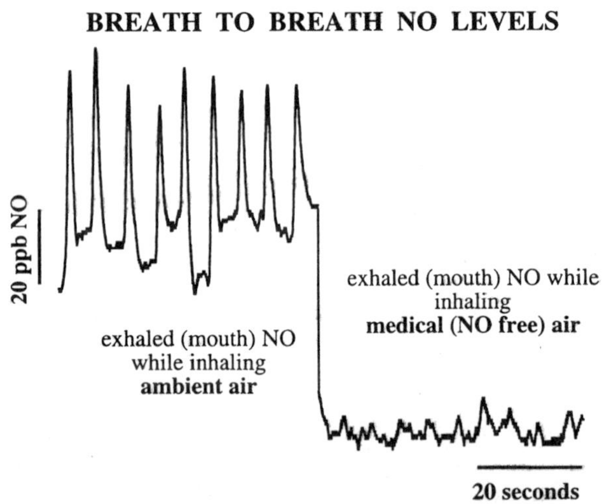

Figure 2. In the left part of this actual tracing the volunteer is breathing ambient air through a mask (non-rebreathing mask, #ABK0020, supplied by Sievers, Boulder, CO) which also occludes her nostrils. She then (right part) breaths medical air through a source connected to the mask. The significant drop in the signal reflects the very high levels of NO in the ambient air in Paris, France, where the experiment took place (note the scale difference between Figures 2 and 3). The rapid response time of this NO analyzer (a Sievers 280 model) permits breath-by-breath analysis of NO levels.

with nitrogen flow is accepted by the analyzer's software as "0". It is important not to use room or ambient air to calibrate the NOA. This is because in some areas the levels of NO in the ambient air can be significantly elevated (Figure 2); for example, in Paris, France, ambient NO levels are often > 100 ppb. A commercially available pressurized gas tank with a known amount of NO (100–1000 ppb, for example 600 ppb) is then connected to the gas inlet. After equilibration, the recorded value is accepted as "600ppb" and a calibration curve is constructed by the analyzer's software. Thereafter, values are reported in ppb instead of mV.

There are two methods we use for collecting exhaled air for NO measurement, one using sample bags to collect a large volume of gas and obtain the mean NO concentration and another to aspirate the NO directly from the airways to the NOA. The former gives mean NO levels but permits transfer of the specimen to the NOA. The NO levels are stable in nitrogen-flushed bags for one hour.[13] The direct method permits a breath-by-breath measure of NO.

Gas sampling bag acquisition

Exhaled breath air (or air from a respirator in intubated patients or air aspirated from a bronchoscopy tube) is collected into a CHEMWARE FEP 1.2 litre gas collecting bag (Curtin Matheson Scientific Inc., Houston, TX) using a TYGON 25 tubing. Ambient air is immediately sealed off using a septum (SEPTA thermogreen LB-1). An INTERMEDIC PE 90 tube (Clay Adams, Parsippang, NJ) is connected to the gas inlet of the analyzer. The valve that seals the bag is opened by rotation. The sample is suctioned into the NOA, using the vacuum created by the analyzer's pump. The signal is then recorded (as ppb NO), and this reflects the mean NO level of the sample.

Direct acquisition of exhaled NO by cannulation

This method allows the on-line 'breath to breath' recording of NO levels. A Tygon tube (8 cm long, external diameter 8 mm) is inserted approximately 1 cm inside the mouth, left or right nostril. The tube is connected to the NOA and the exhaled air is suctioned into the NOA. If properly performed, there should be no sense of suction on the mucosa. The signal is recorded during normal breathing or during breath holding (Figure 3).

Most authors agree that in normal volunteers the predominant source of NO in the breath is the nose[13] and the epithelium of the paranasal sinuses (see Chapter 20).[14] However, in inflammatory states like an exacerbation of asthma, NO is produced in large amounts and the portion of NO originating from the lower airways is increased,[15–17] and can be used to monitor the status of asthmatic attacks in such patients. Exhaled NO might also be used to monitor patients with

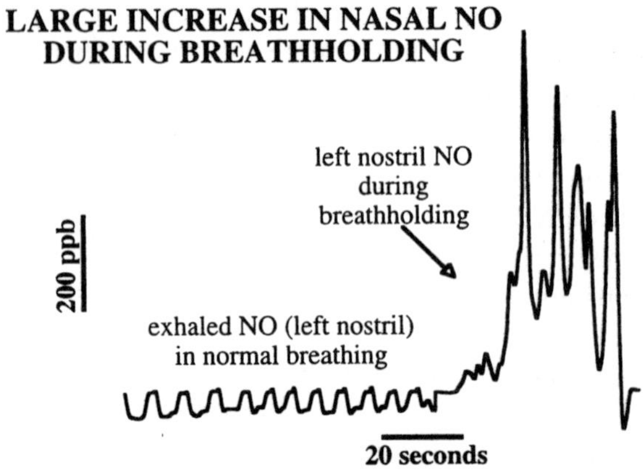

Figure 3. Actual tracing of 'on-line' breath-to-breath NO levels in the nose during normal respiration and apnea. A volunteer is breathing ambient air while at the same time occluding her right nostril externally. A tube is inserted into her left nostril and air is suctioned into the NOA. The breath to breath variations of NO levels in the exhaled nasal air are evident. During breathholding there is a dramatic increase in NO levels. This is due to accumulation of NO as well as decreased dilution with alveolar air, which contains much less NO (see Figure 2).

systemic diseases like cirrhosis, since patients with decompensated cirrhosis have significantly higher exhaled NO levels than healthy volunteers or patients with compensated cirrhosis.[18] Recently, exhaled NO was proposed to serve as a marker for the very common clinical problem of organic nitrate tolerance.[19]

There are several controversies in the measurement of exhaled NO: (1) Should the subject breathe medical air? (2) Should exhalation be against a forced resistance to close the palate? (3) Should nasal NO be excluded? (4) Should the sample be collected in bags or by direct acquisition using a mask that occludes the nose? (5) Should the measure be reported as ppb or corrected for ventilation (V_{NO})? Although 'the jury is out', it appears that NO in the breath must be measured using medical air, to achieve standardization of lower airway NO levels. Nasal air can be excluded by masks which occlude the nose ('non-rebreathing mask' #ABK0020, supplied by Sievers, Boulder, CO, USA) or by having the subject exhale against resistance (which closes the soft palate) or both. The choice of bag or direct measurement depends on what one wants to measure. The use of V_{NO} is another means of correcting for differences in ventilation and is important in measuring NO in exercise.

NO Measurement in Liquid Samples

It is unlikely that free NO, as such, exists in blood or protein-containing body fluids. When a liquid sample is injected in a purge vessel, NO will rapidly escape from the liquid into the headspace; 20 times more NO will finally be in the gas than in the liquid phase (partition coefficient > 20). NO_2^- and NO_3^- are non-volatile and remain in the liquid phase.[6,20] NO alone can be aspirated from the headspace into the NOA, and this accounts for the specificity of the assay in liquid samples. Although the escape of NO into the headspace is very fast (in equilibrium within 1 minute), this can be accelerated by bubbling the liquid sample with a constant flow of an inert gas like helium (see Figure 1). This is commonly referred to as stripping.[21] Helium or another inert gas flows into the purge vessel through a porous frit at a constant rate. The optimal flow rate is above 8–10 ml/min and this can be judged by interposing a flowmeter in the circuit. Many authors recommend that the valve between the purge vessel and the analyzer remains closed for the first 15–30 seconds of the stripping. Theoretically, opening of the valve releases NO as a bolus, improving the chemiluminescence signal. However, in our experience opening and closing the valve results in significant changes to pressure and flow, creating confusing artificial changes in the chemiluminescence signal.[22]

NO stripping of proteinaceous fluids can be challenging, since these fluids tend to form bubbles very easily. These bubbles can be aspirated into the analyzer, coat the 'window' separating the chamber from the photomultiplier tube and decrease the sensitivity of the assay. We use the anti-foaming agent FG-10 (Dow Corning, Midland MI) at 0.1 ml per 6 ml of acid + reducing agent solution. FG-10 alone or in combination with acids or reducing agents does not give a chemiluminescence signal. Although plasma and serum can be measured, the less protein the easier the conduct of the experiment. Whole blood cannot be reliably measured.

As discussed earlier, the amount of free NO in biological fluids is very small, if any, and stripping alone would not reflect at all the amount of NO present since most of it would be already oxidized to NO_2^- or NO_3^-. In an acidic environment and with strong reducing agents (such as KI, NaI or vanadium(III) chloride), these oxides can be reduced back to NO,[23–25] which would then escape to the headspace and be measured by the NO assay. In body fluids, under these reducing conditions, other bioactive NO adducts, like nitrosothiols and nitrosoamines, are reduced to NO as well.[26] Thus, measuring NO_2^- and NO_3^- may overestimate the amount of NO which was biologically available. Furthermore, NO_3^- is abundant in certain foods, beverages and in most water. Therefore, if one systematically studies plasma NO_x in humans, a diet with defined NO_3 content should be considered.

Nitrite (NO_2^-) Assay

In the presence of acid, NO_2^- is converted to the nitrosonium ion (NO^+) (Eq. 7):

$$NO_2^- + 2H^+ \rightarrow NO^+ + H_2O \qquad (7)$$

NO^+ reacts rapidly with nucleophiles like iodide (NaI or KI) or VCl_3, to form NO (e.g. Eqs. 8a,b):

$$NO^+ + I^- \rightarrow ONI \tag{8a}$$

$$2ONI \rightarrow 2NO + I_2 \tag{8b}$$

It is important to remember that KI + acetic acid will not reduce NO_3^-. Thus, the assay preferentially detects NO_2^- and is probably not ideal for use in blood samples as it ignores NO_3^-, the predominant NO_x in blood.

Calibration is an essential part of the assay and frequent calibration curves should be obtained to ensure that the sensitivity of the analyzer is stable. To calibrate the machine, one can prepare serial dilutions of a standard 10^{-5} M NO_2 solution ($NaNO_2$). This should be freshly prepared and protected from light. 3 ml of KI (1 M) and 3 ml of acetic acid (1 M) are added in the purge vessel and bubbled with helium for 5 min, followed by injection of the samples with known amount of NO_2 (1–100 μl at a time, in 5 μl increments). The recorded value in millivolts (for maximal peak height) or 'millivolts . seconds' (for area under the curve) is the amount of signal that a known amount of NO_2 produces. A calibration curve is then prepared: on the *X*-axis the amount of NO_2 given is plotted (number of moles of NO_2 given). On the *Y*-axis, the recorded value is plotted (mV or ppb observed).

If the analyzer is also used to measure NO in breath or ambient air (levels of 0–200 ppb), one can either measure the NO_2 using the same ppb scale or measure using mV. Whether in mV or ppb, one must construct a calibration curve relating the dose administered to the dose observed. Once the calibration curve is generated, one can calculate a regression equation and then use this formula to calculate the quantity of NO_2 in samples. In Figure 4, calibration curves are shown for NO_2^- and NO_3^-. The ability of the assay to measure tiny amounts of NO oxidation products is evident. As stated above, if one is using the NOA also to measure gas NO, the NOA will be calibrated to ppb. Therefore the value that will be read in the recorder after the sample is injected (in the NO_2^- assay) will be ppb NO. One can then enter this value as x in the regression equations shown in Figure 4. Solving for y will give the amount of NO in moles. As can also be seen in Figure 3, one can use the peak height value (Figures 4A and 4C) or the area under the curve (AUC) value (in ppb . seconds) (Figures 4B and 4D).

3 ml of KI (1 M) and 3 ml of acetic acid (1 M) are injected into the purge vessel under constant bubbling with helium. Although one can also use KI and HCl, we prefer acetic acid as it does not involve the problem of baseline stability due to the chlorine ion in HCl. HCl and KI separately do not cause chemiluminescence but together they cause a large and reproducible spike, which disappears within several minutes. The origin of this spike, which requires the presence of O_3, is unknown, but it may be a chlorine oxide or singlet O_2.[27] When the signal returns to its baseline, 10 μl of the sample are injected to the purge vessel, and the signal

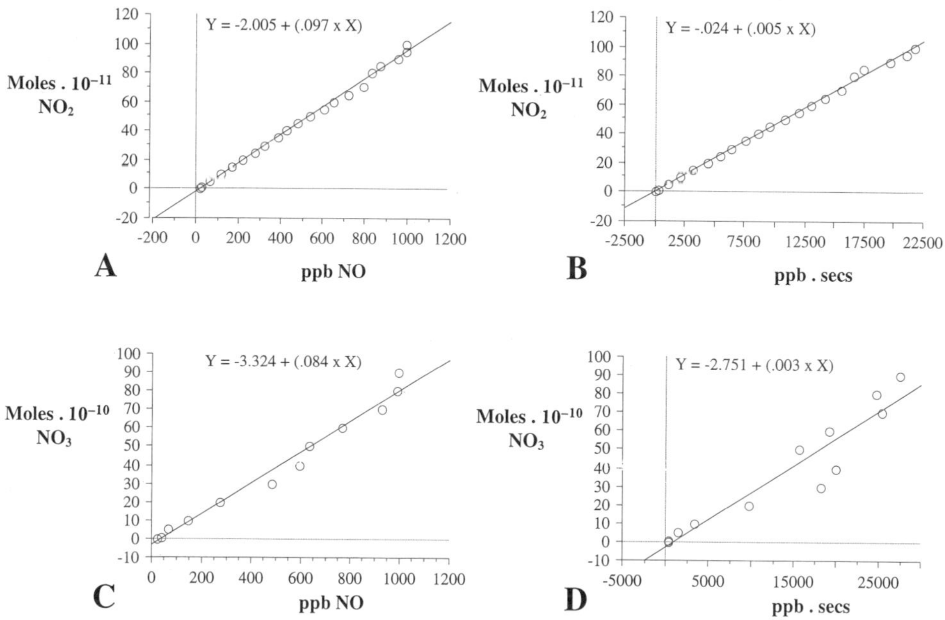

Figure 4. Calibration curves for NO_2^- and NO_3^- that we prepared using standard $NaNO_2$ and $NaNO_3$ solutions. In these experiments the NOA was calibrated for ppb of NO, since we are also using it to measure breath NO. After injecting a known amount of $NaNO_2$ and $NaNO_3$, the signal was read either as ppb NO (peak height of the signal) (A and C), or as ppb . secs (area under the curve) (B and D). This is the amount of NO gas produced from the reduction of the injected NO_2^- and NO_3^-. Using this value as *X*, one can solve the regression equations shown for *Y* and calculate the amount of moles of NO_2^- (A and B) and NO_3^- (C and D).

is recorded. After injection of approximately twenty 10 μl samples, the KI and HCl have to be replaced. This is because they tend to lose their reducing power, and this leads to incomplete and slower reduction of the NO_2^- and also the signals may become slurred. Thus it is ideal to keep specimens small. We also advocate the use of a single dose of NO_2 standard each day to demonstrate that the sensitivity of the assay is unchanged.

Since almost all sources of water are contaminated with NO_2, the background chemiluminescence signal from the control solution has to be subtracted from the sample's signal. To minimize this artifact, use of a relatively NO_2^--free water, such as Omnisolve (Curtin Matheson Scientific, Houston, TX), is advised.

Nitrate (NO_3^-) Assay

This assay is important since NO_3^- is the predominant species of NO_x *in vivo*. Stronger reducing conditions will convert NO_3^- to NO as well. Although in

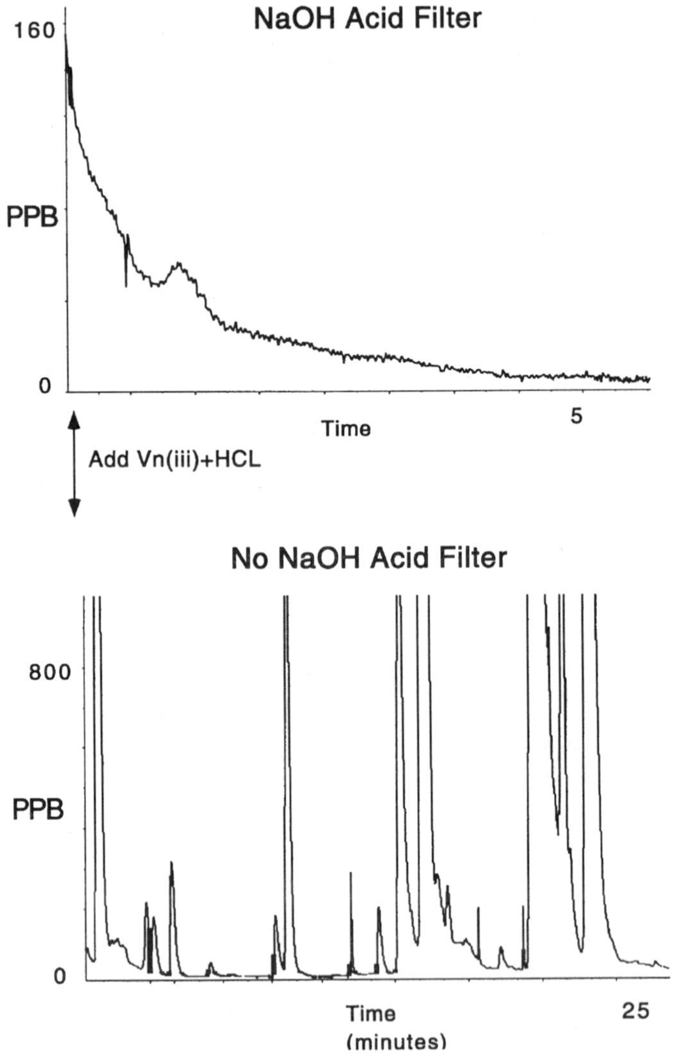

Figure 5. The NaOH filter placed between the purge vessel and the chamber of the NOA prevents the erratic and prolonged fluctuations in the baseline signal caused by the hot vanadium(III) and HCl vapours interacting with ozone.

room temperature, vanadium(III) reduces NO_2^- and only incompletely reduces NO_3^-, by increasing the temperature up to 80–95°C all the NO_3^- is reduced to NO as well.[24] That way both oxides can be measured in one specimen.

The calibration curve is prepared from NO_3 standards (using 10^{-4} M $NaNO_3$) in a manner similar to the NO_2^- assay. The purge vessel is then heated to 90°C and 3 ml of vanadium(III) (0.1 M) and 3 ml of HCl (1 M) are added. Vapours from the heated vanadium(III) are responsible for a large and erratic signal that is then recorded, that may take as much as 1–2 hours to stabilize. The importance of the NaOH filter (that is interposed between the purge vessel and the reaction chamber of the NOA (see Figure 1)) cannot be overemphasized. Figure 5 shows the erratic signal that one obtains when hot vanadium(III) and HCl are injected to the purge vessel. Large and spontaneous fluctuations of the baseline chemiluminescence signal are seen for the 25 minutes that this particular recording lasted. In contrast, when an NaOH filter is interposed, the signal is much smoother and returns to the baseline much more quickly. It can be a major confounding factor and a limitation of this assay. Injecting 3 ml each of vanadium(III)/HCl permits measurement of at least twenty 10 μl samples before being replaced and reduces the amount of time one spends waiting for the signal to return to its baseline.

When the signal returns to its baseline (usually a level of NO close to ambient NO levels), 10 μl of the sample are injected to the purge vessel and the signal value is recorded. This is the total $NO_2^- + NO_3^-$ content of the sample. To determine NO_3^- levels, an aliquot of the sample has to be measured using KI/acetic acid at room temperature (NO_2^- levels) and an aliquot using vanadium(III)/HCl at 90°C (total NO_x). Subtracting the first from the second value will give the NO_3^- levels.

Because time is required for the complete reduction of NO_2^- and NO_3^- to NO, the signals produced tend to be more slurred (broader spikes than when measuring exhaled NO). There are several tips in measuring NO_x that may be helpful:

1. The strong reducing environment can also create problems by reducing certain of the arginine-based NOS inhibitors to NO. Thus, one should not use the assay in samples containing L-N^G monomethyl L-arginine (L-NMMA) or nitro-L-arginine (L-NA), etc.
2. Measurement of too many samples before changing the acid/KI or vanadium(III) should be avoided, as the weakened reduction capacity can lead to incomplete reduction of the NO_x.
3. Always include a standard (a fixed volume of NO_2 or NO_3 standards) and compare to the calibration curve, ensuring that the sensitivity of the NOA has not changed.
4. If the NO peak remains slurred despite fresh acid solutions, consider the possibility that the reducer that controls inflow at the rear of the NOA (Figure 1) is clogged with use — clean it with alcohol and try again.
5. The discovery that animals synthesize NO was based, in part, on the finding that we have more NO in our bodies than we consume; nonetheless, there is substantial NO_3 in many foods (i.e. smoked meats and sausages). If one wishes to measure plasma NO_x in humans carefully, a diet with a defined NO_3 content should be considered.

6. The NO chemiluminescence assay is highly, but not entirely, specific for NO. The few other substances that can give a chemiluminescence signal (like sulphides, amines, H_2S) are either non-volatile, having significantly smaller partition coefficients than NO, or do not occur in biological systems. Nevertheless, most NO analyzers, even with a red filter, will detect ppm levels of hydrogen sulphide. The detection threshold for gas phase NO by the chemiluminescence assay is agreed by most authors to be 20–50 pmol[10,28] or 1 ppb.[13]

Physiological Correlates of NO and its Oxidation Products

Using the chemiluminescence technique, we have previously shown the importance of studying both NO and NO_x, when assessing the effects of conditions like hypoxia on the NOS pathway (Figure 5). Rats exposed to chronic hypoxia develop pulmonary hypertension. Whereas Adnot *et al.* showed that endothelium-dependent vasodilatation is significantly decreased in lungs from chronically hypoxic rats,[29] Isaacson *et al.* showed that NO levels (measured as NO_x in the whole perfused lungs from chronically hypoxic rats, using chemiluminescence) actually increase, when compared with the levels from normoxic lungs.[30] The latter finding presumably reflects an 'attempt' by NO to counterbalance the hypertension. Others have shown that the endothelium-dependent vasodilatation in chronically hypoxic lungs is maintained.[31]

Furthermore, we have recently shown that hypoxia affects both the partitioning of NO and the metabolism of NO_x.[20] Hypoxia favours the survival of NO and thus allows partitioning of NO at the expense of oxidation to NO_2^-. Thus, in a test tube with authentic NO added under normoxic or hypoxic conditions, there will be more NO in the headspace and less NO_2 in the liquid under hypoxic conditions. In rat cultured renal mesangial cells, hypoxia (32 mmHg) does not prevent the induction of iNOS by lipopolysaccharide (LPS); nor does it diminish the accumulation of cGMP. It does — as in the test tube — alter partitioning, however, favouring NO survival and exiting to the headspace and thus reducing accumulation of NO_x in the media.[20] Simply measuring NO in hypoxia might lead to the mistaken conclusion that there is more NO; measuring only NO_2 might lead to the opposite conclusion. Moreover, the chemiluminescence assay cannot accurately assess how a stimulus like hypoxia affects NOS enzymatic activity, since either L-arginine[32] or molecular oxygen[33] (both are substrates for the enzyme) could become rate-limiting factors in hypoxia.

The L-citrulline Assay

A simple and highly specific and sensitive assay to study NOS enzymatic activity has been developed by monitoring the conversion of [^{3}H]arginine to

[^{3}H]citrulline.[34,35] This reaction involves oxidation of a guanidino nitrogen of L-arginine to NO, and the stoichiometric production of L-citrulline, requiring 1.5 equivalents of NADPH, calcium, calmodulin, tetrahydrobiopterin and molecular oxygen. Thus, NOS activity can be studied by measuring the accumulation of the stable end-product, L-citrulline.

Radiolabelled arginine (1 μCi/ml 10 μM [^{3}H]arginine (Amersham) or [^{14}C]arginine is added to homogenized tissue extracts (along with tetrahydrobiopterin (3 μM), FAD (1 μM), FMN (1 μM) calmodulin (100 nM) and NADPH (1 mM), all of which are cofactors for NOS) and after incubation the reaction is terminated by adding EDTA. Cation exchange resin columns tightly bind arginine, which is cationic at pH 5.5, whereas citrulline, which at this pH is neutral, is not retained and flows through. NOS activity can be quantified by simply measuring the radioactivity in the effluent from the column, by elution on silica gel plates and quantification of radioactivity either by liquid scintillation counting or by using a phosphor imager screen. A detailed description of the assay, as well as modifications that allow the use of the assay in intact tissues, can be found elsewhere.[36]

There are several precautions to be kept in mind when one uses the L-citrulline assay. The most important is the possibility that not all L-citrulline comes from NOS. Citrulline can also be derived from arginine with the intermediate production of ornithine, as part of the urea cycle. The two enzymes (arginase and ornithine transcarbamylase) that are needed for this conversion, however, are restricted to tissues that have intact urea cycles such as the liver, brain and kidney.[36] In these tissues special consideration must be given to this, to ensure that the citrulline derived from the urea cycle is not counted as coming from NOS. Endothelial cells lack these enzymes. Interestingly, endothelial cells do have the two other enzymes of the urea cycle (arginosuccinate synthase and arginosuccinase) that allow the recycling of citrulline to arginine.[37] The citrulline assay is very specific and its sensitivity reaches the picomolar range.[36]

Like the chemiluminescence technique, the L-citrulline assay has been used to investigate the effects of hypoxia on NOS activity. Rengasamy *et al.* used the citrulline assay and showed that hypoxia reversibly inhibits the NOS enzymatic activity.[33] They later systematically studied and reported the enzyme's Km values for O_2, in all 3 isoforms of NOS, showing that hypoxia can decrease the activity of NOS by limiting the supply of a substrate, molecular O_2.[38] This does not necessarily disagree with Issacson's results[30] (mentioned above), since Rengasamy measured enzyme activity, whereas Isaacson measured NO accumulation.

A third technique, measurement of NOS protein levels by immunohistochemistry (see below), may explain how these findings can be rationally interpreted. Le Cras *et al.* recently reported that the amount of expressed eNOS was increased in the pulmonary arteries of chronically hypoxic rats (3 weeks hypoxia), and that there was *de novo* expression of eNOS protein in the pulmonary microvascular endothelium.[39] Later, the same group reported that this increase in eNOS expression

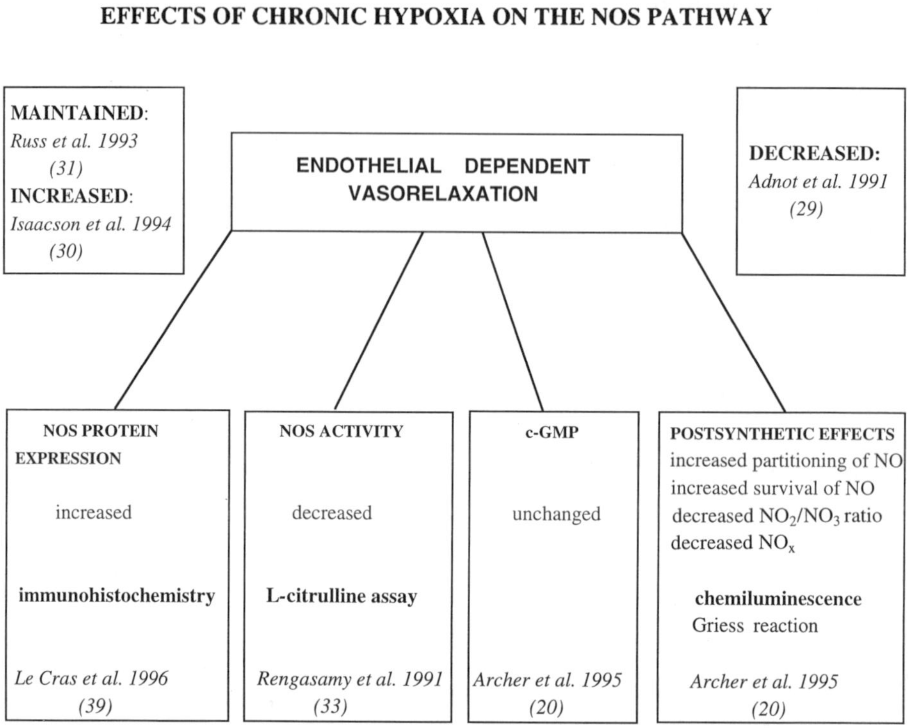

Figure 6. The use of multiple assays to measure NO offers a great advantage in the study of conditions like hypoxia, that affect the NOS pathway at many levels. Chronic hypoxia increases the NO and NO_x levels (measured by chemiluminescence) but at the same time the activity of NOS is decreased (measured by the L-citrulline assay). This apparent paradox appears to be explained at least in part by the finding that the total amount of the expressed NOS protein is increased in chronic hypoxia (shown by immunohistochemistry). Use of only one technique can lead to misleading or confusing results.

was increased over 1–7 days of hypoxia, preceding the vascular remodelling that occurs in the progression of chronically hypoxic pulmonary hypertension. They suggested that this upregulation of the eNOS is the result of hypoxia rather than of the associated haemodynamic changes.[40] Although the activity of NOS might be decreased in hypoxia, the amount of the enzyme is increased. In the above studies, the investigators used a monoclonal antibody to eNOS in their immunohistochemistry experiments.

The work by various groups cited above emphasizes the importance of using a multifaceted approach in the study of conditions like hypoxia, that affect the NOS pathway at multiple levels. Chemiluminescence (NO release), citrulline assay (NOS activity) and immunocytochemistry (visualization of NOS protein in tissue sections) were all important in studying the complex effects of chronic hypoxia in the pulmonary vascular reactivity (see Figure 6).

Immunohistochemistry

Immunohistochemistry allows the detection and localization of relevant antigens at a tissue, cell or subcellular level.[41] It is not a quantitative assay but it gives a semiquantitative idea of the content of the targeted protein. A primary antibody is first used to bind the protein of interest (for example, eNOS). Then a secondary antibody is used to bind to the primary antibody, based on species difference, i.e. if the primary antibody was raised in sheep, the secondary antibody would be an anti-sheep antibody. This is followed by a final staining stage (using peroxidase or fluorescence) to colourize the antigen-antibody complex. Although originally described in endothelial cells, eNOS has also been described in hippocampal neurones[42] and cardiac myocytes of the rat[43] as well as epithelial cells in a lung cancer cell line.[44]

To measure semiquantitatively and to localize eNOS in the lung, heart, brain and kidney of normal rats, we use an anti-eNOS anti-mouse antibody (Transduction Laboratories, Interchim, France) and organs flash-frozen in liquid difluoromethane (−40°C) and stored at −20°C. Sections (10 μm) are thaw-mounted on glass slides (Superfrost, CML, France) and immediately immersed in isotonic paraformaldehyde (4%) in a phosphate buffer (PBS, 1 hour at 10°C). Slides are dehydrated by serial immersions in progressively more concentrated alcohol solutions, air-dried and stored at −80°C. Slides are warmed up at room temperature and air-dried for 10 minutes prior to incubation in a physiologic glycine buffer (pH 7.4, glycine 100 mM) for 1 hour. Then the slides are kept in a PBS buffer containing Tween 20 (0.05%) and bovine serum albumin (0.2%) to facilitate permeabilization of the tissue, facilitating antibody penetration. Following three 10 minute rinses in PBS, sections are dried and incubated overnight with the primary antibody diluted in 0.1 M sodium buffer also containing Tween 20 and albumin (0.01%). Following three 10 minute rinses in PBS, sections are briefly dried and incubated with secondary antibody (eNOS-biotinylated anti-mouse antibodies, Elite Biosys, France). To visualize the antibodies, the slides are rinsed 3 times (10 minutes each) in PBS, briefly dried and then incubated in peroxidase substrate solution for 10 minutes (Sigma Fast DAB (3,3'-diaminobenzidine), Sigma, France). As can be seen in Figure 7, the fine layer of endothelium is clearly stained in reddish brown. The precise 'recipe' used for the process may vary somewhat from laboratory to laboratory. Using similar methodology, tissues can be stained histochemically with antibodies to iNOS or nitrotyrosine (the latter being a marker of peroxynitrite-mediated tissue damage); an example of such staining has recently been illustrated in a rat model of acute pancreatitis.[45]

As stated earlier, exhaled NO measured by chemiluminescence is increased in patients with decompensated cirrhosis. This may be associated with the vasodilatory state that is common in such patients. Morales *et al.* recently showed, using NOS immunohistochemistry, that the expression of both iNOS and eNOS is increased in arteries of cirrhotic rats.[46] Therefore, the use of

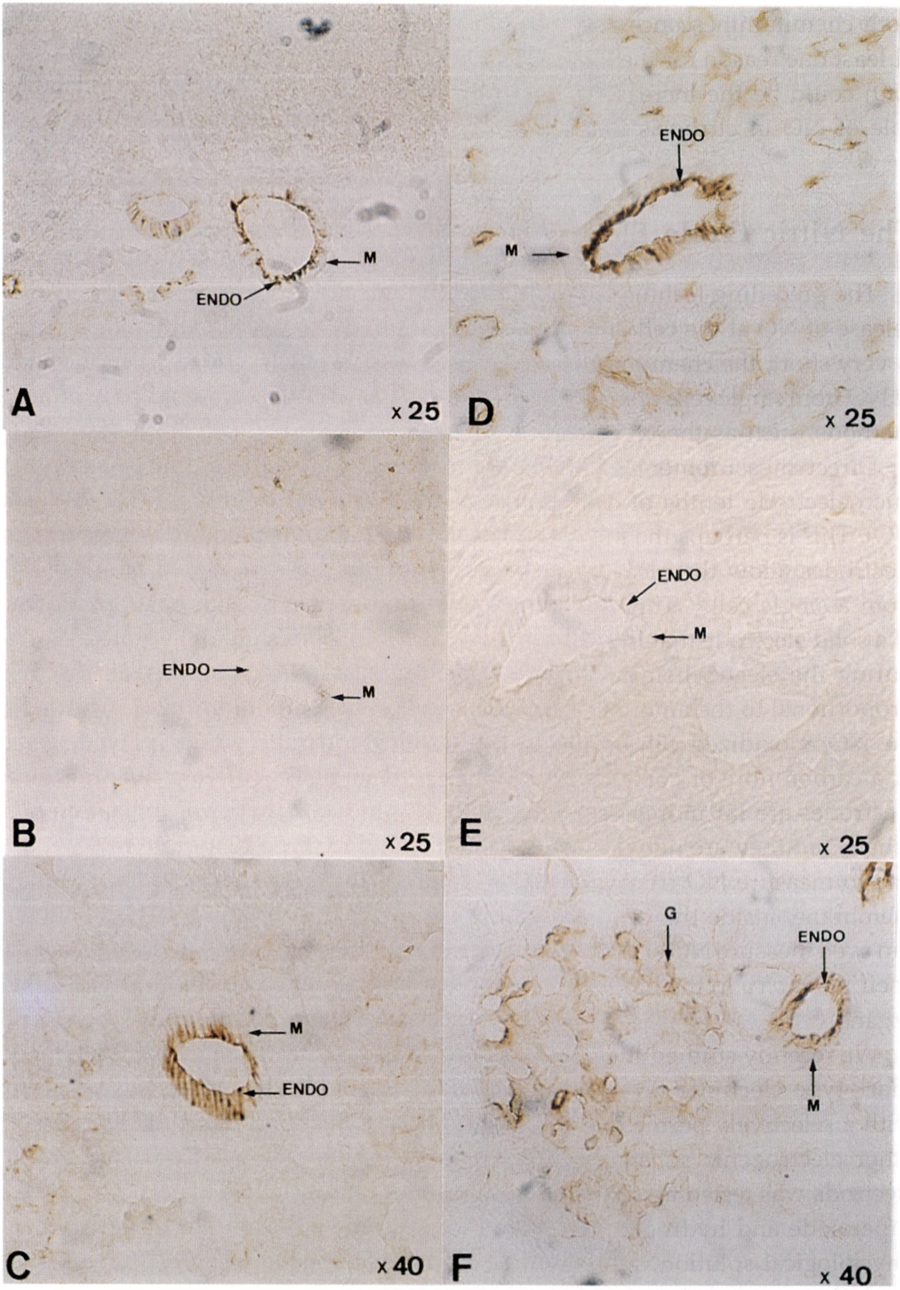

Figure 7. eNOS immunostaining in the endothelium of arteries from the brain (A), myocardium (D), lung (C) and the kidney (F) of the rat. The endothelium is stained reddish brown. There is no staining of the media, suggesting abscence of eNOS in the vascular smooth muscle cells. (B) and (E) serve as controls to (A) and (D) since only the secondary antibody was given. The absence of staining suggests that the assay is specific for eNOS (no cross-reaction of the secondary antibody with other proteins).

both chemiluminescence and immunohistochemistry in cirrhosis suggested that at least one reason for the elevated systemic levels of NO (as reflected by exhaled NO) could be the increased amount of NOS expressed in the arterial bed. The role of NO in cirrhosis is discussed further in Chapter 14.

The Nitric Oxide Electrode

The preceding techniques are obviously limited when one wants to study the release of NO at the cellular level. Because the half life of NO in biological fluids is very short, the chemiluminescence assay is poorly suited to measuring release of NO from single cells or in intact blood vessels. Measuring NO with electrochemical methods, such as the NO electrode, is a very promising technique in this area.

Direct measurement of NO from single cells is possible *in situ* (with the microelectrode tenths of a micrometre distance from the cellular source of the NO). This is perhaps the most important capability of NO microelectrodes. These electrodes allow the measurement of the release of attomoles (10^{-18} mol) of NO from a single cell.[47] This method measures the current produced at the surface of a solid electrode that is being held at a constant voltage (amperometric method) during the electrochemical oxidation of NO. This oxidation current is linearly proportional to the amount of NO present. The surface of the electrode on which the NO is oxidized can be platinum (like in the Clark probe for O_2 detection), or a carbon fibre or coated with a polymeric metalloporphyrin. The porphyrinic electrodes are far more sensitive and have much faster response times than the others, but they are not yet commercially available. Porphyrinic electrodes allow one to measure NO from a cell in 'real time' (with the electrode attached on its membrane, inside the cell or at a fixed distance — 0.2 to 0.5 micrometres). One can even measure NO elaboration from cells cultured on the surface of the electrode itself, or *in vivo* in the bloodstream or in tissue (like in vascular rings or perfused organs).[48,49]

We recently studied the specificity and sensitivity of a commercially available, Clark-type electrode (World Precision Instruments).[50] The electrode is covered with a selectively permeable membrane preferentially to detect NO and exclude other electrogenic substances like hydrogen peroxide or catecholamines. The electrode was tested against a wide variety of redox active agents (like potassium superoxide and hydrogen peroxide) and commonly used chemicals (like Krebs' physiological solution, saline, ethanol and dimethylsulphoxide). It was specific for NO, provided that the membrane coating the platinum tip was intact. We noted that any defects in the coating membrane resulted in detection of other redox agents, like hydrogen peroxide. The electrode had a detection threshold for NO of less than 25 pmol/ml and was linear in its response over the range from 25 pmol/ml to 4 nmol/ml (*in vitro*). However, its sensitivity may be inadequate to detect physiological levels of NO in normal conditions. Using a whole perfused rat lung, we injected boluses of saturated NO solution and caused a dose-dependent decrease in pulmonary artery pressure (pulmonary artery vasoconstriction was achieved with hypoxic gas ventilation of the lung). A bolus of 0.1 μl saturated NO solution (0.2 nmol) caused vasodilatation but was not detected by the NO electrode.

The size of the Clark electrodes does not allow for the delicate single cell measurements that can be performed with the porphyrinic electrodes. Because of its specificity and its reasonable sensitivity, however, it may prove to be a useful tool to monitor the dose of administered NO in isolated organs or during patch-clamp studies.

Conclusions

- To study the role of NO in complex conditions — like hypoxia — that affect the NOS pathway at many levels, a multifaceted approach is required.
- If one wishes to know whether NOS protein is present in the tissue of interest, or if the levels of the enzyme change as a direct response to the condition studied or as part of a counteregulatory mechanism, *immunohistochemical detection* of NOS offers a great advantage.
- One often needs to know how NOS function is affected, during the conditions studied, as a result of direct or indirect changes in the substrates or the cofactors of NOS. The *citrulline assay* has become the standard assay for NOS activity in crude and purified enzyme preparations.
- To measure NO and NO oxidation products directly and to determine how their metabolism is regulated at the post-synthetic level during *in vivo* (NO in exhaled breath) or in physiological experiments (perfused organs, arterial rings), *chemiluminescence* is the assay of choice.
- As research moves more and more to the cellular and subcellular level, the *porphyrinic NO electrodes* will be a very important addition to our armamentarium of NO assays.

References

1. Fontijn A, Sabadell AJ, Ronco RJ. Homogenous chemiluminescent measurement of nitric oxide with ozone. Implications for continuous selective monitoring of gaseous air polutants. *Anal Chem* 1970; **42**: 575–579.
2. Zafiriou OC, McFarland M. Determination of trace levels of nitric oxide in aqueous solution. *Analyt Chem* 1980; **52**: 1662–1667.
3. Ignarro LJ, Buga GM, Wood KS *et al*. Endothelium-derived relaxing factor produced and released from artery and vein is nitric oxide. *Proc Natl Acad Sci USA* 1987; **84**: 9265–9269.

4. Palmer RMJ, Ferrige AG, Moncada S. Nitric oxide release accounts for the biological activity of endothelium-derived relaxing factor. *Nature* 1987; **327**: 524–526.
5. Clyne MAA, Thrush BA, Wayne RP. Kinetics of the chemiluminescent reaction between nitric oxide and ozone. *Trans Faraday Soc* 1964; **60**: 359–370.
6. Archer SL, Shultz PJ, Warren JB *et al.* Preparation of standards and measurement of nitric oxide, nitroxyl, and related oxidation products. *Methods: A Companion to Methods in Enzymology* 1995; **7**: 21–34.
7. Gorimar TS. Total nitrogen determination by chemiluminescence. In: *Bioluminescence and Chemiluminescence: Instruments and Applications*, Van Dyke K (Ed). CRC Press, Boca Raton, FL 1985: 78–93.
8. Johnston HS, Crosby HJ. Kinetics of the fast gas phase reaction between ozone and nitric oxide. *J Chem Phys* 1954; **22**: 689–692.
9. Turner GK. Measurement of light from chemical and biochemical reactions. In: *Bioluminescence and Chemiluminescence: Instruments and Applications*, Van Dyke K (Ed). CRC Press, Boca Raton, FL 1985: 43–77.
10. Archer S. Measurement of nitric oxide in biological models. *FASEB J* 1993; **7**: 349–360.
11. Barnes PJ, Belvisi MG. Nitric oxide and lung disease. *Thorax* 1993; **48**: 1034–1043.
12. Borland C, Cox Y, Higenbottam T. Measurement of exhaled nitric oxide in man. *Thorax* 1993; **48**: 1160–1162.
13. Dillon WC, Hampl V, Shultz PJ *et al.* Origins of breath nitric oxide in humans. *Chest* 1996; **110**: 930–938.
14. Lundberg JON, Farkas-Szallasi T, Weitzberg E *et al.* High nitric oxide production in human paranasal sinuses. *Nat Med* 1995; **1**: 370–373.
15. Kharitonov SA, Yates D, Robbins RA *et al.* Increased nitric oxide in exhaled air of asthmatic patients. *Lancet* 1994; **343**: 133–135.
16. Alving K, Weitzberg E, Lundberg JM. Increased amount of nitric oxide in exhaled air of asthmatics. *Eur Respir J* 1993; **6**: 1368–1370.
17. Persson MG, Zetterström O, Agrenius V *et al.* Single-breath nitric oxide measurements in asthmatic patients and smokers. *Lancet* 1994; **343**: 146–147.
18. Matsumoto A, Ogura K, Hirata Y *et al.* Increased nitric oxide in the exhaled air of patients with decompensated liver cirrhosis. *Ann Intern Med* 1995; **123**: 110–113.
19. Husain M, Adrie C, Ichinose F *et al.* Exhaled nitric oxide as a marker for organic nitrate tolerance. *Circulation* 1994; **89**: 2498–2502.
20. Archer SL, Freude KA, Shultz PJ. Effect of graded hypoxia on the induction and function of inducible nitric oxide synthase in rat mesangial cells. *Circ Res* 1995; **77**: 21–28.
21. Aoki T. Continuous flow determination of nitrite with membrane separation/chemiluminescence detection. *Biomed Chromatography* 1990; **4**: 128–130.

22. Hampl V, Walters CL, Archer SL. Determination of nitric oxide by the chemiluminescence reaction with ozone. In: *Methods in Nitric Oxide Research,* Feelisch M, Stamler JS (Eds). Wiley, Chichester 1996: 310–318.
23. Cox RD. Determination of nitrate and nitrite at the parts per billion level by chemiluminescence. *Anal Chem* 1980; **52**: 332–335.
24. Braman RS, Hendrix SA. Nanogram nitrite and nitrate determination in environmental and biological materials by vanadium (III) reduction with chemiluminescence detection. *Anal Chem* 1989; **61**: 2715–2718.
25. Termin A, Hoffman M, Bing RJ. A simplified method for the determination of nitric oxide in biological solutions. *Life Sci* 1992; **51**: 1621–1629.
26. Walters CL, Gillatt PN, Palmer RC, Smith PLR. A rapid method for the determination of nitrate and nitrite by chemiluminescence. *Food Add Contam* 1987; **4**: 133–140.
27. Wallace J, Springer G, Stedman D. Photochemical ozone and nitric oxide formation in air-nitrogen dioxide mixtures containing sulfur dioxide or chlorine. *Atmosph Environ* 1980; **14**: 1147–1157.
28. Brien J, McLaughlin B, Nakatsu K, Marks G. Quantitation of nitric oxide formation from nitrovasodilator drugs by chemiluminescence analysis of headspace gas. *J Pharmacol Methods* 1991; **25**: 19–27.
29. Adnot S, Raffestin B, Eddahibi S *et al.* Loss of endothelium-dependent relaxant activity in the pulmonary circulation of rats exposed to chronic hypoxia. *J Clin Invest* 1991; **87**: 155–162.
30. Isaacson TC, Hampl V, Weir EK *et al.* Increased endothelium-derived nitric oxide in hypertensive pulmonary circulation of chronically hypoxic rats. *J Appl Physiol* 1994; **76**: 933–940.
31. Russ RD, Walker BR. Maintained endothelium-dependent pulmonary vasodilation following chronic hypoxia in the rat. *J Appl Physiol* 1993; **74**: 339–344.
32. Liu S, Crawley DE, Barnes PJ, Evans TW. Endothelium-derived relaxing factor inhibits hypoxic pulmonary vasoconstriction in rats. *Am Rev Respir Dis* 1991; **143**: 32–37.
33. Rengasamy A, Johns RA. Characterization of endothelium-derived relaxing factor/nitric oxide synthase from bovine cerebellum and mechanism of modulation by high and low oxygen tensions. *J Pharmacol Exp Ther* 1991; **259**: 310–316.
34. Bredt DS, Snyder SH. Nitric oxide mediates glutamate-linked enhancement of cGMP levels in the cerebellum. *Proc Natl Acad Sci USA* 1989; **86**: 9030–9033.
35. Bredt DS, Snyder SH. Isolation of nitric oxide synthase, a calmodulin requiring enzyme. *Proc Natl Acad Sci USA* 1990; **87**: 682–685.
36. Bredt DS, Schmidt HW. The citrulline assay. In: *Methods in Nitric Oxide Research.* Feelisch M, Stamler JS (Eds). Wiley, Chichester 1996: 249–255.

37. Hecker M, Sessa WC, Harris HJ *et al.* The metabolism of L-arginine and its significance for the biosynthesis of endothelium derived relaxing factor: cultured endothelial cells recycle L-citrulline to L-arginine. *Proc Natl Acad Sci USA* 1990; **87**: 8612–8616.
38. Rengasamy A, Johns RA. Determination of K_m for oxygen of nitric oxide synthase isoforms. *J Pharmacol Exp Ther* 1996; **276**: 30–33.
39. Le Cras TD, Xue C, Rengasamy A, Johns RA. Chronic hypoxia upregulates endothelial and inducible NO synthase gene and protein expression in rat lung. *Am J Physiol* 1996; **270**: L164–L170.
40. Xue C, Johns RA. Upregulation of nitric oxide synthase correlates temporally with onset of pulmonary vascular remodeling in the hypoxic rat. *Hypertension* 1996; **28**: 743–753.
41. Sternberger L. *Immunocytochemistry*. Wiley, New York 1986.
42. Dinerman JL, Dawson TM, Schell MJ *et al.* Endothelial nitric oxide synthase localized in hippocampal pyramidal cells: implications for synaptic plasticity. *Proc Natl Acad Sci USA* 1994; **91**: 4214–4218.
43. Balligand JL, Kobzik L, Han X *et al.* Nitric oxide-dependent parasympathetic signaling is due to activation of constitutive endothelial (type III) nitric oxide synthase in cardiac myocytes. *J Biol Chem* 1995; **270**: 14582–14586.
44. Shaul PW, North AJ, Wu LC *et al.* Endothelial nitric oxide synthase is expressed in cultured human bronchiolar epithelium. *J Clin Invest* 1994; **94**: 2231–2236.
45. Al-Mufti RA, Williamson RCN, Mathie RT. Increased nitric oxide activity in a rat model of acute pancreatitis. *Gut* 1998; **43**: 564–570.
46. Morales-Ruiz M, Jiménez W, Pérez-Sala D *et al.* Increased nitric oxide synthase expression in arterial vessels of cirrhotic rats with ascites. *Hepatology* 1996; **24**: 1481–1486.
47. Malinski T, Taha Z. Nitric oxide release from a single cell measured *in situ* by a porphyrinic-based microsensor. *Nature* 1992; **358**: 676–678.
48. Malinski T, Czuchajowski L. NO measurements by electrochemical methods. In: *Methods in Nitric Oxide Research*. Feelisch M, Stamler JS (Eds). Wiley, Chichester 1996: 319–339.
49. Guo JP, Murohara T, Buerke M *et al.* Direct measurement of nitric oxide release from vascular endothelial cells. *J Appl Physiol* 1996; **81**: 774–779.
50. Tristani-Firouzi M, DeMaster EG, Quast E *et al.* Utility of a NO electrode for monitoring the administration of NO in biological systems. *J Lab Clin Med* 1998; **131**: 281–285.

Chapter 9

Hyperaemia and Autoregulation

U Pohl

Introduction

The coordinated regulation of blood flow must ultimately serve to match the oxygen supply of distal tissues with their oxygen demand. The requirements of this regulation are enormous, since minimal and maximal blood flows can differ by up to 1000-fold. Such a broad dynamic range can only be achieved through the coordinated action of local control mechanisms and central neurogenic influences that adjust cardiac output and systemic haemodynamics. Nevertheless, the local mechanisms that control organ blood flow lead to adjustments in such a way as to avoid extremes of flow: *viz* 'luxury perfusion' and tissue oedema on the one hand and local ischaemia on the other. In recent years it has become evident that the release of vasoactive factors by the endothelium is an important component of this local blood flow regulatory system. This chapter will focus on one such mediator, NO. Its specific role in the local adaptation of blood flow to tissue metabolism will be analysed in situations that reflect the capacities of the regulatory system under conditions of stress, namely reactive and active hyperaemia and autoregulation.

Mechanisms Controlling Local Blood Flow

Under resting conditions, blood flow in most organs generally represents 20% or less of its potential maximum, thus indicating that the resistance vasculature is normally under the influence of a continuous constrictor stimulus. Sympathectomy reduces peripheral vascular resistance only by about 20% of the maximal reduction seen during exercise,[1] however, so that mechanisms other than sympathetic vasoconstriction contribute in an important way to the maintenance of 'resting tone'. Although there is a contribution from vasoconstrictor hormones, this resting tone is mainly myogenic in nature (i.e. intrinsic to vascular smooth muscle) and results partly from spontaneous rhythmic depolarizations of smooth muscle and partly from the depolarizing effects of vessel distension caused by a sustained transmural pressure gradient. *Acute* increases in transmural pressure lead to further vasoconstriction (the 'Bayliss effect' or 'myogenic response').[2] Conversely, acute decreases of pressure induce myogenic dilatation, which is an important mechanism that counteracts cerebral hypoperfusion during postural changes.

The accumulation of vasodilator metabolites such as CO_2, adenosine and lactate constitutes the most important mechanism that allows the adaptation of flow to increased demands. Tissue acidosis, moderate hyperkalaemia and low pO_2 act as additional dilator stimuli which are closely linked to the metabolic status of the tissue.[1] Most metabolites are able to relax vascular smooth muscle directly, without the action of intermediate mediators. Moreover, some of them inhibit transmitter release from sympathetic nerve endings, thus enhancing local vasodilatation. No single metabolite can be identified as the dominant mediator of dilatation; rather they act in a cooperative manner. Indeed, it is questionable whether the accumulation of metabolites alone can effect maximal blood flow as large arterioles and upstream feeding arteries must dilate simultaneously with small resistance arterioles to maximize vascular conductance,[3] and it is not evident how metabolites could efficiently control such remote upstream events. As draining veins rarely run in the vicinity of arterial supply vessels,[4] metabolites would have to diffuse over relatively long distances in order to dilate proximal vessels.

To coordinate the behaviour of upstream and downstream vessels additional mechanisms are therefore necessary to achieve maximal conductance. Two principal local control mechanisms have been characterized. The first is flow-dependent dilatation[3,5–7] which in many vascular beds is thought to be mediated by shear stress-induced NO release. Downstream dilatation of arterioles by metabolites will increase upstream flow as a consequence of Poiseuille's law, and the associated increase in shear stress will subsequently stimulate NO-mediated dilatation in upstream vessels. The resultant effect will be a coordinated dilatation along the vascular tree that results in adaptation of vascular conductance to the new ambient flow. Since this is a positive feedback mechanism, it is theoretically possible that 'preferential flow channels' could be established while other vessels running parallel are underperfused, thereby leading to a 'steal phenomenon'. To offset this, additional coordinating mechanisms, especially between different branches of a vessel, are desirable. The second mechanism that contributes to the coordination of the spatial distribution of flow is electrotonic propagation of local changes in membrane potential.[8] Conduction of such signals from capillaries to upstream resistance vessels and between arterioles and upstream vessels have been described and have considerable spatial length constants.[9–11] All coordinating mechanisms potentially enlarge the dynamic range of blood flow changes after transient ischaemia or during exercise.

Contribution of Nitric Oxide to Local Blood Flow Control

Endothelium-derived NO contributes continuously to the control of basal tone, since inhibition of NO synthesis reduces resting diameters of arterioles and small arteries by about 6–15%.[3,12] Moreover, NO acts as a potent mediator of endothelium-dependent dilatation induced by acetylcholine (ACh), platelet-derived compounds, peptides such as bradykinin and other signal molecules such as histamine.[13] At sufficiently high concentrations of these agonists, NO release can induce maximal dilatation of arterioles, being at least as potent as adenosine. NO causes dilatation, not only directly, but also in a synergistic manner with other agents such as prostaglandins and endothelium-derived hyperpolarizing factor (EDHF) which are co-released from the endothelium.[12]

The continuous 'basal' release of NO varies with local haemodynamic conditions, as it is augmented by increases in longitudinal shear stress.[7,14,15] Shear-induced augmentation of NO release represents an important mechanism that opposes pressure-induced myogenic constrictions,[15,16] which would otherwise tend to reduce tissue perfusion in a self-augmenting, positive feedback loop. NO can thus be expected to play an important role in hyperaemic and autoregulatory responses. The contribution of NO to local blood flow regulation has generally been analysed by using substituted analogues of the substrate L-arginine, such as N^G-nitro-L-arginine (L-NNA)[17] and N^G-monomethyl-L-arginine (L-NMMA)[18] as inhibitors of NO synthase (NOS). These compounds inhibit all three isoforms of NOS.[19] The following sections focus mainly on the role of NO synthesized by the endothelial isoform, eNOS.

Reactive Hyperaemia

Reactive hyperaemia results from the dilator response to transient vascular occlusion. It is greatest in magnitude and longest in duration in organs with high metabolic demands, such as the myocardium, and is modified by occlusion duration[20] and temperature,[21] thus reflecting the role of metabolites in determining the post-occlusive changes in flow and vascular tone. Occlusion also alters intravascular pressure and, since the hyperaemic response can be modified by changes in intravascular pressure,[22] a pressure-sensitive myogenic component must be a key contributing factor. Reactive hyperaemia can be characterized by the peak flow achieved after the release of the occlusion and by the overall duration of the increase in flow. While the former is thought to be dependent partly on myogenic relaxation, the latter is thought to be entirely dependent on metabolite accumulation. The hyperaemic response is further characterized by the repayment of flow debt, which can be calculated from the product of basal flow and occlusion time. The flow repayment is a fractional term defined by the total hyperaemic flow divided by flow debt. In the working heart, this fraction takes a value of 2 or 3, indicating that the refilling of exhausted energy stores needs additional supply.

The effects of eNOS inhibitors on reactive hyperaemia have been investigated in a variety of organs, with virtually all studies reporting an absolute decrease either in the duration of the hyperaemic response or a related parameter.[23–27] Some studies also describe an additional reduction in the hyperaemic peak flow that is independent of the duration of the occlusion.[28–32] However, relative changes, i.e. the flow increase expressed as a percentage of basal flow, or the flow repayment, are not always significantly altered by eNOS inhibition.[28,30,33] Since the hyperaemic response is still evident in the absence of normal NO activity, NO cannot be regarded as the primary causal factor or mediator of reactive hyperaemia. NO is nevertheless a potent modulator of the absolute magnitude of the reactive hyperaemic dilator response.

Theoretically, NO could modulate the hyperaemic response through a variety of mechanisms. Firstly, if NO augmented cell respiration, eNOS inhibition would reduce tissue oxygen demands and hence the magnitude and duration of the post-occlusive dilatation. However, recent experiments with isolated mitochondria and in canine hearts *in vivo* suggest, albeit not unanimously, that NO (or peroxynitrite), can inhibit the respiratory chain in a cGMP-independent manner.[34–37] If this action of NO were an important contributor to reactive hyperaemia, its withdrawal (i.e. treatment with L-NNA or N^G-nitro-L-arginine methyl ester [L-NAME]) should enhance oxidative metabolism, and hence increase dilatation and flow repayment, but this contrasts with the actual experimental findings. Secondly, NO could either be the ultimate mediator of the vascular effects of metabolites or enhance the sensitivity of vascular smooth muscle to their action. To date, there is little evidence for such a role. In a recent study, for example, Ishisaka and Kuo[38] found that the vasodilatation of isolated arterioles induced by acidosis was not altered by inhibitors of NO synthase but was susceptible to K_{ATP} channel inhibition. There is, however, some evidence to suggest that NO can enhance the effect of vasodilators that act by increasing intracellular cAMP levels.[12] The vasodilator effect of at least some metabolites, such as adenosine, or other ischaemia-related vasodilators such as prostacyclin, could therefore be amplified by NO and consequently be less pronounced after eNOS inhibition. Thus, it is of particular interest that combined blockade of eNOS and cyclooxygenase, or of eNOS and adenosine receptors, may inhibit reactive hyperaemia to a substantially greater extent than predicted by the effects of the individual inhibitors when administered in isolation.[23,30]

Finally, as NO release is augmented under conditions of high flow, reactive hyperaemia may be associated with increased release of NO.[24] Intravital microscopic studies in rat cremaster arterioles have revealed an endothelium-dependent post-occlusive dilatation that significantly contributed to the duration of the hyperaemic response, at least after short-term occlusions.[27] The time characteristics of this response suggest that it may also underlie the 'delayed' dilatation that is observed in medium-sized and large coronary arterioles during reactive hyperaemia.[39] As discussed above, shear-induced release of NO increases vascular conductance through the coordinated dilatation of downstream and upstream vessels. Even if metabolite concentrations were to increase in the close vicinity of large feed vessels during the occlusion, this would be desirable because they may be intrinsically less sensitive to metabolites (e.g.) because of differences in α-receptor populations.[40] As NO release opposes pressure-induced myogenic constriction, the rapid increase in pressure that follows release of an occlusion could potentially result in an exaggerated vasoconstriction under conditions of NOS inhibition that shortens the duration of the hyperaemic response and reduces peak flow. Acute pressure elevations thus induce a much stronger and more rapid increase in the coronary resistance of isolated rabbit hearts when NO activity is inhibited, in spite of enhanced accumulation of metabolites as evidenced by lactate release.[41] This indicates that strong myogenic responses can override metabolic dilatation.[42]

Moreover, recent intravital studies have shown that eNOS inhibition unmasks myogenic constrictions in large arterioles that normally do not constrict in response to increases in pressure,[43,44] so that myogenic control is then shifted to vessels which are not tightly controlled by metabolites. Measurements of segmental resistance in cat gastrocnemius muscle *in situ* similarly reveal an enhanced myogenic response after eNOS inhibition, a substantial part of which is located in larger arterioles (> 25 μm).[45] Enhanced myogenic constriction could thus contribute to the reduction of the hyperaemic response that follows loss of NO activity in spite of high levels of metabolites.

Taken together, these results suggest that NO is an important modulator of reactive hyperaemia but is not the primary mediator. The available data are consistent with the view that this modulating effect of NO is attributable mainly to attenuation of myogenic responses in the initial phase after occlusion release and the augmenting effect of shear stress-induced NO release on vascular conductance. NO may additionally enhance the vasodilator effect of some metabolites and prostaglandins.

Active Hyperaemia

Organ blood flow usually increases with increased tissue activity, a response known as 'active' or 'functional' hyperaemia. In exercising skeletal muscle, myocardium and secreting exocrine glands, blood flow increases almost linearly with the metabolic activity of the tissue. Active hyperaemia is therefore thought to result principally from an increase in the tissue concentration of metabolites. Studies investigating the role of NO in active hyperaemia have been performed mainly in skeletal muscle,[28,46–54] myocardium[25,26,29,55,56] and brain.[57–62] In heart and skeletal muscle the results are equivocal. While some studies demonstrate a decrease in active hyperaemia of up to 40% after eNOS inhibition, others report no apparent change.

In a study on 26 patients, cardiac pacing to 140 bpm induced a mean 50% increase in coronary blood flow, whereas after intracoronary L-NMMA infusion this was attenuated to 23%.[56] Interestingly, this reduction was observed only in patients who had no signs of impaired endothelial function. In an intravital microscopic study in dog hearts, L-NAME also reduced responses to adenosine and cardiac pacing by 15–35%.[55] A quantitatively similar diminution of active hyperaemia has also been reported in the canine diaphragm, a skeletal muscle with high oxidative capacity[50]. Significant reductions in active hyperaemia after eNOS inhibition have also been found in healthy volunteers immediately after static handgrip exercise for 3 min[49] and in the hindlimbs of rats undergoing treadmill exercise.[51] In the latter model, heart failure due to myocardial infarction resulted in the loss of the NO-dependent component of active hyperaemia.[52] In chronically instrumented dogs, L-NAME reduced active hindlimb hyperaemia during moderate treadmill exercise, but at higher workloads active hyperaemia (~ 2.5-fold increase) was not altered by L-NAME.[28] Measurements of muscle perfusion by a microsphere technique in dogs during exercise (heart rate increasing from 100 to 200 bpm) have demonstrated significant reductions in active hyperaemia in one particular locomotor muscle (flexor digitorum longus), but not in others.[53] After eNOS inhibition, however, the same absolute flows were associated with elevated perfusion pressures, implying that vascular resistances were significantly higher.

Other studies nevertheless suggest that loss of normal NO activity may be without effect on active hyperaemia. Forearm blood flow increased about 5-fold in healthy subjects undergoing wrist exercise, irrespective of the presence or absence of eNOS inhibition.[46] A similar 3-fold increase in blood flow was observed in an electrically stimulated canine gastrocnemius muscle preparation *in situ* before and after eNOS inhibition with L-NNA.[48] Intravital studies in rabbit tenuissimus muscle also failed to reveal an NO-dependent component of hyperaemic responses in small (< 25 μm) arterioles,[47] and active hyperaemia in the coronary bed may be NO-independent in awake dogs during treadmill exercise or cardiac pacing, respectively.[26,29] In the latter studies, however, the associated increase of arterial pressure evoked by L-NNA may have increased coronary flow and, in spite of an apparently unchanged hyperaemic response, an increase in oxygen extraction could have reflected impaired exercise-induced vasodilatation after acute blockade of eNOS.[26] After *chronic* inhibition of eNOS for 7 days with L-NNA there was also no difference in exercise-induced increases of coronary blood flow (about 3-fold) in awake dogs.[25]

The reasons for this high degree of variability are not entirely clear. There does seem, however, to be a trend towards apparently unchanged active hyperaemia at low rates of exercise or small increases in blood flow following eNOS inhibition, suggesting that enhanced metabolic dilatation of small arterioles may then compensate for impaired responses in larger vessels. Moreover, measurements of total forearm or hindlimb blood flow may mask changes in specific muscle groups. Recent studies by Hirai *et al.*[51] have clearly demonstrated that the effect of eNOS inhibitors is most pronounced in muscles with high oxidative capacity and hence the highest requirements for blood supply during exercise. In such muscles, the NO-dependent augmenting effect of upstream vessel dilatation on vascular conductance may be a prerequisite for the maintenance of an adequate oxygen supply, as recently suggested by intravital microscopy in the hamster cremaster muscle.[63] Such experiments also suggest that the control of the smallest arterioles is likely to be dominated by metabolic factors, rather than NO. Studies on active coronary hyperaemia in the presence of chronic inhibition of eNOS further indicate that alternative mediators may compensate for the lack of NO, since the cyclooxygenase inhibitor diclofenac reduced coronary flow after, but not before, the inhibition of eNOS.[25] Interestingly, prostacyclin is co-released with NO from the endothelium and its release may be directly influenced by the presence of

NO[64] (see also Chapter 2). While some experiments clearly demonstrate a significant contribution of NO to active hyperaemia, the apparent lack of effect of NO in certain situations may reflect methodological considerations or, alternatively, indicate that other factors can compensate for the lack of NO. More studies are needed to clarify these issues.

The modulatory role of NO is therefore complex, as the direct effects of certain metabolites may be enhanced, and flow-dependent coordination and adaptation of upstream and downstream vessels becomes increasingly important with increasing blood flow and oxygen demand, (e.g.) in hamster cremaster muscle[63] and in canine myocardium.[55] In skeletal and heart muscle, the contribution of NO to dilatation may be further enhanced by the rhythmic compression of blood vessels (e.g. by myocardial contraction) as pulsatile flow enhances NO release and blood flow.[65–67] Aside from its amplifying effects on vascular conductance and metabolite-induced dilatation, in some circumstances NO may also serve as a primary mediator during the initial phase of active hyperaemia. NO, probably derived from nerves that mediate sympathetic cholinergic dilatation,[68] could thus play a role in the initial phase of active hyperaemia in rat skeletal muscle. It should be emphasized, however, that in this situation the functional role of sympathetic dilatation may lie more in the prevention of increases of blood pressure at the onset of exercise rather than augmentation of the O_2 supply to skeletal muscle. NO release could also be enhanced proportionally to muscle activity by 'spillover' of ACh from neuromuscular junctions, as recent studies demonstrate not only that there is a close spatial correlation between small resistance vessels and neuromuscular junctions[69] but also an ACh-mediated effect on 'conducted dilatation' in exercising skeletal muscle.[70] Responses evoked by stimulation of the arteriolar endothelium by ACh may propagate upstream and thereby elicit a locally spreading dilatation that is NO-dependent. The extent to which NO produced by endothelial or neuronal NOS situated in myocytes or peripheral nerve endings contributes to the hyperaemic response nevertheless remains to be characterized in detail.

A role for NO as a mediator of active hyperaemia has also been established in the brain. Cerebral arteriolar dilatation by glutamate, the predominant excitatory amino acid in the brain, is dependent on the formation of NO,[71–73] thus explaining observations that cerebral vasodilatation during seizures, a condition characterized by the enhanced release of excitatory amino acids, is NO-dependent.[57,58] Cortical hyperaemia associated with peripheral (sciatic) nerve stimulation in rats has also been shown to be mediated by NO.[62,74] Finally, NOS inhibitors alleviate cerebral vasodilatation during spreading depression.[75,76] In some studies, the use of NOS inhibitors with relative specificity for the neuronal isoform suggest that the source of NO mediating cerebral hyperaemia is neuronal rather than endothelial.[59,60] There is no indication that NO affects other factors of flow control such as cerebral glucose utilization or oxygen consumption other than through its vasodilator action.[77,78]

Taken together, current observations suggest that NO modulates the active hyperaemic response in a similar fashion as in reactive hyperaemia. However, in certain situations NO may be replaced by other factors and may be of lesser importance at low levels of exercise. In contrast to reactive hyperaemia, NO can act as a true mediator of active hyperaemia, at least in the brain. The cellular source of this NO is most likely not endothelial in origin.

Autoregulation

Autoregulation is the mechanism whereby the steady-state blood flow entering an organ remains virtually unchanged over a wide range of perfusion pressures. Since this is achieved by appropriate changes in arteriolar resistance, the result is not simply the maintenance of blood flow but also regulation of capillary hydrostatic pressure. Autoregulation occurs in many organ circulations, including skeletal muscle and the heart, and is classically attributed to the kidney and the cerebral circulation, where it has been studied intensively.[79]

The pressure range over which flow exhibits effective autoregulation is generally about 60–160 mmHg. Any decrease in pressure within this range leads to a transient decrease of blood flow which is, however, brought back to baseline levels shortly afterwards. Opposite changes follow increases in pressure. This behaviour is thought to depend on two principal mechanisms, one of which is the vascular reaction to changes in transmural pressure ('myogenic response'). Studies demonstrating arteriolar constriction upon increases of arterial pressure or elevations of venous pressure, which have opposite effects on tissue supply, support this view.[79] The cellular basis of the myogenic response is the stretch-dependent opening or closing of non-specific cation channels which leads to secondary changes in membrane potential and the open state probability of voltage-dependent Ca^{2+} channels.[80] While it seems clear that the resulting change of intracellular free calcium ($[Ca^{2+}]_i$) is necessary for the initial phase of the response, secondary mechanisms such as alterations in the activity of protein kinase C or of the myosin light chain phosphatase are also thought to contribute.[81] NO does not affect the pressure-induced increase in $[Ca^{2+}]_i$ in isolated resistance arteries but nevertheless strongly attenuates myogenic constriction.[82] This implies that it acts principally by altering secondary mechanisms (see Chapter 4).

The second principal control mechanism underlying autoregulation is metabolic, such that a sustained increase in tissue activity results in increased levels of metabolites that 'resets' the autoregulated flow to a new, higher level. According to this theory, an increase in pressure will induce an initial increase in flow which leads to enhanced washout of metabolites. This results in secondary vasoconstriction and hence normalization of blood flow to the previous resting level.[79] The relative contributions of the myogenic and metabolic components may vary between vascular beds and in different haemodynamic situations. Nevertheless, autoregulation involves, in principle, the same basic regulatory mechanisms of blood flow as reactive hyperaemia. This implies a similar influence of NO on both processes.

Griffith and Edwards performed the earliest study of the effects of NO on autoregulation.[83] These authors demonstrated that in the isolated rabbit ear, which normally shows no autoregulation, a distinct autoregulatory range was unmasked when eNOS was inhibited by L-NMMA or NO was scavenged by haemoglobin, consistent with the idea that NO opposes the myogenic response. Pressure-induced vasoconstriction at constant flow results in enhanced longitudinal wall shear stress. In an organ such as the skin, where resistance arteries show such weak myogenic responses that they can only be observed after additional activation by a vasoconstrictor, NO activity may normally be sufficient to suppress autoregulatory behaviour completely.[83] This situation may apply particularly when the metabolic activity of the tissue is so low that accumulation of metabolites cannot attenuate myogenic constriction.

In many organs, however, a distinct autoregulatory range can be defined even in the presence of NO, and inhibition of eNOS either enhances autoregulation[41,84,85] or — especially in the brain and kidney — has no additional effect other than to reduce the plateau at which flow is regulated.[86–89] In isolated perfused guinea pig and rabbit heart, eNOS inhibition strongly augmented the initially weak coronary autoregulation,[41,84] presumably by enhancing the myogenic response which is normally suppressed by shear stress-dependent NO release in isolated resistance arteries.[15,16] Inhibition of NO synthesis by L-NAME similarly enhances the normally weak autoregulation of the cat mesentery *in situ*.[85] In the rabbit heart, the enhanced autoregulatory response that follows inhibition of NO synthesis occurs in the face of increased lactate release and augmented oxygen extraction, indicating that under certain conditions myogenic constriction can override the dilatory influence of metabolites.[41] In this study, enhanced autoregulation was not due simply to increased basal tone, which is known to augment the myogenic response.[90]

By contrast, studies in the heart, brain, and kidney *in situ* have failed to demonstrate a major role for NO in the suppression of autoregulatory responses. In the myocardium of awake dogs, the closed loop autoregulatory gain remained unchanged following NOS inhibition, but the lower autoregulatory break point increased from 45 to about 60 mmHg.[29] A similar shift to higher pressures has also been observed in a study on renal autoregulation in anaesthetized dogs.[86] These findings suggest that the ability of resistance arteries to compensate for a decrease in perfusion pressure may be reduced by exaggerated myogenic activity. At any given pressure level, enhanced myogenic tone reduced vascular conductance and thereby presumably exhausted the compensatory effect of metabolic dilatation, analogous to the situation in the isolated rabbit heart discussed above. There have been a number of studies in brain and kidney, however, which did not detect any effect of eNOS inhibition on the autoregulatory behaviour, except for reduction in the plateau level of the pressure-flow relationship by up to 30%.[87–89,91,92] In some studies,[89,93] the upper end of the autoregulated range was extended towards higher pressures.

There has as yet been no systematic investigation as to why inhibition of NO synthesis has such apparently divergent effects on autoregulation in various organs. One explanation would be that NO opposes myogenic constriction to different extents in organs. Although haemodynamic studies in awake rats and comparisons of vessels from different species suggest that NO activity can indeed vary between organs, species, and artery types,[94,95] eNOS inhibitors reduce basal flow in the brain and kidney while having little effect on autoregulation in these organs.[61,88] This suggests that it is more likely that myogenic reactivity differs between organs. Indeed, isolated mesenteric vessels show less reactivity than coronary vessels,[90] and in the myocardium, subendocardial vessels respond more weakly than subepicardial vessels, even when isolated from the tissue.[96] Both renal and cerebral arterioles normally show a marked myogenic behaviour.[97,98]

It may therefore be proposed that the loss of NO activity has most pronounced functional effects in vessels with inherently weak myogenic behaviour, such as skin or mesenteric resistance arteries. In such vessels, a shift in the balance between myogenic constriction and an opposing NO-dependent dilatation enhances autoregulation when NO release is impaired. In contrast, in organs in which vessels already show strong myogenic responses, inhibition of NO activity is quantitatively less important and does not shift this balance. Rather, it reduces overall vascular conductivity, so that the dominant effect is a reduction in basal flow. It must be emphasized, however, that further studies are necessary to confirm the generality of this hypothesis. Indeed, interventions in a given tissue or organ may lead to unaltered overall autoregulatory behaviour, simply because compensatory adaptations allow the relative contributions of the basic underlying control mechanisms to change. In the kidney, for example, inhibition of NOS reduces medullary, but not cortical blood flow,[99] alters the tubuloglomerular feedback mechanism[100] (which is, however, modulated by NO derived from neuronal NOS), and affects Na^+ retention.[99]

Summary

The continuous matching of blood flow to local cellular oxygen requirements is the main goal of local vascular control mechanisms. Their relative importance, as well as a specific role of NO as mediator or modulator, can be studied best in situations which are characterized by rapid changes in oxygen consumption or haemodynamic parameters. Such situations include reactive and active hyperaemia, and autoregulation of blood flow following rapid changes of perfusion pressure. The available data demonstrate that specific inhibition of NO synthase reduces hyperaemic responses and enhances autoregulation of blood flow at the expense of tissue oxygenation, at least in certain organs. The data are best explained by a loss of flow-dependent dilatation and enhanced myogenic activity of resistance vessels after eNOS inhibition and are in agreement with studies on isolated vessels. In contrast, there is little evidence that NO interferes specifically with the control of vascular tone through direct metabolic effects. The role of NO in the adaptation of local flow to tissue demands is therefore not that of a mediator of metabolic control, but rather as a modulator of vascular conductance. As an exception, NO acts as a direct mediator of active hyperaemia in the brain, but here it is of neuronal rather than vascular origin. Why the influence of NO on autoregulation in different organs is so divergent is a question that remains unresolved.

References

1. Shepherd JT. Circulation to skeletal muscle. In: *The Cardiovascular System. Peripheral Circulation and Organ Blood Flow, Part 1*, Shepherd JT, Abboud FM (Eds). Am Physiol Soc, Bethesda, Maryland 1983: 319–370.
2. Johansson B. Myogenic tone and reactivity: definitions based on muscle physiology. *J Hypertension* 1989; **7 (Suppl)**: S5–S8.
3. Griffith TM, Edwards DH, Davies RLI *et al.* EDRF coordinates the behaviour of vascular resistance vessels. *Nature* 1987; **329**: 442–445.
4. Chilian WM, Eastham CL, Marcus ML. Microvascular distribution of coronary vascular resistance in beating left ventricle. *Am J Physiol* 1986; **251**: H779–H788.
5. Pohl U, Holtz J, Busse R, Bassenge E. Crucial role of endothelium in the vasodilator response to increased flow *in vivo*. *Hypertension* 1986; **8**: 37–44.
6. Smiesko V, Kozik J, Dolezel S. Role of endothelium in the control of arterial diameter by blood flow. *Blood Vessels* 1985; **22**: 247–251.
7. Melkumyants AM, Balashov SA, Khayutin VM. Control of arterial lumen by shear stress on endothelium. *News Physiol Sci* 1995; **10**: 204–210.
8. Xia J, Little TL, Duling BR. Cellular pathways of the conducted electrical response in arterioles of hamster cheek pouch *in vitro*. *Am J Physiol* 1995; **269**: H2022–H2030.
9. Segal SS, Damon DN, Duling BR. Propagation of vasomotor responses coordinates arteriolar resistances. *Am J Physiol* 1989; **256**: H832–H837.
10. Song H, Tyml K. Evidence for sensing and integration of biological signals by the capillary network. *Am J Physiol* 1993; **265**: H1235–H1242.
11. Frame MDS, Sarelius IH. L-arginine-induced conducted signals alter upstream arteriolar responsivity to L-arginine. *Circ Res* 1995; **77**: 695–701.
12. de Wit C, von Bismarck P, Pohl U. Synergistic action of vasodilators that increase cGMP and cAMP in the hamster cremaster microcirculation. *Cardiovasc Res* 1994; **28**: 1513–1518.
13. Furchgott RF. The role of endothelium in the responses of vascular smooth muscle to drugs. *Annu Rev Pharmacol Toxicol* 1984; **24**: 175–197.

14. Hecker M, Mülsch A, Bassenge E, Busse R. Vasoconstriction and increased flow — two principal mechanisms of shear stress-dependent endothelial autacoid release. *Am J Physiol* 1993; **265**: H828–H833.
15. Pohl U, Herlan K, Huang A, Bassenge E. EDRF-mediated, shear-induced dilation opposes myogenic vasoconstriction in small rabbit arteries. *Am J Physiol* 1991; **261**: H2016–H2023.
16. Kuo L, Chilian WM, Davis MJ. Interaction of pressure-induced and flow-induced responses in porcine coronary resistance vessels. *Am J Physiol* 1991; **261**: H1706–H1715.
17. Mülsch A, Busse R. N^G-nitro-L-arginine (N^5-[imino(nitroamino)methyl]-L-ornithine) impairs endothelium-dependent dilations by inhibiting cytosolic nitric oxide synthesis from L-arginine. *Naunyn Schmiedebergs Arch Pharmacol* 1990; **341**: 143–147.
18. Rees DD, Palmer RMJ, Hodson HF, Moncada S. A specific inhibitor of nitric oxide formation from L-arginine attenuates endothelium-dependent relaxation. *Br J Pharmacol* 1989; **96**: 418–424.
19. Pohl U. Detection and contribution of nitric oxide to endothelium regulation of microvascular tone. In: *Nitric Oxide,* Maines M (Ed). Academic Press Inc, Orlando 1996: 253–263.
20. Johnson PC, Burton KS, Henrich H, Henrich U. Effect of occlusion duration on reactive hyperaemia in sartorius muscle capillaries. *Am J Physiol* 1996; **230**: 715–719.
21. Yonce LR. Effect of temperature on reactive hyperaemia in skeletal muscle. *Am J Physiol* 1962; **202**: 1055–1058.
22. Patterson GC. Role of intravascular pressure in the causation of reactive hyperaemia in the human forearm. *Clin Sci* 1956; **15**: 17–25.
23. Yamabe H, Okumura K, Ishizaka H *et al.* Role of endothelium-derived nitric oxide in myocardial reactive hyperemia. *Am J Physiol* 1992; **263**: H8–H14.
24. Kostic MM, Schrader J. Role of nitric oxide in reactive hyperemia of the guinea pig heart. *Circ Res* 1992; **70**: 208–212.
25. Puybasset L, Bea ML, Ghaleh B *et al.* Coronary and systemic hemodynamic effects of sustained inhibition of nitric oxide synthesis in conscious dogs. Evidence for cross talk between nitric oxide and cyclooxygenase in coronary vessels. *Circ Res* 1996; **79**: 343–357.
26. Altman JD, Kinn J, Duncker DJ, Bache RJ. Effect of inhibition of nitric oxide formation on coronary blood flow during exercise in the dog. *Cardiovasc Res* 1994; **28**: 119–124.
27. Koller A, Kaley G. Role of endothelium in reactive dilation of skeletal muscle arterioles. *Am J Physiol* 1990; **259**: H1313–H1316.
28. O'Leary DS, Dunlap RC, Glover KW. Role of endothelium-derived relaxing factor in hindlimb reactive and active hyperemia in conscious dogs. *Am J Physiol* 1994; **266**: R1213–R1219.

29. Smith TP, Canty JM. Modulation of coronary autoregulatory responses by nitric oxide — evidence for flow-dependent resistance adjustments in conscious dogs. *Circ Res* 1993; **73**: 232–240.
30. Engelke KA, Halliwill JR, Proctor DN *et al.* Contribution of nitric oxide and prostaglandins to reactive hyperemia in human forearm. *J Appl Physiol* 1996; **81**: 1807–1814.
31. Minkes RK, Santiago JA, McMahon TJ, Kadowitz PJ. Role of K^{+}_{ATP} channels and EDRF in reactive hyperemia in the hindquarters vascular bed of cats. *Am J Physiol* 1995; **269**: H1704–H1712.
32. Meredith IT, Currie KE, Anderson TJ *et al.* Postischemic vasodilation in human forearm is dependent on endothelium-derived nitric oxide. *Am J Physiol* 1996; **270**: H1435–H1440.
33. Ward ME, Magder SA, Hussain SNA. Role of endothelium-derived relaxing factor in reactive hyperemia in canine diaphragm. *J Appl Physiol* 1993; **74**: 1606–1612.
34. Xie YW, Shen W, Zhao G *et al.* Role of endothelium-derived nitric oxide in the modulation of canine myocardial mitochondrial respiration *in vitro*. Implications for the development of heart failure. *Circ Res* 1996; **79**: 381–387.
35. Sherman AJ, Davis CA, Klocke FJ *et al.* Blockade of nitric oxide synthesis reduces myocardial oxygen consumption *in vivo*. *Circulation* 1997; **95**: 1328–1334.
36. Bernstein RD, Ochoa FY, Xu X *et al.* Function and production of nitric oxide in the coronary circulation of the conscious dog during exercise. *Circ Res* 1996; **79**: 840–848.
37. Sadoff JD, Scholz PM, Weiss HR. Endogenous basal nitric oxide production does not control myocardial oxygen consumption or function. *Proc Soc Exp Biol Med* 1996; **211**: 332–338.
38. Ishizaka H, Kuo L. Acidosis-induced coronary arteriolar dilation is mediated by ATP-sensitive potassium channels in vascular smooth muscle. *Circ Res* 1996; **78**: 50–57.
39. Kanatsuka H, Sekiguchi N, Sato K *et al.* Microvascular sites and mechanisms responsible for reactive hyperemia in the coronary circulation of the beating canine heart. *Circ Res* 1992; **71**: 912–922.
40. Anderson KM, Faber JE. Differential sensitivity of arteriolar α_1- and α_2-adrenoceptor constriction to metabolic inhibition during rat skeletal muscle contraction. *Circ Res* 1991; **69**: 174–184.
41. Pohl U, Lamontagne D, Bassenge E, Busse R. Attenuation of coronary autoregulation in the isolated rabbit heart by endothelium derived nitric oxide. *Cardiovasc Res* 1994; **28**: 414–419.
42. Meininger GA, Mack CA, Fehr KL, Bohlen HG. Myogenic vasoregulation overrides local metabolic control in resting rat skeletal muscle. *Circ Res* 1987; **60**: 861–870.

43. Pohl U, de Wit C. Interaction of nitric oxide with myogenic and adrenergic vasoconstrictor processes in the control of microcirculatory blood flow. *Pflügers Arch* 1996; **432**: R107–R110.
44. de Wit C, Jahrbeck B, Schäfer C *et al.* Nitric oxide opposes myogenic pressure responses predominantly in large arterioles *in vivo. Hypertension* 1998; **31**: 787–794.
45. Ekelund U, Björnberg J, Grände PO *et al.* Myogenic vascular regulation in skeletal muscle *in vivo* is not dependent of endothelium-derived nitric oxide. *Acta Physiol Scand* 1992; **144**: 199–207.
46. Wilson JR, Kapoor S. Contribution of endothelium-derived relaxing factor to exercise-induced vasodilation in humans. *J Appl Physiol* 1993; **75**: 2740–2744.
47. Persson MG, Gustafsson LE, Wiklund NP *et al.* Endogenous nitric oxide as a modulator of rabbit skeletal muscle microcirculation *in vivo. Br J Pharmacol* 1990; **100**: 463–466.
48. Barclay JK, Woodley NE. Nitric oxide synthase inhibitors do not alter functional hyperemia in canine skeletal muscle. *Can J Physiol Pharmacol* 1994; **72**: 1035–1041.
49. Endo T, Imaizumi T, Tagawa T *et al.* Role of nitric oxide in exercise-induced vasodilation of the forearm. *Circulation* 1994; **90**: 2886–2890.
50. Hussain SN, Stewart DJ, Ludemann JP, Magder S. Role of endothelium-derived relaxing factor in active hyperemia of the canine diaphragm. *J Appl Physiol* 1992; **72**: 2393–2401.
51. Hirai T, Visneski MD, Kearns KJ *et al.* Effects of NO synthase inhibition on the muscular blood flow response to treadmill exercise in rats. *J Appl Physiol* 1994; **77**: 1288–1293.
52. Hirai T, Zelis R, Musch TI. Effects of nitric oxide synthase inhibition on the muscle blood flow response to exercise in rats with heart failure. *Cardiovasc Res* 1995; **30**: 469–476.
53. Shen W, Lundborg M, Wang J *et al.* Role of EDRF in the regulation of regional blood flow and vascular resistance at rest and during exercise in conscious dogs. *J Appl Physiol* 1994; **77**: 165–172.
54. Chowienczyk PJ, Watts GF, Cockcroft JR, Ritter JM. Impaired endothelium-dependent vasodilation of forearm resistance vessels in hypercholesterolaemia. *Lancet* 1992; **340**: 1430–1432.
55. Jones CJ, Kuo L, Davis MJ *et al.* Role of nitric oxide in the coronary microvascular responses to adenosine and increased metabolic demand. *Circulation* 1995; **91**: 1807–1813.
56. Quyyumi AA, Dakak N, Andrews NP *et al.* Contribution of nitric oxide to metabolic coronary vasodilation in the human heart. *Circulation* 1995; **92**: 320–326.
57. Faraci FM, Breese KR, Heistad DD. Nitric oxide contributes to dilatation of cerebral arterioles during seizures. *Am J Physiol* 1993; **265**: H2209–H2212.

58. Pereira de Vasconcelos A, Baldwin RA, Wasterlain CG. Nitric oxide mediates the increase in local cerebral blood flow during focal seizures. *Proc Natl Acad Sci USA* 1995; **92**: 3175–3179.
59. Faraci FM, Brian JE Jr. 7-Nitroindazole inhibits brain nitric oxide synthase and cerebral vasodilatation in response to N-methyl-D-aspartate. *Stroke* 1995; **26**: 2172–2175.
60. Zhang F, Xu S, Iadecola C. Role of nitric oxide and acetylcholine in neocortical hyperemia elicited by basal forebrain stimulation: evidence for an involvement of endothelial nitric oxide. *Neuroscience* 1995; **69**: 1195–1204.
61. Brian JE Jr., Faraci FM, Heistad DD. Recent insights into the regulation of cerebral circulation. *Clin Exp Pharmacol Physiol* 1996; **23**: 449–457.
62. Ngai AC, Meno JR, Winn HR. L-NNA suppresses cerebrovascular response and evoked potentials during somatosensory stimulation in rats. *Am J Physiol* 1995; **269**: H1803–H1810.
63. Hester RL, Eraslan A, Saito Y. Differences in EDNO contribution to arteriolar diameters at rest and during functional dilation in striated muscle. *Am J Physiol* 1993; **265**: H146–H151.
64. Doni MG, Whittle BJR, Palmer RMJ, Moncada S. Actions of nitric oxide on the release of prostacyclin from bovine endothelial cells in culture. *Eur J Pharmacol* 1988; **151**: 19–25.
65. Lamontagne D, Pohl U, Busse R. The mechanical activity of the heart stimulates basal EDRF release in the coronary vascular bed. *J Cardiovasc Pharmacol* 1991; **17 (Suppl 1)**: S89–S94.
66. Pohl U, Busse R, Kuon E, Bassenge E. Pulsatile perfusion stimulates the release of endothelial autacoids. *J Appl Cardiol* 1986; **1**: 215–235.
67. Hutcheson IR, Griffith TM. Release of endothelium-derived relaxing factor is modulated both by frequency and amplitude of pulsatile flow. *Am J Physiol* 1991; **261**: H257–H262.
68. Davisson RL, Shaffer RA, Johnson AK, Lewis SJ. Stimulation of lumbar sympathetic nerves may produce hindlimb vasodilation *via* the release of pre-formed stores of nitrosyl factors. *Neuroscience* 1996; **72**: 881–887.
69. Pierzga JM, Segal SS. Spatial relationships between neuromuscular junctions and microvessels in hamster cremaster muscle. *Microvasc Res* 1994; **48**: 50–67.
70. Segal SS, Welsh DG. Conducted vasodilation in skeletal muscle: Roles for motor and sympathetic innervation during exercise [Abstract]. *Int J Microcirc Clin Exp* 1996; **16**: 36.
71. Faraci FM, Breese KR. Nitric oxide mediates vasodilatation in response to activation of N-methyl-D-aspartate receptors in brain. *Circ Res* 1993; **72**: 476–480.
72. Meng W, Tobin JR, Busija DW. Glutamate-induced cerebral vasodilation is mediated by nitric oxide through N-methyl-D-aspartate receptors. *Stroke* 1995; **26**: 857–862.

73. Northington FJ, Tobin JR, Koehler RC, Traystman RJ. *In vivo* production of nitric oxide correlates with NMDA-induced cerebral hyperemia in newborn sheep. *Am J Physiol* 1995; **269**: H215–H221.
74. Northington FJ, Matherne GP, Berne RM. Competitive inhibition of nitric oxide synthase prevents the cortical hyperemia associated with peripheral nerve stimulation. *Proc Natl Acad Sci USA* 1992; **89**: 6649–6652.
75. Colonna DM, Meng W, Deal DD, Busija DW. Nitric oxide promotes arteriolar dilation during cortical spreading depression in rabbits. *Stroke* 1994; **25**: 2463–2470.
76. Wahl M, Schilling L, Parsons AA, Kaumann A. Involvement of calcitonin gene-related peptide (CGRP) and nitric oxide (NO) in the pial artery dilatation elicited by cortical spreading depression. *Brain Res* 1994; **637**: 204–210.
77. Iadecola C, Pelligrino DA, Moskowitz MA, Lassen NA. Nitric oxide synthase inhibition and cerebrovascular regulation. *J Cereb Blood Flow Metab* 1994; **14**: 175–192.
78. Takahashi S, Cook M, Jehle J *et al.* Lack of effects of inhibition of nitric oxide synthesis on local glucose utilization in the rat brain. *J Neurochem* 1995; **65**: 414–419.
79. Johnson PC. The myogenic response. In: *The Cardiovascular System. Vascular Smooth Muscle*, Bohr DF, Somlyo AP, Sparks HV (Eds). Am Physiol Soc, Bethesda, Maryland 1980: 409–442.
80. Kirber MT, Walsh JV, Singer JJ. Stretch-activated ion channels in smooth muscle: a mechanism for the initiation of stretch induced contraction. *Pflügers Arch* 1988; **412**: 339–345.
81. D'Angelo G, Meininger GA. Transduction mechanisms involved in the regulation of myogenic activity. *Hypertension* 1994; **23**: 1096–1105.
82. Bolz SS, Jacobi J, de Wit C *et al.* Effects of nitric oxide (NO) on pressure-induced vasoconstriction and intracellular calcium changes in isolated arterioles [Abstract]. *Int J Microcirc Clin Exp* 1996; **16**: 33.
83. Griffith TM, Edwards DH. Myogenic autoregulation of flow may be inversely related to endothelium-derived relaxing factor activity. *Am J Physiol* 1990; **258**: H1171–H1180.
84. Ueeda M, Silvia SK, Olsson RA. Nitric oxide modulates coronary autoregulation in the guinea pig. *Circ Res* 1992; **70**: 1296–1303.
85. Macedo MP, Lautt WW. Autoregulatory capacity in the superior mesenteric artery is attenuated by nitric oxide. *Am J Physiol* 1996; **271**: G400–G404.
86. Majid DSA, Navar LG. Suppression of blood flow autoregulation plateau during nitric oxide blockade in canine kidney. *Am J Physiol* 1992; **262**: F40–F46.
87. Baumann JE, Persson PB, Ehmke H *et al.* Role of endothelium-derived relaxing factor in renal autoregulation in conscious dogs. *Am J Physiol* 1992; **263**: F208–F213.

88. Beierwaltes WH, Sigmon DH, Carretero OA. Endothelium modulates renal blood flow but not autoregulation. *Am J Physiol* 1992; **262**: F943–F949.
89. Wang Q, Paulson OB, Lassen NA. Is autoregulation of cerebral blood flow in rats influenced by nitro-L-arginine, a blocker of the synthesis of nitric oxide? *Acta Physiol Scand* 1992; **145**: 297–298.
90. Nyborg NC, Baandrup U, Mikkelsen EO, Mulvany MJ. Active, passive and myogenic characteristics of isolated rat intramural coronary resistance arteries. *Pflügers Arch* 1987; **410**: 664–670.
91. Buchanan JE, Phillis JW. The role of nitric oxide in the regulation of cerebral blood flow. *Brain Res* 1993; **610**: 248–255.
92. Thompson BG, Pluta RM, Girton ME, Oldfield EH. Nitric oxide mediation of chemoregulation but not autoregulation of cerebral blood flow in primates. *J Neurosurg* 1996; **84**: 71–78.
93. Talman WT, Dragon DN. Inhibition of nitric oxide synthesis extends cerebrovascular autoregulation during hypertension. *Brain Res* 1995; **672**: 48–54.
94. Gardiner SM, Compton AM, Kemp PA, Bennett T. Regional and cardiac haemodynamic effects of N^G-nitro-L-arginine methyl ester in conscious, Long Evans rats. *Br J Pharmacol* 1990; **101**: 625–631.
95. Collins P, Chappell SP, Griffith TM *et al.* Differences in basal endothelium-derived relaxing factor activity in different artery types. *J Cardiovasc Pharmacol* 1986; **8**: 1158–1162.
96. Kuo L, Davis MJ, Chilian WM. Myogenic activity in isolated subepicardial and subendocardial coronary arterioles. *Am J Physiol* 1988; **255**: H1558–H1562.
97. Harder DR, Gilbert R, Lombard JH. Vascular muscle cell depolarization and activation in renal arteries on elevation of transmural pressure. *Am J Physiol* 1987; **253**: F778–F781.
98. Halpern W, Osol G. Influence of transmural pressure of myogenic responses of isolated cerebral arteries of the rat. *Ann Biomed Eng* 1985; **13**: 287–293.
99. Nakanishi K, Mattson DL, Cowley AW. Role of renal medullary blood flow in the development of L-NAME hypertension in rats. *Am J Physiol* 1995; **268**: R317–R323.
100. Thorup C, Persson AEG. Inhibition of locally produced nitric oxide resets tubuloglomerular feedback mechanism. *Am J Physiol* 1994; **267**: F606–F611.

Part 2

Peripheral Vascular Effects of Nitric Oxide

Chapter 10

The Role of Nitric Oxide in Cerebrovascular Regulation and Stroke

Costantino Iadecola

Introduction

NO is a ubiquitous molecular messenger that contributes to a wide variety of biological processes and, as discussed elsewhere in this volume (see Chapters 3 and 5), is synthesized by a family of haem-containing NO synthase (NOS) enzymes that include endothelial (eNOS), neuronal (nNOS) and inducible (iNOS) isoforms.[1] Subsequent to its identification as a vascular relaxing factor released from endothelial cells in response to acetylcholine (ACh) or bradykinin,[2,3] NO was found to be produced in brain during activation of glutamate receptors;[4] the enzyme responsible, nNOS, was then isolated and cloned from rat cerebellum.[5,6] NO is a potent vasodilator of cerebral blood vessels,[7] thus raising the possibility that this agent could also be important in cerebrovascular regulation.[8] Consequently, numerous studies have addressed the role of NO in the control of cerebral blood flow (CBF). In addition, the observation that NO is produced during activation of glutamate receptors led to the demonstration that NO was also involved in glutamate-mediated neurotoxicity, an important component of cerebral ischaemic damage.[9] In this chapter, the evidence for and against a role for NO in cerebrovascular regulation and ischaemic brain damage will be critically reviewed and discussed in the light of recent concepts concerning the physiology and pathophysiology of the cerebral circulation.

Experimental Approaches

The study of the role of NO in the regulation of the cerebral circulation has largely been based on the effects of pharmacological inhibitors of NOS. One important class of such compounds consists of arginine analogues that act as competitive inhibitors with little or no isoform selectivity.[10] One of these agents, N^G-nitro-L-arginine (L-NA), is a potent inhibitor of the constitutive isoforms (eNOS and nNOS), but also causes enzyme inactivation.[11] More recently, relatively selective NOS inhibitors have been introduced which preferentially inhibit one of the NOS isoforms within a certain concentration range (Table 1, and see Chapter 7). NOS inhibitors can be administered systemically, applied topically to the brain surface, infused through brain microdialysis fibres or microinjected into the brain.[12–14] Systemic administration of a single dose of L-NA or its methyl ester (L-NAME) leads to a protracted but incomplete inhibition of brain NOS as determined *ex vivo* by the arginine/citrulline isotopic conversion assay of Bredt and Snyder.[15,16] The relatively selective nNOS inhibitor 7-nitroindazole (7-NI) leads to a more transient inhibition.[17] Repeated administrations of L-NA over several days inhibit brain NOS activity almost completely.[16] Topical superfusion of L-NA for 45–60 min blocks NOS activity by 80–90% within the field of superfusion.[18,19]

Several studies have questioned the specificity of NOS inhibitors. L-NA may inhibit K^+ channels in the cerebral microcirculation,[20] and L-NA and

Table 1. Selected pharmacological and non-pharmacological approaches for studying the role of NO in cerebrovascular regulation and stroke.

Approach	Tool	Target
Enzyme inhibition	**NO synthase inhibitors**	
	N^G-nitro-L-arginine	NOS
	N^G-nitro-L-arginine-methyl ester	NOS
	N^G-monomethyl-L-arginine	NOS
	7-nitroindazole	nNOS
	AR17477	nNOS
	S-MIN	nNOS
	Aminoguanidine	iNOS
	Guanylate cyclase inhibitors	
	Methylene blue	sGC
	LY83583	sGC
	ODQ	sGC
Gene deletion	Knockout mice	nNOS
		iNOS
		eNOS
Downregulation of NOS expression	Antisense oligonucleotides	nNOS
		iNOS
		eNOS

NOS, NO synthase; sGC, soluble guanylate cyclase; *S*-MIN, *S*-methyl-isothioureido-L-norvaline; ODQ, 1H-[1,2,4]oxadiazolo[4,3,-a]quinoxalin-1-one.

N^G-monomethyl-L-arginine (L-NMMA) have been proposed to activate cyclooxygenase and to scavenge NO *via* free radical production.[21] L-NAME may have anti-muscarinic receptor effects,[22] while L-NAME and L-NMMA interact with iron in several enzymatic systems and may enhance NO synthesis is isolated arterial rings.[23,24] 7-NI produces endothelium-dependent relaxation in isolated arteries[25] and acts as a preferential nNOS inhibitor only when administered systemically.[13] These potential limitations have to be taken into account when using these compounds in cerebrovascular studies, and careful control experiments are necessary to exclude the possibility that the effects observed are not secondary to pharmacological actions unrelated to NOS inhibition.

Soluble guanylate cyclase (sGC) is recognized as the mediator of many of the biological effects of NO (see also Chapters 1, 3 and 4).[26] Therefore, inhibitors of this enzyme have also been used to investigate the role of NO in cerebrovascular regulation. Until recently, however, selective inhibitors have not been available. The sGC inhibitors methylene blue and LY83583, for example, also scavenge

NO through the formation of reactive O_2 species,[7,27] and methylene blue may inhibit NOS directly.[28] A potent and relatively selective inhibitor of sGC, 1H-[1,2,4]oxadiazolo[4,3-a]quinoxalin-1-one (ODQ), has recently been introduced.[29] Initial experiments in the cerebral circulation indicate that ODQ will be a useful tool to study the involvement of the NO/cGMP system in cerebrovascular regulation.[30]

Homologous recombination technology has been used to generate mice lacking selected NOS genes. Thus, 'knockout' mice for nNOS, eNOS and iNOS have been produced.[31–33] Although the validity of this approach in the study of complex biological phenomena is controversial, these mice have proven to be a powerful investigative tool to explore the role of NO in cerebrovascular regulation and cerebral ischaemic damage.[34–37] Different isoforms of NOS can be selectively downregulated with antisense oligonucleotides that block expression of NOS genes.[38,39] However, this approach has not been used extensively due to the inherent difficulties related to the delivery of antisense oligonucleotides *in vivo* and their toxic effects.[39]

The Physiology of Cerebrovascular Regulation

The cerebral circulation is endowed with complex regulatory systems that allow the brain to adjust its own blood supply in a finely controlled manner. Major factors regulating CBF include neural activity, perivascular nerves, arterial pressure (autoregulation), arterial pCO_2 and pO_2, and blood glucose. A large number of pharmacological agents can also have profound effects on the cerebral circulation, mediated either by direct actions on vascular smooth muscle or, indirectly, by release of vasoactive agents from endothelial cells. The interaction among these complex regulatory influences determines the tone of cerebral blood vessels, and hence flow (Figure 1).

Basal release of NO from the endothelium is an important determinant of resting cerebrovascular tone.[40] In brain, endothelial NO is released by shear stress[41] and by a wide variety of agonists, although there may be marked differences in endothelium-dependent responses to agonists between large cerebral arteries and cerebral arterioles (Table 2). Shear stress-induced vasodilatation is an important factor in coordinating regional vascular resistance.[42] In addition, abluminal release of NO-dependent agonists from perivascular nerves or nerve terminals of intrinsic pathways may influence vascular tone, and luminal release of endothelial NO influences platelet and leucocyte aggregation.[43] Although there is increasing evidence that NO is involved in many of the regulatory systems controlling the cerebral circulation, it must be pointed out that it has not been determined to what extent the effects of NO are mediated by NO itself or by primary or secondary products of the reaction of NO with other chemical species. Nitrosothiols (*S*-NO), peroxynitrite and metal-nitrosyl products are examples of such NO-containing

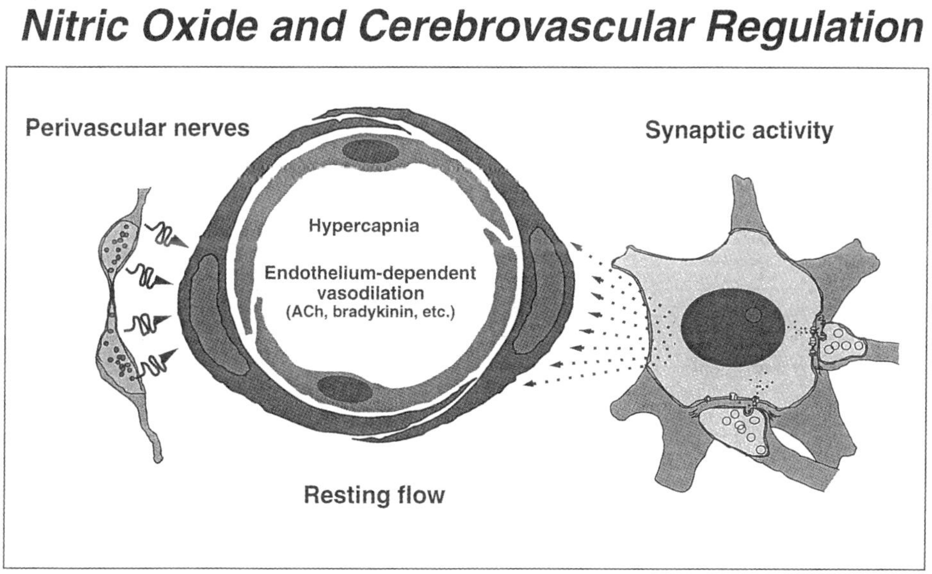

Figure 1. Major roles of NO in cerebrovascular regulation. NO participates in the maintenance of resting cerebral blood flow (CBF) and in the increases in CBF produced by synaptic activity, activation of vasodilator perivascular nerves, hypercapnia and pharmacological agents that release NO from the cerebral endothelium.

Table 2. Agents that dilate cerebral blood vessels through endothelial NO.

Agent	Cerebral Arterial Segment
Acetylcholine	Large cerebral arteries and arterioles
Bradykinin	Large cerebral arteries
Substance P	Large cerebral arteries
Arginine-vasopressin	Large cerebral arteries
Oxytocin	Large cerebral arteries
Endothelin (Endothelin B receptor)	Large cerebral arteries
Histamine	Large cerebral arteries
ATP	Large cerebral arteries
ADP	Cerebral arterioles
Calcium ionophore (A23187)	Large cerebral arteries
Norepinephrine	Cerebral arterioles

Data from Ref. 40.

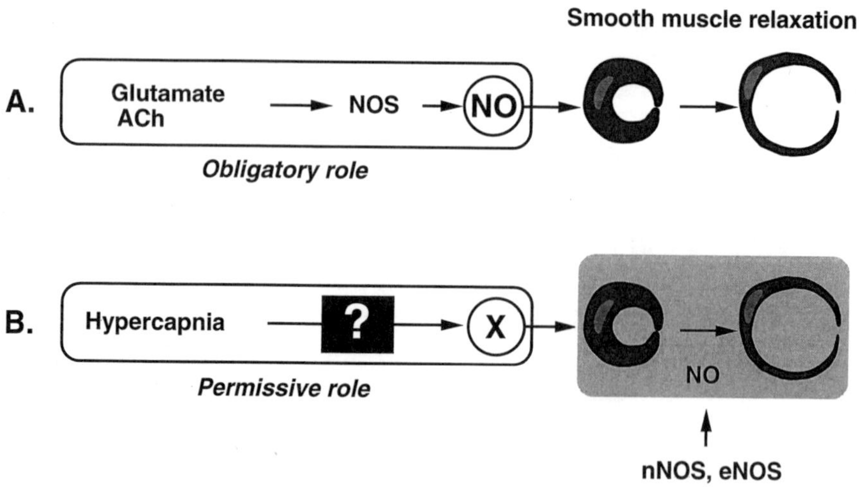

Figure 2. Permissive and obligatory roles of NO in cerebrovascular regulation. NO, or a NO-containing compound, could either act directly on vascular smooth muscles to produce relaxation (obligatory role; A) or could facilitate the vascular relaxation produced by other agents (permissive role; B). NO seems to play an obligatory role in the vasodilatation produced by glutamate or acetylcholine, and a permissive role in the vasodilatation produced by hypercapnia.

species.[44] To simplify presentation, however, NO will be considered to be the biologically active agent. In addition, it will be assumed that most of the cerebrovascular effects of NO are mediated through activation of sGC and cGMP production.[45] However, NO can also exert vascular effects through cGMP-independent mechanisms involving reactions with transition metals and thiols in enzymes, receptors and ion channels.[46] A further possibility is that NO is not the final agent that acts on vascular smooth muscle to produce dilatation, but that its presence is merely required for the response to occur (Figure 2). NO may thus play both 'obligatory' and 'permissive' roles in the regulation of cerebral vascular tone, as discussed below.

Neural Activity

Local neuronal activity is an important determinant of regional CBF. In the resting state, CBF differs markedly among different brain regions, reflecting differences in local synaptic activity. For example, CBF is lower in white matter tracts, where synaptic activity is relatively low, compared with more active grey matter regions.[47] In addition, if the function of a certain brain region is enhanced, local CBF increases in proportion to the degree of activation.[47] Similarly, CBF rises during the massive neuronal firing associated with seizures or arousal.[47,48] Although the observation that brain activation increases local blood flow was made more than a century ago, the mechanisms and mediators responsible for the flow increase remain to be fully elucidated.[45] The discovery that NO is released in brain during synaptic activity led to the suggestion that this agent could be an important link between neural activity and CBF.[8] This hypothesis is supported by several studies in which NO production was blocked by systemic administration or topical application of NOS inhibitors.[10] Thus, NOS inhibitors attenuate the increase in cortical CBF produced by vibrissal or sciatic nerve stimulation in the rat.[49–53] Inhibition of NO synthesis also attenuates the CBF increase produced by seizures,[54–57] and the vascular response produced by increased synaptic activity in cerebellum.[19,58–61] NO may thus play a role in functional hyperaemia in many regions of the brain. Topical application of glutamate receptor agonists to the cerebral or cerebellar cortex produces a vasodilatation that is attenuated by NOS inhibitors,[18,61–64] suggesting that glutamate released during synaptic activity could be one of the factors linking synaptic activity to NO release.

In most instances, near-complete NOS inhibition attenuates the increases in flow produced by neural activity only by 50%,[19,49,51,58] suggesting that NO is responsible for only a component of the vascular response, and that other factors must also be involved. The parallel fibres (PF) of the cerebellar cortex, a relatively simple and well characterized neural system, have allowed a more precise evaluation of the relative contribution of NO to functional hyperaemia. The PF are axons of granule neurones that ascend toward the cerebellar surface, bifurcate, and then run in opposite directions along the major axis of the cerebellar folia, making synaptic contacts with Purkinje cells and interneurones.[65] Stimulation of the PF releases glutamate, which activates receptors on Purkinje cells, molecular layer interneurones and glia of both the ionotropic (N-methyl-D-aspartate [NMDA], α-amino-3-hydroxy-5-methyl-4-isoxazole propionic acid [AMPA] and kainate) and metabotropic (glutamate) types. PF-stimulated activation of glutamate receptors produces an increase in cerebellar blood flow[65] that is attenuated by approximately 50% by NOS inhibition.[19,58] The NO-independent component of the flow increase is attenuated by the adenosine receptor inhibitor 8-sulphophenyltheophylline.[59] These observations suggest that both NO and adenosine participate in the increases in blood flow elicited by activation of the PF system. It is likely, however, that in more complex brain regions, where there may be numerous transmitters and modulators, other factors also participate. These may include neuropeptides as well as vasoactive ions such as K^+ and H^+ released during neural activity.[45] The view that NO is not the only factor regulating CBF during synaptic activity is supported by the fact that mice with a null mutation of the nNOS gene have a normal cortical CBF response to somatosensory activation.[37] Alternative mechanisms may therefore compensate if one of the components contributing to functional hyperaemia is absent — i.e. NO in the case of nNOS knockout mice.

It must be pointed out that some studies have failed to demonstrate a major role for NO in the increase in CBF produced by neural activity. In one study, the NOS inhibitor L-NA (30 mg/kg; *i.v.*) did not affect the increase in cortical

CBF or cerebral glucose utilization produced by vibrissal stimulation in anaesthetized rats.[66] Similar results were obtained in another study in which the vibrissae were stimulated in awake rats and in which brain NOS was substantially inhibited by systemic or intracisternal administration of L-NAME.[67] In another study, the increase in CBF associated with arousal (induced by stimulation of the brainstem reticular formation) was not attenuated by topical application of L-NA (1 mM).[68] These discordant results suggest that there are experimental conditions under which NO is not involved in the mechanisms coupling synaptic activity to blood flow. The balance of evidence indicates, however, that in most instances NO contributes to the increase in CBF produced by functional activation. NO is probably responsible only for a component of the vasodilatation, and other agents also participate in the response.

Neurogenic Mechanisms

Perivascular nerves and selected central neural pathways have important effects on the cerebral circulation. Cerebral blood vessels are closely associated with nerves originating from two major sources, one extrinsic and one intrinsic to the brain. Extrinsic nerves originate from cranial autonomic ganglia and their fibres, and terminals form a dense plexus in the adventitial-medial border of cerebral arteries and arterioles up to a diameter of 50–100 μm.[69] These nerves contain a wide variety of neurotransmitters and neuropeptides.[69] Intrinsic nerves originate from central neurones that send projections to cerebral blood vessels, mainly small arterioles and capillaries.[69] Although these projections come in close contact with blood vessels, they lack the morphological features of true synapses.[70]

Neurogenic factors increase CBF by direct neurovascular mechanisms, rather than indirectly through stimulation of neuronal activity followed by release of vasodilator metabolites, and are therefore typically not associated with increased O_2 or glucose utilization in the regions in which CBF is increased. For example, stimulation of vasodilator fibres travelling through the facial nerve increases CBF without changes in O_2 consumption.[71] Similarly, activation of selected neural centres in brain increases CBF independently of O_2 or glucose consumption. Stimulation of fibres traversing the region of the cerebellar fastigial nucleus increases CBF globally without associated changes in glucose utilization.[69] Stimulation of the cholinergic nuclei of the basal forebrain releases ACh from perivascular terminals and increases CBF in the ipsilateral cerebral cortex.[70,72] Like the fastigial nucleus, the vasodilatation is not associated with increased cortical glucose utilization.[72] Activation of the peri-aqueductal grey matter, a site involved in the integration of the defence reaction, also increases cortical CBF, an effect in part independent of O_2 consumption.[73]

There is evidence that NO is involved in the neurogenic regulation of CBF. NOS-containing nerves surround cerebral arteries and arterioles.[74,75] These nerves originate from the sphenopalatine ganglion, one of the cranial autonomic ganglia, and innervate the anterior and middle cerebral territories more densely than branches of the basilar artery.[74,75] Chemical or electrical stimulation of these nerves *in vitro* produces vasodilatation that is blocked by NOS inhibitors,[76,77] and stimulation of the sphenopalatine ganglion *in vivo* increases CBF, an effect that is attenuated by systemic administration of NOS inhibitors.[78] NOS-containing neurones also have close associations with intracerebral blood vessels. In the cerebral cortex and striatum the processes of NOS-containing neurones make close contacts with local arterioles and capillaries.[74]

NO is also involved in the cerebrovascular responses evoked by activation of central pathways. Topical application of L-NA to the cerebral cortex substantially attenuates the increases in CBF produced by fastigial stimulation.[68] Systemic administration or topical application of NOS inhibitors attenuates the increase in CBF produced by activation of the basal forebrain by 40–70%[18,79–81] and of the peri-aqueductal grey matter by 40–50%.[82] The cellular sources of NO in these responses have not been elucidated. In basal forebrain stimulation, the relatively selective nNOS inhibitor 7-NI does not attenuate the response, whereas the constitutive NOS inhibitor L-NA does so.[18] This observation may imply that the source of NO is not in neurones but in endothelial cells. Collectively, the data suggest that ACh released in cerebral cortex during activation of the basal forebrain cholinergic pathways induces NO production in cerebral endothelial cells.

Autoregulation

CBF is relatively independent of changes in arterial pressure over a certain range (see also Chapter 9).[83] Increases in arterial pressure produce vasoconstriction of cerebral resistance vessels, whereas decreases in arterial pressure produce vasodilatation. The corresponding changes in cerebrovascular resistance counteract the changes in cerebral perfusion pressure so that flow tends to remain constant,[83] although a small increase may occur within the autoregulatory range.[48] At perfusion pressures that exceed the ability of the cerebrovascular bed to autoregulate (150–180 mmHg), CBF increases passively with arterial pressure, a process known as 'breakthrough'.[48] Conversely, if arterial pressure decreases below the lower limit of autoregulation (60–80 mmHg), CBF falls resulting in cerebral hypoperfusion. There is increasing evidence that autoregulation is a direct consequence of the intrinsic ability of cerebral arteries to constrict when intraluminal pressure increases.[83] The cellular mechanisms underlying this myogenic response in cerebral vessels involve, *inter alia*, ATP-sensitive K^+ channels and cytochrome P_{450} metabolites.[84–86] Other mechanisms that modulate cerebrovascular tone, e.g. anaesthesia, drugs, functional activation, neurogenic influences and arterial pCO_2, also have profound effects on cerebrovascular autoregulation.[83]

Several workers have addressed the role of NO in cerebral autoregulation. In these studies, arterial pressure was either increased by infusion of pressor agents or decreased, usually by controlled haemorrhage, before and after administration of NOS inhibitors. A number of reports in rat, dog and monkey have found that NO is not involved in autoregulation,[87–90] whereas other studies suggest that NO is responsible for the vasodilatation that occurs at low arterial pressures[91,92] or that NO contributes to the breakthrough of autoregulation that occurs at high arterial pressures.[93] While mice with a null mutation of nNOS have normal autoregulation,[34] eNOS knockouts have a relative failure of autoregulation at low arterial pressures.[35] Although the mechanisms of this effect have not been studied in detail, the more severe flow reduction during hypotension could be attributed to a shift to the right of the pressure-flow relationship as a consequence of the chronic hypertension that occurs in eNOS null mice.[32] A similar complication arises in studies using NOS inhibitors which produce hypertension and cerebral vasoconstriction, as these haemodynamic effects can have a profound influence on the pressure-flow relationships of cerebral blood vessels. In studies of autoregulation, therefore, it is desirable to carry out additional tests with agents that promote vasoconstriction and induce hypertension to a degree equivalent to that found with NOS inhibitors. Although the investigation of the role of NO in autoregulation is thus complicated by the effects of NOS inhibitors on resting cerebrovascular tone, the available evidence indicates that NO is not responsible for the intrinsic ability of cerebral blood vessels to react to changes in intramural pressure. Under certain conditions, however, NO may exert a modulatory influence on the myogenic tone of cerebral blood vessels.

Cerebrovascular Response to Hypercapnia

Arterial pCO_2 is one of the most potent dilator stimuli in the cerebral circulation, with hypercapnia producing vasodilatation, and hypocapnia producing vasoconstriction.[48] The mechanisms underlying these effects of CO_2 have not been entirely elucidated. Important studies in the late 1970s suggested that the cerebrovascular response to changes in arterial pCO_2 is mediated, at least in part, by attendant changes in perivascular pH.[94] Furthermore, the CBF response to CO_2 is not abolished by endothelial damage, blockade of local neuronal activity with tetrodotoxin, ablation of vasodilator perivascular nerves or destruction of local neurones.[63,95–97]

Taken together, these observations suggest that dilatation is mediated largely by direct effects of pCO_2/pH on vascular smooth muscle. However, there is now evidence that NO is required for the full expression of hypercapnic vasodilatation, and that its role may be to facilitate rather than directly mediate smooth muscle relaxation. In rats, rabbits, cats and monkeys, inhibition of NO synthesis attenuates the vasodilator response to hypercapnia,[12,89,98–100] with the contribution of NO being evident in moderate but not severe hypercapnia, and therefore dependent on the magnitude of the elevation in arterial pCO_2.[98,101,102] The attenuation of the hypercapnic vasodilatation produced by NOS inhibition can be completely reversed by an exogenous source of NO.[81,103,104] Because it has been observed that the cerebral dilatation produced by topical application of H^+ is attenuated by NOS inhibition, it has been suggested that NO production increases in hypercapnia following activation of NOS through the accumulation of H^+ ions.[105] However, the factors influencing NO production in hypercapnia, and its cellular source, remain to be fully established. Hypercapnia increases tissue levels of cGMP within the brain, and this response is absent in mice lacking the nNOS gene, suggesting mediation by neuronally-derived NO.[37,106] Furthermore, relaxations to CO_2 are preserved in isolated endothelium-denuded cerebral arteries,[107] and measurements of cGMP in isolated vessels have demonstrated that hypercapnia does not increase cGMP in vascular tissue.[108] The results of these *in vivo* and *in vitro* experiments suggest that the site of the cGMP increase during hypercapnia (and hence the source of NO) is the brain parenchyma rather than the vasculature.

Observations that mice lacking the eNOS or nNOS gene have a normal cerebrovascular response to CO_2[37,109] confirm that multiple agents and pathways may be involved, and suggest that they can compensate for each other. Indeed, additional mechanisms that have been implicated in the response to CO_2 involve prostanoids and ATP-sensitive K^+ channels.[20,98,110] Studies using the sGC inhibitors methylene blue or LY83583 do not support the involvement of cGMP in hypercapnic vasodilatation,[20,111] although a complicating factor in such studies is that methylene blue and LY83583 both generate reactive O_2 species that scavenge NO.[7,27] Indeed, the recently introduced selective sGC inhibitor ODQ has provided evidence for the role of cGMP in the cerebrovascular response to CO_2.[112]

Permissive and Obligatory Roles of NO

The previous section has reviewed evidence that NO may be required to facilitate the action of other dilator agents, rather than to mediate cerebrovascular vasodilatation directly. This 'permissive' action may be related to maintenance of basal levels of cGMP in vascular smooth muscle, as a cell-permeant cGMP analogue also reverses the attenuation of the hypercapnic vasodilatation produced by NOS inhibitors.[103] In contrast, the NO-independent dilator papaverine does not counteract the effect of NOS inhibition, suggesting that the reversal of the attenuation of the response by NO donors/cGMP is not simply due to non-specific vascular relaxation.[103] Inhibition of prostaglandin synthesis with indomethacin also attenuates hypercapnic vasodilatation, but its effect is not counteracted by NO donors.[103] The latter observation attests further to the specificity of the action of NO donors on the attenuation of the hypercapnic vasodilatation produced by NOS inhibitors. Recent evidence suggests that the source of NO responsible for its permissive effect is neuronal in origin.[81,104] Other responses in which NO may facilitate rather than mediate smooth muscle relaxation directly include the vasodilatation produced by α_2-adrenoceptors, volatile anaesthetics and K^+ application.[113–115] The mechanisms of the permissive role of NO/cGMP have not been elucidated, but could include direct effects on smooth muscle ion channels or the contractile apparatus.

In other situations, NOS activation and NO production seem to be absolutely required for vasodilatation to occur. The increase in CBF produced by ACh, either topically applied or released in cerebral cortex by stimulation of the basal forebrain cholinergic system, is attenuated by L-NA but the attenuation is not reversed by NO donors or cGMP analogues.[81] Similarly, the glutamate-dependent hyperaemia produced by stimulation of the cerebellar PF is attenuated by NOS inhibitors, an effect that is not reversed by NO donors/cGMP.[116] These observations are consistent with the hypothesis that the cerebrovascular effects of ACh and glutamate are specifically mediated by NO synthesis involving eNOS and nNOS, respectively. During these responses, and in contrast to the situation in hypercapnia, NO appears to act directly on vascular smooth muscle to effect dilatation.

Cerebrovascular Responses to Hypoxia and Hypoglycaemia

The structural and functional integrity of the brain is critically dependent on a continuous supply of O_2 and glucose *via* the cerebral circulation. Therefore, under conditions in which the supply is compromised, such as hypoxia and hypoglycaemia, CBF increases to maintain a constant delivery of these substrates.

Hypoxia is a potent vasodilator stimulus for the cerebral circulation, and isolated cerebrovascular smooth muscle cells relax when exposed to low pO_2,[117] suggesting that they possess an intrinsic ability to react to changes in pO_2. Although the cellular basis of this response has not been firmly established, recent evidence suggests that the relaxation may be secondary to smooth muscle hyperpolarization initiated by the opening of ATP-sensitive K^+ channels.[84,118] Hypoxia also causes extracellular accumulation of adenosine, a potent vasodilator, so that adenosine receptor antagonists attenuate the cerebrovascular response to hypoxia.[119] Furthermore, because hypoxia increases neural activity and glucose metabolism,[120] vasodilator factors released by active neurones may also contribute. Recent observations suggest that activation of O_2-sensitive neurones in the rostral ventrolateral medulla play a role in the CBF elevations produced by hypoxia, presumably by acting *via* intrinsic neural pathways that diffusely innervate the brain.[121,122] Therefore, mechanisms intrinsic and extrinsic to the vascular wall participate in the cerebrovascular response to hypoxia.

Although NO may contribute to hypoxic relaxation in isolated preconstricted cerebral arteries,[123,124] *in vivo* data do not suggest a major role for NO in the cerebrovascular changes produced by moderate hypoxia in the rat.[87,125] In severe hypoxia ($pO_2 < 40$ mmHg), however, NO has been reported to be responsible for ~ 50% of the dilator response.[63] It is probable that the NO-dependent component of this response is secondary to activation of NMDA receptors, as it can be attenuated by the NMDA receptor antagonist MK-801.[63] These data, therefore, suggest that NO participates in hypoxic vasodilatation only with reductions in pO_2 that are sufficiently severe to produce neuronal activation. Studies in dog and rat, however, have failed to demonstrate an NO-dependent component even at pO_2 levels lower than 40 mmHg,[126,127] whereas studies in piglet and cat have observed a NO-dependent component at more moderate levels of hypoxia.[128,129] The apparent discrepancies could result from differences in species, anaesthesia and experimental protocols and need to be addressed in future studies.

Hypoglycaemia is also a potent vasodilator stimulus for the cerebral circulation.[130] In the unanaesthetized rat, neither chronic nor acute NOS inhibition affects the increase in CBF produced by insulin- or 2-deoxyglucose-induced hypoglycaemia (plasma glucose 2.5–3.0 mM).[130] In contrast, in pentobarbitone-anaesthetized piglets, in which more severe hypoglycaemia was produced by insulin administration, NO was found to contribute to the increases in CBF, but only in selected brain regions.[131] Therefore, NO is not likely to be a major factor in the cerebrovascular effects of hypoglycaemia, although a contribution may become evident in severe hypoglycaemia.

Cerebral Ischaemia

Early observations that NO contributes to glutamate neurotoxicity in neuronal cultures suggested that NO may be involved in the mechanisms activated during cerebral ischaemia.[9] This possibility was supported by a study in which inhibition of NO synthesis was found to reduce focal cerebral ischaemic damage in mice.[132] Since then there has been great interest in the role of NO in ischaemic brain injury. Most studies have focused on the role of NO in focal cerebral ischaemia produced by occlusion of the rat middle cerebral artery,[133] and the present discussion will be restricted to this model.

Focal cerebral ischaemia increases NO production in the ischaemic brain,[134] an effect probably secondary to activation of glutamate receptors in the injured parenchyma.[135] Cerebral ischaemia also upregulates NOS. While eNOS and nNOS are increased in endothelial cells and neurones within hours of the induction of ischaemia,[136,137] iNOS is upregulated predominantly in inflammatory cells and blood vessels beginning several hours after ischaemia, with or without reperfusion.[138–141] Studies on the role of NO in ischaemic brain injury initially led to conflicting results.[10] The recent introduction of isoform-specific NOS inhibitors and the use of mice with null mutations of the NOS genes have allowed a more clear evaluation of the role of NO, and suggest that the role of NO in ischaemic brain damage can be either protective or destructive depending on the stage of evolution of ischaemic brain damage and on the cellular compartment producing NO (Figure 3). Thus, in the early stage of cerebral ischaemia (< 2 hours after induction of ischaemia), endothelial NO may be beneficial because it promotes vasodilatation and inhibits leucocyte and platelet aggregation. These actions of NO lead to an improvement of blood flow in the ischaemic brain and ameliorate damage.[133] However, at later times (> 2 hours), the vascular and anti-aggregatory effects of NO are no longer beneficial and neurotoxic effects of NO, derived mainly from activation of nNOS, predominate and exacerbate damage. More than 6 hours after ischaemia, inflammatory cells begin to infiltrate the post-ischaemic brain. At this time, iNOS expression in these cells contributes to the further progression of the damage by producing large amounts of NO.[133] (Further discussion on the

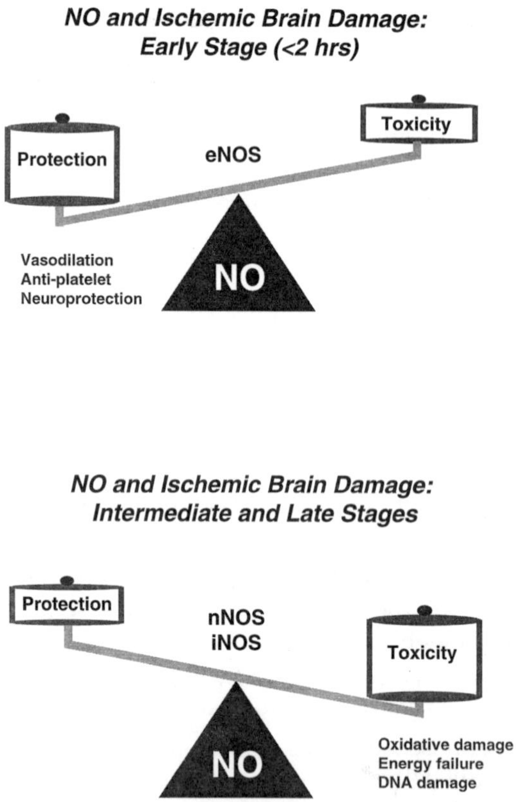

Figure 3. Model of the role of NO in cerebral ischaemia. In the early stages after ischaemia, NO, owing to its vasodilator and anti-platelet effects, is beneficial to the ischaemic tissue and helps to maintain flow. NO may also have a direct neuroprotective effect by downregulating NMDA receptors.[142] In the intermediate and late stages of cerebral ischaemia, the haemodynamic and vascular effects of NO are no longer beneficial. At this time (> 2 hours after ischaemia) large amounts of NO produced by nNOS and, later, by iNOS contribute to tissue injury through induction of oxidative damage, energy failure and DNA damage (see Table 3).

Table 3. Mechanisms contributing to NO toxicity.

NO Reactivity	Biochemical Effect	Cellular Effect
Reaction with $.O_2^-$	Peroxynitrite formation	
Fe/S Clusters	IRE-BP → iron release	*Oxidative damage*
Haem	↑ COX-2	
S-nitrosylation	↓ Creatine kinase ↓ GAPDH	
Fe/S Clusters	↓ Aconitase ↓ Complex I and II	*Energy failure*
	↑ PARS → ATP depletion	
Fe/Tyrosyl radical	↓ Ribonucleotide reductase	*DNA damage*
Base deamination	DNA strand breaks	

NO promotes oxidative damage by combining with superoxide anion ($.O_2$) to form peroxynitrite, a strong oxidant.[143] Furthermore, NO induces loss of iron from the cell, thereby making iron available for free radical generation *via* the Fenton reaction.[144] The mechanisms of the effect are likely to involve binding of NO to the Fe-sulphur cluster (4Fe/4S) of the iron response element binding protein (IRE-BP), a homologue of the mitochondrial enzyme aconitase. NO, or a NO oxidation product, triggers iron loss from the cluster leading to formation of IRE-BP 3Fe/4S which, in turn, leads to release of iron from intracellular stores.[144] In addition, NO activates the prostaglandin synthesizing enzyme cyclooxygenase-2 (COX-2) by binding to the haem group of the enzyme. COX-2, in addition to prostanoids, produces free radicals which contribute to oxidative damage.

NO can inhibit cellular energy production by different mechanisms. NO, or derived species, inhibit the glycolytic enzyme glyceraldehyde-3-phosphate dehydrogenase (GAPDH) and the mitochondrial enzymes aconitase, complex I and complex II.[144] Furthermore, NO may limit ATP formation by inhibiting creatine kinase, an enzyme that catalyzes the transfer of high energy phosphates between phosphocreatine and ATP.[145] NO impairs DNA replication by inhibiting ribonucleotide reductase, the rate-limiting enzyme for DNA synthesis.[144] The effect is mediated by the NO-induced loss of the iron/tyrosyl radical located in the R2 subunit of the enzyme. NO also leads to DNA damage by base deamination.[146] DNA damage activates poly (ADP-ribose) synthetase (PARS) which, in turn, results in cellular depletion of energy-rich substrates.[147]

role of NO in ischaemia-reperfusion injury more generally may be found in Chapter 17.) The mechanisms that contribute to NO-mediated neurotoxicity include oxidative damage, damage to DNA and inhibition of energy-producing enzymes (Table 3).

Conclusions

NO is involved both in the physiology and the pathophysiology of the cerebral circulation. From the standpoint of normal regulation, NO contributes to fundamental responses such as hypercapnic vasodilatation, functional hyperaemia and neurogenic vasodilatation. As for cerebral ischaemia and stroke, NO may be protective or destructive depending on the stage of evolution of ischaemic brain damage. However, NO is only one of a myriad of molecular signals that interact in a highly coordinated fashion to regulate the cerebral circulation in health and disease. Studies focusing on interactions between signalling pathways will ultimately provide a more comprehensive understanding of the role of NO in such complex physiological and pathological processes.

Acknowledgments

Supported by NIH grants (NS31318, NS34179, NS35806) and by grants from the American Heart Association. C.I. is an Established Investigator of the American Heart Association. The editorial assistance of Ms. Karen MacEwan is gratefully acknowledged.

References

1. Marletta, MA. Nitric oxide synthase: aspects concerning structure and catalysis. *Cell* 1994; **78**: 927–930.
2. Furchgott RF, Zawadzki JV. The obligatory role of endothelial cells in the relaxation of arterial smooth muscles by acetylcholine. *Nature* 1980; **228**: 373–376.
3. Palmer RMJ, Ferrige AG, Moncada S. Nitric oxide release accounts for the biological activity of endothelium-derived relaxing factor. *Nature* 1987; **327**: 524–526.
4. Garthwaite J, Charles SL, Chess-Williams R. Endothelium-derived relaxing factor release on activation of NMDA receptors suggests role as intercellular messenger in the brain. *Nature* 1988; **336**: 385–388.
5. Bredt DS, Snyder SH. Isolation of nitric oxide synthetase, a calmodulin-requiring enzyme. *Proc Natl Acad Sci USA* 1990; **87**: 682–685.
6. Bredt DS, Hwang PM, Glatt CE *et al*. Cloned and expressed nitric oxide synthase structurally resembles cytochrome P-450 reductase. *Nature* 1991; **351**: 714–718.
7. Marshall JJ, Wei EP, Kontos HA. Independent blockade of cerebral vasodilation from acetylcholine and nitric oxide. *Am J Physiol* 1988; **255**: H847–H854.
8. Gally JA, Montague PR, Reeke GNJ, Edelman GM. The NO hypothesis: possible effects of a short-lived, rapidly diffusible signal in the development and function of the nervous system. *Proc Natl Acad Sci USA* 1990; **87**: 3547–3551.
9. Dawson VL, Dawson TM, London ED *et al*. Nitric oxide mediates glutamate neurotoxicity in primary cortical cultures. *Proc Natl Acad Sci USA* 1991; **88**: 6368–6371.

10. Iadecola C, Pelligrino DA, Moskowitz MA, Lassen NA. State of the art review: Nitric oxide synthase inhibition and cerebrovascular regulation. *J Cereb Blood Flow Metab* 1994; **14**: 175–192.
11. Klatt P, Schmidt K, Brunner F, Mayer B. Inhibitors of brain nitric oxide synthase. Binding kinetics, metabolism, and enzyme inactivation. *J Biol Chem* 1994; **269**: 1674–1680.
12. Iadecola C. Does nitric oxide mediate the increases in cerebral blood flow elicited by hypercapnia? *Proc Natl Acad Sci USA* 1992; **89**: 3913–3916.
13. Fabricius M, Rubin I, Bundgaard M, Lauritzen M. NOS activity in brain and endothelium: relation to hypercapnic rise of cerebral blood flow in rats. *Am J Physiol* 1996; **271**: H2035–H2044.
14. Northington FJ, Tobin JR, Koehler RC, Traystman RJ. *In vivo* production of nitric oxide correlates with NMDA-induced cerebral hyperemia in newborn sheep. *Am J Physiol* 1995; **269**: H215–H221.
15. Iadecola C, Xu X, Zhang F *et al.* Prolonged inhibition of brain nitric oxide synthase by short-term systemic administration of nitro-L-arginine methyl ester. *Neurochem Res* 1994; **19**: 501–505.
16. Dwyer MA, Bredt DS, Snyder SH. Nitric oxide synthase: irreversible inhibition by L-N^G-nitroarginine in brain *in vitro* and *in vivo*. *Biochem Biophys Res Commun* 1991; **176**: 1136–1141.
17. MacKenzie GM, Rose S, Bland WP *et al.* Time course of inhibition of brain nitric oxide synthase by 7-nitro indazole. *Neuroreport* 1994; **5**: 1993–1996.
18. Zhang F, Xu X, Iadecola C. Role of nitric oxide and acetylcholine in neocortical hyperemia elicited by basal forebrain stimulation: evidence for an involvement of endothelial nitric oxide. *Neuroscience* 1995; **69**: 1195–1204.
19. Iadecola C, Li J, Ebner TJ, Xu S. Nitric oxide contributes to functional hyperemia in cerebellar cortex. *Am J Physiol* 1995; **268**: R1153–R1162.
20. Kontos HA, Wei EP. Arginine analogues inhibit responses mediated by ATP-sensitive K^+ channels. *Am J Physiol* 1996; **271**: H1498–H1506.
21. Rosenblum WI, Nishimura H, Nelson GH. L-NMMA in brain microcirculation of mice is inhibited by blockade of cyclooxygenase and by superoxide dismutase. *Am J Physiol* 1992; **262**: H1343–H1349.
22. Buxton IL, Cheek DJ, Eckman D *et al.* N^G-nitro L-arginine methyl ester and other alkyl esters of arginine are muscarinic receptor antagonists. *Circ Res* 1993; **72**: 387–395.
23. Archer SL, Hampl V. N^G-monomethyl-L-arginine causes nitric oxide synthesis in isolated arterial rings: trouble in paradise. *Biochem Biophys Res Commun* 1992; **188**: 590–596.
24. Peterson DA, Peterson DC, Archer S, Weir EK. The non specificity of specific nitric oxide synthase inhibitors. *Biochem Biophys Res Commun* 1992; **187**: 797–801.

25. Medhurst AD, Greenlees C, Parsons AA, Smith SJ. Nitric oxide synthase inhibitors 7- and 6-nitroindazole relax smooth muscle *in vitro*. *Eur J Pharmacol* 1994; **256**: R5–R6.
26. Ignarro LJ. Heme-dependent activation of guanylate cyclase by nitric oxide: a novel signal transduction mechanism. *Blood Vessels* 1991; **28**: 67–73.
27. Kontos HA, Wei EP. Hydroxyl radical-dependent inactivation of guanylate cyclase in cerebral arterioles by methylene blue and by LY83583. *Stroke* 1993; **24**: 427–434.
28. Mayer B, Brunner F, Schmidt K. Inhibition of nitric oxide synthesis by methylene blue. *Biochem Pharmacol* 1993; **45**: 367–374.
29. Garthwaite J, Southam E, Boulton CL *et al.* Potent and selective inhibition of nitric oxide-sensitive guanylyl cyclase by 1H-[1,2,4]oxadiazolo[4,3-a] quinoxalin-1-one. *Mol Pharmacol* 1995; **48**: 184–188.
30. Sobey CG, Faraci FM. Effects of a novel inhibitor of guanylyl cyclase on dilator responses of mouse cerebral arterioles. *Stroke* 1997; **28**: 837–842.
31. Huang PL, Dawson TM, Bredt DS *et al.* Targeted disruption of the neuronal nitric oxide synthase gene. *Cell* 1993; **75**: 1273–1286.
32. Huang PL, Huang Z, Mashimo H *et al.* Hypertension in mice lacking the gene for endothelial nitric oxide synthase. *Nature* 1995; **377**: 239–242.
33. MacMicking JD, Nathan C, Hom G *et al.* Altered responses to bacterial infection and endotoxic shock in mice lacking inducible nitric oxide synthase. *Cell* 1995; **81**: 641–650.
34. Huang A, Huang PL, Panahian N *et al.* Effects of cerebral ischemia in mice deficient in neuronal nitric oxide synthase. *Science* 1994; **265**: 1883–1885.
35. Huang A, Huang PL, Ma J *et al.* Enlarged infarcts in endothelial nitric oxide synthase knockout mice are attenuated by nitro-L-arginine. *J Cereb Blood Flow Metab* 1996; **16**: 981–987.
36. Iadecola C, Zhang F, Casey R *et al.* Delayed reduction in ischemic brain injury and neurological defects in mice lacking the inducible nitric oxide synthase gene. *J Neurosci* 1997; **17**: 9157–9164.
37. Irikura K, Huang PL, Ma J *et al.* Cerebrovascular alteration in mice lacking neuronal nitric oxide synthase gene expression. *Proc Natl Acad Sci USA* 1995; **92**: 6823–6827.
38. Rosenblum WI, Murata S. Antisense evidence for two functionally active forms of nitric oxide synthase in brain microvascular endothelium. *Biochem Biophys Res Commun* 1996; **224**: 535–543.
39. Ross ME, Iadecola C. Nitric oxide synthase expression in cerebral ischemia: neurochemical, immunocytochemical and molecular approaches. In: *Methods in Enzymology. Nitric Oxide: Physiological and Pathological Processes*, Packer L (Ed). Academic Press, San Diego 1996; **269**: 408–426.
40. Faraci FM, Heistad DD. Regulation of the cerebral circulation: role of endothelium and potassium channels. *Physiol Rev* 1998; **78**: 53–97.

41. Ngai AL, Winn HR. Modulation of cerebral arteriolar diameter by intraluminal flow and pressure. *Circ Res* 1995; **77**: 832–840.
42. Griffith TM. Temporal chaos in the microcirculation. *Cardiovasc Res* 1996; **31**: 342–358.
43. Lindauer U, Dreier J, Angstwurm K *et al.* Role of nitric oxide synthase inhibition in leukocyte-endothelium interaction in the rat pial microvasculature. *J Cereb Blood Flow Metab* 1996; **16**: 1143–1152.
44. Stamler JS. Redox signaling: Nitrosylation and related target interactions of nitric oxide. *Cell* 1994; **78**: 931–936.
45. Iadecola C. Regulation of the cerebral microcirculation during neural activity: Is nitric oxide the missing link? *Trends Neurosci* 1993; **16**: 206–214.
46. Stamler JS, Toone EJ, Lipton SA, Sucher NJ. (S)NO signals: translocation, regulation and consensus motif. *Neuron* 1997; **18**: 691–696.
47. Edvinsson L, MacKenzie ET, McCulloch J. *Cerebral Blood Flow and Metabolism.* Raven Press, New York 1993: 3–39.
48. Heistad D, Kontos H. Cerebral circulation. In: *Handbook of Physiology: The Cardiovascular System III*, Abboud FM, Shepherd JT (Eds). American Physiological Society, Bethesda 1983: 137–182.
49. Dirnagl U, Lindauer U, Villringer A. Role of nitric oxide in the coupling of cerebral blood flow to neural activation in rats. *Neurosci Lett* 1993; **149**: 43–46.
50. Cholet N, Bonvento G, Seylaz J. Effect of neuronal NO synthase inhibition on the cerebral vasodilatory response to somatosensory stimulation. *Brain Res* 1996; **708**: 197–200.
51. Irikura K, Maynard KI, Moskowitz MA. The importance of nitric oxide synthase inhibition to the attenuated vascular responses induced by topical L-nitroarginine during vibrissal stimulation. *J Cereb Blood Flow Metab* 1994; **14**: 45–48.
52. Ngai AC, Meno JR, Winn HR. L-NNA suppresses cerebrovascular response and evoked potentials during somatosensory stimulation in rats. *Am J Physiol* 1995; **269**: H1803–H1810.
53. Northington FJ, Matherne GP, Berne RM. Competitive inhibition of nitric oxide synthase prevents the cortical hyperemia associated with peripheral nerve stimulation. *Proc Natl Acad Sci USA* 1992; **89**: 6649–6652.
54. Pereira de Vasconcelos A, Baldwin RA, Wasterlain CG. Nitric oxide mediates the increase in local cerebral blood flow during focal seizures. *Proc Natl Acad Sci USA* 1995; **92**: 3175–3179.
55. Faraci FM, Breese KR, Heistad DD. Nitric oxide contributes to dilatation of cerebral arterioles during seizures. *Am J Physiol* 1993; **265**: H2209–H2212.
56. Pelligrino, DA, Gay RL 3rd, Baughman VL, Wang Q. NO synthase inhibition modulates NMDA-induced changes in cerebral blood flow and EEG activity. *Am J Physiol* 1996; **271**: H990–H995.

57. Montecot C, Borredon J, Seylaz J, Pinard E. Nitric oxide of neuronal origin is involved in cerebral blood flow increase during seizures induced by kainate. *J Cereb Blood Flow Metab* 1997; **17**: 94–99.
58. Akgören N, Fabricius M, Lauritzen M. Importance of nitric oxide for local increases of blood flow in rat cerebellar cortex during electrical stimulation. *Proc Natl Acad Sci USA* 1994; **91**: 5903–5907.
59. Li J, Iadecola C. Nitric oxide and adenosine mediate vasodilation during functional activation in cerebellar cortex. *Neuropharmacology* 1994; **33**: 1453–1461.
60. Iadecola C, Xu S, Yang G. 7-Nitroindazole attenuates vasodilation from cerebellar parallel fiber stimulation but not acetylcholine. *Am J Physiol* 1996; **270**: R914–R919.
61. Yang G, Iadecola C. Glutamate microinjections in cerebellar cortex reproduce cerebral vascular effects of parallel fiber stimulation. *Am J Physiol* 1996; **271**: R1568–R1575.
62. Faraci FM, Breese KR. Nitric oxide mediates vasodilation in response to activation of N-methyl-D-aspartate receptors in brain. *Circ Res* 1993; **72**: 476–480.
63. Pelligrino DA, Wang Q, Koenig HM, Albrecht RF. Role of nitric oxide, adenosine, N-methyl-D-aspartate receptors, and neuronal activation in hypoxia-induced pial arteriolar dilation in rats. *Brain Res* 1995; **704**: 61–70.
64. Meng W, Tobin JR, Busija DW. Glutamate-induced cerebral vasodilation is mediated by nitric oxide through N-methyl-D-aspartate receptors. *Stroke* 1995; **26**: 857–863.
65. Iadecola C, Li J, Yang G, Xu S. Neural mechanisms of blood flow regulation during synaptic activity in cerebellar cortex. *J Neurophysiol* 1996; **75**: 940–950.
66. Wang Q, Kjaer T, Jorgensen MB *et al.* Nitric oxide does not act as a mediator coupling cerebral blood flow to neural activity following somatosensory stimuli in rats. *Neurol Res* 1993; **15**: 33–36.
67. Adachi K, Takahashi S, Melzer P *et al.* Increases in local cerebral blood flow associated with somatosensory activation are not mediated by NO. *Am J Physiol* 1994; **267**: H2155–H2162.
68. Iadecola C. Nitric oxide participates in the cerebrovasodilation elicited from cerebellar fastigial nucleus. *Am J Physiol* 1992; **263**: R1156–R1161.
69. Iadecola C. Intrinsic and extrinsic neural regulation of the cerebral circulation. In: *Stimulated Cerebral Blood Flow*, Schmiedek P, Einhäupl K, Kirsch C (Eds). Springer-Verlag, Heidelberg 1992: 19–36.
70. Vaucher E, Hamel E. Cholinergic basal forebrain neurons project to cortical microvessels in the rat: electron microscopic study with anterogradely transported Phaseolus vulgaris leucoagglutinin and choline acetyltransferase immunocytochemistry. *J Neurosci* 1995; **15**: 7427–7441.

71. Goadsby PJ. Characteristics of facial nerve elicited cerebral vasodilation determined by laser Doppler flowmetry. *Am J Physiol* 1991; **260**: R255–R262.
72. Sato A, Sato Y. Regulation of regional cerebral blood flow by cholinergic fibers originating in the basal forebrain. *Neurosci Res* 1992; **14**: 242–274.
73. Nakai M, Maeda M. Vasodilatation and enhanced oxidative metabolism of the cerebral cortex provoked by the periaqueductal gray matter in anaesthetized rats. *Neuroscience* 1996; **72**: 1133–1140.
74. Iadecola C, Beitz AJ, Renno W *et al.* Nitric oxide synthase-containing neural processes on large cerebral arteries and cerebral microvessels. *Brain Res* 1993; **606**: 148–155.
75. Nozaki K, Moskowitz MA, Maynard KI *et al.* Possible origin and distribution of immunoreactive nitric oxide synthase-containing nerve fibers in cerebral arteries. *J Cereb Blood Flow Metab* 1993; **13**: 70–79.
76. Lee TJ, Sarwinski SJ. Nitric oxidergic neurogenic vasodilation in the porcine basilar artery. *Blood Vessels* 1991; **28**: 407–412.
77. Toda N, Okamura T. Mechanism underlying the response to vasodilator nerve stimulation in isolated dog and monkey cerebral arteries. *Am J Physiol* 1990; **259**: H1511–H1517.
78. Morita-Tsuzuki Y, Hardebo JE, Bouskela E. Inhibition of nitric oxide synthase attenuates the cerebral blood flow response to stimulation of postganglionic parasympathetic nerves in the rat. *J Cereb Blood Flow Metab* 1993; **13**: 993–997.
79. Raszkiewicz JL, Linville DG, Kerwin JFJ *et al.* Nitric oxide synthase is critical in mediating basal forebrain regulation of cortical cerebral circulation. *J Neurosci Res* 1992; **33**: 129–135.
80. Adachi T, Inanami O, Sato A. Nitric oxide (NO) is involved in increased cerebral cortical blood flow following stimulation of the nucleus basalis of Meynert in anesthetized rats. *Neurosci Lett* 1992; **139**: 201–204.
81. Iadecola C, Zhang F. Permissive and obligatory roles of NO in cerebrovascular responses to hypercapnia and acetylcholine. *Am J Physiol* 1996; **271**: R990–R1001.
82. Nakai M, Maeda M. Nitrergic cerebral vasodilatation provoked by the periaqueductal grey. *Neuroreport* 1996; **7**: 2571–2574.
83. Paulson OB, Strandgaard S, Edvinsson L. Cerebral autoregulation. *Cerebrovasc Brain Metab Rev* 1990; **2**: 162–192.
84. Kitazono T, Faraci FM, Taguci H, Heistad DD. Role of potassium channels in cerebral blood vessels. *Stroke* 1995; **26**: 1713–1723.
85. Harder DR, Narayanan J, Gebremedhin D, Roman RJ. Transduction of physical forces by the vascular wall. Role of phospholipase C and cytochrome P450 metabolites of arachidonic acid. *Trends Cardiovasc Med* 1995; **5**: 7–14.
86. Faraci FM, Heistad DD. Regulation of the cerebral circulation: role of endothelium and potassium channels. *Physiol Rev* 1998; **78**: 53–97.

87. Buchanan JE, Phillis JW. The role of nitric oxide in the regulation of cerebral blood flow. *Brain Res* 1993; **610**: 248–255.
88. Saito S, Wilson DA, Hanley DF, Traystman RJ. Nitric oxide synthase does not contribute to cerebral autoregulatory phenomenon in anesthetized dogs. *J Auton Nerv Syst* 1994; **49**: S73–S76.
89. Thompson BG, Pluta RM, Girton ME, Oldfield EH. Nitric oxide mediation of chemoregulation but not autoregulation of cerebral blood flow in primates. *J Neurosurg* 1996; **84**: 71–78.
90. Wang Q, Paulson OB, Lassen NA. Is autoregulation of cerebral blood flow in rats influenced by nitro-L-arginine, a blocker of the synthesis of nitric oxide. *Acta Physiol Scand* 1992; **145**: 297–298.
91. Kobari M, Fukuuchi Y, Tomita M *et al.* Role of nitric oxide in regulation of cerebral microvascular tone and autoregulation of cerebral blood flow in cats. *Brain Res* 1994; **667**: 255–62.
92. Tanaka K, Fukuuchi Y, Gomi S *et al.* Inhibition of nitric oxide synthesis impairs autoregulation of local cerebral blood flow in the rat. *Neuroreport* 1993; **4**: 267–270.
93. Talman WT, Dragon DN. Inhibition of nitric oxide synthesis extends cerebrovascular autoregulation during hypertension. *Brain Res* 1995; **672**: 48–54.
94. Kontos HA, Raper AJ, Patterson JL. Analysis of vasoactivity of local pH, pCO_2 and bicarbonate on pial vessels. *Stroke* 1977; **8**: 358–360.
95. Iadecola C, Arneric S, Baker H *et al.* Role of local neurons in the cerebrocortical vasodilation elicited from cerebellum. *Am J Physiol* 1987; **252**: R1082–R1091.
96. Wang Q, Pelligrino DA, Koenig HM, Albrecht RF. The role of endothelium and nitric oxide in rat arteriolar dilatory responses to CO_2 *in vivo. J Cereb Blood Flow Metab* 1994; **14**: 944–951.
97. Iadecola C, Zhang F, Xu X. Role of nitric oxide synthase-containing vascular nerves in cerebrovasodilation elicited from cerebellum. *Am J Physiol* 1993; **264**: R738–R746.
98. Faraci FM, Breese KR, Heistad DD. Cerebral vasodilation during hypercapnia. Role of glibenclamide-sensitive potassium channels and nitric oxide. *Stroke* 1994; **25**: 1679–1683.
99. Sandor P, Komjati K, Reivich M, Nyary I. Major role of nitric oxide in the mediation of regional CO_2 responsiveness of the cerebral and spinal cord vessels of the cat. *J Cereb Blood Flow Metab* 1994; **14**: 49–58.
100. McPherson RW, Kirsch JR, Ghaly RF, Traystman RJ. Effect of nitric oxide synthase inhibition on the cerebral vascular response to hypercapnia in primates. *Stroke* 1995; **26**: 682–687.
101. Iadecola C, Zhang F. Nitric oxide-dependent and independent components of cerebrovasodilation elicited by hypercapnia. *Am J Physiol* 1994; **266**: R546–R552.

102. Wang Q, Pelligrino DA, Paulson OB, Lassen NA. Comparison of the effects of N^G-nitro-L-arginine and indomethacin on the hypercapnic cerebral blood flow increase in rats. *Brain Res* 1994; **641**: 257–264.
103. Iadecola C, Zhang F, Xu X. SIN-1 reverses attenuation of hypercapnic cerebrovasodilation by nitric oxide synthase inhibitors. *Am J Physiol* 1994; **267**: R228–R235.
104. Okamoto H, Hudetz AG, Roman RJ *et al.* Neuronal NOS-derived NO plays permissive role in cerebral blood flow response to hypercapnia. *Am J Physiol* 1997; **272**: H559–H566.
105. Niwa K, Lindauer U, Villringer A, Dirnagl U. Blockade of nitric oxide synthesis in rats strongly attenuates the CBF response to extracellular acidosis. *J Cereb Blood Flow Metab* 1993; **13**: 535–539.
106. Irikura K, Maynard KI, Lee WS, Moskowitz MA. L-NNA decreases cortical hyperemia and brain cGMP levels following CO_2 inhalation in Sprague-Dawley rats. *Am J Physiol* 1994; **267**: H837–H843.
107. Toda N, Ayajiki K, Okamura T. Hypercapnia relaxes cerebral arteries and potentiates neurally-induced relaxation. *J Cereb Blood Flow Metab* 1996; **16**: 1068–1074.
108. You JP, Wang Q, Zhang W *et al.* Hypercapnic vasodilatation in isolated rat basilar arteries is exerted via low pH and does not involve nitric oxide synthase stimulation or cyclic GMP production. *Acta Physiol Scand* 1994; **152**: 391–397.
109. Ma J, Meng W, Ayata C *et al.* L-NNA-sensitive regional cerebral blood flow augmentation during hypercapnia in type III NOS mutant mice. *Am J Physiol* 1996; **271**: H1717–H1719.
110. Busija DW, Heistad DD. Factors involved in the physiological regulation of the cerebral circulation. *Rev Physiol Biochem Pharmacol* 1984; **101**: 161–211.
111. Busija DW, Leffler CW, Wagerle LC. Mono-L-arginine-containing compounds dilate piglet pial arterioles via an endothelium-derived relaxing factor-like substance. *Circ Res* 1990; **67**: 1374–1380.
112. Yang G, Iadecola C. Activation of cerebellar climbing fibers increases cerebellar blood flow. Role of glutamate receptors, nitric oxide, and cGMP. *Stroke* 1998; **29**: 499–508.
113. Bryan RM Jr, Steenberg ML, Eichler MY *et al.* Permissive role of NO in α_2-adrenoceptor-mediated dilations in rat cerebral arteries. *Am J Physiol* 1995; **269**: H1171–H1174.
114. Smith JJ, Hudetz AG, Bosnjak ZJ, Kampine JP. The role of nitric oxide in cerebrocortical laser-Doppler flow response to halothane in the rat. *J Neurosurg Anesth* 1995; **7**: 187–195.
115. Dreier JP, Körner K, Görner A *et al.* Nitric oxide modulates the CBF response to increased extracellular potassium. *J Cereb Blood Flow Metab* 1995; **15**: 914–919.

116. Yang G, Iadecola C. Obligatory role of NO in glutamate-dependent hyperemia evoked from cerebellar parallel fibers. *Am J Physiol* 1997; **272**: R1155–R1161.
117. Madden JA, Vadula MS, Kurup VP. Effects on hypoxia and other vasoactive agents on pulmonary and cerebral artery smooth muscles. *Am J Physiol* 1992; **263**: L384–L393.
118. Faraci FM, Heistad DD. Regulation of the cerebral circulation: role of endothelium and potassium channels. *Physiol Rev* 1998; **78**: 53–97.
119. Phillis JW. Adenosine in the control of the cerebral circulation. *Cerebrovasc Brain Metab Rev* 1989; **1**: 26–54.
120. Pulsinelli WA, Duffy TE. Local cerebral glucose metabolism during controlled hypoxemia in the rat. *Science* 1979; **204**: 626–629.
121. Golanov EV, Reis DJ. Contribution of oxygen-sensitive neurons of the rostral ventrolateral medulla to hypoxic cerebral vasodilatation in the rat. *J Physiol (Lond)* 1996; **495**: 201–216.
122. Underwood MD, Iadecola C, Reis DJ. Lesions of the rostral ventrolateral medulla reduce the cerebrovascular response to hypoxia. *Brain Res* 1994; **635**: 217–223.
123. Pohl U, Busse R. Hypoxia stimulates release of endothelium-derived relaxant factor. *Am J Physiol* 1989; **256**: H1595–H1600.
124. Taguci H, Faraci F, Kitazono T, Heistad DD. Relaxation of the carotid artery to hypoxia is impaired in Watanabe heritable hyperlipidemic rabbits. *Arterioscler Thromb Vasc Res* 1995; **15**: 1641–1645.
125. Pelligrino DA, Koenig HM, Albrecht RF. The role of nitric oxide (NO) in cerebral arteriolar relaxation during hypoxia in the rat [Abstract]. *Anesthesiology* 1993; **79**: A757.
126. Kozniewska E, Oseka M, Stys T. Effects of endothelium-derived nitric oxide on cerebral circulation during normoxia and hypoxia in the rat. *J Cereb Blood Flow Metab* 1992; **12**: 311–317.
127. McPherson RW, Koehler RC, Traystman RJ. Hypoxia, α_2-adrenergic, and nitric oxide-dependent interactions on canine cerebral blood flow. *Am J Physiol* 1994; **266**: H476–H482.
128. Ishimura N, Kitaguchi K, Tatsumi K, Furuya H. Nitric oxide involvement in hypoxic dilation of pial arteries in the cat. *Anesthesiology* 1996; **85**: 1350–1356.
129. Armstead WM. Opioids and nitric oxide contribute to hypoxia-induced pial arterial vasodilation in newborn pigs. *Am J Physiol* 1995; **268**: H226–H232.
130. Horinaka N, Artz N, Jehle J *et al.* Examination of potential mechanisms in the enhancement of cerebral blood flow by hypoglycemia and pharmacological doses of deoxyglucose. *J Cereb Blood Flow Metab* 1997; **17**: 54–63.
131. Ichord RN, Helfaer MA, Kirsch JR *et al.* Nitric oxide synthase inhibition attenuates hypoglycemic cerebral hyperemia in piglets. *Am J Physiol* 1994; **266**: H1062–H1068.

132. Nowicki JP, Duval D, Poignet H, Scatton B. Nitric oxide mediates neuronal death after focal cerebral ischemia in the mouse. *Eur J Pharmacol* 1991; **204**: 339–340.
133. Iadecola C. Bright and dark sides of nitric oxide in ischemic brain damage. *Trends Neurosci* 1997; **20**: 132–139.
134. Malinski T, Bailey F, Zhang ZG, Chopp M. Nitric oxide measured by a porphyrinic microsensor in rat brain after transient middle cerebral artery occlusion. *J Cereb Blood Flow Metab* 1993; **13**: 355–358.
135. Lin SZ, Chiou AL, Wang Y. Ketamine antagonizes nitric oxide release from cerebral cortex after middle cerebral artery ligation in rats. *Stroke* 1996; **27**: 747–752.
136. Zhang ZG, Chopp M, Zaloga C *et al.* Nitric oxide synthase expression following focal cerebral ischemia in rat [Abstract]. *J Cereb Blood Flow Metab* 1993; **13 (Suppl 1)**: S164.
137. Zhang ZG, Chopp M, Zaloga C *et al.* Cerebral endothelial nitric oxide synthase expression after focal cerebral ischemia in rats. *Stroke* 1993; **24**: 2016–2022.
138. Iadecola C, Zhang F, Xu X. Inhibition of inducible nitric oxide synthase ameliorates cerebral ischemic damage. *Am J Physiol* 1995; **268**: R286–R292.
139. Iadecola C, Zhang F, Xu X *et al.* Inducible nitric oxide synthase gene expression in brain following focal cerebral ischemia. *J Cereb Blood Flow Metab* 1995; **15**: 378–384.
140. Iadecola C, Xu X, Zhang F *et al.* Marked induction of calcium-independent nitric oxide synthase activity after focal cerebral ischemia. *J Cereb Blood Flow Metab* 1995; **14**: 52–59.
141. Iadecola C, Zhang F, Casey R *et al.* Inducible nitric oxide synthase gene expression in vascular cells following transient focal cerebral ischemia. *Stroke* 1996; **27**: 1373–1380.
142. Fagni L, Olivier M, Lafon CM, Bockaert J. Involvement of divalent ions in the nitric oxide-induced blockade of N-methyl-D-aspartate receptors in cerebellar granule cells. *Mol Pharmacol* 1995; **47**: 1239–1247.
143. Beckman JS, Beckman TW, Chen J *et al.* Apparent hydroxyl radical production by peroxynitrite: implications for endothelial injury from nitric oxide and superoxide. *Proc Natl Acad Sci USA* 1990; **87**: 1620–1624.
144. Gross SS, Wolin MS. Nitric oxide: pathophysiological mechanisms. *Ann Rev Physiol* 1995; **57**: 737–769.
145. Gross WL, Bak MI, Ingwall JS *et al.* Nitric oxide inhibits creatine kinase and regulates rat heart contractile reserves. *Proc Natl Acad Sci USA* 1996; **93**: 5604–5609.
146. Nguyen T, Brunson D, Crespi CL *et al.* DNA damage and mutation in human cells exposed to nitric oxide *in vitro*. *Proc Natl Acad Sci USA* 1992; **89**: 3030–3034.
147. Zhang J, Dawson VL, Dawson TM, Snyder SH. Nitric oxide activation of poly(ADP-ribose) synthetase in neurotoxicity. *Science* 1994; **263**: 687–689.

Chapter 11

The Physiological and Pathophysiological Effects of Nitric Oxide in the Coronary Circulation

Peter Groves

Introduction

The coronary arterial circulation comprises vessels of diminishing calibre which begin on the epicardial surface of the heart and plunge through the sub-epicardium into the myocardium and ultimately into the sub-endocardial layer. The major epicardial arteries serve as conductance vessels and give rise to intramural side-branches which lead to the arterioles where much of the coronary vascular resistance resides. Diastolic coronary blood flow is determined by a combination of the driving pressure and the total resistance of the coronary vascular bed. The epicardial conductance vessels offer little resistance to flow (< 10% of total coronary resistance), while 50% of total coronary vascular resistance is proximal to arterioles 100 μm in diameter and about 25% proximal to larger arterioles 200 μm in diameter.[1] The regulation of coronary blood flow is therefore intimately related to factors that influence vascular tone within the coronary microcirculation, including intrinsic myogenic behaviour and superimposed neural, humoral and metabolic factors.[2] The control of myocardial perfusion is thus complex and multi-factorial, and NO plays an important modulatory role at different anatomical and functional levels.

Physiological Role of Nitric Oxide in the Coronary Circulation

Basal Release of NO and Coronary Vasomotor Tone

Basal NO production has been demonstrated in isolated coronary microvessels[3] but the direct measurement of NO production from vascular beds *in vivo* has proved more difficult. Using the oxidation of oxyhaemoglobin to methaemoglobin, Kelm and Schrader[4] detected the production of NO from the coronary circulation of isolated guinea pig hearts, and these findings were corroborated by the demonstration of elevated levels of cGMP in platelets during passage through the coronary circulation of isolated rabbit[5] and guinea pig hearts.[6] More recently, Zellabos *et al.*[7] demonstrated basal NO production in the coronary circulation of conscious dogs *in vivo* by measuring plasma nitrate + nitrite concentrations.

The role of basal NO activity in regulating coronary vascular tone has been studied *in vitro* and *in vivo* through the administration of L-arginine analogues which specifically inhibit the formation of NO in a concentration-dependent fashion. Systemic doses of N^G-monomethyl-L-arginine (L-NMMA), which increased blood pressure and decreased heart rate in awake dogs, led to a 5–8% reduction in circumflex coronary artery diameter.[8,9] Similar changes were reported in anaesthetized dogs by Woodman and Dusting,[10] but in all these studies changes in systemic haemodynamics may have led to secondary alterations in coronary tone that were unrelated to inhibition of NO synthesis. Sudhir *et al.*,[11] however, administered N^G-nitro-L-arginine methyl ester (L-NAME) directly into the coronary circulation of anaesthetized dogs and reported epicardial vasoconstriction, which was partially reversed by L-arginine, in the absence of changes in blood pressure or heart rate. Lefroy *et al.*[12] reported the effects of L-NMMA administered selectively into the human coronary circulation at doses that did not influence systemic haemodynamics. They identified a small but significant constriction in the distal (~ 6%) but not proximal left anterior descending (LAD) coronary artery. More recently, other investigators have shown consistent vasoconstriction of both proximal and distal epicardial vessels following the inhibition of basal NO release.[13,14]

The influence of basal NO release has also been investigated in the coronary microcirculation. In the buffer-perfused rabbit coronary circulation, L-NMMA led to a marked increase in coronary vascular resistance, implying a high basal release of NO,[15] but in the absence of haemoglobin (which scavenges NO) the physiological effects of these observations are uncertain. In the canine coronary circulation, Chu *et al.*[9] showed a small reduction in blood flow at the highest dose of systemic L-NMMA, while Woodman and Dusting[10] and Sudhir *et al.*[11] found no change in coronary vascular resistance and coronary blood flow velocity under blockade of basal NO synthesis. Jones *et al.*[16] identified a differential role for NO in vessels of different calibre within the microvasculature of dogs, with the administration of L-NAME leading to constriction of small coronary arteries (100–300 μm in diameter) but dilatation of arterioles. One interpretation of these data is that basal NO activity is greater in small arteries than arterioles and thereby aids in maintaining the normal distribution of coronary vasodilator reserve available to metabolic and autoregulatory stimuli that dominate at the arteriolar level (see below). Studies in the human coronary circulation also indicate a role for NO in the regulation of basal microvascular tone. Lefroy *et al.*[12] used coronary venous oxygen saturation as an index of coronary blood flow and observed a small decrease (6%) with intra-coronary L-NMMA. Using similar doses of L-NMMA and a sensitive Doppler flow wire technique to monitor changes in coronary blood flow velocity, Shiode *et al.*[13] found a 16% decrease and Nishikawa and Ogawa[14] a 33% reduction in coronary blood flow following inhibition of basal NO release. Current evidence therefore supports an important physiological role for NO in the regulation of basal coronary tone in vessels ranging from small arteries in the proximal microcirculation up to the largest epicardial conductance vessels.

The Stimulated Release of NO and Coronary Vasomotor Tone

Receptor-mediated release

Indirect evidence of the role of NO in the coronary circulation has come from the intracoronary administration of agents which are known to stimulate endothelial NO production. Acetylcholine (ACh) normally stimulates endothelial NO release, and leads to a dose-dependent dilatation of the epicardial coronary arteries and

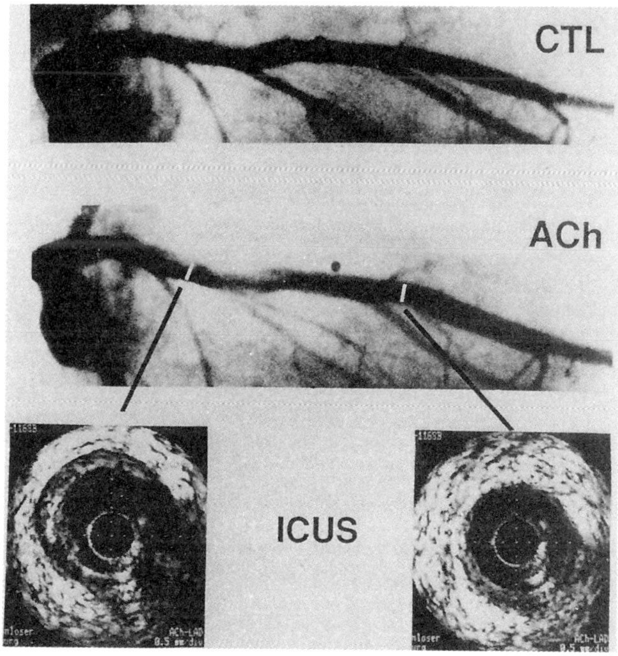

Figure 1. The influence of intracoronary acetylcholine (ACh) on human epicardial coronary artery tone. This angiogram shows a normal appearance within the left anterior descending coronary artery at baseline (CTL) but segmental vasoconstriction and vasodilatation during ACh infusion. Intravascular ultrasound imaging (ICUS) shows that vasoconstriction in response to ACh occurs in coronary segments with atheromatous thickening while vasodilatation is apparent in segments with normal wall architecture. Reproduced and modified from Zeiher *et al.*[75] with the kind permission of *Circulation*.

increase in coronary blood flow through a reduction in coronary vascular resistance, but leads to paradoxical vasoconstriction in vessel segments where endothelial function is locally impaired[17] (Figure 1). The epicardial response to ACh depends on location and on the dose of agent administered, with dilator effects being observed more consistently in distal than proximal epicardial vessels.[12,18] These results could reflect either specific differences in stimulated NO production or non-specific differences in the responsiveness of proximal and larger coronary artery segments to vasoactive stimuli.[19] Vessel segments which dilate at low doses of ACh constrict at higher infusion concentrations, probably due to direct effects on vascular smooth muscle cells.[18] The effects of ACh are nevertheless not uniquely NO-mediated throughout the coronary circulation. In anaesthetized dogs, L-NMMA led to complete inhibition of ACh-induced dilatation in arterioles with a diameter $> 120\ \mu m$ but only partial inhibition in arterioles $< 120\ \mu m$.[20] Similarly, while dilatation of epicardial coronary arteries was completely inhibited by substituted L-arginine analogues,[10] increases in coronary blood flow were only partially attenuated.[10,21] In the human coronary circulation, L-NMMA inhibited ACh-induced

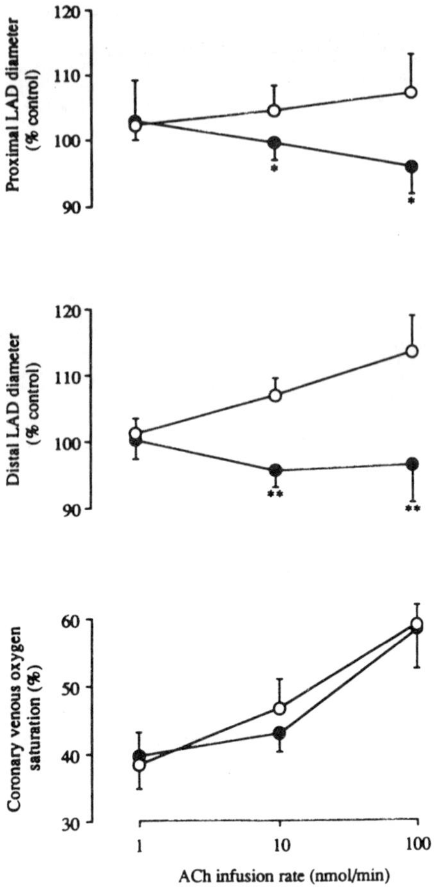

Figure 2. The influence of ACh on human epicardial coronary diameters and coronary blood flow (coronary venous oxygen saturation) in six patients before (open circles) and after (closed circles) intracoronary L-NMMA. Inhibition of NO synthesis converts vasodilatation to vasoconstriction but has no significant influence on the coronary blood flow response to ACh. *P < 0.05, **P < 0.01 compared with before L-NMMA. Reproduced and modified from Lefroy *et al.*[12] with the kind permission of *Circulation*.

dilatation of distal epicardial coronary segments but had no effect on ACh-induced increases in coronary blood flow,[12,13] implying that the effects of ACh on resistance vessels are at least partly, if not entirely, independent of NO synthesis (Figure 2). ACh induces a transient endothelium-dependent hyperpolarization in small coronary arteries,[22] perhaps as a result of the opening of potassium channels,[23] and it is possible therefore that ACh leads to the release of relaxing factors other than NO within coronary resistance vessels (see Chapters 1 and 2). One candidate is the putative endothelium-derived hyperpolarizing factor (EDHF) which recent evidence suggests could be an epoxy-eicosanoid metabolite of arachidonic acid produced by the cytochrome P_{450} mono-oxygenase pathway or an endogenous cannabinoid.[24] The importance of NO-independent mechanisms of relaxation appears to increase from proximal to distal vessels within the coronary circulation.[25]

5-hydroxytryptamine (5-HT), which is released *in vivo* from activated platelets, produces effects on human epicardial coronary arteries which parallel those of ACh, being dependent on the dose of agent used, location of vessel studied, as well as the clinical characteristics of the patient.[26,27] Thus, in patients with angiographically normal coronary arteries, 5-HT evokes dilatation at low concentration but constriction at high concentration, while in patients with coronary artery disease, constriction is observed at all doses and is more pronounced in distal than proximal vessels.[26,27] It appears, therefore, that the biphasic effects of 5-HT involve endothelial NO release at low concentrations and direct smooth muscle constriction at high concentrations, as in the case of ACh. Substance P is also an endothelium-dependent vasodilator but is without direct vasoconstrictor action. Intracoronary substance P consequently leads to human epicardial coronary vasodilatation, which is greater in distal than proximal vessels[28] but, unlike ACh and 5-HT, this action is preserved in diseased vessels.[29] Substance P also increases coronary blood flow, indicating a vasodilator effect on coronary resistance vessels, although the specific role of NO in mediating this response has not been investigated.[28] The exogenous administration of vasoactive agents gives only limited information about underlying physiological mechanisms, but these studies corroborate a role for NO as an endogenous regulator of coronary tone *in vivo*, with receptor-mediated effects being more prominent in conductance than in resistance vessels.

Flow-mediated release

Dilatation in response to an increase in flow was first reported in the isolated canine femoral artery[30] and may contribute to coronary autoregulation by counteracting myogenic vasoconstriction[31] (see also Chapters 2 and 9). Flow-dependent dilatation has been demonstrated *in vitro* in isolated perfused coronary conductance[32] and resistance vessels[33] and *in vivo* following brief coronary occlusion in animals[34] when the extent of vessel dilatation was found to be directly proportional to the increase in flow.[35] Flow-dependent dilatation has also been demonstrated in human epicardial vessels in response to pharmacologically-induced vasodilatation of the distal coronary bed[36,37] (Figure 3). As in other circulations, the dilatation induced by flow is endothelium-dependent, both in conductance[38] and resistance coronary vessels[33] (see also Chapter 2). In isolated sub-epicardial porcine arterioles, flow-dependent dilatation is more rapid (10–30 seconds to reach steady state)[39] than in epicardial vessels (60 seconds),[40] and the extent of the increase in vessel diameter is also greater (~ 30% versus ~ 10%). In resistance vessels the phenomenon is completely abolished by inhibition of NO production with L-NMMA,[39] while

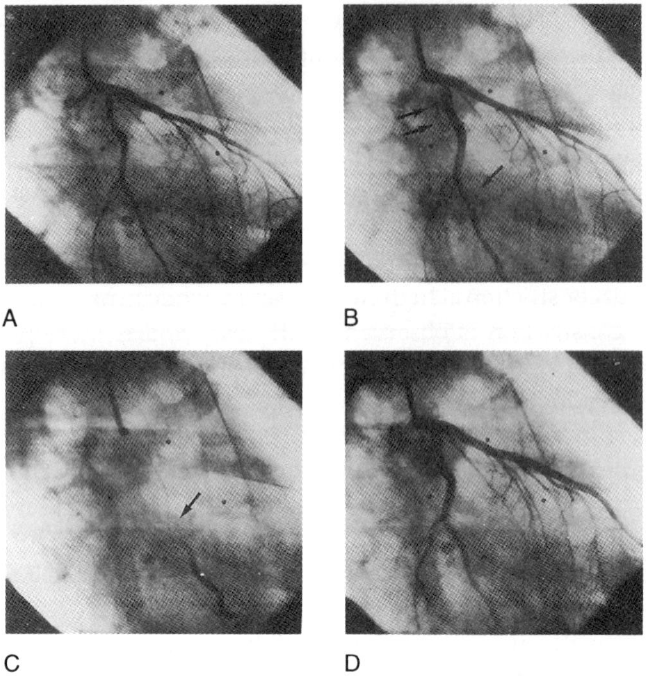

Figure 3. Flow-dependent dilatation in the human coronary circulation. This left coronary angiogram shows smooth coronary arteries at baseline (Panel A) and flow-dependent dilatation of the proximal circumflex (two arrows) following injection of papaverine into the distal vessel through an infusion catheter (single arrow; Panel B). The position of the infusion catheter is confirmed in Panel C (single arrow; with contrast). Dilatation of the whole of the left coronary artery following nitroglycerine injection into the main stem is shown in Panel D). Reproduced and modified from Drexler *et al.*[36] with the kind permission of *Circulation*.

in epicardial vessels the role of NO is less clearly defined. Chu *et al.*[9] demonstrated that peak epicardial coronary dilatation following transient coronary artery occlusion was attenuated but not abolished by L-NMMA in the dog, while Shiode *et al.*[13] found that flow-dependent dilatation in human epicardial vessels secondary to adenosine-induced distal vasodilatation was not influenced by inhibition of NO synthesis. The latter observation is consistent with the findings of Canty and Schwartz,[40] who identified differential effects of NO in the epicardial artery response to transient increases in flow induced by changes in pulse frequency compared to those induced by sustained flow. Thus, flow-dependent dilatation following pacing-induced increases in coronary flow was markedly attenuated by L-NAME, while epicardial coronary artery dilatation induced by increases in coronary flow at a constant heart rate was unaffected.[40]

Taken together, these findings provide evidence that NO-mediated effects predominate in resistance vessels, whereas epicardial responses may be partially NO independent. Consequently, the mechanisms underlying flow-dependent dilatation may differ between conductance and resistance vessels in the coronary circulation. While prostaglandins influence basal coronary tone, they do not appear to be involved in flow-dependent responses.[35] Nevertheless, Groves *et al.*[41] recently demonstrated an important role for endogenous bradykinin in mediating flow-dependent dilatation in human epicardial vessels by employing intracoronary administration of a selective bradykinin B_2 receptor antagonist. Bradykinin promotes endothelial cell hyperpolarization, as does the direct influence of shear stress resulting from flow, and may also stimulate the release of an EDHF.[42,43] The role of alternative endothelium-derived autacoids in mediating flow-dependent dilatation therefore remains to be fully evaluated, particularly *in vivo*.

The coronary response to increased metabolic demand

Changes in coronary flow are closely linked to myocardial oxygen consumption and metabolic demand during physiological stress. Thus, during exercise or mental activity, coronary arteries dilate and flow increases, indicating a reduction in overall vascular resistance.[44,45] The mechanisms responsible for this matching process are incompletely understood, but changes in vascular tone are likely to occur in response to a combination of endothelium-dependent, metabolic and myogenic mechanisms, whose relative importance differs between epicardial and resistance vessels (see Chapter 9). Although it had been suspected that NO has an important role to play in adaptive physiological changes, Bernstein *et al.*[46] were the first to demonstrate an increase in the NO production by the coronary vasculature during exercise *in vivo*. During exercise, epicardial coronary vasodilatation is largely the result of β-adrenoceptor activation and release of endothelium-derived vasodilators. The importance of the latter mechanism was demonstrated by Berdeaux *et al.*,[47] who described paradoxical vasoconstriction with exercise following endothelial denudation in chronically instrumented dogs with the restoration of a vasodilatory response following endothelial regrowth. Epicardial vasodilatation is prevented by a flow-limiting stenosis[48] and by the administration of the NO synthesis inhibitor nitro-L-arginine,[49] emphasizing the importance of increased NO release secondary to increases in flow and shear stress. The contribution of NO to metabolic epicardial dilatation in the human coronary circulation has also been studied by employing pacing-induced tachycardia to increase myocardial oxygen demand. Quyyumi *et al.*[50] reported human epicardial coronary vasodilatation (~ 9%) during control pacing which was converted to vasoconstriction (–6%) after intracoronary L-NMMA (Figure 4), while Nishikawa and Ogawa[14] demonstrated a blunted dilator response. These results are consistent with the hypothesis that flow-induced increases in endothelial NO production, particularly in response to an increase in pulse frequency, are responsible for epicardial coronary dilatation in the face of increased metabolic demand.

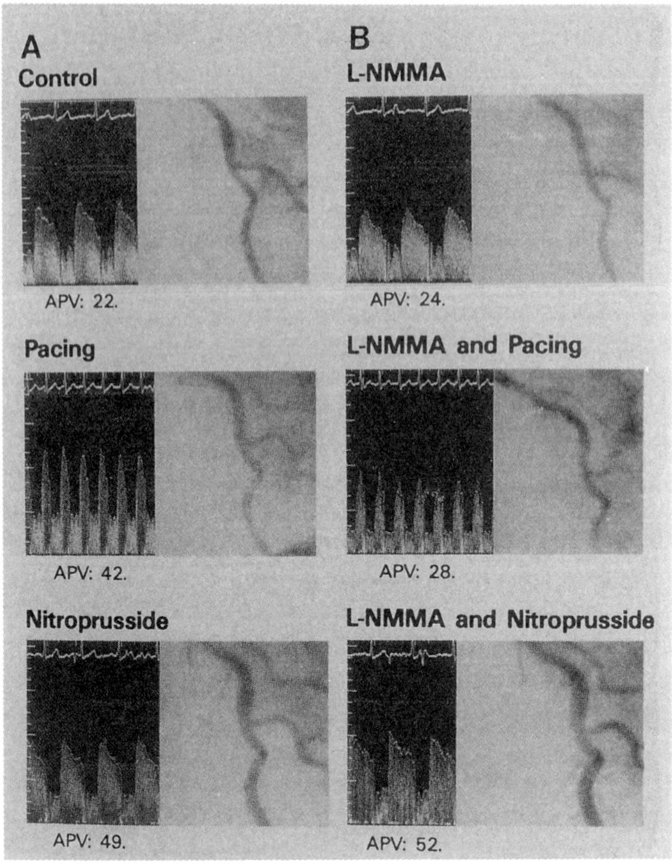

Figure 4. The influence of inhibition of NO synthesis on human coronary responses following pacing-induced tachycardia and an increase in metabolic demand. Data from a single patient without risk factors for coronary artery disease, showing coronary flow velocity signals on the left and angiograms of the left anterior descending coronary artery on the right before (Panel A) and after (Panel B) L-NMMA. Inhibition of NO synthesis led to vasoconstriction at baseline and a blunting of the increase in coronary blood flow that occurs in response to pacing-induced tachycardia, but had no effect on the vasodilatory effects of sodium nitroprusside. Reproduced and modified from Quyyumi *et al.*[50] with the kind permission of *Circulation*.

Alterations in coronary flow are largely secondary to changes in distal resistance, and it is therefore the microvasculature that is primarily involved in adapting coronary flow to the instantaneous myocardial oxygen requirement. Metabolic microvascular dilatation results mainly from the accumulation of local metabolites such as adenosine, prostaglandins, lactate, phosphate and potassium within the interstitium.[51] Studies in the conscious dog have consistently shown that systemic[49,52] and local[53] inhibition of NO synthesis have no effect on changes in coronary blood flow during acute exercise and may even enhance the blood flow response.[54] In human studies, Nishikawa and Ogawa[14] also reported no influence of the inhibition of NO synthesis on pacing-induced increases in coronary blood flow, while Quyyumi *et al.*[50] found that administration of high doses of intracoronary L-NMMA during pacing to 140 bpm attenuated the expected increase in coronary blood flow (Figure 4). Others have implicated NO in coronary microvascular adjustments during autoregulation,[55] hypoxia,[56] reactive hyperaemia,[57] adenosine administration[58] and during the response to increased myocardial oxygen consumption,[59] and it is therefore likely that NO has variable functional importance at different levels within the microvasculature. Metabolite and pressure-induced changes in tone predominate in coronary microvessels $< 150\ \mu m$ in diameter[60] while NO-mediated flow-induced changes in tone are greatest in larger microvessels.[61] Thus, tonic endothelium-derived NO activity in small coronary arteries may shift resistance into smaller arterioles and promote the potential for downstream metabolic microvascular vasodilatation with the subsequent increases in coronary flow leading to upstream flow-dependent vasodilatation[16]. With pharmacological inhibition of NO synthesis or endothelial dysfunction, an alteration in the distribution of resistance may therefore account for a reduction in coronary flow reserve and autoregulatory range (see below).

Neurogenic coronary vasodilatation

Stimulation of parasympathetic nerves, either by direct vagal stimulation[62] or by baroreceptor[63] or chemoreceptor activation,[64] leads to coronary vasodilatation and an increase in coronary blood flow. Although partly attributable to activation of muscarinic cholinergic receptors on vascular smooth muscle, Broten *et al.*[21] demonstrated that parasympathetic coronary vasodilatation also results from the endothelium-dependent release of NO. Since cholinergic nerve fibres terminate at the adventitia/media border, it thus appears that neurally-released ACh may traverse the vessel wall and activate endothelial muscarinic receptors, leading to the release of NO (see also Chapter 19).

β-adrenergic stimulation also dilates coronary conductance and resistance vessels, although the involvement of NO in these responses is controversial. β-adrenergic vasodilatation in isolated canine epicardial vessels has been reported to be endothelium-independent[65] as well as partially endothelium-dependent,[66] while in conscious dogs, inhibition of NO synthesis antagonizes increases in coronary flow evoked by β_2-adrenergic stimulation.[67] It was concluded, however, that the underlying mechanism of NO release was receptor- rather than flow-dependent, as inhibition of β-adrenergic vasodilatation by L-NAME could be demonstrated even when increases in coronary blood flow were prevented.[67] More recently it has been shown that β-adrenergic dilatation of resistance coronary vessels is associated with the opening of K_{ATP} channels[68] and it may be that this is a primary event leading to endothelial cell hyperpolarization, increased Ca^{2+} entry into cells, and thereby increased NO production.

Nitric Oxide and Coronary Pathophysiology

Hypercholesterolaemia and Atherosclerosis

Cohen *et al.*[69] were the first to demonstrate *in vitro* that responses to endothelium-dependent agonists were reduced in epicardial coronary artery rings from hypercholesterolaemic pigs, and subsequent *in vivo* studies confirmed that hypercholesterolaemia impairs endothelial function in the human coronary circulation.[70] Flow-mediated dilatation is attenuated in the coronary circulation in animal models of atherosclerosis[71] as well as in the human coronary circulation[72] at an early stage in the disease process. Moreover, endothelial dysfunction is not simply limited to epicardial vessels in patients with coronary disease, as endothelium-dependent microvascular responses are also abrogated in hypercholesterolaemia, thus implying a generalized disorder.[70] Dysfunction can thus be demonstrated before there is histological evidence of atherosclerosis, and coronary flow reserve may be significantly impaired in the absence of flow-limiting coronary stenosis.[73] As noted above, intracoronary infusion of ACh[17] and 5-HT[26] leads to paradoxical vasoconstriction in patients with coronary artery disease, which can be attributed to direct constrictor effects on vascular smooth muscle in the presence of diminished NO activity. ACh responses may be variable in different segments of the same epicardial artery, suggesting heterogeneity in the extent of endothelial dysfunction[74,75] (Figure 1). Sympathetic stimulation provoked by cold pressor testing,[44] exercise[76] or mental stress[77] also evokes vasoconstriction in patients with coronary artery disease. Zeiher and coworkers were thus able to correlate the extent of endothelial dysfunction with the atherosclerotic plaque load in human epicardial vessels.[75,78]

It has become increasingly recognized that the mechanisms underlying endothelial dysfunction in hypercholesterolaemia and atherosclerosis are highly complex (see Chapters 2, 18 and 19). Functional studies employing endothelium-dependent relaxation of isolated vascular segments as an index of NO activity and *in vivo* studies in the human coronary circulation superficially imply reduced NO production by atherosclerotic coronary vessels.[79,80] Direct assay, however, indicates that NO synthase (NOS) activity may paradoxically be augmented,[81,82] and that functional impairment may therefore result from enhanced inactivation of NO by oxygen-derived free radicals (ODFR). Increased generation of superoxide anion ($O_2^{\cdot -}$) has been reported in the early stages of atherosclerosis when derived largely from the endothelium, but as lipid rich atheromatous plaques develop, infiltration with inflammatory cells and accumulation of oxidized-LDL also contribute.[83,84] The interaction of $O_2^{\cdot -}$ with NO leads to the formation of the strong oxidant peroxynitrite at a rate that exceeds the scavenging capability of superoxide dismutase (see Chapter 18). Peroxynitrite is itself able to oxidize LDL and effect structural alterations in cellular proteins by oxidation and nitrosylation of sulphydryl groups and nitration of tyrosine residues, and may itself reduce coronary vascular reactivity to ACh and NO donors.[85] The hypothesis that increased NO inactivation by ODFR contributes to the genesis of endothelial dysfunction has been further strengthened by the demonstration of improved endothelium-dependent vasodilatation in patients with hypercholesterolaemia and atherosclerosis if treated with the anti-oxidant ascorbic acid.[86,87] Oxidized-LDL may also attenuate NO-mediated vasodilatation by impairing membrane signal transduction, reducing transcription of NOS mRNA, and causing post-transcriptional destabilization of mRNA.[88]

Although L-arginine levels within the endothelial cell are high, its bioavailability may become a rate-limiting step for NO synthesis in hypercholesterolaemic animals[89] as dietary L-arginine supplementation restores endothelium-dependent responses in cholesterol-fed rabbits.[90] Drexler *et al.*[91] demonstrated an analogous phenomenon in the human coronary microcirculation, but observed that beneficial effects were not apparent in epicardial vessels when early atheromatous changes were present on angiography. The paradox that the supply of L-arginine is rate-limiting under conditions where saturation concentrations are present within the endothelial cell itself may reflect the influence of other compounds such as L-glutamine[92] or tetrahydrobiopterin[93] on the substrate affinity between NO synthase and L-arginine. The recent demonstration that tetrahydrobiopterin supplementation in hypercholesterolaemia improves functional NO activity and counteracts the rate limiting role of L-arginine implies that a deficiency of this cofactor may have an important bearing on NO metabolism.[93] The reduced functional NO activity that accompanies hypercholesterolaemia can also be reversed by aggressive lipid lowering therapy, which improves endothelial dysfunction in atherosclerotic coronary arteries within months,[94,95] vascular reactivity in the forearm within weeks,[96] and the deleterious effects of LDL-cholesterol on vasomotor function within minutes following apheresis.[97] In addition to the increased inactivation of NO by $O_2^{\cdot -}$, the excess release of endothelium-derived constrictor factors, which functionally antagonize the smooth muscle effects of NO, may also partially account for impaired endothelium-dependent coronary vasodilatation in patients with coronary artery disease. In this regard, plasma endothelin levels may be elevated, and endothelin-1-like immunoreactivity was found to be increased in endothelium and smooth muscle cells of patients with advanced atherosclerosis.[98]

The concept of endothelial dysfunction as an initiator of the disease process has gained momentum with the recognition that abnormalities are apparent long before intimal thickening is detectable, and are also present in those individuals with only risk factors for the disease.[99] The balance of cause and effect thus remains unresolved.

Hypertension

Hypertension is accompanied by structural changes within the arterial wall which result in thickening of the intima and media and increased extracellular matrix production. In common with other risk factors for atherosclerosis, alterations in endothelial function have also been widely reported (see Chapter 21), and suggest that impaired tonic functional NO activity can be normalized with anti-hypertensive therapy.[100] Reduced ACh vasodilatory responses in hypertensive patients[101,102] and even their offspring[103] have been reported by some investigators, although this is by no means a consistent finding;[104] reduced ACh-induced vasorelaxation cannot be unequivocally attributed to alterations in NO release, however, as discussed above.

Controversy thus exists regarding the influence of hypertension on ACh-induced vasodilatation in the human coronary circulation. While some investigators have reported vasoconstriction of epicardial vessels[105] and impaired resistance vessel vasodilatation[106] following intracoronary ACh, this has not been borne out in all studies.[107,108] The presence of coronary endothelial dysfunction and impairment of NO-mediated vasodilatation in hypertension is, however, corroborated by the recent finding of depressed epicardial and microvascular dilator responses to substance P.[109] The use of L-NMMA in this study identified that the impaired coronary endothelium-dependent vasodilator responses to both ACh and substance P in patients with hypertension is NO-dependent, although it is unclear at what level the abnormality is manifest.[109]

The presence of structural change within coronary vessels as well as ventricular hypertrophy in response to chronically elevated arterial pressure also has a bearing on pathophysiological changes in coronary blood flow in hypertension. The recent finding, for instance, that epicardial coronary arteries fail to increase in size in patients with hypertension and are relatively unresponsive to exogenous nitrovasodilators implies that their failure to adapt to increased myocardial oxygen demand is related to structural changes as well as intrinsic abnormalities of vasomotor function.[110]

Heart Failure

Systemic vasoconstriction is a characteristic feature of advanced congestive cardiac failure and is associated with increased sympathetic tone and activation of the renin-angiotensin system. The endothelium plays an important role in regulating tissue perfusion in heart failure syndromes and several clinical studies have described the importance of peripheral endothelial dysfunction in contributing to impaired arterial vasodilatation.[111,112] Since many of the peripheral features of heart failure are adaptive, compensatory changes in response to falling cardiac output, it cannot be assumed that abnormalities in endothelial function are present in the coronary circulation. Indeed, coronary flow may be preserved and coronary resistance reduced.[113] Nevertheless, some experimental and human studies have identified abnormalities in NO-mediated coronary vasodilatation. In a pacing-induced canine model of heart failure, Wang *et al.*[114] demonstrated that reactive dilatation of epicardial coronary vessels becomes markedly impaired with the development of cardiac decompensation and that this is associated with a reduction in basal and ACh-stimulated release of NO in large coronary arteries and microvessels. These observations are consistent with those of Treasure *et al.*[115] who found that ACh-induced dilatation of the coronary microvasculature was impaired in patients with dilated cardiomyopathy. Zhao *et al.*[116] identified that vagally-mediated coronary vasodilatation is also depressed in conscious dogs in pacing-induced heart failure, with NO-mediated responses to carotid chemoreflex or cardiac receptor activation being reduced. O'Murchu *et al.*,[117] however, described early enhancement of endothelial responses in a pacing-induced model of heart failure, with a paradoxical increase in NO activity and augmentation of ACh and α_2-adrenergic pathways of coronary endothelium-dependent vasodilatation. These early alterations in coronary vascular responses may contribute to the homeostatic mechanisms responsible for preserving coronary flow in the face of a falling cardiac output, and illustrate the likely variation in NO-mediated responses in the coronary bed according to the duration and severity of the heart failure syndrome. While early changes are likely to be adaptive, it remains unclear whether a reduction in NO-mediated coronary vasodilatation contributes specifically to the development of heart failure or merely represents a secondary phenomenon.

Summary

Current evidence supports an important physiological role for NO in the regulation of coronary vascular tone in vessels ranging in size from small arteries in the proximal microcirculation up to the largest epicardial conduit vessels. Increased coronary endothelial NO release occurs in response to a variety of stimuli, including receptor and neurogenic activation and alterations in coronary blood flow. The consequences of receptor-mediated NO release are more evident in conduit than in resistance vessels, while NO-mediated flow-dependent dilatation predominates in resistance vessels and may be partially NO-independent in epicardial arteries. In small coronary arteries, tonic NO activity shifts resistance into smaller arterioles and thereby enhances the potential for downstream metabolic vasodilatation, with subsequent increases in coronary flow leading to flow-dependent and NO-mediated vasodilatation in more proximal vessels.

Abnormal NO-mediated responses have been identified in the coronary circulation in a variety of disease states, including hypercholesterolaemia, atherosclerosis, hypertension and heart failure and may contribute to the early pathogenesis of vascular disease. While the relationships between cause and effect have yet to be clearly defined, it is already apparent that modification of lifestyle and the ability to modulate NO synthesis pharmacologically may provide therapeutic options for the future prevention and treatment of coronary vascular disease.

References

1. Chilian WM, Eastham CL, Marcus ML. Microvascular distribution of coronary vascular resistance in beating left ventricle. *Am J Physiol* 1986; **251**: H779–H788.
2. Jain M, Dai HB, Carrozza JP *et al.* Intrinsic tone as potential vascular reserve in conductance and resistance vessels. *Circulation* 1996; **94**: 1083–1088.
3. Seyedi N, Xu X, Nasjletti A, Hintze TH. Coronary kinin generation mediates nitric oxide release after angiotensin receptor stimulation. *Hypertension* 1995; **26**: 164–170.
4. Kelm M, Schrader J. Control of coronary vascular tone by nitric oxide. *Circ Res* 1990; **66**: 1561–1575.
5. Pohl U, Busse R. EDRF increase cyclic GMP in platelets during passage through the coronary vascular bed. *Circ Res* 1989; **65**: 1798–1803.
6. Kelm M, Schrader J. Nitric oxide release from the isolated guinea pig heart. *Eur J Pharmacol* 1988; **155**: 317–321.
7. Zellabos GA, Bernstein RD, Thompson CI *et al.* Pharmacodynamics of plasma nitrate/nitrite as an indication of nitric oxide formation in conscious dogs. *Circulation* 1995; **91**: 2982–2988.
8. Chu A, Chambers DE, Lin CC *et al.* Nitric oxide modulates epicardial coronary vasomotor tone in awake dogs. *Am J Physiol* 1990; **258**: H1250–H1254.
9. Chu A, Chambers DE, Lin CC *et al.* Effects of inhibition of nitric oxide formation on basal vasomotion and endothelium-dependent responses of the coronary arteries in awake dogs. *J Clin Invest* 1991; **87**: 1964–1968.
10. Woodman OL, Dusting GJ. N-Nitro-L-arginine causes coronary vasoconstriction and inhibits endothelium-dependent vasodilatation in anaesthetized greyhounds. *Br J Pharmacol* 1991; **103**: 1407–1410.
11. Sudhir K, MacGregor JS, Amidon TM *et al.* Differential contribution of nitric oxide to regulation of vascular tone in coronary conductance and resistance arteries: intravascular ultrasound studies. *Am Heart J* 1994; **127**: 858–865.
12. Lefroy DC, Crake T, Uren NG *et al.* Effect of inhibition of nitric oxide synthesis on epicardial coronary artery caliber and coronary blood flow in humans. *Circulation* 1993; **88**: 43–54.
13. Shiode N, Morishima N, Nakayama K *et al.* Flow-mediated vasodilation of human epicardial coronary arteries: effect of inhibition of nitric oxide synthesis. *J Am Coll Cardiol* 1996; **27**: 304–310.
14. Nishikawa Y, Ogawa S. Importance of nitric oxide in the coronary artery at rest and during pacing in humans. *J Am Coll Cardiol* 1997; **29**: 85–92.
15. Amezcua JL, Palmer RMJ, de Souza BM, Moncada S. Nitric oxide synthesized from L-arginine regulates vascular tone in the coronary circulation of the rabbit. *Br J Pharmacol* 1989; **97**: 1119–1124.

16. Jones CJH, Kuo L, Davis MJ *et al.* Role of nitric oxide in the coronary microvascular responses to adenosine and increased metabolic demand. *Circulation* 1995; **91**: 1807–1813.
17. Ludmer PL, Selwyn AP, Shook TL *et al.* Paradoxical vasoconstriction induced by acetylcholine in atherosclerotic coronary arteries. *N Engl J Med* 1986; **315**: 1046–1051.
18. Newman CM, Maseri A, Hackett DR *et al.* Response of angiographically normal and atherosclerotic left anterior descending coronary arteries to acetylcholine. *Am J Cardiol* 1990; **66**: 1070–1076.
19. Brown BG, Bolson EL, Dodge HT. Dynamic mechanisms in human coronary stenosis. *Circulation* 1984; **70**: 917–922.
20. Komaru T, Lamping KG, Eastham CL *et al.* Effect of an arginine analogue on acetylcholine-induced coronary microvascular dilatation in dogs. *Am J Physiol* 1991; **261**: H2001–H2007.
21. Broten TP, Miyashiro JK, Moncada S, Feigl EO. Role of endothelium-derived relaxing factor in parasympathetic coronary vasodilation. *Am J Physiol* 1992; **262**: H1579–H1584.
22. Feletou M, Vanhoutte PM. Endothelium-dependent hyperpolarization of canine coronary smooth muscle. *Br J Pharmacol* 1988; **93**: 515–524.
23. Chen G, Suzuki H, Weston AH. Acetylcholine releases endothelium-derived hyperpolarizing factor and EDRF from rat blood vessels. *Br J Pharmacol* 1988; **95**: 1165–1174.
24. Mombouli J-V, Vanhoutte PM. Endothelium-derived hyperpolarizing factor(s): updating the unknown. *Trends Pharmacol Sci* 1997; **18**: 252–256.
25. Hoeffner V, Boulanger C, Vanhoutte PM. Proximal and distal coronary arteries respond differently to basal EDRF but not to NO. *Am J Physiol* 1989; **256**: H828–H831.
26. Golino P, Piscione F, Wilterson JT *et al.* Divergent effects of serotonin on coronary artery dimensions and blood flow in patients with coronary atherosclerosis and control patients. *N Engl J Med* 1991; **324**: 641–648.
27. McFadden EP, Clarke JG, Davies GJ *et al.* Effect of intracoronary serotonin on coronary vessels in patients with stable angina and patients with variant angina. *N Engl J Med* 1991; **324**: 648–654.
28. Crossman DC, Larkin SW, Fuller RW *et al.* Substance P dilates epicardial coronary arteries and increases coronary blood flow in humans. *Circulation* 1989; **80**: 475–484.
29. Crossman DC, Larkin SW, Dashwood MR *et al.* Responses of atherosclerotic human coronary arteries *in vivo* to the endothelium-dependent vasodilator substance P. *Circulation* 1991; **84**: 2001–2010.
30. Schretzenmayr A. Über kreislaufregulatorische Vorgänge an den grossen Arterien bei der Muskelarbeit. *Pflügers Arch* 1933; **232**: 743–748.

31. Pohl U, Herlan K, Huang A, Bassenge E. EDRF-mediated shear-induced dilation opposes myogenic vasoconstriction in small rabbit arteries. *Am J Physiol* 1991; **261**: H2016–H2023.
32. Gerova M, Gero J, Barta E *et al.* Neurogenic and myogenic control of conduit coronary arteries: a possible interference. *Basic Res Cardiol* 1981; **76**: 503–507.
33. Kuo J, Davis MJ, Chilian WM. Endothelium-dependent, flow-induced dilation of isolated coronary arteries. *Am J Physiol* 1990; **259**: H1063–H1070.
34. Hintze TH, Vatner SF. Reactive dilation of large coronary arteries in conscious dogs. *Circ Res* 1984; **54**: 50–57.
35. Holtz J, Förstermann U, Pohl U *et al.* Flow-dependent, endothelium-mediated dilation of epicardial coronary arteries in conscious dogs: effects of cyclooxygenase inhibition. *J Cardiovasc Pharmacol* 1984; **6**: 1161–1169.
36. Drexler H, Zeiher AM, Wollschläger H *et al.* Flow-dependent coronary artery dilatation in humans. *Circulation* 1989; **80**: 466–474.
37. Nabel EG, Selwyn AP, Ganz P. Large coronary arteries in humans are responsive to changing blood flow: an endothelium dependent mechanism that fails in patients with atherosclerosis. *J Am Coll Cardiol* 1990; **16**: 349–356.
38. Inoue T, Tomoike H, Hisano K, Nakamura M. Endothelium determines flow-dependent dilation of the epicardial coronary artery in dogs. *J Am Coll Cardiol* 1988; **11**: 187–191.
39. Kuo L, Chilian WM, Davis MJ. Interaction of pressure-and flow-induced responses in porcine coronary resistance vessels. *Am J Physiol* 1991; **261**: H1706–H1715.
40. Canty JM, Schwartz JS. Nitric oxide mediates flow dependent epicardial coronary vasodilation to changes in pulse frequency but not mean flow in conscious dogs. *Circulation* 1994; **89**: 375–384.
41. Groves P, Kurz S, Just H, Drexler H. Role of endogenous bradykinin in human coronary vasomotor control. *Circulation* 1995; **92**: 3424–3430.
42. Nakashima M, Mombouli JV, Taylor AA, Vanhoutte PM. Endothelium-dependent hyperpolarization caused by bradykinin in human coronary arteries. *J Clin Invest* 1993; **92**: 2867–2871.
43. Kauser K, Stekiel WJ, Rubanyi G, Harder DR. Mechanism of action of EDRF on pressurized arteries: effect on K^+ conductance. *Circ Res* 1989; **65**: 199–204.
44. Nabel EG, Ganz P, Gordon JB *et al.* Dilation of normal and constriction of atherosclerotic coronary arteries caused by the cold pressure test. *Circulation* 1988; **77**: 43–52.
45. Gordon JB, Ganz P, Nabel EG *et al.* Atherosclerosis influences the vasomotor response of epicardial coronary arteries to exercise. *J Clin Invest* 1989; **83**: 1946–1952.

46. Bernstein RD, Ochoa FY, Xu X *et al.* Function and production of nitric oxide in the coronary circulation of the conscious dog during exercise. *Circ Res* 1996; **79**: 840–848.
47. Berdeaux A, Ghaleh B, Dubois-Rande JL *et al.* Role of vascular endothelium in exercise-induced dilation of large epicardial coronary arteries in conscious dogs. *Circulation* 1994; **89**: 2799–2808.
48. Schwartz JS, Baran KW, Bache RJ. Effect of stenosis on exercise induced dilation of large coronary arteries. *Am Heart J* 1990; **119**: 520–524.
49. Wang J, Wolin MS, Hintze TH. Chronic exercise enhances endothelium mediated dilation of epicardial coronary artery in conscious dogs. *Circ Res* 1993; **73**: 829–838.
50. Quyyumi AA, Dakak N, Andrews NP *et al.* Contribution of nitric oxide to metabolic coronary vasodilation in the human heart. *Circulation* 1995; **92**: 320–326.
51. Feigl EO. Coronary physiology. *Physiol Rev* 1983; **63**: 1–205.
52. Shen W, Lundborg M, Wang J *et al.* Role of EDRF in the regulation of regional blood flow and vascular resistance at rest and during exercise in conscious dogs. *J Appl Physiol* 1994; **77**: 165–172.
53. Dunker DJ, Bache RJ. Inhibition of nitric oxide production aggravates myocardial hypoperfusion during exercise in the presence of a coronary artery stenosis. *Circ Res* 1994; **74**: 629–640.
54. Altman JD, Kinn J, Duncker DJ, Bache RJ. Effect of inhibition of nitric oxide formation on coronary blood flow during exercise in the dog. *Cardiovasc Res* 1994; **28**: 119–124.
55. Smith TP Jr, Canty JM Jr. Modulation of coronary autoregulatory responses by nitric oxide. Evidence for flow-dependent resistance adjustments in conscious dogs. *Circ Res* 1993; **73**: 232–240.
56. Park KH, Rubin LE, Gross SS, Levi RR. Nitric oxide is a mediator of hypoxic coronary vasodilatation. Relation to adenosine and cyclooxygenase-derived metabolites. *Circ Res* 1992; **71**: 992–1001.
57. Kostic MM, Schrader J. Role of nitric oxide in reactive hyperaemia of the guinea pig heart. *Circ Res* 1992; **70**: 208–212.
58. Parent R, Pare R, Lavallée M. Contribution of nitric oxide to dilation of resistance coronary vessels in conscious dogs. *Am J Physiol* 1992; **262**: H10–H16.
59. Ueda M, Silvia SK, Olsson RA. Nitric oxide modulates coronary autoregulation in the guinea pig. *Circ Res* 1992; **70**: 1296–1303.
60. Kanatsuka H, Lamping KG, Eastham CL *et al.* Comparison of the effects of increased myocardial oxygen consumption and adenosine on the coronary microvascular resistance. *Circ Res* 1989; **65**: 1296–1305.
61. Griffith TM, Edwards DH, Davies RL *et al.* Endothelium-derived relaxing factor (EDRF) and resistance vessels in an intact vascular bed: a

microangiographic study of the rabbit isolated ear. *Br J Pharmacol* 1988; **93**: 654–662.
62. Reid JV, Ito BR, Huang AH *et al.* Parasympathetic control of transmural coronary blood flow in dogs. *Am J Physiol* 1985; **249**: H337–H343.
63. Ito BR, Feigl EO. Carotid baroreceptor reflex coronary vasodilation in the dog. *Circ Res* 1985; **56**: 486–495.
64. Ito BR, Feigl EO. Carotid chemoreceptor reflex parasympathetic coronary vasodilation in the dog. *Am J Physiol* 1985; **249**: H1167–H1175.
65. Béa M-L, Ghaleh B, Giudicelli J-F, Berdeaux A. Lack of importance of NO in β-adrenoceptor-mediated relaxation of large epicardial canine coronary arteries. *Br J Pharmacol* 1994; **111**: 981–982.
66. Rubanyi G, Vanhoutte PM. Endothelium-removal decreases relaxations of canine coronary arteries caused by β-adrenergic agonists and adenosine. *J Cardiovasc Pharmacol* 1985; **7**: 139–144.
67. Parent R, Al-Obaidi M, Lavallée M. Nitric oxide formation contributes to β-adrenergic dilation of resistance coronary vessels in conscious dogs. *Circ Res* 1993; **73**: 241–251.
68. Ming Z, Parent R, Lavallée M. β_2-adrenergic dilation of resistance coronary vessels involves K_{ATP} channels and nitric oxide in conscious dogs. *Circulation* 1997; **95**: 1568–1576.
69. Cohen RA, Zitnay KM, Haudenschild CC, Cunningham LD. Loss of selective endothelial cell vasoactive functions caused by hypercholesterolemia in pig coronary arteries. *Circ Res* 1988; **63**: 903–910.
70. Zeiher AM, Drexler H, Wollschläger H, Just H. Endothelial dysfunction of the coronary microvasculature is associated with coronary blood flow regulation in patients with early atherosclerosis. *Circulation* 1991; **84**: 1984–1992.
71. McLenachan JM, Williams JK, Fish D *et al.* Loss of flow-mediated endothelium-dependent dilation occurs early in the development of atherosclerosis. *Circulation* 1991; **84**: 1273–1278.
72. Cox DA, Vita JA, Treasure CB *et al.* Atherosclerosis impairs flow-mediated dilation of coronary arteries in humans. *Circulation* 1989; **80**: 458–465.
73. Pitkänen OP, Raitakari OT, Niinikoski H *et al.* Coronary flow reserve is impaired in young men with familial hypercholesterolemia. *J Am Coll Cardiol* 1996; **28**: 1705–1711.
74. el-Tamimi H, Mansour M, Wargovich TJ *et al.* Constrictor and dilator responses to intracoronary acetylcholine in adjacent segments of the same coronary artery in patients with coronary artery disease. Endothelial function revisited. *Circulation* 1994; **89**: 45–51.
75. Zeiher AM, Schächlinger V, Hohnloser SH *et al.* Coronary atherosclerotic wall thickening and vascular reactivity in humans. Elevated high-density lipoprotein levels ameliorate abnormal vasoconstriction in early atherosclerosis. *Circulation* 1994; **89**: 2525–2532.

76. Gage JE, Hess OM, Murakami T *et al.* Vasoconstriction of stenotic coronary arteries during dynamic exercise in patients with classic angina pectoris: reversibility by nitroglycerin. *Circulation* 1986; **73**: 865–876.
77. Yeung AC, Vekshtein VL, Krantz DS *et al.* The effect of atherosclerosis on the vasomotor response of coronary arteries to mental stress. *N Engl J Med* 1991; **325**: 1551–1556.
78. Zeiher AM, Drexler H, Wollschläger H, Just H. Modulation of coronary vasomotor tone in humans. Progressive endothelial dysfunction with different early stages of coronary atherosclerosis. *Circulation* 1991; **83**: 391–401.
79. Chester AH, O'Neil GS, Moncada S *et al.* Low basal and stimulated release of nitric oxide in atherosclerotic epicardial coronary arteries. *Lancet* 1990; **336**: 897–900.
80. Quyyumi AA, Dakak N, Mulcahy D *et al.* Nitric oxide activity in the atherosclerotic human coronary circulation. *J Am Coll Cardiol* 1997; **29**: 308–317.
81. Minor RL Jr, Myers PR, Guerra R Jr *et al.* Diet-induced atherosclerosis increases the release of nitrogen oxides from rabbit aorta. *J Clin Invest* 1990; **86**: 2109–2116.
82. Lang D, Smith JA, Lewis MJ. Induction of a calcium-independent NO synthase by hypercholesterolaemia in the rabbit. *Br J Pharmacol* 1993; **198**: 290–293.
83. Ohara Y, Peterson TE, Harrison DG. Hypercholesterolemia increases endothelial superoxide anion production. *J Clin Invest* 1993; **91**: 2546–2551.
84. White CR, Brock TA, Chang LY *et al.* Superoxide and peroxynitrite in atherosclerosis. *Proc Natl Acad Sci USA* 1994; **91**: 1044–1048.
85. Villa LM, Salas E, Darley-Usmar VM *et al.* Peroxynitrite induces both vasodilatation and impaired vascular relaxation in the isolated perfused rat heart. *Proc Natl Acad Sci USA* 1994; **91**: 12383–12387.
86. Ting HH, Timimi FK, Haley EA *et al.* Vitamin C improves endothelium-dependent vasodilation in forearm resistance vessels of humans with hypercholesterolemia. *Circulation* 1997; **95**: 2617–2622.
87. Levine GN, Frei B, Koulouris SN *et al.* Ascorbic acid reverses endothelial vasomotor dysfunction in patients with coronary artery disease. *Circulation* 1996; **93**: 1107–1113.
88. Liao JK, Shin WS, Lee WY, Clark SL. Oxidized low-density lipoprotein decreases the expression of endothelial nitric oxide synthase. *J Biol Chem* 1995; **270**: 319–324.
89. Rubanyi GM. Reversal of hypercholesterolemia-induced endothelial dysfunction by L-arginine. *Circulation* 1991; **83**: 1118–1120.
90. Girerd XJ, Hirsch AT, Cooke JP *et al.* L-arginine augments endothelium-dependent vasodilation in cholesterol-fed rabbits. *Circ Res* 1990; **67**: 1301–1308.

91. Drexler H, Zeiher AM, Meinzer K, Just H. Correction of endothelial dysfunction in coronary microcirculation of hypercholesterolaemia patients by L-arginine. *Lancet* 1991; **338**: 1546–1550.
92. Arnal JF, Munzel T, Venema RC *et al*. Interactions between L-arginine and L-glutamine change endothelial NO production. An effect independent of NO synthase substrate availability. *J Clin Invest* 1995; **95**: 2565–2572.
93. Stroes E, Kastelein J, Cosentino F *et al*. Tetrahydrobiopterin restores endothelial function in hypercholesterolemia. *J Clin Invest* 1997; **99**: 41–46.
94. Leung WH, Lau CP, Wong CK. Beneficial effect of cholesterol-lowering therapy on coronary endothelium dependent relaxation in hypercholersterolaemia patients. *Lancet* 1993; **341**: 1496–1500.
95. Anderson TJ, Meredith IT, Yeung AC *et al*. The effect of cholesterol-lowering and antioxidant therapy on endothelium-dependent coronary vasomotion. *N Engl J Med* 1995; **332**: 488–493.
96. O'Driscoll G, Green D, Taylor RR. Simvastatin, an HMG-coenzyme A reductase inhibitor, improves endothelial function within 1 month. *Circulation* 1997; **95**: 1126–1131.
97. Tamai O, Matsuoka H, Itabe H *et al*. Single LDL apheresis improves endothelium-dependent vasodilatation in hypercholesterolemic humans. *Circulation* 1997; **95**: 76–82.
98. Winkles JA, Alberts GF, Brogi E, Libby P. Endothelin-1 and endothelin receptor mRNA expression in normal and atherosclerotic human arteries. *Biochem Biophys Res Commun* 1993; **191**: 1081–1088.
99. Reddy KG, Nair RN, Sheehan HM, Hodgson JM. Evidence that selective endothelial dysfunction may occur in the absence of angiographic or ultrasound atherosclerosis in patients with risk factors for atherosclerosis. *J Am Coll Cardiol* 1994; **23**: 833–843.
100. Calver A, Collier J, Vallance P. Forearm blood flow responses to a nitric oxide synthase inhibitor in patients with treated essential hypertension. *Cardiovasc Res* 1994; **28**: 1720–1725.
101. Linder L, Kiowski W, Buhler FR, Lüscher TF. Indirect evidence for release of endothelium-derived relaxing factor in human forearm circulation *in vivo*. Blunted response in essential hypertension. *Circulation* 1990; **81**: 1762–1767.
102. Panza JA, Quyyumi AA, Brush JE Jr, Epstein SE. Abnormal endothelium-dependent vascular relaxation in patients with essential hypertenson. *N Engl J Med* 1990; **323**: 22–27.
103. Taddei S, Virdis A, Mattei P *et al*. Defective L-arginine-nitric oxide pathway in offspring of essential hypertensive patients. *Circulation* 1996; **94**: 1298–1303.
104. Cockcroft JR, Chowienczyk PJ, Benjamin N, Ritter JM. Preserved endothelium-dependent vasodilatation in patients with essential hypertension. *N Engl J Med* 1994; **330**: 1036–1040.

105. Treasure CB, Manoukian SV, Klein JL *et al.* Epicardial coronary artery responses to acetylcholine are impaired in hypertensive patients. *Circ Res* 1992; **71**: 776–781.
106. Treasure CB, Klein JL, Vita JA *et al.* Hypertension and left ventricular hypertrophy are associated with impaired endothelium-mediated relaxation in human coronary resistance vessels. *Circulation* 1993; **87**: 86–93.
107. Vita JA, Treasure CB, Nabel EG *et al.* Coronary vasomotor response to acetylcholine relates to risk factors for coronary artery disease. *Circulation* 1990; **81**: 491–497.
108. Zeiher AM, Drexler H, Saurbier B, Just H. Endothelium-mediated coronary blood flow modulation in humans. Effects of age, atherosclerosis, hypercholesterolemia and hypertension. *J Clin Invest* 1993; **92**: 652–662.
109. Quyyumi AA, Mulcahy D, Andrews NP *et al.* Coronary vascular nitric oxide activity in hypertension and hypercholesterolemia. Comparison of acetylcholine and substance P. *Circulation* 1997; **95**: 104–110.
110. Nitenberg A, Antony I. Epicardial coronary arteries are not adequately sized in hypertensive patients. *J Am Coll Cardiol* 1996; **27**: 115–123.
111. Kubo SH, Rector TS, Bank AJ *et al.* Endothelium-dependent vasodilation is attenuated in patients with heart failure. *Circulation* 1991; **84**: 1589–1596.
112. Drexler H, Hayoz D, Münzel T *et al.* Endothelial function in chronic congestive heart failure. *Am J Cardio*l 1992; **69**: 1596–1601.
113. Nitenberg A, Foult JM, Blanchet F, Zouioueche S. Multifactorial determinants of reduced coronary flow reserve after dipyridamole in dilated cardiomyopathy. *Am J Cardiol* 1985; **55**: 748–754.
114. Wang J, Seyedi N, Xu X-B *et al.* Defective endothelium-mediated control of coronary circulation in conscious dogs after heart failure. *Am J Physiol* 1994; **266**: H670–H680.
115. Treasure CB, Vita JA, Cox DA *et al.* Endothelium-dependent dilation of the coronary microvasculature is impaired in dilated cardiomyopathy. *Circulation* 1990; **81**: 772–779.
116. Zhao G, Shen W, Xu X *et al.* Selective impairment of vagally mediated, nitric oxide-dependent coronary vasodilation in conscious dogs after pacing-induced heart failure. *Circulation* 1995; **91**: 2655–2663.
117. O'Murchu B, Miller VM, Perrella MA, Burnett JC Jr. Increased production of nitric oxide in coronary arteries during congestive heart failure. *J Clin Invest* 1994; **93**: 165–171.

Chapter 12

The Effect of Nitric Oxide on Myocardial Function and Dysfunction

Bernard D Prendergast and Ajay M Shah

Background

The first report of a role for NO in modulating myocardial contraction was published as recently as 1991.[1] Since then, there has been a profusion of information concerning its physiological and pathophysiological roles within the heart. This review summarizes recent research in this rapidly growing field. Space constraints make it impossible to quote all original articles, but most of these can be accessed from the manuscripts cited here or from recent comprehensive reviews.[2,3]

NO can potentially influence cardiac function indirectly secondary to alterations in: (i) arterial tone and therefore afterload, (ii) venous return and capacitance, and thus ventricular filling volume, thereby modulating stroke volume via the Frank-Starling mechanism, and (iii) coronary vascular tone. In addition to these effects, NO has direct actions on cardiac myocytes — the focus of the present review. The precise effects on myocardial function in any given situation are likely to depend upon many factors: the cellular source of NO, the amount released (isoform-dependent), stimuli such as coronary flow rate and heart rate (which alter fluid shear stress), the target tissue (e.g. atrial or ventricular myocyte), interactions with autonomic and other stimuli, and the presence of immune activation or disease.

Nitric Oxide Synthase in the Heart

All three NO synthase isoforms (nNOS, iNOS, eNOS) can be expressed in the heart, albeit in a cell-specific manner. nNOS ('neuronal' NOS) appears to be expressed only in nervous tissue: pre- and post-ganglionic fibres innervating the sinoatrial and atrioventricular nodes, subepicardial neuronal cells, and intrinsic cardiac neurons.[4] iNOS ('inducible' NOS) expression occurs in infiltrating inflammatory cells, coronary microvascular and endocardial endothelium, vascular smooth muscle, fibroblasts and cardiac myocytes, in response to cytokine stimulation or in disease states.[2,3] eNOS ('endothelial-type' NOS) is constitutively expressed in several cell types, the highest level being in microvascular and endocardial endothelium;[5,6] lower levels are detectable in cardiac myocytes and in sinoatrial and atrioventricular nodal cells.[3,5,6]

Subcellular Actions of Nitric Oxide in Myocardium

Although the general subcellular actions of NO are considered elsewhere in this book, specific aspects pertinent to the heart are discussed here (see Figure 1). Many of the cardiac actions of NO appear to be mediated through guanosine 3',5'-cyclic monophosphate (cGMP).[7] The major myocardial target proteins for cGMP are cGMP-dependent protein kinases (PKG) and cGMP-regulated cyclic nucleotide

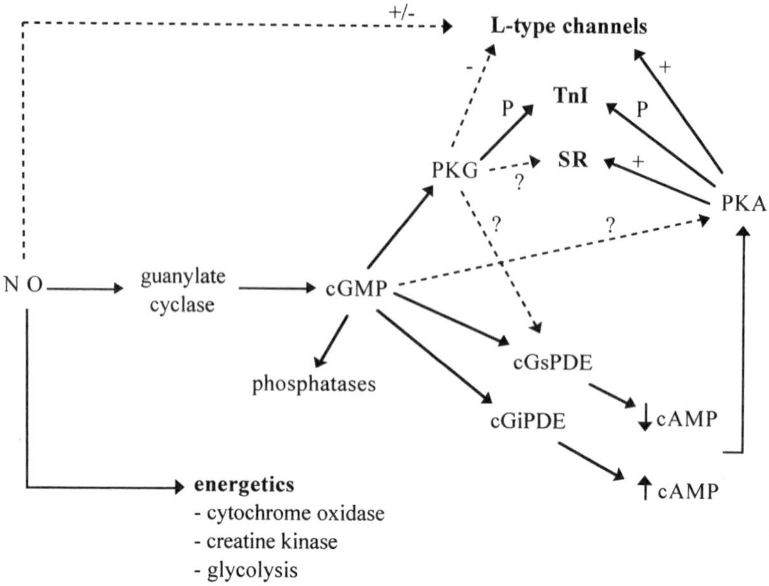

Figure 1. Subcellular actions of NO in myocardium. TnI, troponin I; SR, sarcoplasmic reticulum; P, phosphorylation; cGMP, guanosine 3',5'-cyclic monophosphate; cAMP, adenosine 3',5'-cyclic monophosphate; PKG, cGMP-dependent protein kinase; PKA, cAMP-dependent protein kinase; cGsPDE, cGMP-stimulated phosphodiesterase; cGiPDE, cGMP-inhibited phosphodiesterase.

phosphodiesterases (PDE), although cGMP-activated phosphatases and cGMP-regulated ion channels could also be important. Downstream signalling cascades are potentially complex due to the presence of multiple PKG and PDE isoforms (with different cGMP sensitivities), cross-reaction between PKG and adenosine 3',5'-cyclic monophosphate (cAMP)-dependent kinases, and the likely subcellular compartmentation of cGMP and its target proteins. Reported cardiac actions of cGMP include: (i) modulation of sarcolemmal Ca^{2+} influx, (ii) reduction in myofilament Ca^{2+} sensitivity, (iii) changes in the action potential, (iv) altered sarcoplasmic reticular function, (v) modulation of cell volume, and (vi) reduction in oxygen consumption. These actions are affected by factors such as tissue, species, amount and source of NO, cyclic nucleotide content, and pre-existing levels of PDE activity and protein phosphorylation.

Some data support cGMP-independent actions of NO in myocardium, including modulation of sarcolemmal Ca^{2+} influx secondary to *S*-nitrosylation of Ca^{2+} channels,[8] changes in mitochondrial respiration, and in glycolytic enzyme and creatine kinase activity (see *NO and myocardial energetics* below). The large amounts of NO generated by iNOS may exert poorly reversible, toxic effects, e.g. on mitochondrial enzymes and nucleic acids. Excessive NO production *per se* or in the context of increased oxygen-derived free radical production (e.g. during myocardial reperfusion) leads to peroxynitrite ($OONO^-$) formation,[9] which may have deleterious effects on myocardium.[10] Peroxynitrite formation depends critically upon the redox balance between NO and superoxide and on endogenous antioxidant status.

Endothelial and Neuronal Nitric Oxide Synthases in Myocardium: Physiological and Pathophysiological Roles

Several different functional roles have been suggested for NO generated by eNOS in cardiac endothelial cells and myocytes, while nNOS is implicated in the regulation of heart rate. Many of these effects are relatively subtle (e.g. selective changes in myocardial relaxation, the force-frequency relation, or the Frank-Starling response) and may easily be missed on cursory assessment of contractile function, as with the measurement of force alone. This may account, at least in part, for some of the variable results discussed below. The physiological stimuli that regulate eNOS activation in cardiac cells and the associated downstream signalling cascades remain poorly defined.

Myocardial Relaxation and Diastolic Function

Experimental studies in several different species indicate a role for endothelium-derived NO in modulating basal myocardial relaxation and diastolic tone (i.e. in the absence of myocardial stimulation by other agonists) (Figure 2). Smith and colleagues showed in isolated ferret papillary muscles that substance P-induced stimulation of NO release from endocardial endothelium caused selective effects on myocardial relaxation.[1] The onset of isometric relaxation occurred earlier, thereby abbreviating contractile duration, but with only minor reduction in peak force or rate of force development. Similar effects were observed with substance P in cat papillary muscle,[11] and with the exogenous NO donor, sodium nitroprusside, or a cGMP analogue in ferret papillary muscle.[12] These effects of NO/cGMP differ from most other inotropic interventions, e.g. cAMP- or Ca^{2+}-mediated actions, which substantially alter early systolic parameters (e.g. maximal rate of force development).

In isolated ejecting ('working') guinea-pig hearts, studied under conditions of constant loading and heart rate, both sodium nitroprusside and specific agonists for release of NO (substance P, bradykinin) induced a dose-dependent acceleration and earlier onset of left ventricular (LV) pressure fall without any change in peak LV pressure or LV dP/dt_{max}.[13,14] These LV relaxant effects were inhibited by haemoglobin and were independent of changes in coronary flow, implying a direct effect of NO on cardiac myocytes in the intact heart. Although the cellular source of NO was not definitively established in these studies, it is most likely

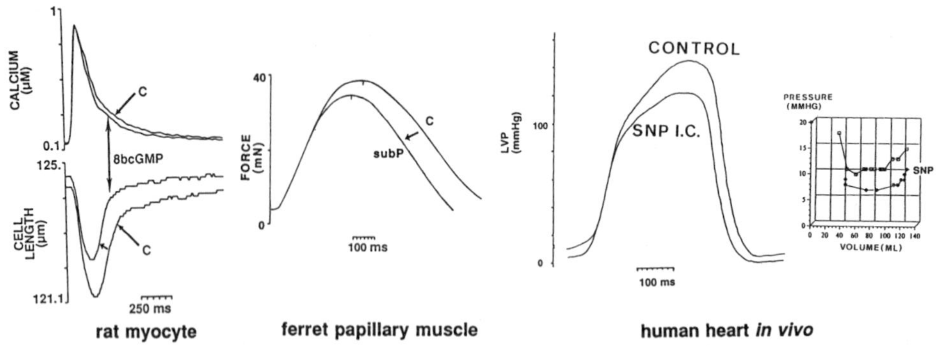

Figure 2. Effects of NO on myocardial relaxation and diastolic function. Left panel: effect of 8bromo-cGMP (8bcGMP 50 µM) on cell shortening (bottom) and cytosolic calcium (top) in an isolated indo-1 loaded rat cardiac myocyte (Reference 15). Middle panel: effect of substance P (subP) on isolated ferret papillary muscle (Reference 1). Right panel: effect of a 5 min low-dose bicoronary sodium nitroprusside (SNP) infusion on high fidelity left ventricular pressure (LVP) in a human subject with normal cardiac function (Reference 18). The inset shows a downward shift in the diastolic pressure-volume relation. c, control trace in each case.

to have been the coronary microvascular endothelial cells. These data support the hypothesis that endothelium-derived NO selectively modulates LV relaxation in the isolated whole heart.

In isolated rat ventricular myocytes, 8-bromo-cGMP induced a similar pattern of effects, i.e. an earlier onset of isotonic twitch relaxation and reduction in twitch shortening, without significant reduction in shortening velocity.[15] The underlying mechanism appeared to be a reduction in myofilament response to Ca^{2+}, since no change in cytosolic Ca^{2+} transient was observed (Figure 2). In intact myocytes tetanized by high frequency stimulation following thapsigargin-induced inhibition of the sarcoplasmic reticulum Ca^{2+} ATPase, 8-bromo-cGMP decreased steady-state tetanic shortening without reducing peak Ca^{2+} — providing further support for this hypothesis. These effects were inhibited by KT5823, a PKG inhibitor.[15] A previous study in skinned porcine cardiac fibres also reported a PKG-induced reduction in myofilament Ca^{2+} sensitivity.[16]

Interestingly, 8-bromo-cGMP also induces a Ca^{2+}-independent increase in diastolic length both in electrically stimulated and quiescent rat myocytes, which may reflect a cGMP-induced reduction in intrinsic diastolic tone.[15] Possible underlying mechanisms include a reduction in myofilament response to Ca^{2+}, and thus in force produced by diastolic cross-bridge cycling, or a cGMP-induced change in properties of sarcomeric proteins such as titin which contribute to diastolic tone.[17] NO- or cGMP-induced changes in diastolic properties have not been adequately addressed in isolated muscle or isolated heart studies (but see *NO and the Frank-Starling response* below). However, recent human studies indicate a similar effect of NO.

Paulus *et al.* studied the effects of 5 minute low-dose bicoronary sodium nitroprusside infusion in subjects with normal coronary arteries undergoing diagnostic cardiac catheterization.[18] Nitroprusside infusion resulted in an earlier onset of LV relaxation (reduced time to LV dP/dt_{min}) and a reduction in peak and end-systolic LV pressure, but no change in LV dP/dt_{max} or isovolumic relaxation. Perhaps the most striking effect, however, was a reduction in LV end-diastolic pressure with a slight increase in end-diastolic volume, i.e. a downward shift in the end-diastolic pressure-volume relation, consistent with an increase in LV end-diastolic distensibility (Figure 2). This was unrelated to systemic vasodilatation, since right atrial infusion of an identical dose of nitroprusside failed to reproduce the effects, nor was it accounted for by changes in ventricular interaction or pericardial pressure, based on estimation of transmural LV end-diastolic pressure. Despite the reduction in end-systolic pressure, no change in ejection fraction or stroke volume was observed, presumably because improved diastolic distensibility compensated for the decline in systolic function. The same group reported similar changes in LV function after coronary endothelial stimulation with substance P in normal human subjects and cardiac transplant recipients free of rejection or graft vasculopathy using a similar protocol.[19] These studies provide the first evidence for a paracrine coronary endothelial influence of NO on myocardial contractile function and diastolic distensibility *in vivo*, presumably mediated through cGMP.

The most relevant physiological stimuli for release of NO from coronary endothelial cells *in vivo* are likely to be flow-induced shear stress and the cyclical mechanical deformation of the endothelium that occurs during the cardiac cycle. Augmented release of NO during increases in heart rate and/or coronary flow, e.g. during exercise, would enhance both ventricular relaxation and diastolic distensibility. The consequence would be to increase ventricular filling and improve subendocardial coronary perfusion, secondary to an increased diastolic interval, a greater driving pressure for filling, and reduced extravascular compressive forces. This could help to optimize global pump function by matching the pattern of ventricular function to coronary flow and/or heart rate. The anatomical level at which such regulation may occur is uncertain: the role of eNOS in capillary and venular endothelium is likely to be different from that in arteries and arterioles, since shear forces and mechanical stimuli for NO release will vary among these sites.

Other Effects of NO on Basal Myocardial Contraction

Some authors have failed to observe any effect of NO (or of NOS inhibition) on basal myocardial function. Weyrich *et al.* reported that NO donors had no effect on isometric force in cat or rat papillary muscle or on rat myocyte shortening.[20] Crystal & Gurevicius found no effect of intracoronary nitroprusside or of NOS inhibitors on regional segment shortening in open-chest dogs when coronary blood flow was maintained constant.[21] Neither of these studies adequately characterized effects on relaxation or diastolic properties. Brady *et al.* found that both authentic NO in solution and NO donor drugs exerted slight negative inotropic effects in guinea-pig myocytes[22] (consistent with the data in rat myocytes),[15] but noted no change in relaxation.

In rat myocytes, Kojda *et al.* reported that NO donors exerted significant positive inotropic effects at low doses but were negatively inotropic at high doses.[23] The former was related to cGMP-mediated inhibition of PDE with consequent increase in cAMP levels, while the latter was attributed to activation of PKG. Similar biphasic effects were reported in cat papillary muscle.[11] Campbell *et al.* reported both enhancement and reduction of basal Ca^{2+} current by NO donors in ferret ventricular myocytes, apparently *via* direct effects of NO on Ca^{2+} channels.[8] However, the very high (millimolar) doses of drug used in this study are of doubtful physiological relevance. In general, biphasic effects of NO involving cGMP-mediated modulation of PDE activities have been noted more often in preparations pre-stimulated with β-adrenergic agonists than in those investigated under basal conditions (see below).

NO and the Frank-Starling Response

In view of the fact that a change in myofilament responsiveness to Ca^{2+} is involved in both the Frank-Starling response and some of the effects of NO,[2] and the knowledge that diastolic function influences the manifestation of the Frank-Starling response in the whole heart, the potential interaction between NO and the Frank-Starling response has recently been investigated. In isolated ejecting guinea-pig hearts, inhibition or inactivation of endogenous NO, with N^G-monomethyl-L-arginine (L-NMMA) or haemoglobin respectively, significantly attenuated preload-induced increases in cardiac output independently of changes in coronary flow.[24] Interestingly, L-NMMA effects were only apparent at higher preloads, with no reduction in cardiac output at lower preload. L-NMMA effects were abolished by concurrent treatment with L-arginine, and were not mimicked by similar or greater coronary vasoconstriction using prostaglandin $F_{2\alpha}$. Thus, this study suggests that endogenous NO facilitates the Frank-Starling response. The cellular source of this NO (i.e. whether endothelial cell or cardiac myocyte) is uncertain. The underlying mechanism of the above interaction may involve either an NO-induced increase in diastolic chamber distensibility (thus facilitating ventricular filling when preload is raised) and/or a length-dependent increase in NO release or activity.[24] The latter possibility is intriguing since eNOS is localized in plasmalemmal caveolae, which are postulated to be sites where function of proteins may be modulated by many signals including mechanical forces.[25] The possibility arises that eNOS activation in the heart could be directly responsive to the mechanical signal of stretch during the cardiac cycle.

NO and the Force-frequency Relationship

Two groups have reported an interaction between endogenous NO and the contractile response to altered heart rate (the force-frequency relation). The negative force-frequency relation normally observed in isolated hamster papillary muscle was altered to a positive one by L-NMMA.[26] These effects of L-NMMA were mimicked by methylene blue, an inhibitor of guanylate cyclase, and blunted by nitroprusside. In rat ventricular myocytes, both a NOS inhibitor and putative guanylate cyclase inhibitors (methylene blue, LY83583) augmented the positive frequency-shortening relation.[27] An increase in frequency was associated with increased myocyte nitrite production, which was inhibited by the Ca^{2+} chelator 1,2-bis(o-aminophenoxy)ethane-N,N,N',N'-tetraacetic acid (BAPTA). This finding suggests that eNOS activation was increased at higher frequencies due to the associated increase in Ca^{2+} transient amplitude. The implication from these studies is that endogenous NO may inhibit contraction at higher frequencies. Although a cGMP-dependent signalling pathway is implicated, the downstream actions remain to be elucidated.

Inotropic Interactions of NO with the β-adrenergic Pathway

Convincing evidence indicates that NO can influence the inotropic response to β-adrenergic stimulation. Balligand *et al.* reported that the inotropic response of adult rat ventricular myocytes to the β-agonist, isoproterenol, was enhanced by about 30% following treatment with a NOS inhibitor; this was associated with a decrease in myocyte NO production.[28] They concluded that endogenous myocyte NO inhibits the inotropic response to β-adrenergic agonists, an hypothesis strengthened by the demonstration that eNOS is expressed in rat myocytes.[5] In isolated rat hearts, Ebihara & Karmazyn reported that the inotropic response to isoproterenol was reduced by NO donors.[29] Analogous data were obtained in open-chest anaesthetized and autonomically blocked dogs, in which the LV dP/dt_{max} response to intracoronary β-agonists was augmented by an intracoronary NOS inhibitor, N^G-nitro-L-arginine methyl ester (L-NAME), apparently independently of changes in systemic loading.[30] However, in chronically instrumented dogs, NOS inhibition with L-NAME had no effect on β-adrenergic contractile response.[31] It remains to be established if these conflicting results relate to the differences in experimental conditions.

Recent human studies support an interaction between NO and the β-adrenergic system. In patients with severe non-ischaemic dilated cardiomyopathy, intracoronary L-NMMA potentiated the increase in LV dP/dt_{max} in response to dobutamine infusion by about 30%.[32] Using a different approach, Bartunek *et al.* compared the effects of intracoronary substance P infusion with or without concurrent intravenous dobutamine in a group of cardiac transplant recipients and patients with dilated cardiomyopathy.[33] They reported that substance P induced potent cardiodepressant effects in the presence of the β-agonist, indicating an interaction between endothelium-derived NO and β-adrenergic response. No patients with normal LV function were included in either of these studies.

Based on studies in isolated myocytes, the subcellular mechanism underlying these interactions is likely to involve modulation of cAMP PDE activity (Figure 1). An exogenous NO donor, SIN-1, decreased β-adrenergic stimulated L-type Ca^{2+} current in frog[34] and guinea-pig cardiac myocytes,[35] *via* stimulation of a cAMP PDE and subsequent reduction in cAMP levels.

Elevation of cGMP can also inhibit a cAMP PDE with a consequent rise in cAMP levels and potential positive inotropic effects (Figure 1). Indeed, in the myocyte studies mentioned above,[34,35] low concentrations of SIN-1 *increased* cAMP-stimulated Ca^{2+} current, whereas higher concentrations resulted in a decrease. These data are similar to those of Kojda *et al.* in rat myocytes,[23] and Mohan *et al.* in cat papillary muscle,[11] discussed above, except that no β-adrenergic stimulation was employed in those studies. We have also observed biphasic effects of NO donors on the inotropic response to dobutamine in isolated guinea-pig hearts (Prendergast & Shah, unpublished data).

The above studies provide compelling evidence that inotropic interactions may occur between NO and the β-adrenergic system. Several issues, however, remain unclear. What are the circumstances under which these interactions result in enhancement rather than reduction in β-response? Do the effects of endothelial NO or exogenous NO differ from those of myocyte-derived NO? Are there species differences? What is the influence of myocardial disease processes? What happens when iNOS is expressed?

Mediation of Cholinergic Responses

It has long been known that cholinergic agonists (e.g. acetylcholine) elevate myocardial cGMP, but most studies in the 'pre-NO' era concluded that this was not responsible for its myocardial actions.[7] Indeed, cholinergic activation results in cGMP-independent effects, such as G-protein-mediated changes in K^+ current, phosphatidylinositol hydrolysis, and reduced adenylyl cyclase activity. The appreciation that acetylcholine activates eNOS in the vascular endothelium led to a resurgence of interest in myocardial cholinergic signalling. Balligand *et al.* reported that carbachol-induced decreases in spontaneous beating frequency of neonatal rat myocytes were inhibited either by haemoglobin or L-NMMA.[28] The same group reported that cholinergic inhibition of isoproterenol-stimulated increases in Ca^{2+} current and contraction in adult rat myocytes was mediated by intra-myocyte NO.[5] Similar data were reported with respect to cholinergic modulation of Ca^{2+} current in rabbit sinoatrial and atrioventricular nodal cells,[3,36] with implications for regulation of heart rate and atrioventricular conduction (see below). In anaesthetized closed-chest dogs, vagal inhibition of β-adrenergic positive inotropic response was markedly attenuated by L-NMMA,[37] suggesting that NO mediated this effect, although neither the cell type nor the NOS isoform responsible was defined. In contrast to these reports, Méry *et al.* showed convincingly that eNOS played no role in cholinergic inhibition of β-stimulated Ca^{2+} current in frog myocytes (nor did NOS inhibitors affect the response to β-agonists *per se*).[38] Although L-NMMA blocked carbachol-induced rises in cGMP in rabbit papillary muscles, it had no effect on carbachol inhibition of forskolin-induced rises in tension.[39] In guinea-pig myocytes, cholinergic inhibition of isoprenaline-stimulated Ca^{2+} current was unaffected either by NOS inhibitors or NO donors.[40]

NO and Heart Rate

Both nNOS and eNOS have been implicated in chronotropic and dromotropic regulation. nNOS, expressed in neurones which innervate the sinoatrial and atrioventricular nodes, may do so by modifying nerve firing rate. In isolated canine intrinsic neurones, NOS inhibitors reduced spontaneous firing rate while NO donors increased it.[41] The spontaneous beating rate of adult guinea-pig ventricular myocytes in co-culture with primary stellate ganglion or 'intrinsic cardiac' neurones was also increased by NO donors or L-arginine.[42] In Langendorff-perfused adult rat hearts, a NOS inhibitor modulated norepinephrine release during electrical sympathetic nerve stimulation.[4]

In contrast, the potential influence of eNOS on heart rate results from its expression within sinoatrial and atrioventricular nodal cells, where it may mediate cholinergic effects. Thus, cholinergic inhibition of isoproterenol-stimulated Ca^{2+} current in rabbit sino-atrial cells or in atrioventricular nodal cells involves intra-myocyte generation of NO.[3,36] In adult cells, NO does not appear to influence basal heart rate, i.e. in the absence of autonomic stimulation.[32] In neonatal cells, however, NO may lower spontaneous beating rate.[3] It remains to be established whether there are species differences in these effects, as seems to be the case for cholinergic modulation of inotropic responses.

Cardiac Angiotensin-converting Enzyme and NO

The level of angiotensin-converting enzyme (ACE) activity in the microvasculature may modulate NO production by endothelial cells, since ACE has kininase activity which degrades bradykinin and other kinins that are potent agonists for eNOS activation.[43] Bradykinin itself can be generated by endothelial cells *via* a local kallikrein-kinin pathway, and there is now evidence that such a pathway exists in the human heart. In human coronary microvascular preparations, ACE inhibitors increased agonist-induced nitrite production, an effect blocked by a bradykinin receptor antagonist (HOE140) or by L-NAME — indicating that local kinin generation led to eNOS activation.[44] Recent studies have suggested possible functional effects of this pathway. In isolated ejecting guinea-pig hearts, the ACE inhibitor captopril induced LV relaxant effects similar to those described earlier with NO, which were attributable to an endogenous bradykinin-NO pathway.[45] In human subjects with LV hypertrophy, acute intracoronary infusion of an ACE inhibitor improved relaxation and chamber distensibility,[46] although whether this involved bradykinin or NO was not investigated. ACE inhibitors were also shown to decrease oxygen consumption *via* an increase in kinin-stimulated NO production in human coronary microvascular preparations.[44]

NO and Myocardial Energetics

Several reports suggest cGMP-independent effects of NO on muscle energetics. In striated muscle, studies with NOS inhibitors have shown that NO reduces oxygen consumption independently of any negative inotropic effect, possibly *via* effects on mitochondrial cytochrome oxidase.[47] However, other studies have shown no effect of NO on oxygen consumption under basal conditions.[48] Interestingly, eNOS was recently reported to be present in mitochondria isolated from cardiac myocytes and other cells.[49] In isolated rat hearts, an NO donor inhibited phosphoryl transfer by creatine kinase, secondary to a reversible nitrosothiol modification.[50] In isolated rabbit hearts subjected to low-flow ischaemia, NOS inhibitors were reported to stimulate glycolysis from exogenous glucose and thus improve functional recovery.[51] Under conditions where peroxynitrite formation is enhanced, e.g. during hypoxia-reoxygenation (ischaemia-reperfusion), this may cause relatively irreversible reduction in myocardial oxygen consumption, in contrast to the 'physiological' reversible decrease induced by NO.[52]

Chronic Effects of eNOS on Myocardial Function

Most studies to date have addressed only the acute influences of NO on myocardial function. Does eNOS have long-term effects on cardiac structure and function? The expression of eNOS may be chronically modulated, e.g. increasing with chronic exercise. It is possible that increased NO could contribute to LV dilatation by increasing diastolic distensibility, e.g. in trained athletes. On the other hand, chronic reduction in eNOS activity could contribute to the diastolic dysfunction observed in conditions such as LV hypertrophy and heart failure.[53] A recent study demonstrated that expression of eNOS in rat cardiac myocytes was decreased either by *in vitro* exposure to cAMP or by *in vivo* treatment with the PDE inhibitor milrinone.[3] Intriguingly, similar changes in eNOS were not observed in cultured coronary microvascular endothelial cells. These data suggest that eNOS levels and activity could be chronically altered by β-adrenergic stimulation, e.g. in chronic heart failure.

Since NO has smooth muscle anti-proliferative activity, it is of interest whether it influences myocardial growth and hypertrophy. Although this has not been addressed directly, it has been reported that chronic treatment with nitrates reduced LV remodelling following myocardial infarction in dogs, independently of changes in loading.[54] Likewise, chronic L-arginine treatment reduced the development of hypertrophy in spontaneously hypertensive rats independently of changes in blood pressure.[55]

Ischaemia-reperfusion

Myocardial function during ischaemia-reperfusion (I-R) may be affected by NO-related impairment of coronary perfusion, secondary to increased vascular tone and increased neutrophil adhesion to endothelial cells. NO donors and L-arginine have been shown to ameliorate reperfusion injury *in vivo* by inhibiting these processes and improving perfusion.[56]

In addition to these indirect effects, NO might influence myocardial function independently of changes in coronary perfusion. Several recent studies have addressed this possibility. In isolated ejecting rat hearts subjected to I-R injury, post-ischaemic function was worsened by a NOS inhibitor present throughout the protocol, independently of changes in coronary flow.[57] In two studies in rabbits *in vivo*, treatment with NOS inhibitors during ischaemia increased infarct size.[58,59] In rat myocytes, 8-bromo cGMP pre-treatment completely inhibited the abnormalities of relaxation otherwise observed at reoxygenation following brief hypoxia.[60] An NO donor also protected rat myocytes against reoxygenation injury following prolonged anoxia.[61] These studies suggest beneficial effects of NO/ cGMP in I-R injury. In contrast, a NOS inhibitor present throughout ischaemia and reperfusion improved post-ischaemic function in Langendorff-perfused rat hearts[62] and reduced infarct size in rabbits *in vivo*.[63] A NOS inhibitor also improved *in vivo* contractile function in piglet hearts after a 2 hour hypoxic period on cardiopulmonary bypass,[64] suggesting that NO was harmful. These apparently contradictory data may be explained on the basis that NO has both beneficial and harmful actions, the net effect depending on their relative magnitude and on the timing of NO release (i.e. whether during ischaemia or reperfusion). Beneficial actions of NO include a reduction in Ca^{2+} influx and myofilament response to Ca^{2+} (thus reducing cytosolic Ca^{2+} overload), and a decrease in oxygen demand.[2,3,47] Harmful actions are probably related to peroxynitrite formation upon reaction of NO with superoxide radical,[9,10] and may be particularly important immediately upon reperfusion when free radical production is greatest.[10] This concept is supported by recent studies. In Langendorff-perfused rat hearts subjected to brief hypoxia at constant coronary flow, post-hypoxic LV function was enhanced by an NO donor present during hypoxia, but was worsened if the drug was administered during reoxygenation.[65] Likewise, recovery of function following ischaemia with cardioplegia in rat hearts was improved by L-arginine administered during ischaemia, but was worsened by the drug if given during reperfusion.[66]

NO has also been suggested to have a protective role in a canine model of myocardial stunning, independently of changes in coronary flow,[67] and to contribute to the anti-arrhythmic effects of ischaemic preconditioning.[68] The topic of preconditioning is discussed in Chapter 17.

Activation of eNOS by Inflammatory Cytokines

In addition to induction of iNOS (see below), cytokines may exert acute NO-mediated effects (within minutes), secondary to activation of eNOS. High concentrations of human tumour necrosis factor (TNF)-α, interleukin (IL)-2 or IL-6 decreased isometric force of hamster papillary muscles within 2–5 minutes, an effect inhibited by L-NMMA.[69] TNF-α similarly acutely decreased the shortening of rabbit myocytes and increased diastolic length, secondary to NO production.[70] Interestingly, the latter effects appeared to be due to a reduction in myofilament response to Ca^{2+} (since cytosolic Ca^{2+} was unaltered), an identical mechanism to that previously reported with a cGMP analogue in rat myocytes.[15]

Inducible Nitric Oxide Synthase in Myocardium: Pathophysiological Roles

The presence of Ca^{2+}-independent NOS activity in cardiac myocytes was first demonstrated by Schulz *et al.* in cells isolated from rats injected with endotoxin.[71] Subsequent studies showed that *in vitro* exposure to cytokines or *in vivo* treatment with endotoxin induces iNOS mRNA and protein in cardiac myocytes.[72–75] iNOS is similarly expressed in other cardiac cells, including coronary microvascular endothelium, vascular smooth muscle, and macrophages.[3] There is growing evidence that iNOS is also expressed in the heart in many conditions associated with inflammation and elevation of cytokine levels (e.g. myocarditis, allograft rejection) (Table 1). However, the precise roles of iNOS in these situations remain to be clearly defined. It is likely to have both beneficial and harmful actions, which will vary according to the disease and its severity or stage, and the temporal and spatial distribution of iNOS activity. For example, restricted expression of iNOS (predominantly in infiltrating macrophages) is beneficial in viral myocarditis,[76] whereas widespread expression (in cardiac myocytes and the vasculature) is harmful in endotoxaemia.[3]

Detailed discussion of the regulation of iNOS expression and activity is beyond the scope of this chapter. In general, the major mechanism is transcriptional regulation, but post-translational factors such as mRNA stability, substrate and co-factor availability are also important. iNOS-mediated co-induction of GTP

Table 1. Potential roles of myocardial iNOS in pathological conditions.

Viral myocarditis	Anti-viral activity Cytoprotection Reduced vascular permeability
Autoimmune myocarditis	Myocyte necrosis/apoptosis Contractile depression
Septic shock	Systolic depression Diastolic dilatation
Transplant rejection	Cytotoxicity Contractile depression
Non-rejecting cardiac allografts	Systolic and diastolic dysfunction Altered response to β-adrenergic agonists
Cardiomyopathy/heart failure	Unknown

cyclohydrolase I (and thus tetrahydrobiopterin synthesis), L-arginine transporters, and arginosuccinate synthase (which catalyses the conversion of L-citrulline to L-arginine), have been reported in coronary microvascular endothelial cells.[3]

Viral Myocarditis/Antiviral Activity/Other Beneficial Effects of iNOS

It is not surprising that iNOS has beneficial effects, given that the cytokines that induce it (e.g. TNF-α, IL-1β) are involved as mediators of natural immunity. Expression of iNOS in macrophages leads to suppression of viral replication and reactivation, and promotes bactericidal activity.[77,78] Other beneficial actions of iNOS include cytoprotection, decreased leucocyte adhesion, anti-platelet activity, and reduced vascular permeability.[3]

In a murine model of coxsackie B3 viral myocarditis, treatment with NOS inhibitors, either early (day 0–14) or late (day 14–35), resulted in higher viral titres, greater cardiac damage (necrosis and fibrosis), and increased mortality.[76] iNOS expression in these animals occurred predominantly in infiltrating inflammatory cells, with little detectable expression in cardiac myocytes or endothelial cells. iNOS knockout mice were also highly susceptible to viral myocarditis and lethal infection with facultative intracellular and opportunistic pathogens.[79,80]

Autoimmune Myocarditis

In contrast to viral myocarditis, iNOS appears to be harmful in autoimmune myocarditis, probably because of its widespread expression in cardiac myocytes and coronary microvessels, as well as in macrophages.[81] In an experimental rat model, induced by injection with porcine myosin, myocarditis was associated with necrosis and increased production of NO, superoxide and peroxynitrite.[81] Treatment with a selective iNOS inhibitor, aminoguanidine, attenuated these changes and improved haemodynamic parameters.[81,82]

Septic Shock

Following the demonstration that endotoxin induces iNOS in cardiac myocytes,[71] it was speculated that this may contribute to intrinsic cardiac depression in septic shock. Indeed, Brady *et al.* reported that depressed contractile function of myocytes isolated from endotoxin-treated guinea-pigs was reversed by NOS inhibitors.[83] Balligand *et al.* found that basal contractile function of rat myocytes expressing iNOS, following treatment with macrophage-conditioned medium, was no different from control; however, the inotropic response to isoproterenol was blunted, an effect reversed by NOS inhibitors.[73] In isolated ferret papillary muscles exposed to IL-1β, contractile duration was markedly abbreviated, an effect blocked by L-NMMA.[84] Schulz *et al.* demonstrated that the combination of IL-1β and TNF-α depressed contractile function of isolated ejecting rat hearts over a period of 2 hours, and that this was prevented by cycloheximide or a NOS inhibitor (although only over a limited dose range).[85] However, in other studies in endotoxin-treated guinea-pigs, NOS inhibitors did not affect either the baseline contractile depression or the response to isoproterenol in isolated myocytes, atrial muscle or isolated hearts.[86]

Studies performed *in vivo* have failed convincingly to demonstrate a role for treatment with NOS inhibitors in septic shock. In a rat model of chronic sepsis, iNOS expression was not directly responsible for the haemodynamic effects of systemic toxaemia, while systemic treatment with NOS inhibitors worsened perfusion in several beds and impaired overall cardiac performance.[87] Although a NOS inhibitor reduced hypotension in endotoxin-treated dogs, cardiac output and oxygen delivery were impaired.[88] In pigs injected with endotoxin, pre-treatment with a NOS inhibitor reduced intrinsic myocardial systolic dysfunction (assessed by the end-systolic pressure volume relation), but impaired overall cardiac function.[89] These *in vivo* studies indicate that: (a) NO only accounts for part of the myocardial depression of septic shock (an effect which itself may be species-dependent), and (b) treatment with non-selective NOS inhibitors does not improve overall cardiac function. This may in part be related to the fact that NO may have some beneficial actions in sepsis, e.g. by contributing to the diastolic ventricular dilatation that forms an important compensatory response in septic shock. Whether selective iNOS inhibitors have any role in the alleviation of septic shock remains to be established.

Transplant Rejection

iNOS expression in inflammatory cells, the microvasculature and cardiac myocytes has been demonstrated in several animal models of heterotopic cardiac allograft rejection.[90,91] Induction of iNOS may lead to direct cytotoxicity and apoptosis of cardiac myocytes. A close temporal relationship was reported between iNOS expression, enzyme activity, peroxynitrite formation and myocyte apoptosis during cardiac rejection in a rat model of heterotopic transplantation.[92] In one study, continuous infusion of the iNOS-selective inhibitor, aminoguanidine, improved cardiac graft survival and *ex vivo* contractile function in papillary muscles.[93]

The situation in human cardiac transplantation is less clear. No published studies have adequately examined the role of iNOS during acute rejection. In a recent study, cardiac iNOS was detectable (at mRNA and protein level) in routine surveillance endomyocardial biopsy specimens, but was not related to the histological grade of rejection or serum nitrite/nitrate levels.[94] These authors reported that iNOS expression was associated with systolic and diastolic dysfunction as assessed by echocardiography.[94] However, in another study in cardiac transplant recipients who had detectable NOS expression on endomyocardial biopsy but no evidence of rejection, intracoronary infusion of L-arginine had no effect on LV function.[95] Since L-arginine would be expected to augment the effects of iNOS, one interpretation of this result is that iNOS did not affect basal contractile function in these subjects. A subsequent study from the same group indicates that although resting contractile function is apparently the same in cardiac transplant recipients with and without significant iNOS expression, the response to β-adrenergic stimulation is altered.[96]

Cardiomyopathy and Heart Failure

A number of reports suggest a possible link between iNOS induction and some forms of human cardiomyopathy. De Belder *et al.* documented the presence of Ca^{2+}-independent NOS (i.e. iNOS) activity in endomyocardial biopsies obtained from patients with idiopathic dilated cardiomyopathy, post-partum cardiomyopathy and healing myocarditis, but not in patients with ischaemic heart disease.[97,98] In contrast, Haywood *et al.* demonstrated expression of iNOS mRNA and protein in myocardium of patients with heart failure, irrespective of aetiology.[99] In patients with dilated cardiomyopathy, iNOS expression in cardiac myocytes is reported to be accompanied by TNF-α expression in the microvasculature,[100] suggesting a potential paracrine pathway for the induction of iNOS. These observational studies have all involved relatively small numbers of patients, however, and it remains to be seen whether iNOS has any pathophysiological role in cardiomyopathy and heart failure. Given that heart failure is usually accompanied by neurohumoral activation, it is of interest that iNOS expression in cardiac myocytes is potentiated by cAMP or angiotensin II.[101]

Conclusions

The data reviewed in this chapter clearly demonstrate the involvement of NO in the regulation of myocardial function, both under physiological and pathological circumstances. NO has different actions on myocardium depending on factors such as its cellular source, the NOS isoform responsible, interactions with the β-adrenergic pathway and free radicals, and the state of the target myocardium. Under basal conditions, eNOS is involved in the regulation of myocardial relaxation and diastolic tone, modulates the force-frequency relation and the Frank-Starling response, and may affect myocardial energetics (the latter independently of cGMP). eNOS also modulates the inotropic response to β-adrenergic stimulation, generally decreasing its magnitude, but in some situations (e.g. at very low doses) may augment the β-response. It has been suggested that eNOS mediates cholinergic effects, such as the antagonism of β-adrenergic inotropic and chronotropic responses, but this appears to be species-dependent. Both nNOS in nerves and eNOS in myocytes may be involved in the regulation of heart rate. Although most studies to date have focused on acute influences on myocardial function, chronic effects on cardiac structure and function may turn out to be equally important.

In pathological situations, NO appears to be something of a double-edged sword. Thus during myocardial I-R, eNOS has beneficial effects (e.g. anti-platelet, anti-neutrophil, anti-Ca^{2+} overload) but may also be harmful, particularly through the formation of peroxynitrite during reperfusion. Decreased eNOS activity could contribute to diastolic dysfunction in conditions such as LV hypertrophy, whereas in heart failure NO may contribute to systolic dysfunction secondary to interactions with the activated β-adrenergic system.

iNOS is expressed in several cell types in situations where there are increased levels of cytokines, its effects depending upon the context, and the temporal and spatial restriction of its expression. Beneficial effects of iNOS include anti-viral and anti-bacterial activity, decreased leucocyte adhesion, and anti-platelet activity. Macrophage iNOS expression is crucial for host defence in viral myocarditis. In contrast, widespread induction of iNOS in cardiac myocytes and the coronary vasculature in autoimmune myocarditis contributes to myocardial damage and contractile depression. iNOS also contributes to myocardial depression in septic shock. Although this isoform is expressed in conditions such as cardiac transplantation, cardiomyopathy and heart failure, it remains to be established whether it has any pathophysiological role in these situations.

A vast amount of new information concerning the roles of NO in the heart has been obtained in the few years since the first paper on NO and myocardial function was published. These studies raise many questions with respect to its physiological and pathophysiological functions and potential clinical and therapeutic relevance, which will keep cardiovascular researchers occupied for many more years.

Acknowledgements

The work in the authors' laboratory is supported by the UK Medical Research Council and the British Heart Foundation.

References

1. Smith JA, Shah AM, Lewis MJ. Factors released from the endocardium of the ferret and pig modulate myocardial contraction. *J Physiol (Lond)* 1991; **439**: 1–14.
2. Shah AM. Paracrine modulation of heart cell function by endothelial cells. *Cardiovasc Res* 1996; **31**: 847–867.
3. Kelly RA, Balligand J-L, Smith TW. Nitric oxide and cardiac function. *Circ Res* 1996; **7**: 363–380.
4. Schwarz P, Diem R, Dun NJ, Förstermann U. Endogenous and exogenous nitric oxide inhibits norepinephrine release from rat heart sympathetic nerves. *Circ Res* 1995; **77**: 841–848.
5. Balligand JL, Kobzik L, Han X *et al.* Nitric oxide-dependent parasympathetic signaling is due to activation of constitutive endothelial (type III) nitric oxide synthase in cardiac myocytes. *J Biol Chem* 1995; **270**: 14582–14586.
6. Seki T, Hagiwara H, Naruse K *et al. In situ* identification of messenger RNA of endothelial type nitric oxide synthase in rat cardiac myocytes. *Biochem Biophys Res Commun* 1996; **218**: 601–605.
7. Lincoln TM, Cornwell TL. Intracellular cyclic GMP receptor proteins. *FASEB J* 1993; **7**: 328–338.
8. Campbell DL, Stamler JS, Strauss HC. Redox modulation of L-type calcium channels in ferret ventricular myocytes. *J Gen Physiol* 1996; **108**: 277–293.
9. Beckman JS, Beckman TW, Chen J *et al.* Apparent hydroxyl radical production by peroxynitrite: implications for endothelial injury from nitric oxide and superoxide. *Proc Natl Acad Sci USA* 1990; **87**: 1620–1624.
10. Wang P, Zweier JL. Measurement of nitric oxide and peroxynitrite generation in the postischemic heart. Evidence for peroxynitrite-mediated reperfusion injury. *J Biol Chem* 1996; **271**: 29223–29230.
11. Mohan P, Brutsaert DL, Paulus WJ, Sys SU. Myocardial contractile response to nitric oxide and cGMP. *Circulation* 1996; **93**: 1223–1229.
12. Shah AM, Lewis MJ, Henderson AH. Effects of 8-bromo-cyclic GMP on contraction and on inotropic response of ferret cardiac muscle. *J Mol Cell Cardiol* 1990; **23**: 55–64.
13. Grocott-Mason R, Fort S, Lewis MJ, Shah AM. Myocardial relaxant effect of exogenous nitric oxide in the isolated ejecting heart. *Am J Physiol* 1994; **266**: H1699–H1705.

14. Grocott-Mason RM, Anning PB, Evans H *et al.* Modulation of left ventricular relaxation in isolated ejecting heart by endogenous nitric oxide. *Am J Physiol* 1994; **267**: H1804–H1813.
15. Shah AM, Spurgeon H, Sollott SJ *et al.* 8-bromo cyclic GMP reduces the myofilament response to calcium in intact cardiac myocytes. *Circ Res* 1994; **74**: 970–978.
16. Pfitzer G, Rüegg JC, Flockerzi V, Hofmann F. cGMP-dependent protein kinase decreases calcium sensitivity of skinned cardiac fibers. *FEBS Lett* 1982; **149**: 171–175.
17. Keller TCS. Structure and function of titin and nebulin. *Curr Opin Cell Biol* 1995; **7**: 32–38.
18. Paulus WJ, Vantrimpont PJ, Shah AM. Acute effects of nitric oxide on left ventricular relaxation and diastolic distensibility in man. *Circulation* 1994; **89**: 2070–2078.
19. Paulus WJ, Vantrimpont PJ, Shah AM. Paracrine coronary endothelial control of left ventricular function in humans. *Circulation* 1995; **92**: 2119–2126.
20. Weyrich AS, Ma X, Buerke M *et al.* Physiological concentrations of nitric oxide do not elicit an acute negative inotropic effect in unstimulated cardiac muscle. *Circ Res* 1994; **75**: 692–700.
21. Crystal GJ, Gurevicius J. Nitric oxide does not modulate myocardial contractility acutely in *in situ* canine hearts. *Am J Physiol* 1996; **270**: H1568–H1576.
22. Brady AJB, Warren JB, Poole-Wilson PA *et al.* Nitric oxide attenuates cardiac myocyte contraction. *Am J Physiol* 1993; **265**: H176–H182.
23. Kojda G, Kottenberg K, Nix P *et al.* Low increase in cGMP induced by organic nitrates and nitrovasodilators improves contractile response of rat ventricular myocytes. *Circ Res* 1996; **78**: 91–101.
24. Prendergast BD, Sagach VF, Shah AM. Basal release of nitric oxide augments the Frank-Starling response in the isolated heart. *Circulation* 1997; **96**: 1320–1329.
25. Shaul PW, Smart EJ, Robinson LJ *et al.* Acylation targets endothelial nitric-oxide synthase to plasmalemmal caveolae. *J Biol Chem* 1996; **271**: 6518–6522.
26. Finkel MS, Oddis CV, Mayer OH *et al.* Nitric oxide synthase inhibitor alters papillary muscle force-frequency relationship. *J Pharmacol Exp Ther* 1995; **272**: 945–952.
27. Kaye DM, Wiviott SD, Balligand JL *et al.* Frequency-dependent activation of a constitutive nitric oxide synthase and regulation of contractile function in adult rat ventricular myocytes. *Circ Res* 1996; **78**: 217–224.
28. Balligand JL, Kelly RA, Marsden PA *et al.* Control of cardiac muscle function by an endogenous nitric oxide signaling system. *Proc Natl Acad Sci USA* 1993; **90**: 347–351.

29. Ebihara Y, Karmazyn M. Inhibition of β- but not α_1-mediated adrenergic responses in isolated hearts and cardiomyocytes by nitric oxide and 8-bromo cyclic GMP. *Cardiovasc Res* 1996; **32**: 622–629.
30. Keaney JF, Hare JM, Balligand JL *et al.* Inhibition of nitric oxide synthase augments myocardial contractile responses to β-adrenergic stimulation. *Am J Physiol* 1996; **271**: H2646–H2652.
31. Parent R, Al-Obaidi M, Lavallee M. Nitric oxide formation contributes to β-adrenergic dilation of resistance coronary vessels in conscious dogs. *Circ Res* 1993; **73**: 241–251.
32. Hare JM, Loh E, Creager MA, Colucci W. Nitric oxide inhibits the positive inotropic response to β-adrenergic stimulation in humans with left ventricular dysfunction. *Circulation* 1995; **92**: 2198–2203.
33. Bartunek J, Shah AM, Vanderheyden M, Paulus WJ. Dobutamine enhances cardiopressant effects of receptor-mediated coronary endothelial stimulation. *Circulation* 1997; **95**: 90–96.
34. Méry PF, Pavoine C, Belhassen L *et al.* Nitric oxide regulates cardiac Ca^{2+} current. *J Biol Chem* 1993; **268**: 26286–26295.
35. Wahler GM, Dollinger S. Nitric oxide donor SIN-1 inhibits mammalian cardiac calcium current through cGMP-dependent protein kinase. *Am J Physiol* 1995; **268**: C45–C54.
36. Han X, Shimoni Y, Giles WR. An obligatory role for nitric oxide in autonomic control of mammalian heart rate. *J Physiol (Lond)* 1994; **476**: 309–314.
37. Hare JM, Keaney JF, Balligand JL *et al.* Role of nitric oxide in parasympathetic modulation of β-adrenergic myocardial contractility in dogs. *J Clin Invest* 1995; **95**: 360–366.
38. Méry P-F, Hove-Madsen L, Chesnais J-M *et al.* Nitric oxide synthase does not participate in negative inotropic effect of acetylcholine in frog heart. *Am J Physiol* 1996; **270**: H1178–H1188.
39. Hui J, Tabatabaei A, MacLeod KM. L-NMMA blocks carbachol-induced increases in cGMP levels but not decreases in tension in the presence of forskolin in rabbit papillary muscles. *Cardiovasc Res* 1995; **30**: 372–376.
40. Zakharov SI, Pieramici S, Kumar GK *et al.* Nitric oxide synthase activity in guinea-pig ventricular myocytes is not involved in muscarinic inhibition of cAMP-regulated ion channels. *Circ Res* 1996; **78**: 925–935.
41. Armour JA, Smith FM, Losier AM *et al.* Modulation of intrinsic neuronal activity by nitric oxide donors induces cardiodynamic changes. *Am J Physiol* 1995; **268**: R403–R413.
42. Horackova M, Armour JA, Hopkins DA, Huang MH. Nitric oxide modulates signaling between cultured adult peripheral cardiac neurons and cardiomyocytes. *Am J Physiol* 1995; **269**: C504–C510.
43. Linz W, Wiemer G, Gohlke P *et al.* Contribution of kinins to the cardiovascular actions of angiotensin-converting enzyme inhibitors. *Pharmacol Rev* 1995; **47**: 25–49.

44. Zhang XP, Xie YW, Nasjletti A *et al.* ACE inhibitors promote nitric oxide accumulation to modulate myocardial oxygen consumption. *Circulation* 1997; **95**: 176–182.
45. Anning PB, Grocott-Mason RM, Lewis MJ, Shah AM. Enhancement of left ventricular relaxation in the isolated heart by an angiotensin converting-enzyme inhibitor. *Circulation* 1995; **92**: 2660–2665.
46. Friedrich S, Lorell B, Rousseau M *et al.* Intracardiac angiotensin-converting enzyme inhibition improves diastolic function in patients with left ventricular hypertrophy due to aortic stenosis. *Circulation* 1994; **90**: 2761–2771.
47. Shen W, Xu X, Ochoa M *et al.* Role of nitric oxide in the regulation of oxygen consumption in conscious dogs. *Circ Res* 1994; **75**: 1086–1095.
48. Saeki A, Recchia FA, Senzaki H, Kass DA. Minimal role of nitric oxide in basal coronary flow regulation and cardiac energetics of blood-perfused isolated canine heart. *J Physiol (Lond)* 1996; **491**: 455–463.
49. Bates TE, Loesch A, Burnstock G, Clark JB. Mitochondrial nitric oxide synthase: a ubiquitous regulator of oxidative phosphorylation? *Biochem Biophys Res Commun* 1996; **218**: 40–44.
50. Gross WL, Bak MI, Ingwall JS *et al.* Nitric oxide regulates rat heart contractile reserve and inhibits creatinine kinase. *Proc Natl Acad Sci USA* 1996; **93**: 5604–5609.
51. Depré C, Vanoverschelde J-L, Goudemant J-F *et al.* Protection against ischemic injury by nonvasoactive concentrations of nitric oxide synthase inhibitors in the perfused rabbit heart. *Circulation* 1995; **92**: 1911–1918.
52. Xie Y-W, Wolin MS. Role of nitric oxide and its interaction with superoxide in the suppression of cardiac muscle mitochondrial respiration. Involvement in response to hypoxia/reoxygenation. *Circulation* 1996; **94**: 2580–2586.
53. Shah AM, Prendergast B, Grocott-Mason RM *et al.* The influence of endothelium-derived nitric oxide on myocardial contractile function. *Int J Cardiol* 1995; **50**: 225–231.
54. McDonald KM, Francis GS, Matthews J *et al.* Chronic oral nitrate therapy prevents long-term ventricular remodelling in the dog. *J Am Coll Cardiol* 1993; **21**: 514–522.
55. Matsuoka H, Nakata M, Kohno K *et al.* Chronic L-arginine administration attenuates cardiac hypertrophy in spontaneously hypertensive rats. *Hypertension* 1996; **27**: 14–18.
56. Lefer DJ, Nakanishi K, Johnston WE *et al.* Anti-neutrophil and myocardial protecting actions of a novel nitric oxide donor after acute myocardial ischemia and reperfusion in dogs. *Circulation* 1993; **88**: 2337–2350.
57. Beresewicz A, Karwatowska-Prokopczuk E, Lewartowski B, Cedro-Ceremuzynska K. A protective role of nitric oxide in isolated ischaemic/reperfused rat heart. *Cardiovasc Res* 1995; **30**: 1001–1008.

58. Williams MW, Taft CS, Ramnauth S *et al.* Endogenous nitric oxide (NO) protects against ischaemic-reperfusion injury in the rabbit. *Cardiovasc Res* 1995; **30**: 79–86.
59. Hoshida S, Yamashita N, Igarashi J *et al.* Nitric oxide synthase protects the heart against ischemia-reperfusion injury in rabbits. *J Pharmacol Exp Ther* 1995; **274**: 413–418.
60. Shah AM, Silverman HS, Griffith EJ *et al.* Cyclic GMP prevents delayed relaxation at reoxygenation following brief hypoxia in isolated cardiac myocytes. *Am J Physiol* 1995; **268**: H2396–H2404.
61. Schlüter KD, Weber M, Schraven E, Piper HM. NO donor SIN-1 protects against reoxygenation-induced cardiomyocyte injury by a dual action. *Am J Physiol* 1994; **267**: H1461–H1466.
62. Naseem SA, Kontos MC, Rao PS *et al.* Sustained inhibition of nitric oxide by N^G-nitro-L-arginine improves myocardial function following ischaemia/reperfusion in isolated perfused rat heart. *J Mol Cell Cardiol* 1995; **27**: 419–426.
63. Patel VC, Yellon DM, Singh KJ *et al.* Inhibition of nitric oxide limits infarct size in the *in situ* rabbit heart. *Biochem Biophys Res Commun* 1993; **194**: 234–238.
64. Matheis G, Sherman MP, Buckberg GD *et al.* Role of L-arginine-nitric oxide pathway in myocardial reoxygenation injury. *Am J Physiol* 1992; **262**: H616–H620.
65. Draper NJ, Shah AM. Beneficial effects of a nitric oxide donor on recovery of contractile function following brief hypoxia in isolated rat heart. *J Mol Cell Cardiol* 1997; **29**: 1195–1205.
66. Engelman DT, Watanabe M, Maulik N *et al.* Critical timing of nitric oxide supplementation in cardioplegic arrest and reperfusion. *Circulation* 1996; **94 (Suppl II)**: II407–II411.
67. Hasebe N, Shen YT, Vatner SF. Inhibition of endothelium-derived relaxing factor enhances myocardial stunning in conscious dogs. *Circulation* 1993; **88**: 2862–2871.
68. Vegh A, Szekeres L, Parratt J. Preconditioning of the ischaemic myocardium; involvement of the L-arginine nitric oxide pathway. *Br J Pharmacol* 1992; **107**: 648–652.
69. Finkel MS, Oddis CV, Jacob TD *et al.* Negative inotropic effects of cytokines on the heart mediated by nitric oxide. *Science* 1992; **257**: 387–389.
70. Goldhaber JI, Kim KH, Weiss JN. Tumour necrosis factor-α causes acute reduction of myofilament calcium responsiveness in isolated myocytes [Abstract]. *Circulation* 1995; **92 (Suppl I)**: I–240.
71. Schulz R, Nava E, Moncada S. Induction and potential biological relevance of Ca^{2+}-independent nitric oxide synthase in the myocardium. *Br J Pharmacol* 1992; **105**: 575–580.

72. Tsujino M, Hirata Y, Imai T *et al.* Induction of nitric oxide synthase gene by interleukin-1β in cultured rat cardiocytes. *Circulation* 1994; **90**: 375–383.
73. Balligand JL, Ungureanu D, Kelly RA *et al.* Abnormal contractile function due to induction of nitric oxide synthesis in rat cardiac myocytes follows exposure to activated macrophage-conditioned medium. *J Clin Invest* 1993; **91**: 2314–2319.
74. Balligand JL, Ungureanu-Longrois D, Simmons WW *et al.* Cytokine-inducible nitric oxide synthase (iNOS) expression in cardiac myocytes. *J Biol Chem* 1994; **269**: 27580–27588.
75. Luss H, Watkins SC, Freeswick PD *et al.* Characterization of inducible nitric oxide synthase expression in endotoxaemic rat cardiac myocytes *in vivo* and following cytokine exposure *in vitro*. *J Mol Cell Cardiol* 1995; **27**: 2015–2029.
76. Hiraoka Y, Kishimoto C, Takada H *et al.* Nitric oxide and murine coxsackievirus B3 myocarditis: aggravation of myocarditis by inhibition of nitric oxide synthase. *J Am Coll Cardiol* 1996; **28**: 1610–1615.
77. Karupiah G, Xie Q, Buller RML *et al.* Inhibition of viral replication by interferon-γ-induced nitric oxide synthase. *Science* 1993; **261**: 1445–1448.
78. Pacelli R, Wink DA, Cook JA *et al.* Nitric oxide potentiates hydrogen peroxide-induced killing of Escherichia coli. *J Exp Med* 1995; **182**: 1469–1479.
79. MacMicking JD, Nathan C, Hom G *et al.* Altered responses to bacterial infection and endotoxic shock in mice lacking inducible nitric oxide synthase. *Cell* 1995; **81**: 641–650.
80. Wei X-Q, Charles IG, Smith A *et al.* Altered immune responses in mice lacking inducible nitric oxide synthase. *Nature* 1995; **375**: 408–411.
81. Ishiyama S, Hiroe M, Nishikawa T *et al.* Nitric oxide contributes to the progression of myocardial damage in experimental autoimmune myocarditis in rats. *Circulation* 1997; **95**: 489–496.
82. Hirono S, Islam O, Nakazawa M *et al.* Expression of inducible nitric oxide synthase in rat experimental autoimmune myocarditis with special reference to changes in cardiac hemodynamics. *Circ Res* 1997; **80**: 11–20.
83. Brady AJB, Poole-Wilson PA, Harding SE, Warren JB. Nitric oxide within cardiac myocytes reduces their contractility in endotoxemia. *Am J Physiol* 1992; **263**: H1963–H1966.
84. Evans HG, Lewis MJ, Shah AM. Interleukin-1β modulates myocardial contraction *via* dexamethasone-sensitive production of nitric oxide. *Cardiovasc Res* 1993; **27**: 1486–1490.
85. Schulz R, Panas DL, Catena R *et al.* The role of nitric oxide in cardiac depression induced by interleukin-1β and tumour necrosis factor-α. *Br J Pharmacol* 1995; **114**: 27–34.
86. Keller RS, Jones JJ, Kim KF *et al.* Endotoxin-induced myocardial dysfunction: is there a role for nitric oxide? *Shock* 1995; **4**: 338–344.
87. Gardiner SM, Kemp PA, March JE, Bennett T. Cardiac and regional haemodynamics, inducible nitric oxide synthase (NOS) activity, and the effects

of NOS inhibitors in conscious, endotoxaemic rats. *Br J Pharmacol* 1995; **116**: 2005–2016.

88. Mitaka C, Hirata Y, Ichikawa K *et al.* Effects of nitric oxide synthase inhibitor on hemodynamic change and O_2 delivery in septic dogs. *Am J Physiol* 1995; **268**: H2017–H2023.
89. Herbertson MJ, Werner HA, Walley KR. Nitric oxide synthase inhibition partially prevents decreased LV contractility during endotoxemia. *Am J Physiol* 1996; **270**: H1979–H1984.
90. Lancaster JR, Langrehr JM, Bergonia HA *et al.* EPR detection of heme and nonheme iron-containing protein nitrosylation by nitric oxide during rejection of rat heart allograft. *J Biol Chem* 1992; **267**: 10994–10998.
91. Yang X, Chowdhury N, Cai B *et al.* Induction of myocardial nitric oxide synthase by cardiac allograft rejection. *J Clin Invest* 1994; **94**: 714–721.
92. Szabolcs M, Michler RE, Yang X *et al.* Apoptosis of cardiac myocytes during cardiac allograft rejection. Relation to induction of nitric oxide synthase. *Circulation* 1996; **94**: 1665–1673.
93. Worrall NK, Lazenby WD, Misko TP *et al.* Modulation of *in vivo* alloreactivity by inhibition of inducible nitric oxide synthase. *J Exp Med* 1995; **181**: 63–70.
94. Lewis NP, Tsao PS, Rickenbacher PR *et al.* Induction of nitric oxide synthase in the human cardiac allograft is associated with contractile dysfunction of the left ventricle. *Circulation* 1996; **93**: 720–729.
95. Paulus WJ, Kastner S, Vanderheyden M *et al.* Myocardial contractile effects of L-arginine in the human allograft. *J Am Coll Cardiol* 1997; **29**: 1332–1338.
96. Paulus WJ, Kästner S, Pujadas P *et al.* Left ventricular contractile effects of inducible nitric oxide synthase in the human allograft. *Circulation* 1997; **96**: 3436–3442.
97. de Belder AJ, Radomski MW, Why HJF *et al.* Nitric oxide synthase in human myocardium. *Lancet* 1993; **341**: 84–85.
98. de Belder AJ, Radomski MW, Why HJ *et al.* Myocardial calcium-independent nitric oxide synthase activity is present in dilated cardiomyopathy, myocarditis, and postpartum cardiomyopathy but not in ishaemic or valvar heart disease. *Br Heart J* 1995; **74**: 426–430.
99. Haywood GA, Tsao PS, von der Leyen HE *et al.* Expression of inducible nitric oxide synthase in human heart failure. *Circulation* 1996; **93**: 1087–1094.
100. Habib FM, Springall DR, Davies GJ *et al.* Tumour necrosis factor and inducible nitric oxide synthase in dilated cardiomyopathy. *Lancet* 1996; **347**: 1151–1155.
101. Ikeda U, Yamamoto K, Ichida M *et al.* Cyclic AMP augments cytokine-stimulated nitric oxide synthesis in rat cardiac myocytes. *J Mol Cell Cardiol* 1996; **28**: 789–795.

Chapter 13

Regulation of Skeletal Muscle Microcirculation by Nitric Oxide

Akos Koller and Gabor Kaley

Introduction

As described elsewhere in this volume, it was established in 1980 that the endothelium plays a major role in mediating the responses of large vessels to different stimuli through the release of an endothelium-derived relaxing factor, now characterized as NO or a closely related NO-containing compound.[1–3] Using optical methods (light-dye impairment), *in vivo* studies on the function of skeletal muscle arteriolar endothelium were reported nine years later, and revealed that acetylcholine (ACh) no longer evoked arteriolar dilatation after optical damage, whereas adenosine and sodium nitroprusside (SNP) were still effective.[4,5] Similar results were obtained in studies using L-arginine analogues such as N^G-nitro-L-arginine (L-NA), thus confirming the important role of NO in the control of skeletal muscle blood flow.[6] The endothelium can thus be a source of dilator factors that modulate the resistance of the skeletal muscle microcirculation, a local mechanism of blood flow regulation that had not previously been recognized.[7]

Skeletal Muscle Microvascular Control

Arterioles

NO released from cultured microvascular endothelial cells causes arteriolar dilatation in an *in vivo* skeletal muscle preparation coupled to the cells in a bioassay system.[8] In isolated arterioles of rat gracilis muscle 35–250 μm diameter, removal of endothelium by an air perfusion technique abolishes, and administration of L-arginine analogues attenuates, the dilatation evoked by ACh.[9] Thus, NO release from the arteriolar endothelium participates in the response to ACh in skeletal muscle. Furthermore, L-arginine (but not D-arginine) evokes a concentration- and endothelium-dependent arteriolar dilatation, suggesting that nitric oxide synthase (NOS) activity can be substrate-dependent,[9] a finding that may have significance under pathological conditions.

Venules

Using an *in vitro* bioassay system, Falcone and Meininger demonstrated that ACh and bradykinin elicit NO release from a donor skeletal muscle venule and dilatation of an endothelium-denuded arteriole coupled in series with it.[10] In venules, low concentrations of ACh (10^{-9}–10^{-6} M) elicited dilatation, whereas high concentrations (10^{-5}–10^{-4} M) caused smaller dilatations or even constriction, with both types of response being abolished by endothelial denudation.[11] In the presence of indomethacin to inhibit the cyclooxygenase enzyme or SQ29548, a thromboxane A_2/prostaglandin H_2 (PGH_2) receptor antagonist, the higher concentrations of ACh caused further dilatation (up to 75%) instead of constriction.

Preliminary studies have shown that a suicide substrate of cytochrome P_{450}, 17-octadecynoic acid (17-ODYA), converts ACh-induced constriction to dilatation in the presence of indomethacin, providing evidence that ACh may also stimulate the synthesis of a vasodilator cytochrome P_{450} metabolite. Furthermore, the nonselective K^+ channel blocker tetraethylammonium (TEA) decreases L-NA and indomethacin-insensitive dilatations to ACh, suggesting a role for membrane hyperpolarization, whereas dilatation to SNP is retained in the presence of both 17-ODYA and TEA. ACh thus appears to stimulate the release of both dilator (NO and a hyperpolarizing factor) and constrictor (PGH_2) factors from the venular endothelium. Release/synthesis of such agents may be interdependent, since L-NA significantly reduces both dilator and constrictor responses to ACh, thus suggesting that NO production may be necessary for the release of other autacoids.[11] These findings emphasize that the function of arteriolar and venular endothelium may differ, and that during systemic application of vasoactive agents consideration should be given to the fact that the circulatory responses observed are the sum of both arteriolar and venular events.

Myogenic and Shear Stress-related Mechanisms of Local Microcirculatory Control

The role of myogenic and metabolic mechanisms in the regulation of local blood flow is well established (Figure 1). Formerly, it was assumed that these two local mechanisms, together with neural and humoral factors that modulate arteriolar tone, accounted for the regulation of blood flow in skeletal muscle.[7] There is accumulating evidence that release of NO by the arteriolar and venular endothelium makes it necessary to re-evaluate this view.

Role of NO in the Response to Changes in Shear Stress

As in other vessel types, increases in blood flow stimulate an endothelium-dependent dilatation in skeletal muscle microvessels.[12–16] That the underlying signal to the endothelium is longitudinal wall shear stress has been confirmed by experiments in which increases in perfusate viscosity elicit dilatation in the presence of constant perfusate flow and pressure in skeletal muscle arterioles.[17] Such studies reveal the presence of a negative feedback mechanism that reduces wall shear stress, facilitates increases in blood flow, and prevents inappropriately high hydraulic power losses in the circulation (see also Chapter 2). These effects may be important in the signal-amplifying process that permits large increases in blood flow during reactive hyperaemia or exercise, and may therefore coordinate the behaviour of the arteriolar and venular sides of the skeletal muscle microcirculation (see also Chapter 9).

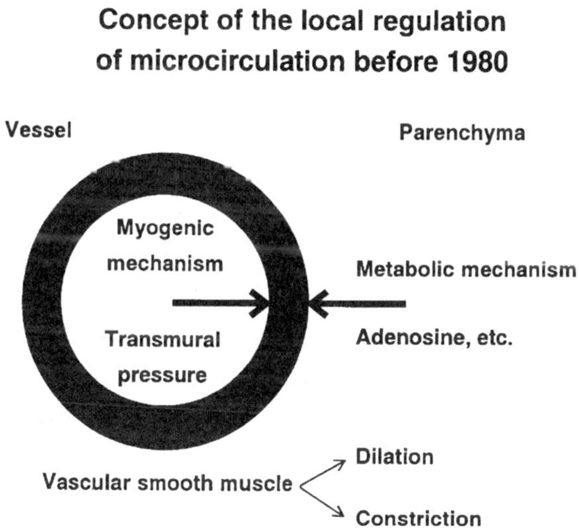

Figure 1. Before 1980, myogenic and metabolic mechanisms were considered as the two primary local regulators of microvascular resistance and blood flow in skeletal muscle. The myogenic mechanism is sensitive to changes in transmural pressure and is thought to be responsible for autoregulation, whereas the metabolic mechanism regulates blood flow in accordance with tissue O_2 demands.

It is now established that NO plays a central role in the mediation of flow-dependent dilatation in arterioles from several skeletal muscle types, including rat gracilis and spinotrapezius,[18,19] as well as in venules of 100–200 μm diameter isolated from the porcine myocardium[20] and rat gracilis muscle.[21] Indeed, observations that L-arginine analogues elevate systemic pressure by as much as 30–40%,[22] and that their local application induces arteriolar constriction,[6,8] suggest that basal vascular tone is continuously modulated by NO release. Furthermore, in rat spinotrapezius muscle, increases in venular flow velocity and shear rate induced by a localized occlusion technique elicit dilatation of adjacent parallel arterioles, a response that appears to be dependent on NO release from the venular wall and its subsequent diffusion.[13,23] ATP and bradykinin released from stores within endothelial cells in response to shear stress may also trigger the synthesis of NO.[24] This pathway may amplify NO-mediated dilatation in the skeletal muscle microcirculation during treatment with angiotensin-converting enzyme (ACE) inhibitors that decrease the metabolism of bradykinin.[25,26] Furthermore, recent studies by Hornig *et al.*[27] have demonstrated that ACE inhibitors enhance flow-dependent dilatation in the human forearm through a bradykinin-dependent mechanism that involves NO and prostaglandins. Although the primary stimulus for the continuous release of NO in the skeletal muscle microcirculation is wall shear stress, changes in the boundary layer mass transport of endothelium-dependent agonists close to the glycocalyx may also be important during increases in flow (see Chapter 2). In addition to effects mediated by direct activation of the soluble guanylate cyclase enzyme (see Chapters 1, 3 and 4), several other mechanisms potentially contribute to the effects of NO on tone in skeletal muscle arterioles. Thus, NO may influence the response to, or interfere with the release of, noradrenaline at peripheral nerve terminals[24,28] and may interact chemically with this neurotransmitter.[29] Shear stress-related release of NO from arteriolar and venular endothelium will facilitate other NO-related functions such as the regulation of platelet aggregation, vascular permeability and leucocyte adhesion (see Chapters 15, 17 and 19).

Signal transduction mechanisms leading to release of NO by shear stress

The intima is directly exposed to the forces generated by blood flow, so that changes in shear stress may distort the endothelial cell membrane and thereby initiate a signalling cascade involving changes in membrane potential and phospholipid metabolism.[30] The transduction mechanisms and intracellular signalling pathways subsequently leading to the release of NO are not yet completely elucidated. Previous work has suggested that endothelial cell hyperpolarization secondary to activation of several classes of transmembrane ion channels is involved in flow-induced NO production.[31] One of these is a K_{Ca} channel coupled to a pertussis toxin-sensitive G protein which is activated by shear stress and promotes K^+ efflux and hyperpolarization.[32,33] As the endothelial cell is electrically inexcitable, this will promote Ca^{2+} entry and lead to activation of eNOS.[34,35] Inhibition of endothelial K_{Ca} channels consequently inhibits dilatation of skeletal muscle arterioles to increases in perfusate flow.[36] Recently, however, Gooch and Frangos[37] have challenged this idea by demonstrating that elevations in $[Ca^{2+}]_i$ may not always be essential to sustain enhanced NO synthesis in response to increases in shear stress and that release of NO in response to flow may be independent of alterations in membrane potential.

The endothelial cytoskeleton contributes to the maintenance of membrane integrity, provides structural stability to the cell by binding to focal adhesion points and is associated with a variety of enzymes (see 38, 39 for review). In perfused isolated aortic segments, the microfilament and microtubular elements of the endothelial cytoskeleton both appear to be involved in the mechanotransduction pathways leading to flow-induced release of NO, but not in the agonist-induced response to ACh.[40] The effects of nocodazole, an agent that interferes with microtubule polymerization, have also been investigated in isolated skeletal muscle arterioles.[41] Dilatation to ACh or SNP was not affected by nocodazole, whereas flow-dependent dilatation was abolished.[41] Recent observations also suggest an important role for integrins in shear stress-induced responses.[42] Integrins may serve as mechanotransducers of shear forces to initiate signals through tyrosine kinase activation and are co-localized with cytoskeletal proteins at focal adhesion sites which provide a tensile link between the surface of the cell and the cell matrix. Focal adhesions on the abluminal surface of endothelial cells increase their rates of association/dissociation with the extracellular matrix in response to changes in luminal shear stress. Deformation of the endothelial cytoskeleton may consequently be an important intermediate step in the shear stress-induced release of NO.

Interaction of Myogenic and Shear Stress-related Mechanisms

Although NO and other endothelial mediators are not directly involved in the generation of myogenic tone in arterioles and venules from skeletal muscle and other organs, pressure-induced constriction is generally enhanced following loss of NO activity.[43–49] In isolated porcine coronary arterioles, Kuo *et al.*[47] demonstrated that myogenic tone is attenuated by flow-induced dilatation, and that, conversely, flow-induced dilatation is opposed by myogenic constriction. Analogous protocols have demonstrated that interactions between pressure- and flow-induced responses modulate the tone of rat skeletal muscle arterioles.[50] Increases in intraluminal pressure (from 0 to 100 mmHg) elicited greatest constriction in the absence of perfusion, and constriction was attenuated by up to ~50% and pressure-diameter relationships shifted upwards as perfusate flow was increased. By contrast, flow-diameter relationships were shifted downwards on increasing pressure from 60 to 100 mmHg so as to stimulate myogenic constriction. Such reductions in flow-induced dilatation were also seen when basal arteriolar diameter was adjusted to equivalent levels following constriction with phenylephrine. There is therefore accumulating evidence that the interaction between pressure and shear stress is an important local determinant of microvascular resistance in a variety of different species and vascular beds.

Theoretically, the pressure-induced myogenic response is a positive feedback mechanism which, if not opposed, could lead to collapse of the circulation through regenerative constriction, whereas flow-dependent dilatation is based on the negative feedback regulation of shear stress. Therefore, when systemic pressure increases, the initial vascular response will be constriction. At the same time, however, the increased pressure drop across the vascular network will increase flow velocity and shear stress, thus promoting dilatation. These two opposing mechanisms will reach a dynamic equilibrium that establishes the tone of the vessel and serves to stabilize pressure and flow. Similar arguments lead to the conclusion that any intervention that influences the tone of blood vessels will be modulated by the changes in pressure and flow that the stimulus itself initiates.

Hypertension

Hypertension is characterized not only by elevated peripheral resistance and augmented arteriolar myogenic constriction,[51] but also by increases in haemodynamic forces that might be expected to affect endothelial function.[52] Indeed, there is well-documented evidence for impaired NO-mediated relaxation in the conduit arteries of subjects with essential hypertension[53] as well as in experimental models of hypertension.[54–58] In isolated gracilis muscle arterioles from normotensive (WKY) rats (~ 70 μm diameter), dilatation to increases in shear stress over the range 0–60 dyne/cm^2 was significantly greater than in arterioles of spontaneously hypertensive rats (SHR).[59,60] The diameter of WKY arterioles increased in a sigmoidal fashion as a function of wall shear stress, but in SHR this relationship was shifted significantly to the right. Inhibition of prostaglandin synthesis resulted in a rightward shift of the diameter/shear stress relationship in both strains of rat (by ~ 15 dyne/cm^2), whereas inhibition of NOS caused a rightward shift of similar magnitude in arterioles from WKY but not SHR. Thus, shear stress-induced dilatation of arterioles is reduced in hypertension as a consequence of impaired NO-mediated dilatation, resulting in elevated tone and an increase in shear stress. Figure 2 depicts the shear stress-arteriolar diameter relationships for control, exercise-trained (see below) and hypertensive rats, and illustrates differences in the maintained range of shear stress. Prostaglandin-dependent mechanisms may also be affected by hypertension in rat skeletal muscle, with altered arachidonic acid metabolism resulting in enhanced production of PGH_2.[61] The extent to which endothelial dysfunction contributes directly to the development of hypertension (or is itself a secondary phenomenon) is likely to remain an area of active research and may ultimately lead to new therapeutic strategies (see Chapter 21).

Gender Differences in Nitric Oxide Expression

Epidemiological studies have demonstrated that the female cardiovascular system is less vulnerable to atherosclerosis, hypertension and other vascular disorders, and this may in part depend on a beneficial effect of oestrogens on lipid profile.[62] Recent evidence, however, suggests that oestrogens also exert a direct dilator action mediated by activation of K_{Ca} channels present in vascular smooth muscle,[63] and that there are gender differences in the bioassayable endothelium-derived NO from rat aorta.[64] Indeed, Rosselli *et al.*[65] found that circulating [$NO_2^- + NO_3^-$] increases with follicular development, thus providing indirect evidence for oestrogen-mediated NO release. Such observations have led to the hypothesis that upregulation of NO synthesis by oestrogens is responsible for a lower incidence of hypertension in pre-menopausal females.

Investigations with isolated gracilis muscle arterioles ~ 55 μm in diameter have shown that diameters in normal males are 10 ~ 15% smaller than in females over the pressure range 60–140 mmHg, thus implying greater basal tone.[66] Furthermore, L-NA causes an approximately 2-fold greater downward shift in the pressure-diameter relationships of arterioles from females compared to males, thus suggesting that NO is responsible for the observed differences. Dilatations to substance P and increases in perfusate flow are significantly greater (by ~ 30%) in arterioles from female rats and less susceptible to inhibition by L-NA than arterioles from males.[67] Furthermore, NO-dependent responses mediated *via* the endothelium, but not those induced by an exogenous source of NO, are markedly attenuated in gracilis muscle arterioles from hypertensive male rats as compared to hypertensive females. Indeed, endothelium-dependent arteriolar dilatation to the Ca^{2+} ionophore, A23187, which involves release of NO, is preserved in hypertensive female rats, whereas a PGH_2-dependent constriction develops in arterioles from hypertensive males.[68] Observations that arterioles from ovariectomized female rats and ovariectomized females with oestrogen replacement behave in a fashion similar to those from males and normal females, respectively, suggest that enhancement of NO activity is oestrogen-mediated, a conclusion supported by recent studies that show induction of eNOS by oestrogens.[67–69] Ovariectomy has also been shown to attenuate increases in forearm blood flow induced by ACh in humans, without affecting responses to SNP.[70] Three months of oestrogen replacement therapy normalized responses to ACh, before any changes occurred in the serum lipid pattern.[70]

Thus, there is a clear gender difference in the modulation of pressure-induced myogenic constriction of arterioles between females and males under both normotensive and hypertensive conditions. This is most likely due to the enhanced release of NO caused by oestrogens. The beneficial effects of oestrogen replacement therapy after the menopause may therefore reflect, at least in part, enhanced NO synthesis and not merely its cholesterol lowering properties.

Reactive Hyperaemia

Although reactive hyperaemia has been investigated since the beginning of the 20th century, not all contributing mechanisms have been clearly identified (see also Chapter 9). Immediately after the release of an occlusion, restoration of blood flow would be expected to stimulate endothelial release of NO and prostaglandins through shear-dependent mechanisms. This hypothesis is supported by evidence that peak reactive dilatation in skeletal muscle after brief arteriolar occlusion is significantly reduced by endothelial impairment, whereas the hyperaemic response after the release of long occlusions is dominated by metabolic mechanisms, and dilatation during the occlusion itself elicited by myogenic mechanisms.[71,72] While the contribution of prostaglandins in the early and late phases of reactive hyperaemia seems well-established,[71–73] the role of NO remains more controversial because the reduction in baseline blood flow that generally follows administration of NOS inhibitors complicates the interpretation of responses during reactive hyperaemia. Both peak and total excess flow are significantly reduced in the canine diaphragm and hindlimb following administration of NOS inhibitors,[74,75] whereas in the human forearm NO has a relatively modest effect on the peak dilator response, but no effect on the total excess flow.[76] However, the contribution of NO to reactive hyperaemia in the human forearm becomes clearly evident in the presence of an inhibitor of prostaglandin synthesis, when inhibition of NOS then causes a reduction in peak flow and total flow excess.[76] These findings suggest that NO- and prostaglandin-mediated mechanisms may operate synergistically and to some extent compensate for each other.

Some studies suggest that the endothelium is sensitive to changes in the partial pressure of O_2 and may respond by releasing prostaglandins and/or NO.[77–81] In theory, such a mechanism could promote increases in blood flow and serve to restore the O_2 supply before vascular smooth muscle and parenchymal cells are affected by low O_2 levels. Although there is some evidence that changes in O_2 tension can affect flow-induced dilatation,[82] the role of hypoxia needs further investigation.[81] The release and interaction of reactive O_2 metabolites with NO may also be relevant, especially after the release of occlusions of longer duration.[83,84]

Exercise Hyperaemia

Multiple redundant vasoactive mechanisms may similarly contribute to functional hyperaemia, making it difficult to define the role of shear stress-induced NO release clearly (see also Chapter 9). Nevertheless, exercise results in large intermittent increases in shear stress in the skeletal muscle circulation and other vascular beds, and in rats a short-term daily exercise regimen leads to a sustained enhancement of endothelial NO synthesis that is likely to make a significant contribution to the hyperaemic response of skeletal muscle during exercise.[85] It is important to note that increased sympathetic nerve activity functionally attenuates the dilatation of skeletal muscle arterioles that occurs during exercise,[86] thereby tending to increase local pressures, flow velocity and thus shear stress. Therefore, it is possible that even if flow measurements do not indicate major quantitative increases in blood flow in vessels of a given skeletal muscle, endothelial shear stress may still increase during exercise.

In isolated arterioles from the gracilis muscle ~ 45 μm in diameter, flow-induced dilatation is significantly greater (by ~ 36% at maximum flow) in the vessels of exercised rats (EX; after ~ 3 weeks of daily exercise by running on a treadmill) than in those of sedentary rats (SED).[85] Calculations of wall shear stress indicated that the increase in diameter of EX arterioles compared to baseline/SED was nearly double for the same stimulus (~ 50 dynes/cm^2). In both SED and EX vessels, removal of endothelium or inhibition of NO and prostaglandin synthesis eliminated flow-dependent dilatation. Inhibition of eNOS shifted flow-diameter relationships to the right, but had a significantly greater effect in arterioles from EX than SED rats (Figure 2). Similar observations have been made in gracilis muscle venules.[87] By contrast, mesenteric arterioles do not undergo adaptation, presumably because they do not experience increases in shear stress during exercise.[88] Adaptation

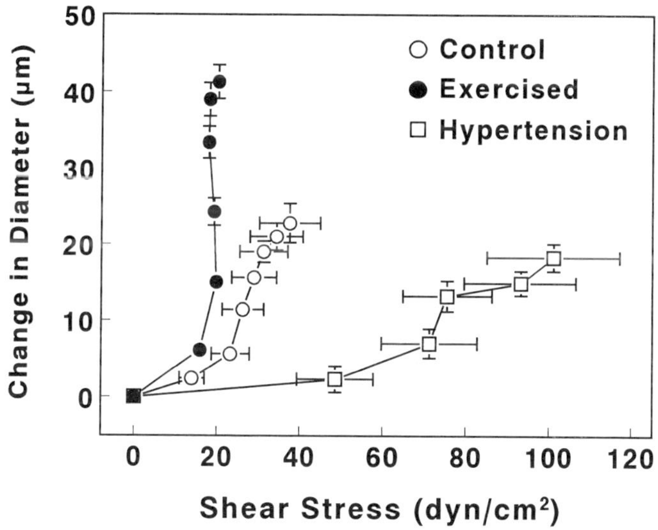

Figure 2. Effect of hypertension and exercise on wall shear stress-diameter relationships in skeletal muscle arterioles.

following exercise training is likely to reflect upregulation of NOS gene expression, as demonstrated in the aorta of dogs[89] and skeletal muscle arterioles of rats (unpublished observations). Indeed, dilatation evoked by ACh and L-arginine is also enhanced by short-term exercise training.[90]

Other Possible Roles of Nitric Oxide in Skeletal Muscle

Because NO is highly diffusible, it affects not only arteriolar smooth muscle but the endothelium itself, blood components and parenchymal tissues. The effects of NO on platelet aggregation, the role of shear stress-induced release of NO in modulating the permeability and hydraulic conductivity of the endothelium,[91,92] and the role of NO in the development of atherosclerosis[93] and reperfusion injury[84] are all areas of intense investigation (see Chapters 17–19). In addition to these mechanisms, there are other possible roles for NO in the skeletal muscle circulation.

Effects of NO on Endothelium

High concentrations of L-arginine analogues not only blunt the arteriolar responses to NO-dependent agonists but also those to arachidonic acid, the precursor of prostaglandins.[94] Studies by Salvemini *et al.*[95] and others have shown that NO directly stimulates cyclooxygenase, the key enzyme involved in the production of prostaglandins (see also Chapter 2). Thus, NO production may be required to maintain optimal synthesis of prostaglandins. Under certain conditions, endothelial function may also be affected by reactive O_2 metabolites (such as superoxide anion) generated during the production of constrictor factors (PGH_2, endothelin) or related to a reduced release of dilator factors.[96] Thus, superoxide dismutase attenuates the constrictor response to ACh in skeletal muscle venules, a response that is mediated by PGH_2.[97] Even under control conditions it thus seems that in venular endothelium there is a continuous production of reactive O_2 metabolites that interferes with the NO and prostaglandin pathways (see also Chapter 19).

Effects of NO on Parenchymal Tissue Function

In addition to effects on vascular smooth muscle, *in vivo* and *in vitro* studies both suggest that NO may affect the metabolism of parenchymal tissues and thus play a direct role in the adaptation to different conditions of demand.[98,99] This hypothesis is supported by immunohistochemical staining of eNOS in the skeletal muscle microcirculation using a specific monoclonal antibody. In rat gracilis muscle, NOS is localized exclusively to the vascular endothelium with staining intensity being strongest in arterioles, less strong in capillary endothelium, and weakest in venules.[100] It has been known for some time that NO modulates the contractility of cardiac muscle[101] (see Chapter 12) and recent studies have demonstrated that NO released from blood vessels can significantly affect O_2 uptake and consumption in dog cardiac[98] and skeletal muscle.[102] Moreover, it has been shown that L-arginine reduces O_2 consumption by isolated mitochondria in an L-NA-sensitive manner.[99] Since capillaries have no smooth muscle layer, their diameters cannot increase actively, and therefore it is likely that shear stress increases at the intraluminal surface during increases in blood flow (e.g. during exercise), thereby stimulating NO synthesis. NO released from capillary endothelium could thus play a role in the regulation of skeletal muscle metabolism (Figure 3). If indeed NO reduces O_2 consumption by parenchymal tissues, this would be advantageous during exercise when the intensity and duration of skeletal muscle work is highly dependent on the availability of O_2. By reducing O_2 consumption, this mechanism could enhance skeletal muscle efficiency and couple tissue metabolism more closely to vascular events. The presence of NOS in capillaries may also be relevant to the effects of NO on capillary permeability (see Chapter 15).

Summary

Endothelium-derived NO plays an important role in the regulation of the skeletal muscle microcirculation, although many aspects of its physiology remain to be established. Tonic release of NO is involved in the modulation of arteriolar and venular tone by balancing myogenic constriction. Release of NO can be stimulated

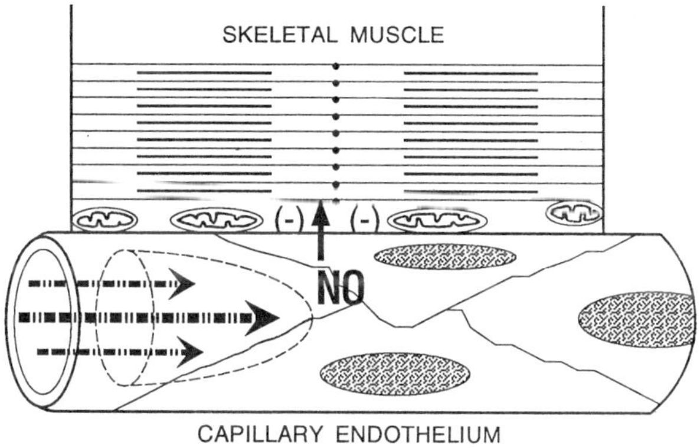

Figure 3. NO released from capillary endothelium in response to increases in shear stress could reach the mitochondria of skeletal muscle fibres and thereby modulate O_2 consumption.

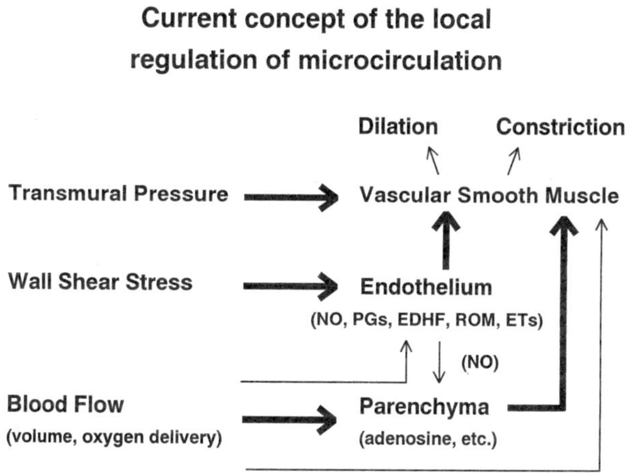

Figure 4. Haemodynamic forces, together with changes in the mass transport of O_2, can affect vascular and parenchymal tissues. The vasoactive factors released from the parenchyma and endothelium modify the intraluminal pressure-induced tone of skeletal muscle arterioles. PG: prostaglandin, EDHF: endothelium-derived hyperpolarizing factor; ROM: reactive oxygen metabolites; ET: endothelin.

by various agents whose action is completely or partly mediated by NO or whose action is modulated by a parallel release of NO. The primary role of NO in the skeletal muscle microcirculation may be to mediate shear stress-induced arteriolar dilatation. However, it is likely to have other functions, including those directly related to the consumption of O_2. These functions may change under various conditions such as exercise and hypertension and are influenced by sex hormones. A new model that incorporates some of the physiological roles now ascribed to NO in the regulation of microcirculatory blood flow in skeletal muscle is given in Figure 4.

Acknowledgement

Professor Kaley is supported by NIH grants HL-43023 and HL-46813.

References

1. Furchgott RF, Zawadzki JV. The obligatory role of endothelial cells in the relaxation of arterial smooth muscle by acetylcholine. *Nature* 1980; **288**: 373–376.
2. Ignarro LJ, GM Buga, Wood KS *et al.* Endothelium-derived relaxing factor produced and released from artery and vein is nitric oxide. *Proc Natl Acad Sci USA* 1987; **84**: 9265–9269.
3. Moncada S, Palmer RMJ, Higgs EA. The discovery of nitric oxide as the endogenous nitrovasodilator. *Hypertension* 1988; **12**: 365–372.
4. Koller A, Messina EJ, Wolin MS, Kaley G. Effects of endothelial impairment on arteriolar dilator responses *in vivo*. *Am J Physiol* 1989; **257**: H1485–H1489.
5. Koller A, Messina EJ, Wolin MS, Kaley G. Endothelial impairment inhibits prostaglandin and EDRF-mediated arteriolar dilation *in vivo*. *Am J Physiol* 1989; **257**: H1966–H1970.
6. Kaley G, Koller A, Rodenburg J *et al.* Regulation of arteriolar tone and responses *via* L-arginine pathway in skeletal muscle. *Am J Physiol* 1992; **262**: H987–H992.
7. Shepherd JT. Circulation to skeletal muscle. In: *Handbook of Physiology, Section 2: The Cardiovascular System*. American Physiological Society, Bethesda 1983: 319–370.
8. Koller A, Seyedi N, Gerritsen ME, Kaley G. EDRF released from microvascular endothelial cells dilates arterioles *in vivo*. *Am J Physiol* 1991; **261**: H128–H133.
9. Sun D, Messina EJ, Koller A *et al.* Endothelium-dependent dilation to L-arginine in isolated rat skeletal muscle arterioles. *Am J Physiol* 1992; **262**: H1211–H1216.
10. Falcone JC, Meininger GA. Arteriolar dilation produced by venule endothelium-derived nitric oxide. *Microcirculation* 1997; **4**: 303–310.
11. Dörnyei G, Kaley G, Koller A. Release of nitric oxide and prostaglandin H_2 to acetylcholine in skeletal muscle venules. *Am J Physiol* 1997; **272**: H2541–H2546.
12. Koller A, Kaley G. Flow velocity-dependent regulation of microvascular resistance *in vivo*. *Microvasc Endothelium Lymphatics* 1989; **5**: 519–529.
13. Smiesko V, Lang DJ, Johnson PC. Dilator response of rat mesenteric arcarding arterioles to increased blood flow velocity. *Am J Physiol* 1989; **257**: H1958–H1965.
14. Koller A, Kaley G. Endothelium regulates skeletal muscle microcirculation by a blood flow velocity sensing mechanism. *Am J Physiol* 1990; **258**: H916–H920.

15. Koller A, Kaley G. Endothelial control of shear stress and resistance in skeletal muscle microcirculation. In: *Flow Dependent Regulation of Vascular Function*, Bevan J, Kaley G, Rubanyi GM (Eds). Oxford University Press, Oxford 1994: 236–260.
16. Hecker M, Mulsch A, Bassenge E, Busse R. Vasoconstriction and increased flow: two principal mechanisms of shear stress-dependent endothelial autacoid release. *Am J Physiol* 1993; **265**: H828–H833.
17. Koller A, Sun D, Kaley G. Role of shear stress and endothelial prostaglandins in flow- and viscosity-induced dilation of arterioles *in vitro*. *Circ Res* 1993; **72**: 1276–1284.
18. Koller A, Sun D, Huang A, Kaley, G. Corelease of nitric oxide and prostaglandins mediates flow-dependent dilation flow of rat gracilis muscle arterioles. *Am J Physiol* 1994; **267**: H326–H332.
19. Boegehold MA. Flow-dependent arteriolar dilation in normotensive rats fed low- or high-salt diets. *Am J Physiol* 1995; **269**: H1407–H1414.
20. Kuo L, Arko F, Chilian WM, Davis MJ. Coronary venular responses to flow and pressure. *Circ Res* 1993; **72**: 607–615.
21. Koller A, Dörnyei, G, Kaley G. Flow-induced responses in skeletal muscle venules: modulation of nitric oxide and prostaglandins. *Am J Physiol* 1998; **275**: H831–H836.
22. Rees DD, Palmer, MJ, Moncada, S. Role of endothelium-derived nitric oxide in the regulation of blood pressure. *Proc Natl Acad Sci USA* 1989; **86**: 3375–3378.
23. Boegehold MA. Shear-dependent release of venular nitric oxide: effect on arteriolar tone in rat striated muscle. *Am J Physiol* 1996; **271**: H387–H395.
24. Burnstock G. Local mechanisms of blood flow control by perivascular nerves and endothelium. *J Hypertens* 1990; **8 (Suppl)**: S95–S106.
25. Koller A, Rodenburg JM, Kaley G. Effects of Hoe-140 and ramiprilat on arteriolar tone and dilation to bradykinin in skeletal muscle of rats. *Am J Physiol* 1995; **268**: H1628–H1633.
26. Busse R, Lamontagne D. Endothelium-derived bradykinin is responsible for the increase in calcium produced by angiotensin-converting enzyme inhibitors in human endothelial cells. *Naunyn-Schmiedebergs Arch Pharmacol* 1991; **344**: 126–129.
27. Hornig B, Kohler C, Drexler H. Role of bradykinin in mediating vascular effects of angiotensin-converting enzyme inhibitors in humans. *Circulation* 1997; **95**: 1115–1118.
28. Ohyanagi M, Nishigaki K, Faber JE. Interaction between microvascular α_1- and α_2-adrenoceptors and endothelium-derived relaxing factor. *Circ Res* 1992; **71**: 188–200.
29. Rubanyi GM, Lorenz RR, Vanhoutte PM. Bioassay of endothelium-derived relaxing factor(s): inactivation by catecholamines. *Am J Physiol* 1985; **249**: H95–H101.

30. Bhagyalakshmi A, Berthiaume F, Reich K, Frangos JA. Fluid shear stimulates membrane phospholipid metabolism in cultured human endothelial cells. *J Vasc Res* 1992; **29**: 443–449.
31. Olesen SP, Clapham DE, Davies PF. Haemodynamic shear stress activates a K^+ current in vascular endothelial cells. *Nature* 1988; **331**: 168–170.
32. Ohno M, Gibbons GH, Dzau VJ, Cooke JP. Shear stress elevates endothelial cGMP. Role of a potassium channel and G protein coupling. *Circulation* 1993; **88**: 193–197.
33. Kuchan MJ, Hanjoong J, Frangos JA. Role of G proteins in shear stress-mediated nitric oxide production. *Am J Physiol* 1994; **267**: C753–C758.
34. Schilling WP. Effect of membrane potential on cytosolic calcium of bovine aortic endothelial cells. *Am J Physiol* 1989; **257**: H778–H784.
35. Schwarz G, Droogmans G, Nilius B. Shear stress induced membrane currents and calcium transients in human vascular endothelial cells. *Pflügers Arch* 1992; **421**: 394–396.
36. Kaley G, Sun D, Huang A, Koller A. Activation of endothelial K_{Ca} channels is essential in the mediation of flow-dependent dilation of arterioles [Abstract]. *FASEB J* 1996; **10**: A571.
37. Gooch KJ, Frangos JA. Flow-and bradykinin-induced nitric oxide production by endothelial cells is independent of membrane potential. *Am J Physiol* 1996; **270**: C546–C551.
38. Wang N, Ingber DE. Control of cytoskeletal mechanics by extracellular matrix, cell shape, and mechanical tension. *Biophys J* 1994; **66**: 2181–2189.
39. Davies PF, Tripathi SC. Mechanical stress mechanisms and the cell; an endothelial paradigm. *Circ Res* 1993; **72**: 239–245.
40. Hutcheson IR, Griffith TM. Mechanotransduction through the endothelial cytoskeleton: mediation of flow- but not agonist-induced EDRF release. *Br J Pharmacol* 1996; **118**: 720–726.
41. Sun D, Huang A, Koller A, Kaley G. Flow-dependent dilation of arterioles is abolished by blockade of endothelial microtubule assembly [Abstract]. *Circulation* 1995; **92 (Suppl)**: I–629.
42. Muller JM, Chilian WM, Davis MJ. Integrin signaling transduces shear stress-dependent vasodilation of coronary arterioles. *Circ Res* 1997; **80**: 320–326.
43. Sun D, Kaley G, Koller A. Characteristics and origin of myogenic response in isolated gracilis muscle arterioles. *Am J Physiol* 1994; **266**: H1177–H1183.
44. Falcone JC, Davis MJ, Meininger GA. Endothelial independence of myogenic response in isolated skeletal muscle arterioles. *Am J Physiol* 1991; **260**: H130–H135.
45. Dörnyei G, Monos E, Kaley G, Koller A. Myogenic responses of isolated rat skeletal muscle venules: modulation by norepinephrine and endothelium. *Am J Physiol* 1996; **271**: H267–H272.

46. Griffith TM, Edwards DH. Myogenic autoregulation of flow may be inversely related to endothelium derived relaxing factor activity. *Am J Physiol* 1990; **259**: H1171–H1180.
47. Kuo L, Chilian WM, Mavis MJ. Interaction of pressure- and flow-induced responses in porcine coronary resistance vessels. *Am J Physiol* 1991; **261**: H1706–H1715.
48. Kuo L, Arko F, Chilian WM, Davis JM. Coronary venular responses to flow and pressure. *Circ Res* 1993; **72**: 607–614.
49. Pohl U, Herlan K, Huang A, Bassenge E. EDRF mediated shear-induced dilation opposes myogenic vasoconstriction in small rabbit arteries. *Am J Physiol* 1991; **261**: H2016–H2023.
50. Sun D, Huang A, Koller A, Kaley G. Flow dependent dilation and myogenic constriction interact to establish the resistance of skeletal muscle arterioles. *Microcirculation* 1995; **2**: 289–295.
51. Huang A, Sun D, Koller A. Endothelial dysfunction augments myogenic arteriolar constriction in hypertension. *Hypertension* 1993; **22**: 913–921.
52. Folkow B. Structural factor in primary and secondary hypertension. *Hypertension* 1990; **16**: 89–101.
53. Panza JA, Casion PR, Kilcoyne CM, Quyyumi AA. Role of endothelium-derived nitric oxide in the abnormal endothelium-dependent vascular relaxation of patients with essential hypertension. *Circulation* 1993; **87**: 1468–1474.
54. Lahera V, Salazar J, Salom MG, Romero JC. Deficient production of nitric oxide induces volume-dependent hypertension. *J Hypertens* 1992; **10 (Suppl)**: S173–S177.
55. Li F, Joshua IG. Decreased arteriolar endothelium-derived relaxing factor production during the development of genetic hypertension. *Clin Exp Hypertens* 1993; **15**: 511–526.
56. Boegehold MA. Reduced influence of nitric oxide on arteriolar tone in hypertensive Dahl rats. *Hypertension* 1992; **19**: 290–295.
57. Chen PY, Sanders PW. Role of nitric oxide synthesis in salt-sensitive hypertension in Dahl/Rapp rats. *Hypertension* 1993; **22**: 812–818.
58. Chu ZM, Beilin LJ. Nitric oxide-mediated changes in vascular reactivity in spontaneously hypertensive rats. *Br J Pharmacol* 1993; **110**: 1184–1188.
59. Koller A, Huang A. Impaired nitric oxide-mediated flow-induced dilation in arterioles of spontaneously hypertensive rats. *Circ Res* 1994; **74**: 416–421.
60. Koller A, Huang A. Shear stress-induced dilation is attenuated in skeletal muscle arterioles of hypertensive rats. *Hypertension* 1995; **25**: 758–763.
61. Huang A, Koller A. Both nitric oxide and prostaglandin-mediated responses are impaired in skeletal muscle arterioles of hypertensive rats. *J Hypertens* 1996; **14**: 887–895.

62. Lerner DJ, Kannel WB. Patterns of coronary heart disease morbidity and mortality in the sexes: 26-year follow-up of the Framingham population. *Am Heart J* 1986; **111**: 383–390.
63. White RE, Darkow DJ, Lang JLF. Estrogen relaxes coronary arteries by opening BK_{Ca} channels through a cGMP-dependent mechanism. *Circ Res* 1995; **77**: 936–942.
64. Kauser K, Rubanyi GM. Gender difference in bioassayable endothelium-derived nitric oxide from rat aortae. *Am J Physiol* 1994; **267**: H2311–H2317.
65. Rosselli M, Imthurm B, Macas E *et al.* Circulating nitrite/nitrate levels increase with follicular development: indirect evidence for estradiol mediated NO release. *Biochem Biophys Res Commun* 1994; **202**: 1543–1552.
66. Huang A, Sun D, Koller A, Kaley G. Gender difference in myogenic tone of rat arterioles is due to estrogen-induced, enhanced release of nitric oxide. *Am J Physiol* 1997; **272**: H1804–H1809.
67. Huang A, Sun D, Koller A, Kaley G. Estrogen enhances flow-induced dilation of arterioles of female rats via increased release of endothelial nitric oxide (NO) [Abstract]. *Circulation* 1997; **96 (Suppl)**: I-355–I-356.
68. Huang A, Sun D, Kaley G, Koller A. Estrogen maintains nitric oxide synthesis in arterioles of female hypertensive rats. *Hypertension* 1997; **29**: 1351–1356.
69. Weiner CP, Lizasoain I, Baylis SA *et al.* Induction of calcium-dependent nitric oxide synthases by sex hormones. *Proc Natl Acad Sci USA* 1994; **91**: 5212–5216.
70. Pinto S, Virdis A, Ghiadoni L *et al.* Endogenous estrogen and acetylcholine-induced vasodilation in normotensive women. *Hypertension* 1997; **29**: 268–273.
71. Carlsson I, Sollevi A, Wennmalm Å. The role of myogenic relaxation, adenosine and prostaglandins in human forearm reactive hyperaemia. *J Physiol (Lond)* 1987; **389**: 147–161.
72. Koller A, Kaley G. Role of endothelium in reactive dilation of skeletal muscle arterioles. *Am J Physiol* 1990; **259**: H1313–H1316.
73. Messina EJ, Weiner R, Kaley G. Arteriolar reactive hyperemia: modification by inhibitors of prostaglandin synthesis. *Am J Physiol* 1977; **232**: H571–H575.
74. Ward ME, Magder SA, Hussain SN. Role of endothelium-derived relaxing factor in reactive hyperemia in canine diaphragm. *J Appl Physiol* 1993; **74**: 1606–1612
75. O'Leary DS, Dunlap RC, Glover KW. Role of endothelium-derived relaxing factor in hindlimb relative and active hyperemia in conscious dogs. *Am J Physiol* 1996; **266**: R1213–R1219.
76. Engelke KA, Halliwill JR, Proctor DN *et al.* Contribution of nitric oxide and prostaglandins to reactive hyperemia in human forearm. *J Appl Physiol* 1996; **81**: 1807–1814.
77. Messina EJ, Sun D, Koller A *et al.* Role of endothelium-derived prostaglandins in hypoxia-elicited arteriolar dilation in rat skeletal muscle. *Circ Res* 1992; **71**: 790–796.

78. Pohl U, Busse R. Hypoxia stimulates release of endothelium-derived relaxing factor. *Am J Physiol* 1989; **256**: H1595–H1600.
79. Park KH, Rubin LE, Gross SS, Levi R. Nitric oxide is a mediator of hypoxic coronary vasodilation. Relation to adenosine and cyclooxygenase-derived metabolites. *Circ Res* 1992; **71**: 992–1001.
80. Kovitz KL, Aleskowich TD, Sylvester JT, Flavahan NA. Endothelium derived contracting and relaxing factors contribute to hypoxic responses of pulmonary arteries. *Am J Physiol* 1993; **265**: H1139–H1148.
81. Xu XP, Pollock J, Tanner MA, Myers PR. Hypoxia activates nitric oxide synthase and stimulates nitric oxide release in porcine coronary resistance arteriolar endothelial cells. *J Cardiovasc Res* 1995; **30**: 841–847.
82. Jimenez AH, Tanner MA, Caldwell WM, Myers PR. Effects of oxygen tension on flow-induced vasodilation in porcine coronary resistance arterioles. *Microvasc Res* 1996; **51**: 365–377.
83. Wolin MS, Rodenburg JM, Messina EJ, Kaley G. Similarities in the pharmacological modulation of reactive hyperemia and vasodilation to hydrogen peroxide in rat skeletal muscle arterioles: effects of probes for endothelium-derived mediators. *J Pharmacol Exp Ther* 1990; **253**: 508–512.
84. Lefer AM, Lefer DJ. The role of nitric oxide and cell adhesion molecules on the microcirculation in ischaemia-reperfusion. *Cardiovasc Res* 1996; **32**: 743–751.
85. Sun D, Huang A, Koller A, Kaley G. Short term daily exercise activity enhances endothelial nitric oxide synthesis in skeletal muscle arterioles of rats. *J Appl Physiol* 1994; **76**: 2241–2247.
86. O'Leary DS, Robinson ED, Butler JL. Is active skeletal muscle functionally vasoconstricted during dynamic exercise in conscious dogs? *Am J Physiol* 1997; **272**: R386–R391.
87. Koller A, Dörnyei G, Kaley G. Exercise training augments flow-dependent dilation in rat skeletal muscle venules [Abstract]. *FASEB J* 1996; **10**: A51.
88. Sun D, Huang A, Koller A, Kaley G. Adaptation of flow-induced dilation of arterioles to daily exercise. *Microvasc Res* 1998; **56**: 54–61.
89. Sessa WC, Pritchard K, Seyedi N *et al.* Chronic exercise in dogs increases coronary vascular nitric oxide production and endothelial nitric oxide synthase gene expression. *Circ Res* 1994; **74**: 349–353.
90. Koller A, Huang A, Sun D, Kaley G. Exercise training augments flow-dependent dilation in rat skeletal muscle arterioles. Role of endothelial nitric oxide and prostaglandins. *Circ Res* 1995; **76**: 544–550.
91. Yuan Y, Granger HJ, Zawieja DC, Chilian WM. Flow modulates coronary venular permeability by a nitric oxide-related mechanism. *Am J Physiol* 1992; **263**: H641–H646.
92. Sill HW, Chang YS, Artman JR *et al.* Shear stress increases hydraulic conductivity of cultured endothelial monolayers. *Am J Physiol* 1993; **268**: H535–H543.

93. Flavahan NA. Atherosclerosis or lipoprotein-induced endothelial dysfunction. Potential mechanisms underlying reduction in EDRF/nitric oxide activity. *Circulation* 1992; **85**: 1927–1938.
94. Koller A, Sun D, Messina EJ, Kaley G. L-Arginine analogs blunt prostaglandin-related dilation of isolated arterioles. *Am J Physiol* 1993; **264**: H1194–H1199.
95. Salvemini D, Misko TP, Masferrer JL *et al.* Nitric oxide activates cyclooxygenase enzymes. *Proc Natl Acad Sci USA* 1993; **90**: 7240–7244.
96. Rubanyi GM, Vanhoutte PM. Oxygen-derived free radicals, endothelium, and responsiveness of vascular smooth muscle. *Am J Physiol* 1986; **250**: H815–H821.
97. Dörnyei G, Kaley G, Koller A. Superoxide is involved in the mediation of acetylcholine-induced constriction of venules [Abstract]. *FASEB J* 1997; **11**: A69.
98. Shen W, Xu X, Ochoa M *et al.* Role of nitric oxide in the regulation of oxygen consumption in conscious dogs. *Circ Res* 1986; **75**: 1086–1095.
99. Kobzik L, Stringer B, Balligand JL *et al.* Endothelial type nitric oxide synthase in skeletal muscle fibers: mitochondrial relationships. *Biochem Biophys Res Commun* 1995; **211**: 375–381.
100. Chander PN, Sun D, Pollock JS *et al.* Localization of endothelial nitric oxide synthase in rat skeletal muscle microvasculature [Abstract]. *Circulation* 1996; **92 (Suppl)**: I–366.
101. Brady AJB, Warren JB, Poole-Wilson PA *et al.* Nitric oxide attenuates cardiac myocyte contraction. *Am J Physiol* 1993; **265**: H176–H182.
102. King CE, Melinshyn MJ, Mewburn JD *et al.* Canine hindlimb blood flow and oxygen uptake after inhibition of EDRF/nitric oxide synthesis. *J Appl Physiol* 1994; **76**: 1166–1171.

Chapter 14

The Hepatic Haemodynamic Effects of Nitric Oxide

Robert T Mathie

Introduction

The liver possesses a microcirculatory structure and control system whose complexity complements the functional diversity of the organ. The hepatic circulation comprises a dual blood supply, whose arterial and portal venous components converge in the liver sinusoids (Figure 1a) and have a common outlet

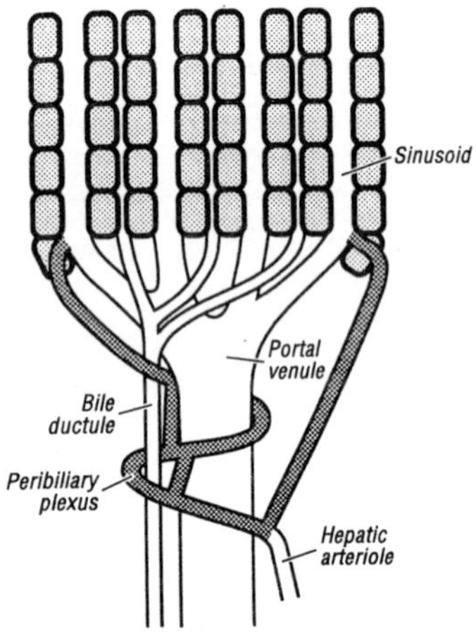

(a)

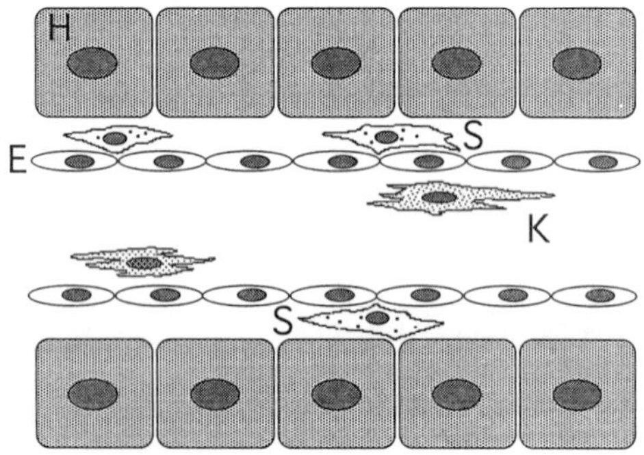

(b)

Figure 1. Diagrammatic representation of the microvascular bed of the liver. (a) The anatomical association between hepatic arterioles, portal venules, hepatic sinusoids and bile ductules. The probable site of action of vasoactive agents, including NO, on the hepatic arterial microcirculation is indicated by shading. (Reproduced from *Surgery of the Liver and Biliary Tract*, 2nd Edition, Blumgart LH (Ed), Churchill Livingstone, Edinburgh 1994, p97; with the kind permission of the publishers.) (b) The hepatic sinusoid, illustrating the association between hepatocytes (H), endothelial cells (E), Kupffer cells (K) and Ito (stellate) cells (S). It is believed that sinusoidal (and thus portal) vascular resistance is regulated by the relaxation and contraction of Ito cells. NO, generated and released under a variety of circumstances by hepatic cells of any of the above types, is closely involved in this controlling mechanism (see text).

via the hepatic vein. Hepatic arterial flow, which accounts for approximately 25% of the total blood supply to the liver, is controlled by mechanisms that possess many features in common with those of other peripheral arteries. By contrast, the liver is unable to regulate portal venous inflow (accounting for the remainder of hepatic blood flow), which is largely a passive consequence of outflow from the intestines, spleen and pancreas. Factors that may regulate intrahepatic portal vascular tone generally have an impact on portal *pressure* rather than on portal blood flow. Haemodynamic inter-relationships between the two bloodstreams of the liver often confound the interpretation of research data obtained for this vascular territory. These complications become further accentuated in a number of disease states, in which profound circulatory disturbances may exist.

NO is implicated in the mechanisms controlling hepatic haemodynamics both in the normal and in the diseased liver, most notably in cirrhosis, but also in septic and haemorrhagic shock, hepatic ischaemia-reperfusion (I-R) injury and liver transplantation. The evidence underlying our current understanding about the extent to which NO contributes to the regulation of the hepatic circulation in health and in these disease states is examined in this chapter.

Physiological Haemodynamic Control

Hepatic Artery

There is no universal consensus that endothelially generated NO modulates hepatic arterial blood flow under normal circumstances, though the balance of evidence indicates that endogenous NO does have an active role. Inhibition of NO synthesis has been observed to increase basal hepatic arterial resistance in pigs[1,2] but not in cats.[3] There is good evidence from *in vitro* perfusion studies of the rabbit liver, however, that the hepatic arterial vasodilatation elicited by acetylcholine and adenosine 5'-triphosphate (ATP) is mediated by NO.[4]

One of the haemodynamic features consequent on the dual nature of the hepatic circulation is the reciprocal change in hepatic arterial blood flow that occurs following variation in portal venous inflow. The phenomenon is known as the 'hepatic arterial buffer response' and is most evident as a marked increase in hepatic arterial blood flow during acute loss of portal flow, a response which is usually identified with the vasodilatory action of adenosine[5] rather than NO.

Indeed, endogenous hepatic arterial NO has been shown to *attenuate* the magnitude of the 'buffer' response.[1,2] The likelihood that NO-mediated vasodilatation of the hepatic artery does not actively contribute to the hyperaemia of the 'buffer' response is given further support by the recent observation that inhibition of hepatic NO synthesis did not alter the vasodilator effect of intra-arterially injected adenosine.[3] Endogenous release of NO within the prehepatic splanchnic vascular bed may exert an indirect controlling influence on hepatic arterial vascular tone *via* the 'buffer' response. Systemically delivered NO synthesis inhibition caused a decrease rather than the predicted increase in hepatic arterial resistance in rats,[6] suggesting that decreased portal venous hepatic inflow arising from splanchnic vasoconstriction resulted in hepatic arterial vasodilatation by the 'buffer' response mechanism.

Little is known about the extent to which NO may participate in the other intrinsic physiological mechanisms that regulate hepatic arterial blood flow. There is evidence, however, that NO opposes autoregulation of flow in the hepatic artery following changes in supply pressure,[1] an effect similar to that established in other vascular beds.[7,8] It has been speculated that NO regulates the peak reactive hyperaemic flow rate observed in the hepatic artery after brief periods (up to 10 seconds) of interrupted flow.[9]

Because of the close apposition of the two vascular beds, the hepatic circulation displays another intriguing haemodynamic relationship between the portal venous and hepatic arterial circulations: a mechanism by which many vasoactive agents contained in the portal vein can result in first-pass haemodynamic effects in the hepatic artery.[10] An example of this 'transhepatic' phenomenon is the vasodilatation of the hepatic arterial circulation evoked by intraportal injection of ATP in the perfused rabbit liver. Although the hepatic arterial vasodilatation elicited by ATP is mediated by NO,[4] the response occurs as a consequence of diffusion of ATP *itself* between the vascular beds, stimulating the local release of NO in the hepatic arterioles following activation of endothelial purine receptors, rather than intrahepatic diffusion of portally-released NO.[11]

Inhibition of NO synthesis was noted to potentiate the vasoconstriction elicited by endothelin (ET)-1 in the hepatic arterial microcirculation of the isolated perfused canine liver,[12] indicating a dynamic balance between the effects of NO and ET-1. However, the direct or indirect haemodynamic relationships that may exist between NO and the action of other vasoactive influences (such as catecholamines, angiotensin, prostaglandins and perivascular nerve activity), increasingly recognized in many other arterial vascular beds,[13,14] have not been extensively investigated in the normal hepatic arterial circulation.

Portal Vein

The role of NO in the regulation of intrahepatic portal vascular resistance has been the recent focus of intense research activity. It is becoming increasingly recognised that endogenous production of NO by the liver contributes importantly to normal portal vascular resistance. Though *in vivo* studies in the pig revealed no evidence in support of a basal NO vasodilatory tone,[1] a contribution by NO to basal portal vascular tone has been identified in the *in situ* perfused rat liver.[15,16] These latter findings have been confirmed in some *in vitro* isolated perfused rat liver experiments[17] but not in others.[18] Intravital fluorescence microscopy studies in rats have revealed modest narrowing of hepatic sinusoids following the intraportal administration of the NO synthesis inhibitor N^G-monomethyl-L-arginine.[19] The soluble guanylate cyclase (sGC) inhibitor methylene blue was found to reduce portal vascular resistance *in vitro*, and to abolish the vasodilatory effect of L-arginine, showing that NO exerts vasodilatation in the liver *via* activation of sGC.[17] However, activation of sGC can also be elicited by carbon monoxide (CO) — generated by the constitutively expressed enzyme haem oxygenase (HO)-2 — and it is thus possible that CO contributes to the regulation of hepatic vascular tone.[20] Data from cell culture experiments show that both sinusoidal endothelial cells and hepatocytes can spontaneously release NO.[17] *In vitro* or *in situ* perfused liver experiments in rat or rabbit models have demonstrated that NO modulates the portal vasoconstrictive effect of prostaglandin $F_{2\alpha}$,[21] noradrenaline[15] and ATP (the latter studied at basal portal vascular tone).[11]

It is increasingly believed that a contribution to the regulation of portal pressure normally takes place *via* the contraction and relaxation of perisinusoidal stellate cells (also referred to as Ito cells), which possess a fat-storing function but which also exhibit smooth muscle characteristics (Figure 1b). The NO donor sodium nitroprusside has been shown to inhibit the sinusoidal constriction elicited by ET-1 in the perfused rat liver,[22] and to induce relaxation in cultured Ito cells.[23] Constitutive production of NO in the liver may thus effectively modulate portal vascular resistance by acting on perisinusoidal Ito cells. A much greater haemodynamic importance for such interactions is evident in disease conditions that involve endotoxaemia, as in septic shock and many cases of end-stage cirrhosis (*vide infra*). During liver injury of this kind, Ito cells undergo 'activation', acquiring an enhanced contractile capability with increased expression of α-smooth muscle actin;[24] at the same time, hepatic synthesis and release of both ET-1 and NO may be substantially increased.

Hepatic Metabolism

The liver possesses two other non-vascular cell types, both recognized many years before Ito cells were identified: the true hepatic parenchymal cells or hepatocytes, and the phagocytic Kupffer cells (see Figure 1b). Following stimulation (by bacterial lipopolysaccharide (LPS) challenge, for example), hepatocytes and Kupffer cells are both capable of synthesizing inducible NO synthase (iNOS). Under such circumstances, these cells can produce NO in quantities sufficient to inhibit several fundamental metabolic functions of the hepatocyte, such as protein synthesis[25] and mitochondrial respiration,[26] the latter effect possibly due to formation of the toxic species peroxynitrite,[27] which is the reaction product of NO with superoxide. Thus, a number of damaging NO-mediated metabolic changes occur in the liver in endotoxaemic shock. Excess NO does not appear to be responsible for the suppression of hepatic gluconeogenesis associated with sepsis, however.[28]

Under normal circumstances, it is unlikely that basal low-level NO release has any direct effect on liver metabolism.

Does liver metabolism regulate hepatic blood flow *via* locally produced NO? It is feasible that basally released NO contributes to the vasodilator effects of metabolites such as adenosine (see Chapter 9), which has a prominent role in the physiological control of hepatic arterial blood flow, as discussed above. However, it has become generally accepted that, within physiological limits, the oxygen demands of the liver do not significantly influence hepatic arterial blood flow.[29] Altered liver metabolism is frequently associated with induced overproduction of NO, as discussed elsewhere, and this excess NO in itself is likely to elicit changes in hepatic vascular resistance. Whether a marked increase in liver metabolism *per se* can evoke NO-mediated changes in hepatic blood flow remains unresearched.

Pathophysiology of Hepatic Haemodynamics

Cirrhosis and Portal Hypertension

The haemodynamics of advanced cirrhosis are characterized by increased cardiac output and hepatic portal venous pressure and resistance, refractoriness to exogenously applied vasopressor agents, and arterial hypotension associated with a peripheral vasodilatation that is most marked in the vasculature of the prehepatic splanchnic bed. The low vascular resistance state is maintained despite increased sympathetic nervous activity and plasma noradrenaline levels,[30] indicating the presence of a potent local vasodilator agent (and/or absence of a potent vasoconstrictor) in the peripheral vasculature. The 'hyperdynamic circulation' of the prehepatic splanchnic vasculature is an important contributor to the maintenance of the portal hypertensive state.[31]

Given a background of evidence that endotoxaemia is frequently associated with cirrhosis and that NO is an important mediator of the vasodilatory effects of endotoxins and cytokines, it was proposed by Vallance and Moncada in 1991 that an endotoxin- or cytokine-induced increase in vascular NO synthesis and release may contribute to the haemodynamic changes observed in the disease.[32] Early wholehearted acceptance of this proposal has gradually given way to a rather more cautious approach and a refinement of the concept, in the light of an emerging number of published studies providing only limited evidence in support of the hypothesis. It is necessary to note, however, that overproduction of NO arising from stimulation of iNOS is only one mechanism by which NO may contribute to the splanchnic vasodilatory syndrome. It is feasible that the primary haemodynamic mechanism involving NO is through enhanced release of constitutively-expressed NO as a *consequence* of a flow-induced increased shear on the vascular endothelium,[33] resulting from the action of other vasodilators such as glucagon or prostacyclin, for example.[31] The present section examines the evidence 'for' and 'against' the hypothesis that overproduction of NO plays a role in the haemodynamic abnormalities of portal hypertension, with emphasis on those studies that have focused on the splanchnic and hepatic circulations. The role of NO in the hepatorenal syndrome associated with portal hypertension is discussed in Chapter 16.

The evidence 'for' increased splanchnic NO in portal hypertension

Early *in vitro* and *in vivo* experimental studies in portal hypertensive rats, such as those reported by the research groups in New Haven,[34] Barcelona[35] and Clichy,[36] established that non-selective NO synthesis inhibition modulates the systemic and splanchnic vasodilatation and the hypo-reactivity of the mesenteric bed to vasoconstriction by catecholamines, vasopressin, potassium chloride and ET-1. Subsequent reports of a relationship between endotoxaemia and increased NO synthesis associated with cirrhosis in human subjects[37,38] and experimental animals[39] have further substantiated the original Vallance/Moncada hypothesis. Forearm resistance vessels of cirrhotic patients showed enhanced endothelium-mediated vasodilatation, consistent with increased endothelial NO synthesis.[40] Similar experiments, but with de-endothelialized vessels, indicated that attenuation of vasoconstrictor responses to noradrenaline was due to increased smooth muscle-derived NO.[41] Elevated levels of NO metabolites in peripheral plasma[40] and exhaled NO[42] in cirrhotic patients have also been documented, suggesting increased vascular synthesis of NO.

Upregulation of the constitutive NO pathway in the endothelium of the mesenteric artery of cirrhotic rats[43,44] and portal vein ligated rabbits[45] has been suggested in the last 2–3 years. NO overproduction arising from induction of iNOS in the mesenteric or systemic vasculature associated with portal hypertension has also been reported. For example, iNOS inhibition by aminoguanidine significantly increased arterial blood pressure and systemic vascular resistance in portal hypertensive but not in control rats.[46] Augmented NO activity in vascular smooth muscle leading to decreased sGC availability might account for the diminished vasodilator responses to sodium nitroprusside observed in the mesenteric bed of cirrhotic rats.[47] Inhibition of NO synthesis potentiated vasoconstrictor responses to ATP in the rat mesentery at basal tone in cirrhotic but not in control preparations,[48] suggesting upregulation of NO synthesis in the vascular smooth muscle of cirrhotic animals. Recent work has shown that the hypo-reactivity to mesenteric nerve stimulation of rats with portal hypertension induced by partial portal vein ligation may result from increased vascular formation of NO.[49]

In the past 2 years, investigators have sought to determine the precise source of increased NO and NOS activity within the splanchnic and other peripheral vasculatures. In portal hypertensive rats, samples of jejunum,[50] superior mesenteric artery,[51] mesenteric tissue[52] and aorta[53] revealed increased Ca^{2+}-dependent NOS activity, while increased production of NO from upregulated constitutive NOS in gastric mucosal blood vessels has been described.[54] Overexpression of both constitutive and inducible NOS mRNA in oesophageal mucosa of portal hypertensive rats has been observed by Tanoue and colleagues.[55] Increased endothelial NOS (eNOS) protein immuno-reactivity[56] and iNOS mRNA expression[57] in cirrhotic rat mesenteric artery have been detected.

One of the very few NO-related studies to examine the splanchnic territory in *human* portal hypertension found evidence for an increase in both constitutive and inducible NOS activity in the gastric mucosa of cirrhotic patients.[58] Our own group has recently observed increased NO metabolites in the portal blood of patients with end-stage cirrhosis compared with non-cirrhotic controls,[59] while workers in Japan have demonstrated a positive correlation between blood flow and serum nitrate concentration in the portal venous circulation of cirrhotic subjects,[60] implicating NO in the hyperdynamic splanchnic circulatory syndrome. NO is also thought to promote the collateralization of the portal system that is a frequent complication of portal hypertension.[61,62]

The evidence 'against' increased splanchnic NO in portal hypertension

For every three studies published that have implicated NO in the abnormal haemodynamics of cirrhosis and portal hypertension, there has been on average one publication that places doubt on the concept. This was evident from the earliest *in vivo* experiments,[63,64] which were unable to detect any evidence for an enhanced splanchnic regulatory role for NO in portal hypertensive animals. A degree of scepticism for the concept of NO involvement has remained in the light of later *in vitro* studies which have failed to demonstrate an increase in endothelial NO-mediated vasodilatation[65] or in the splanchnic tissue enzyme activity of constitutive[66] or inducible[66,67] NOS in portal hypertensive rats. Responses to ATP in the cirrhotic mesentery, studied after NOS inhibition and at raised vascular tone, were converted from vasodilatation to vasoconstriction,[47] findings consistent with diminished endothelial NO formation; these effects were not observed before inhibition of NO synthesis, and so it was considered unlikely that a change in ATP receptors was responsible. Rats with secondary biliary cirrhosis were found to possess a *decrease* in serum concentrations of NO metabolites.[68] Samples of mesenteric artery or liver taken from cirrhotic rats revealed little or no detectable iNOS gene expression.[56,69]

Cirrhotic subjects displayed no evidence of a functionally significant induction of NOS activity in peripheral arterioles or veins.[70] Other clinical studies in cirrhosis have failed to detect evidence of endotoxaemia or raised cytokine levels[71] or to identify any correlation between elevated blood endotoxin and nitrate levels.[72] In a recent series of liver transplant patients, lower Ca^{2+}-dependent NOS activity but similar iNOS activity was reported in the omentum of cirrhotic compared with non-cirrhotic patients.[59]

NO activity in the cirrhotic liver

In contrast to the hyporesponsiveness to vasoconstrictor agents known to occur in other regions of the peripheral vasculature in cirrhosis and portal hypertension,[73] the intrahepatic portal vascular bed of the cirrhotic liver exhibits *hyper*-reactivity to vasoconstriction by noradrenaline[74,75] and sympathetic nerve stimulation.[76] These findings would suggest that NO synthesis is not upregulated in the cirrhotic portal bed,[75,77] and indeed would be consistent with a *reduced* constitutive NOS reported in the liver of portal hypertensive patients[59] and rats.[68] In a recent report, Gupta *et al.*[78] have shown that impaired release of endothelial NO may explain the hyper-reactivity of the cirrhotic portal vascular bed to vasoconstriction. If one accepts that constitutively expressed NO can contribute to a low basal portal vascular resistance, then reduced hepatic synthesis of NO in cirrhosis will be likely to promote the portal hypertensive state by increasing intrahepatic resistance to portal flow.

Such increases in portal vascular resistance due to reduced NO may be potentiated by the interplay of increased ET-1 expression by the endothelium, causing contraction of activated Ito cells[24,79] (see Figure 1b). These effects would probably be modulated by co-existing endotoxaemia, which is known to induce upregulation of hepatic cellular NO synthesis *via* iNOS activation (*vide infra*). Such excess NO might, however, diminish eNOS enzyme expression or activity by negative feedback.[80] Evidence for a decrease in NO production by non-parenchymal liver cells with an associated increase in hepatocyte NO expression has been observed in ethanol-fed rats, but only in those animals manifesting endotoxaemia and histological hepatic injury.[81] Thus, no studies identified to date have implicated the hepatic tissue or vasculature in an upregulation of NO activity in cirrhosis and portal hypertension in the proven absence of endotoxaemia.

The balance of evidence: A role for nitric oxide

The weight of evidence currently favours the view that NO is implicated in the haemodynamic alterations associated with cirrhosis and portal hypertension, though it is increasingly apparent that it is the decompensated end-stage of the cirrhotic condition that is most strongly associated with both endotoxaemia[82] and a hyperdynamic circulation involving NO.[83] Thus, a consensus view is emerging of a prehepatic splanchnic circulation in which the activity of constitutive NOS is upregulated in discrete anatomical locations, such as in the endothelium of the mesenteric artery and in the oesophageal, gastric and jejunal mucosa. By contrast, considerable evidence points to a *decrease* in constitutive NOS synthesis in the liver, that will tend to exacerbate portal hypertension. Inducible NOS has proved to be a more elusive candidate, but there may well be expression of this isoform in discrete regions of the prehepatic splanchnic and peripheral vasculature. In the absence of endotoxaemia, upregulation of hepatic iNOS activity in cirrhosis probably does not occur. Explanations for the discrepant findings of different published studies include: the different animal models of portal hypertension used[84] and their degree of similarity to the human condition; the severity of the disease and the timing of the measurements made;[82] and the different concentrations, selectivity and potency of the NO synthesis inhibitors used to address the question.[85]

In view of such heterogeneity of NO synthesis and release, together with a recognition of anatomical differences in the responsiveness to vasoconstrictors in the portal hypertensive hepatosplanchnic circulation,[86] it is perhaps not surprising that neither non-selective NO synthesis inhibition[87] nor systemically administered iNOS inhibition[88] has succeeded in modulating portal venous pressure. Given that endogenously released NO influences the vasodilatory activity of NO donors such as sodium nitroprusside,[89] it should be recognized that a decrease in constitutive hepatic NOS activity, discussed above, could be the explanation for the enhanced portal hypotensive effect of NO donors in cirrhosis. It is thus possible that combination NO-targeted therapy, comprising the selective inhibition of splanchnic and/or peripheral vascular iNOS together with the administration of liver-specific NO donor prodrugs (such as those currently under development),[90] may play a future role in the pharmacological control of the haemodynamic abnormalities of advanced portal hypertension.

Additional studies are clearly necessary to elucidate further the role of NO in the pathogenesis of cirrhosis and portal hypertension. In particular, there is a need to identify more clearly: (a) the sequence of events and the pathogenic stimuli that induce changes in NO synthesis; (b) the specific anatomical/cellular locations associated with altered synthesis and/or activity of NO and with the message for and expression of each NOS isoform; (c) the interaction of NO with other vasoactive agents such as ET-1; and (d) the relationship of the above processes with the stage of the cirrhotic process and the haemodynamics of the disease. Such investigations will provide further insight into the therapeutic potential of NO donors and selective inhibitors of NOS activity in modulating portal venous and systemic arterial blood pressure.

It is noteworthy that some of the other candidate mediators of the hyperdynamic circulatory syndrome involve NO in their mechanisms of action. Thus, for example, endothelial release of NO contributes to the vasodilator effect of adenosine,[91] substance P[92] and shear stress (Chapter 2). A synergistic haemodynamic relationship between NO and prostacyclin in the splanchnic hyperaemia of portal hypertensive rats has been characterized,[93] and NO is known to stimulate endothelial prostacyclin release (Chapter 2). In addition, it is recognized that the vascular synthesis and release of NO and the activation and regulation of NOS are dependent on G proteins,[94] guanosine 3',5'-cyclic monophosphate (cGMP),[95] protein tyrosine kinases and nuclear transcription factor NF-κB,[96] some or all of which might function abnormally in end-stage cirrhosis.

Circulatory Shock

The hepatic circulation, particularly the portal blood supply and sinusoidal perfusion, is profoundly affected by both haemorrhagic and septic shock. The initial cardiovascular response to haemorrhage is characterized by an intense vasoconstriction of the peripheral vascular system, while the decompensated or refractory phase of both forms of circulatory shock is frequently associated with peripheral vasodilatation, sustained arterial hypotension and hyporesponsiveness to vasopressor agents. Changes in NO activity have been implicated in many of these phenomena,[97] though in this regard substantially more experimental investigation of the liver has focused on the effects of sepsis rather than hypovolaemia.

In endotoxic shock, whether hepatic NO is beneficial or harmful to the organ may depend upon the quantity of NO present, in turn dependent upon whether the time elapsed from the initial stimulus has been sufficient to elicit iNOS synthesis and thus excessive production of NO. Thus, NO has been shown to be important to the preservation of hepatic blood flow 2–3 hours after initiation of endotoxaemia in mice[98] and rabbits (Figure 2).[99] In LPS-treated rats, hepatic damage was increased when a non-specific NO synthesis inhibitor was administered at the time of LPS injection,[100] and it has been suggested that the NO synthesized and released in the first few hours of endotoxaemia may limit liver damage by reacting with oxygen-derived free radicals (ODFR).[101] After 6 hours of endotoxic shock, by contrast, selective inhibition of iNOS prevented circulatory failure and improved liver function (Figure 3).[102] In a more recent study, the liver damage associated with 6 hours exposure to LPS was markedly increased by the constitutive NOS inhibitor N^G-nitro-L-arginine methyl ester (L-NAME), but not by the selective iNOS inhibitor *S*-methylisothiourea.[103] Collectively, these findings are consistent with an iNOS-stimulated overproduction of NO and associated cytotoxicity *via* the formation of peroxynitrite,[104] together with constitutive eNOS-mediated protection of the liver's microvascular blood supply.

It therefore appears likely that in early septic shock conditions, hepatic NO activity has a cytoprotective role. The sustained synthesis of iNOS and the excess NO released by virtually all hepatic cell types during prolonged (at least 6 hours) endotoxaemia, however, may exert damaging effects on hepatic functional status.[105] Nevertheless, NO (even in excess supply) may tend to preserve sinusoidal perfusion, possibly by limiting the effects of endogenously produced vasoconstrictor substances such as ET-1[18] and cyclooxygenase products,[106] the mechanism for which may involve modulation of Ito cell contractility through autocrine release

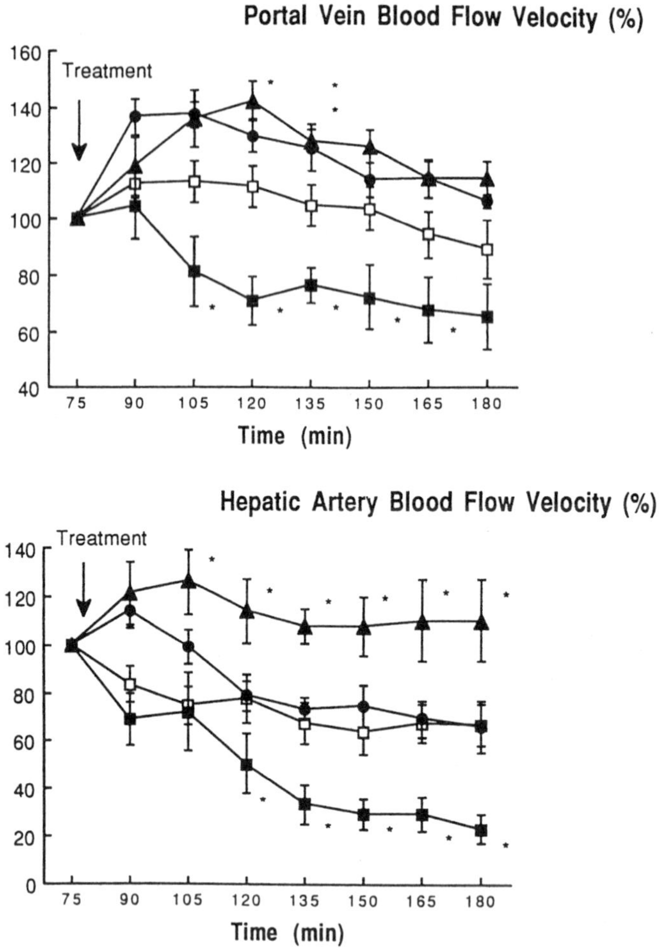

Figure 2. Hepatic blood flow up to 3 h following endotoxin administration. Doppler flowmeter measurements of portal venous and hepatic arterial blood flow velocities (% ± standard error of the mean), in anaesthetized rabbits injected with 1200 μg endotoxin at time zero. At 75 min after endotoxin administration, treatments were injected *i.v.* as follows: saline in 8 rabbits (open squares); the NO synthesis inhibitor N^G-nitro-L-arginine in 5 rabbits (7.5 mg/kg, closed squares); L-arginine in 5 rabbits (600 mg/kg, closed circles); the NO donor linsidomine in 5 rabbits (1 mg, closed triangles). *, significantly different from saline-treated animals ($P < 0.05$). Note the decrease in blood flows following NO synthesis inhibition. (Reproduced from *Shock* 1994; **2:** p200,[99] with the kind permission of the publishers, BioMedical Press.)

of NO.[107] Whether eNOS activity is abrogated in the liver during sepsis remains uncertain; greatly reduced eNOS mRNA concentrations have been reported in the heart, lung and aorta of rats only 4 hours after LPS challenge.[108]

Haemorrhagic shock may result in generalized vascular endothelial dysfunction with reduced eNOS activity and impaired endothelial biosynthesis of NO;[109]

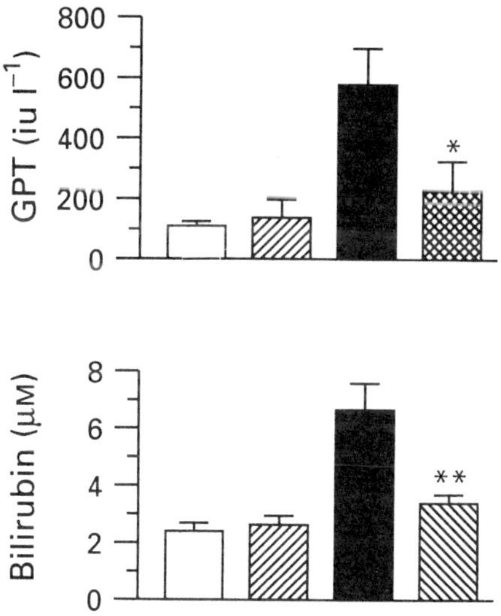

Figure 3. Liver function 6 h after administration of LPS. Effect of the selective iNOS inhibitor 1-amino-2-hydroxy-guanidine on hepatic damage assessed by activity of alanine aminotransferase (GPT) and concentration of bilirubin in the serum of rats treated with *E. coli* LPS (10 mg/kg, *i.v.*) for 6 h and in controls. LPS-treated rats were infused for 4 h either with saline vehicle (0.6 ml/kg/h, solid column, $n = 10$) or 1-amino-2-hydroxy-guanidine (10 mg/kg + 10 mg/kg/h, cross-hatched column, $n = 10$). Similarly, control animals received infusions either of saline vehicle (0.6 ml/kg/h, open column, $n = 4$) or 1-amino-2-hydroxy-guanidine (10 mg/kg + 10 mg/kg/h, hatched column, $n = 10$). The infusion of drug or saline vehicle was started at 2 h after LPS. Data are expressed as mean ± standard error of the mean of n observations. A significant reduction in GPT activity ($^*P < 0.05$) and bilirubin concentration ($^{**}P < 0.01$) was obtained in those LPS-treated animals given the iNOS inhibitor compared with LPS-treated rats given vehicle alone. (Extract of figure reproduced from *Br J Pharmacol* 1996; **118**: p266,[102] with the kind permission of the publishers, MacMillan Press.)

prolonged hypovolaemia induces upregulation of iNOS activity in a variety of organs and in vascular smooth muscle.[109] Little work has been carried out to identify the effects of haemorrhage on NO expression in the liver. It has been observed, however, that after 2 hours of severe haemorrhage in the rat, non-selective inhibition of NO synthesis resulted in markedly increased hepatic injury.[110] By contrast, after nearly 6 hours of severe haemorrhagic shock, there was increased Ca^{2+}-independent NOS activity in the liver, mesentery and aorta,[111] associated with systemic vascular hypo-reactivity and decompensation. An appealing hypothesis, therefore, also appreciated recently by the Pittsburgh group,[112] is that constitutively expressed NO serves to protect the liver from the early injury following severe haemorrhage; on the other hand, more prolonged haemorrhagic shock (of greater than about 6 hours duration) induces iNOS activation and consequently a greatly increased production of NO and probably peroxynitrite, with resulting loss of hepatic vascular tone and tissue integrity.

Given the above evidence, future therapeutic options that target the liver of patients with prolonged decompensated circulatory shock might thus involve selective inhibition of iNOS activity, while preservating or enhancing the functional activity of eNOS and endothelium-derived NO. Preliminary studies by Billiar and co-workers have demonstrated that a novel liver-specific NO donor prodrug can reduce LPS-induced liver injury in mice,[90] by a mechanism presumably associated with hepatic microcirculatory improvement. Exciting opportunities for future drug development in this area are thus indicated.

Hepatic Ischaemia-reperfusion Injury

The damage to the hepatic endothelium and parenchyma that results from ischaemia and reperfusion is caused by a number of inter-related phenomena, including the action of locally liberated ODFR.[113] The reaction of superoxide with NO liberated by the damaged endothelium results in the formation of peroxynitrite, but there is little evidence that peroxynitrite-associated hepatic damage occurs in the early reperfusion period. Only one research group has identified hepatocellular injury associated with peroxynitrite within one hour of reperfusion,[114] abnormalities that were amplified by pre-treatment with LPS. The balance of available evidence, however, strongly suggests that endogenous NO tends to *protect* the liver in the early reperfusion period following hepatic ischaemia.

In recent reports, improvements in liver haemodynamics and tissue function were conferred by supplying an NO donor[115] or L-arginine[116] at the start of reperfusion following an hepatic ischaemic episode of up to 2 hours. Treatment of rats with L-NAME was found to produce a striking increase in the number of non-perfused sinusoids 70 minutes after liver reperfusion.[117] Four hours after a 20 minute period of hepatic ischaemia in the rat, hepatic blood flow and function were poorer in animals given L-NAME than in untreated control animals,[118] observations which further suggest that NO endogenously released by the liver serves to limit early I-R injury. Corroboration of these findings is available from experiments that examined the hepatic haemodynamic response of rats 1 hour after volume resuscitation following severe haemorrhage: inhibition of NO synthesis markedly reduced the improvement in both hepatic arterial and portal venous blood flow.[119] The protective effect of NO may involve its action as a scavenger of ODFR, or may be a reflection of its ability to counteract the vasoconstrictive effect of ET-1.[119] Interestingly, NO has been implicated in the early hepatocellular protection resulting from ischaemic preconditioning in a rat model of hepatic I-R injury;[120] the mechanism of such protection was shown to involve the inhibitory action of NO on ET-1.[121] An attempt to supplement the compromised formation of endogenous hepatic NO with L-arginine or an NO donor may thus be a rational approach to the immediate management of hepatic I-R injury.

The hypothesis that hepatic iNOS activity may be stimulated by inflammatory mediators during the later reperfusion period, leading to the production of cytotoxic levels of peroxynitrite (at least 6 hours after ischaemia), has been investigated to only a very limited extent. There are some data, however, which suggest that neither NO nor peroxynitrite is involved in the hepatocellular injury observed in the rat liver during reperfusion up to 24 hours after a 45 minute period of ischaemia.[122] Similar experiments have demonstrated a *detrimental* influence of NO synthesis inhibition on hepatic blood flow and function, an effect abolished by pre-treatment with L-arginine.[123] The possibility that vasodilatation associated with activation of sGC in the hepatic portal vascular bed may be an induced protective event in the later reperfusion period is supported by a recent report from the Baltimore group: 5 hours after fluid resuscitation from severe haemorrhagic hypotension, blockade of the inducible enzyme HO-1 (responsible for excess production of CO) markedly increased intrahepatic portal vascular resistance, implicating an important pathway for preservation of sinusoidal perfusion at that time.[124] Further work is clearly required to ascertain whether targeted inhibition of iNOS activity during these longer term processes of reperfusion would be protective or harmful to the liver.

Liver Transplantation

Following transplantation, reperfusion of the grafted liver soon results in systemic hypotension[125] with raised portal venous blood flow,[126] and it seems likely that NO and subsequently endotoxin released by the ischaemic viscera contribute to these effects. The release of arginase from the grafted organ severely depletes peripheral plasma levels of L-arginine and NO metabolites, promoting pulmonary hypertension.[127] Taken together with its role in protection from hepatic I-R injury *per se*, infusion of L-arginine may thus be further indicated as a useful strategy in the early reperfusion period following liver transplantation. Although transplantation for cirrhosis leads to a swift normalization of portal venous pressure, the raised cardiac output, splanchnic blood flow and portal-systemic collateralization associated with cirrhosis may persist for several months after the operation.[128] Sustained iNOS expression and excess production of NO in the prehepatic splanchnic vasculature may underlie some of these longer-term effects, but direct proof for this hypothesis is lacking.

The effect of cold preservation of the liver on hepatic NO activity and its sequelae has received little attention. There is some evidence from experimental studies, however, that a decline in hepatocyte protein synthesis after cold storage may be mediated by upregulation of the synthesis and release of NO from activated Kupffer cells.[129] It is increasingly recognized that sinusoidal endothelial function of the liver is optimally preserved by cold storage of the organ in University of Wisconsin solution.[130] The relative importance of altered NO activity in the hepatic dysfunction resulting from cold preservation *per se*, compared with its contribution to the hepatic injury arising from ischaemia and reperfusion during transplantation procedures, has not been investigated.

Overproduction of NO has been implicated in the tissue injury and rejection that may arise in the first several days after liver transplantation.[131–133] In biopsy specimens from patients with chronic liver rejection, immunocytochemical expression of hepatic iNOS was significantly enhanced.[134] Inhibition of NOS expression may partly account for the cytoprotective effects of immunosuppressant drugs such as cyclosporin A and FK506.[97] Cyclosporin-induced systemic hypertension may be the result of altered NO metabolism and/or increased ET-1 production.[135] These effects of immunosuppressant agents confound attempts to determine the long-term consequences of liver transplantation *per se* on hepatic haemodynamics and the role of NO in these changes.

Conclusions

NO constitutively produced by the hepatic endothelium almost certainly contributes to the normal control of hepatic arterial and portal venous vascular tone. Endothelially-derived NO is protective to the liver in the early stages of circulatory shock or local ischaemia-reperfusion injury, an effect at least partly due to modulation of endothelin-mediated vasoconstriction. In prolonged sepsis, there is good evidence for excess production of NO in the liver arising from stimulated iNOS activity, resulting in hepatic tissue damage. Hepatic ischaemia and reperfusion may also cause iNOS stimulation, but the effects of the resulting NO overproduction are poorly established. The hyperdynamic splanchnic circulation of cirrhosis is likely to be mediated, at least in part, by upregulation of NOS activity (particularly the constitutively expressed isoform) in discrete regions of the splanchnic vascular bed. Hepatic iNOS synthesis may occur in end-stage cirrhosis, though probably only in cases where endotoxaemia co-exists. Selective modulation of hepatic iNOS activity, with preservation or enhancement of endothelially-derived NO activity, may provide a future therapeutic option in the control of hepatic haemodynamics and the improvement of hepatic function in the disease states described. Our appreciation of these possibilities may be enhanced with the development of NOS 'knockout' transgenic mouse models, which promise a fundamentally more direct experimental approach with which to examine in detail the effects of NO on hepatic haemodynamics and function in both health and disease.

Acknowledgements

The author is grateful to Professor Anthony Wheatley (University of Otago, New Zealand), Professor James Robotham (Imperial College School of Medicine, London) and Dr Vera Ralevic (University of Nottingham) for helpful critical review of the manuscript.

References

1. Ayuse T, Brienza N, Revelly JP *et al.* Role of nitric oxide in porcine liver circulation under normal and endotoxemic conditions. *J Appl Physiol* 1995; **78**: 1319–1329.
2. Grund F, Sommerschild HT, Winecoff A *et al.* Importance of nitric oxide in hepatic arterial blood flow and total hepatic blood volume regulation in pigs. *Acta Physiol Scand* 1997; **161**: 303–309.
3. Macedo MP, Lautt WW. Potentiation to vasodilators by nitric oxide synthase blockade in superior mesenteric but not hepatic artery. *Am J Physiol* 1997; **272**: G507–G514.
4. Mathie RT, Ralevic V, Alexander B, Burnstock G. Nitric oxide is the mediator of ATP-induced dilatation of the rabbit hepatic arterial vascular bed. *Br J Pharmacol* 1991; **103**: 1602–1606.
5. Lautt WW, Legare DJ, d'Almeida MS. Adenosine as putative regulator of hepatic arterial flow (the buffer response). *Am J Physiol* 1985; **248**: H331–H338.
6. Greenblatt EP, Loeb AL, Longnecker DE. Marked regional heterogeneity in the magnitude of EDRF/NO-mediated vascular tone in awake rats. *J Cardiovasc Pharmacol* 1993; **21**: 235–240.
7. Griffith TM, Edwards DH. Myogenic autoregulation of flow may be inversely related to endothelium-derived relaxing factor activity. *Am J Physiol* 1990; **258**: H1171–H1180.
8. Macedo MP, Lautt WW. Nitric oxide antagonizes pressure-flow autoregulation in the superior mesenteric artery. In: *The Biology of Nitric Oxide, Part 5*, Moncada S, Stamler J, Gross S, Higgs EA (Eds). Portland Press, London 1996: 254.
9. Mathie RT. Hepatic haemodynamics during and after brief occlusion of the hepatic artery. *Hepatol Res* 1997; **8**: 198–206.
10. Richardson PDI. Physiological regulation of the hepatic circulation. *Fed Proc* 1982; **41**: 2111–2116.
11. Browse DJ, Mathie RT, Benjamin IS, Alexander B. The transhepatic action of ATP on the hepatic arterial and portal venous vascular beds of the rabbit: the role of nitric oxide. *Br J Pharmacol* 1994; **113**: 987–993.

12. Faro R, Grassi-Kassisse DM, Donato JL *et al.* Role of endothelin ET_A and ET_B receptors in the arterial vasculature of the isolated canine liver. *J Cardiovasc Pharmacol* 1995; **26**: S204–S207.
13. Navar LG, Inscho EW, Majid DSA *et al.* Paracrine regulation of the renal microcirculation. *Physiol Rev* 1996; **76**: 425–536.
14. Ralevic V, Burnstock G. Interactions between perivascular nerves and endothelial cells in control of local vascular tone. In: *Nervous Control of Blood Vessels*, Bennett T, Gardiner SM (Eds). Harwood Academic, Amsterdam 1996: 135–175.
15. Mittal MK, Gupta TK, Lee F-Y *et al.* Nitric oxide modulates hepatic vascular tone in normal rat liver. *Am J Physiol* 1994; **267**: G416–G422.
16. Shah V, Haddad FG, García-Cardeña G *et al.* Liver sinusoidal endothelial cells are responsible for nitric oxide modulation of resistance in the hepatic sinusoids. *J Clin Invest* 1997; **100**: 2923–2930.
17. Zhang B, Borderie D, Sogni P *et al.* NO-mediated vasodilation in the rat liver. Role of hepatocytes and liver endothelial cells. *J Hepatol* 1997; **26**: 1348–1355.
18. Pannen BHJ, Bauer M, Zhang JX *et al.* A time-dependent balance between endothelins and nitric oxide regulating portal resistance after endotoxin. *Am J Physiol* 1996; **271**: H1953–H1961.
19. Bauer C, Walcher F, Kalweit U *et al.* Role of nitric oxide in the regulation of the hepatic microcirculation *in vivo*. *J Hepatol* 1997; **27**: 1089–1095.
20. Suematsu M, Kashiwagi S, Sano T *et al.* Carbon monoxide as an endogenous modulator of hepatic vascular perfusion. *Biochem Biophys Res Commun* 1994; **205**: 1333–1337.
21. Weidenbach H, Beckh K, Günthör M *et al.* The role of nitric oxide in hemodynamic and metabolic alterations induced by prostaglandin $F_{2\alpha}$ in the perfused rat liver. *Biochim Biophys Acta* 1995; **1245**: 181–186.
22. Zhang JX, Pegoli W, Jr., Clemens MG. Endothelin-1 induces direct constriction of hepatic sinusoids. *Am J Physiol* 1994; **266**: G624–G632.
23. Kawada N, Tran-Thi T-A, Klein H, Decker K. The contraction of hepatic stellate (Ito) cells stimulated with vasoactive substances. Possible involvement of endothelin 1 and nitric oxide in the regulation of the sinusoidal tonus. *Eur J Biochem* 1993; **213**: 815–823.
24. Pinzani M, Milani S, De Franco R *et al.* Endothelin 1 is overexpressed in human cirrhotic liver and exerts multiple effects on activated hepatic stellate cells. *Gastroenterology* 1996; **110**: 534–548.
25. Curran RD, Ferrari FK, Kispert PH *et al.* Nitric oxide and nitric oxide-generating compounds inhibit hepatocyte protein synthesis. *FASEB J* 1991; **5**: 2085–2092.
26. Harbrecht BG, Billiar TR. The role of nitric oxide in Kupffer cell-hepatocyte interactions. *Shock* 1995; **3**: 79–87.
27. Szabó C, Salzman AL. Endogenous peroxynitrite is involved in the inhibition of mitochondrial respiration in immuno-stimulated J774.2 macrophages. *Biochem Biophys Res Commun* 1995; **209**: 739–743.

28. Ou J, Molina L, Kim Y-M, Billiar TR. Excessive NO production dose not account for the inhibition of hepatic gluconeogenesis in endotoxemia. *Am J Physiol* 1996; **271**: G621–G628.
29. Lautt WW. Control of hepatic arterial blood flow: independence from liver metabolic activity. *Am J Physiol* 1980; **239**: H559–H564.
30. Esler M, Dudley F, Jennings G *et al.* Increased sympathetic nervous activity and the effects of its inhibition with clonidine in alcoholic cirrhosis. *Ann Intern Med* 1992; **116**: 446–455.
31. Groszmann RJ. Hyperdynamic state in chronic liver diseases. *J Hepatol* 1993; **17 (Suppl 2)**: S38–S40.
32. Vallance P, Moncada S. Hyperdynamic circulation in cirrhosis: a role for nitric oxide? *Lancet* 1991; **337**: 776–778.
33. Griffith TM. Modulation of blood flow and tissue perfusion by endothelium-derived relaxing factor. *Exp Physiol* 1994; **79**: 873–913.
34. Sieber CC, Groszmann RJ. Nitric oxide mediates hyporeactivity to vasopressors in mesenteric vessels of portal hypertensive rats. *Gastroenterology* 1992; **103**: 235–239.
35. Pizcueta P, Piqué JM, Fernández M *et al.* Modulation of the hyperdynamic circulation of cirrhotic rats by nitric oxide inhibition. *Gastroenterology* 1992; **103**: 1909–1915.
36. Hartleb M, Moreau R, Cailmail S *et al.* Vascular hyporesponsiveness to endothelin 1 in rats with cirrhosis. *Gastroenterology* 1994; **107**: 1085–1093.
37. Guarner C, Soriano G, Tomas A *et al.* Increased serum nitrite and nitrate levels in patients with cirrhosis: relationship to endotoxemia. *Hepatology* 1993; **18**: 1139–1143.
38. Laffi G, Foschi M, Masini E *et al.* Increased production of nitric oxide by neutrophils and monocytes from cirrhotic patients with ascites and hyperdynamic circulation. *Hepatology* 1995; **22**: 1666–1673.
39. Chu C-J, Lee F-Y, Wang S-S *et al.* Hyperdynamic circulation of cirrhotic rats with ascites: role of endotoxin, tumour necrosis factor-α and nitric oxide. *Clin Sci* 1997; **93**: 219–225.
40. Albillos A, Rossi I, Cacho G *et al.* Enhanced endothelium-dependent vasodilation in patients with cirrhosis. *Am J Physiol* 1995; **268**: G459–G464.
41. Ryan J, Jennings G, Dudley F, Chin-Dusting J. Smooth muscle-derived nitric oxide is elevated in isolated forearm veins in human alcoholic cirrhosis. *Clin Sci* 1996; **91**: 23–28.
42. Matsumoto A, Ogura K, Hirata Y *et al.* Increased nitric oxide in the exhaled air of patients with decompensated liver cirrhosis. *Ann Intern Med* 1995; **123**: 110–113.
43. Ros J, Jiménez W, Lamas S *et al.* Nitric oxide production in arterial vessels of cirrhotic rats. *Hepatology* 1995; **21**: 554–560.

44. Atucha NM, Shah V, García-Cardeña G *et al.* Role of endothelium in the abnormal response of mesenteric vessels in rats with portal hypertension and liver cirrhosis. *Gastroenterology* 1996; **111**: 1627–1632.
45. Cahill PA, Foster C, Redmond EM *et al.* Enhanced nitric oxide synthase activity in portal hypertensive rabbits. *Hepatology* 1995; **22**: 598–606.
46. Lopez-Talavera JC, Groszmann RJ. Treatment with aminoguanidine, a specific inhibitor of the inducible nitric oxide synthase, ameliorates the hyperdynamic syndrome in portal hypertensive rats [Abstract]. *Hepatology* 1994; **20**: 143A.
47. Mathie RT, Ralevic V, Moore KP, Burnstock G. Mesenteric vasodilator responses in cirrhotic rats: a role for nitric oxide? *Hepatology* 1996; **23**: 130–136.
48. Ralevic V, Mathie RT, Moore KP, Burnstock G. Vasoconstrictor responsiveness of the rat mesenteric arterial bed in cirrhosis. *Br J Pharmacol* 1996; **118**: 435–441.
49. Sieber CC, Sumanovski LT, Moll-Kaufmann C, Stalder GA. Hyposensitivity to nerve stimulation in portal hypertensive rats: role of nitric oxide. *Eur J Clin Invest* 1997; **27**: 902–907.
50. Kanwar S, Kubes P, Tepperman BL, Lee SS. Nitric oxide synthase activity in portal-hypertensive and cirrhotic rats. *J Hepatol* 1996; **25**: 85–89.
51. Cahill PA, Redmond EM, Hodges R *et al.* Increased endothelial nitric oxide synthase activity in the hyperemic vessels of portal hypertensive rats. *J Hepatol* 1996; **25**: 370–378.
52. Sarela AI, Mathie RT. Nitric oxide release in the splanchnic circulation of cirrhotic rats [Abstract]. *Gut* 1997; **40 (Suppl 1)**: A14.
53. Sogni P, Smith APL, Gadano A *et al.* Induction of nitric oxide synthase II does not account for excess vascular nitric oxide production in experimental cirrhosis. *J Hepatol* 1997; **26**: 1120–1127.
54. Ohta M, Tanoue K, Tarnawski AS *et al.* Overexpressed nitric oxide synthase in portal-hypertensive stomach of rat: a key to increased susceptibility to damage? *Gastroenterology* 1997; **112**: 1920–1930.
55. Tanoue K, Ohta M, Tarnawski AS *et al.* Portal hypertension activates the nitric oxide synthase genes in the esophageal mucosa of rats. *Gastroenterology* 1996; **110**: 549–557.
56. Martin P-Y, Xu DL, Niederberger M *et al.* Upregulation of endothelial constitutive NOS: a major role in the increased NO production in cirrhotic rats. *Am J Physiol* 1996; **270**: F494–F499.
57. Morales-Ruiz M, Jiménez W, Pérez-Sala D *et al.* Increased nitric oxide synthase expression in arterial vessels of cirrhotic rats with ascites. *Hepatology* 1996; **24**: 1481–1486.
58. El-Newihi HM, Kanji VK, Mihas AA. Activity of gastric mucosal nitric oxide synthase in portal hypertensive gastropathy. *Am J Gastroenterol* 1996; **91**: 535–538.

59. Sarela AI, Mihaimeed FMA, Batten JJ *et al.* Hepatic and splanchnic nitric oxide activity in patients with cirrhosis. *Gut* 1999; **44**: 749–753.
60. Shijo H, Yokoyama M, Ota K *et al.* Nitrate kinetics in patients with compensated cirrhosis: correlation with hemodynamics. *Am J Gastroenterol* 1996; **91**: 2190–2194.
61. Lee F-Y, Colombato LA, Albillos A, Groszmann RJ. Administration of N^{ω}-nitro-L-arginine ameliorates portal-systemic shunting in portal-hypertensive rats. *Gastroenterology* 1993; **105**: 1464–1470.
62. Pilette C, Moreau R, Sogni P *et al.* Haemodynamic and hormonal responses to long-term inhibition of nitric oxide synthesis in rats with portal hypertension. *Eur J Pharmacol* 1996; **312**: 63–68.
63. Iwata F, Joh T, Kawai T, Itoh M. Role of EDRF in splanchnic blood flow of normal and chronic portal hypertensive rats. *Am J Physiol* 1992; **263**: G149–G154.
64. Sogni P, Moreau R, Ohsuga M *et al.* Evidence for normal nitric oxide-mediated vasodilator tone in conscious rats with cirrhosis. *Hepatology* 1992; **16**: 980–983.
65. Cawley T, Breslin E, Geraghty J *et al.* Investigations of the function of the vascular endothelium in portal hypertensive rats. *Pharmacology* 1995; **51**: 381–390.
66. Fernández M, García-Pagán JC, Casadevall M *et al.* Evidence against a role for inducible nitric oxide synthase in the hyperdynamic circulation of portal-hypertensive rats. *Gastroenterology* 1995; **108**: 1487–1495.
67. Heinemann A, Stauber RE. The role of inducible nitric oxide synthase in vascular hyporeactivity of endotoxin-treated and portal hypertensive rats. *Eur J Pharmacol* 1995; **278**: 87–90.
68. Zimmermann H, Kurzen P, Klossner W *et al.* Decreased constitutive hepatic nitric oxide synthase expression in secondary biliary fibrosis and its changes after Roux-en-Y choledocho-jejunostomy in the rat. *J Hepatol* 1996; **25**: 567–573.
69. Soubrane O, Lacronique V, Miquerol L *et al.* The role of inducible NO synthase (iNOS) in the hyperdynamic circulation of portal hypertension [Abstract]. *Hepatology* 1995; **22**: 156A.
70. Calver A, Harris A, Maxwell JD, Vallance P. Effect of local inhibition of nitric oxide synthesis on forearm blood flow and dorsal hand vein size in patients with alcoholic cirrhosis. *Clin Sci* 1994; **86**: 203–208.
71. Le Moine O, Soupison T, Sogni P *et al.* Plasma endotoxin and tumor necrosis factor in the hyperkinetic state of cirrhosis. *J Hepatol* 1995; **23:** 391–395.
72. Campillo B, Bories PN, Benvenuti C, Dupeyron C. Serum and urinary nitrate levels in liver cirrhosis: endotoxemia, renal function and hyperdynamic circulation. *J Hepatol* 1996; **25**: 707–714.

73. Ryan J, Sudhir K, Jennings G *et al.* Impaired reactivity of the peripheral vasculature to pressor agents in alcoholic cirrhosis. *Gastroenterology* 1993; **105**: 1167–1172.
74. Grossman HJ, Grossman VL, Bhathal PS. Enhanced vasoconstrictor response of the isolated perfused cirrhotic rat liver to humoral vasoconstrictor substances found in portal venous blood. *J Gastroenterol Hepatol* 1992; **7**: 283–287.
75. Mathie RT, Ralevic V, Burnstock G. Portal vascular responsiveness to sympathetic stimulation and nitric oxide in cirrhotic rats. *J Hepatol* 1996; **25**: 90–97.
76. Zimmermann T, Gardemann A, Machnik G *et al.* Metabolic and hemodynamic responses of bivascularly perfused rat liver to nerve stimulation, noradrenaline, acetylcholine and glucagon in thioacetamide-induced micronodular cirrhosis. *Hepatology* 1992; **15**: 464–470.
77. Mittal MK, Lee FY, Groszmann RJ. Role of nitric oxide in the intrahepatic portal circulation of cirrhotic rats [Abstract]. *Gastroenterology* 1993; **104**: A956.
78. Gupta TK, Toruner M, Chung MK, Groszmann RJ. Endothelial dysfunction and decreased production of nitric oxide in the intrahepatic microcirculation of cirrhotic rats. *Hepatology* 1998; **28**: 926–931.
79. Rockey D. The cellular pathogenesis of portal hypertension: stellate cell contractility, endothelin, and nitric oxide. *Hepatology* 1997; **25**: 2–5.
80. Griscavage JM, Hobbs AJ, Ignarro LJ. Negative modulation of nitric oxide synthase by nitric oxide and nitroso compounds. *Adv Pharmacol* 1995; **34**: 215–234.
81. Nanji AA, Greenberg SS, Tahan SR *et al.* Nitric oxide production in experimental alcoholic liver disease in the rat: role in protection from injury. *Gastroenterology* 1995; **109**: 899–907.
82. Chu C-J, Lee F-Y, Wang S-S *et al.* Hyperdynamic circulation of cirrhotic rats with ascites: role of endotoxin, tumour necrosis factor-α and nitric oxide. *Clin Sci* 1997; **93**: 219–225.
83. Niederberger M, Martin P-Y, Ginès P *et al.* Normalization of nitric oxide production corrects arterial vasodilation and hyperdynamic circulation in cirrhotic rats. *Gastroenterology* 1995; **109**: 1624–1630.
84. Niederberger M, Ginès P, Martin P-Y *et al.* Comparison of vascular nitric oxide production and systemic hemodynamics in cirrhosis versus prehepatic portal hypertension in rats. *Hepatology* 1996; **24**: 947–951.
85. Casadevall M, Panés J, Piqué JM *et al.* Involvement of nitric oxide and prostaglandins in gastric mucosal hyperemia of portal-hypertensive anesthetized rats. *Hepatology* 1993; **18**: 628–634.
86. Moreno L, Martínez-Cuesta MA, Piqué JM *et al.* Anatomical differences in responsiveness to vasoconstrictors in the mesenteric veins from normal and portal hypertensive rats. *Naunyn Schmiedebergs Arch Pharmacol* 1996; **354**: 474–480.

87. Forrest EH, Jones AL, Dillon JF *et al.* The effect of nitric oxide synthase inhibition on portal pressure and azygos blood flow in patients with cirrhosis. *J Hepatol* 1995; **23**: 254–258.
88. Lee F-Y, Wang S-S, Tsai Y-T *et al.* Aminoguanidine corrects hyperdynamic circulation without ameliorating portal hypertension and portal hypertensive gastropathy in anesthetized portal hypertensive rats. *J Hepatol* 1997; **26**: 687–693.
89. Ralevic V, Mathie RT, Alexander B, Burnstock G. N^G-nitro-L-arginine methyl ester attenuates vasodilator responses to acetylcholine but enhances those to sodium nitroprusside. *J Pharm Pharmacol* 1991; **43**: 871–874.
90. Billiar TR, Saavedra JE, Williams DL, Keefer LK. Toward design of a liver-specific nitric oxide donor. In: *The Biology of Nitric Oxide, Part 5,* Moncada S, Stamler J, Gross S, Higgs EA (Eds). Portland Press, London 1996: 35.
91. Smits P, Williams SB, Lipson DE *et al.* Endothelial release of nitric oxide contributes to the vasodilator effect of adenosine in humans. *Circulation* 1995; **92**: 2135–2141.
92. Chu CJ, Lee FY, Wang SS *et al.* Hyperdynamic circulation of cirrhotic rats: role of substance P and its relationship to nitric oxide. *Scand J Gastroenterol* 1997; **32**: 841–846.
93. Fernández M, García-Pagán JC, Casadevall M *et al.* Acute and chronic cyclooxygenase blockage in portal-hypertensive rats: influence in nitric oxide biosynthesis. *Gastroenterology* 1996; **110**: 1529–1535.
94. Moreau R, Lebrec D. Endogenous factors involved in the control of arterial tone in cirrhosis. *J Hepatol* 1995; **22**: 370–376.
95. Niederberger M, Ginès P, Tsai P *et al.* Increased aortic cyclic guanosine monophosphate concentration in experimental cirrhosis in rats: evidence for a role of nitric oxide in the pathogenesis of arterial vasodilation in cirrhosis. *Hepatology* 1995; **21**: 1625–1631.
96. Lee B-S, Kang H-S, Pyun K-H, Choi I. Roles of tyrosine kinases in the regulation of nitric oxide synthesis in murine liver cells: modulation of NF-κB activity by tyrosine kinases. *Hepatology* 1997; **25**: 913–919.
97. Dusting GJ. Nitric oxide in cardiovascular disorders. *J Vasc Res* 1995; **32**: 143–161.
98. Nishida J, McCuskey RS, McDonnell D, Fox ES. Protective role of NO in hepatic microcirculatory dysfunction during endotoxemia. *Am J Physiol* 1994; **267**: G1135–G1141.
99. Pastor CM, Payen DM. Effect of modifying nitric oxide pathway on liver circulation in a rabbit endotoxin shock model. *Shock* 1994; **2**: 196–202.
100. Billiar TR, Curran RD, Harbrecht BG *et al.* Modulation of nitrogen oxide synthesis *in vivo*: N^G-monomethyl-L-arginine inhibits endotoxin-induced nitrate/nitrate biosynthesis while promoting hepatic damage. *J Leukoc Biol* 1990; **48**: 565–569.

101. Harbrecht BG, Billiar TR, Stadler J *et al.* Inhibition of nitric oxide synthesis during endotoxemia promotes intrahepatic thrombosis and an oxygen radical-mediated hepatic injury. *J Leukoc Biol* 1992; **52**: 390–394.
102. Ruetten H, Southan GJ, Abate A, Thiemermann C. Attenuation of endotoxin-induced multiple organ dysfunction by 1-amino-2-hydroxy-guanidine, a potent inhibitor of inducible nitric oxide synthase. *Br J Pharmacol* 1996; **118**: 261–270.
103. Vos TA, Gouw ASH, Klok PA *et al.* Differential effects of nitric oxide synthase inhibitors on endotoxin-induced liver damage in rats. *Gastroenterology* 1997; **113**: 1323–1333.
104. Szabó C, Salzman AL, Ischiropoulos H. Endotoxin triggers the expression of an inducible isoform of nitric oxide synthase and the formation of peroxynitrite in the rat aorta *in vivo*. *FEBS Lett* 1995; **363**: 235–238.
105. Thiemermann C, Ruetten H, Wu C-C, Vane JR. The multiple organ dysfunction syndrome caused by endotoxin in the rat: attenuation of liver dysfunction by inhibitors of nitric oxide synthase. *Br J Pharmacol* 1995; **116**: 2845–2851.
106. Yamamoto S, Burman HP, O'Donnell CP *et al.* Endothelin causes portal and pulmonary hypertension in porcine endotoxemic shock. *Am J Physiol* 1997; **272**: H1239–H1249.
107. Rockey DC, Chung JJ. Inducible nitric oxide synthase in rat hepatic lipocytes and the effect of nitric oxide on lipocyte contractility. *J Clin Invest* 1995; **95**: 1199–1206.
108. Liu SF, Adcock IM, Old RW *et al.* Differential regulation of the constitutive and inducible nitric oxide synthase mRNA by lipopolysaccharide treatment *in vivo* in the rat. *Crit Care Med* 1996; **24**: 1219–1225.
109. Szabó C, Thiemermann C. Invited opinion: Role of nitric oxide in hemorrhagic, traumatic, and anaphylactic shock and thermal injury. *Shock* 1994; **2**: 145–155.
110. Harbrecht BG, Wu B, Watkins SC *et al.* Inhibition of nitric oxide synthase during hemorrhagic shock increases hepatic injury. *Shock* 1995; **4**: 332–337.
111. Thiemermann C, Szabó C, Mitchell JA, Vane JR. Vascular hyporeactivity to vasoconstrictor agents and hemodynamic decompensation in hemorrhagic shock is mediated by nitric oxide. *Proc Natl Acad Sci USA* 1993; **90**: 267–271.
112. Peitzman AB, Billiar TR, Harbrecht BG *et al.* Hemorrhagic shock. *Curr Probl Surg* 1995; **32**: 925–1002.
113. Wendon JA. Local consequences of reperfusion in the liver. In: *Ischaemia-Reperfusion Injury*, Grace PA, Mathie RT (Eds). Blackwell Science, Oxford 1999: 56–64.
114. Ma TT, Ischiropoulos H, Brass CA. Endotoxin-stimulated nitric oxide production increases injury and reduces rat liver chemiluminescence during reperfusion. *Gastroenterology* 1995; **108**: 463–469.

115. Shimamura T, Zhu Y, Zhang S *et al.* Nitric oxide down-regulates endothelin production during ischemia and reperfusion of canine livers [Abstract]. *Hepatology* 1996; **24**: 579A.
116. Shiraishi M, Hiroyasu S, Nagahama M *et al.* Role of exogenous L-arginine in hepatic ischemia-reperfusion injury. *J Surg Res* 1997; **69**: 429–434.
117. Koeppel TA, Thies JC, Schemmer P *et al.* Inhibition of nitric oxide synthesis in ischemia/reperfusion of the rat liver is followed by impairment of hepatic microvascular blood flow. *J Hepatol* 1997; **27**: 163–169.
118. Wang Y, Mathews WR, Guido DM *et al.* Inhibition of nitric oxide synthesis aggravates reperfusion injury after hepatic ischemia and endotoxemia. *Shock* 1995; **4**: 282–288.
119. Pannen BHJ, Bauer M, Nöldge-Schomburg GFE *et al.* Regulation of hepatic blood flow during resuscitation from hemorrhagic shock: role of NO and endothelins. *Am J Physiol* 1997; **272**: H2736–H2745.
120. Peralta C, Hotter G, Closa D *et al.* Protective effect of preconditioning on the injury associated to hepatic ischemia-reperfusion in the rat: role of nitric oxide and adenosine. *Hepatology* 1997; **25**: 934–937.
121. Peralta C, Closa D, Hotter G *et al.* Liver ischemic preconditioning is mediated by the inhibitory action of nitric oxide on endothelin. *Biochem Biophys Res Commun* 1996; **229**: 264–270.
122. Jaeschke H, Schini VB, Farhood A. Role of nitric oxide in the oxidant stress during ischemia/reperfusion injury of the liver. *Life Sci* 1992; **50**: 1797–1804.
123. Kobayashi H, Nonami T, Kurokawa T *et al.* Role of endogenous nitric oxide in ischemia-reperfusion injury in rat liver. *J Surg Res* 1995; **59**: 772–779.
124. Bauer M, Pannen BHJ, Bauer I *et al.* Evidence for a functional link between stress response and vascular control in hepatic portal circulation. *Am J Physiol* 1996; **271**: G929–G935.
125. Kang Y. Hemodynamic changes during intra-abdominal organ transplantation. *Transplant Proc* 1993; **25**: 2583–2587.
126. Henderson JM, Gilmore GT, Mackay GJ *et al.* Hemodynamics during liver transplantation: the interactions between cardiac output and portal venous and hepatic arterial flows. *Hepatology* 1992; **16**: 715–718.
127. Längle F, Steininger R, Waldmann E *et al.* Improvement of cardiac output and liver blood flow and reduction of pulmonary vascular resistance by intravenous infusion of L-arginine during the early reperfusion period in pig liver transplantation. *Transplantation* 1997; **63**: 1225–1233.
128. Wheatley AM, Rieder HU, Kuznetsova LV *et al.* Effect of liver transplantation on hepatic hemodynamics: methodology and mechanisms. *Dig Surg* 1997; **14**: 344–354.
129. Lindell SL, Southard JH, Vreugdenhil P, Belzer FO. Kupffer cells depress hepatocyte protein synthesis on cold storage of the rat liver. *Transplantation* 1994; **58**: 869–874.

130. Gryf-Lowczowski JVD. Techniques for organ preservation. In: *Ischaemia-Reperfusion Injury*, Grace PA, Mathie RT (Eds). Blackwell Science, Oxford 1999: 357–368.
131. Devlin J, Palmer RMJ, Gonde CE *et al.* Nitric oxide generation. A predictive parameter of acute allograft rejection. *Transplantation* 1994; **58**: 592–595.
132. Ioannidis I, Hellinger A, Dehmlow C *et al.* Evidence for increased nitric oxide production after liver transplantation in humans. *Transplantation* 1995; **59**: 1293–1297.
133. Goto M, Yamaguchi Y, Ichiguchi O *et al.* Phenotype and localization of macrophages expressing inducible nitric oxide synthase in rat hepatic allograft rejection. *Transplantation* 1997; **64**: 303–310.
134. Töx U, Burkhard M, Goeser T *et al.* Distribution of inducible nitric oxide synthase in liver tissue after liver transplantation [Abstract]. *Hepatology* 1995; **22**: 361A.
135. Ventura HO, Malik FS, Mehra MR *et al.* Mechanisms of hypertension in cardiac transplantation and the role of cyclosporine. *Curr Opin Cardiol* 1997; **12**: 375–381.

Chapter 15

The Physiological and Pathophysiological Effects of Nitric Oxide on the Intestinal Circulation

Donna-Marie McCafferty and Paul Kubes

Background

Since the identification of EDRF as NO,[1] this molecule has been recognized to be produced by virtually all mammalian cells, including endothelium lining the vasculature, neurones of the central and enteric nervous system, and cells of the immune system.[2–4] In the intestine, the enzyme NO synthase (NOS) is expressed as three distinct isoforms: (1) a constitutive form present primarily in vascular endothelial cells, known as eNOS, (2) a constitutive form found in neurones of the enteric nervous system (nNOS), and (3) an inducible form (iNOS) which is expressed by endothelium,[5] epithelium,[6] and inflammatory cells[3] and is capable of the high output production of NO, in response to cytokines or lipopolysaccharide (LPS) in inflammatory processes. As described elsewhere in this volume, a key difference between the constitutive and the inducible enzyme systems is the dependence of constitutive NOSs (cNOS), but not iNOS, on Ca^{2+}. Transcription is required for the induction of iNOS and is mediated by activation of nuclear transcriptional factors such as NFκB.[7] The NO pathway plays a crucial role in the regulation of homeostasis of many physiological organ systems, including the cardiovascular (central and peripheral), respiratory, gastrointestinal, endocrine and immune systems.[4]

It is the purpose of this chapter to review the role of NO as a regulator in the intestinal vasculature in both physiological and pathophysiological states. There is a growing body of data to suggest that NO has effects on many components of intestinal haemodynamics, including: (1) maintenance of blood flow, (2) inhibition of platelet and leucocyte adhesion and/or aggregation within the vasculature, (3) modulation of mast cell reactivity, (4) antioxidant effects, and (5) microvascular permeability. The role of NO on these factors in gastric, small intestinal and colonic circulation will be discussed.

Nitric Oxide and the Stomach

Perfusion of the mucosal microcirculation is intimately involved in the maintenance of gastric integrity, and NO has been shown to regulate normal gastric vascular homeostasis. A reduction in resting gastric mucosal blood flow was observed after N^G-monomethyl-L-arginine (L-NMMA) administration in rats, using either the hydrogen gas clearance or the radiolabelled microspheres technique to measure flow.[8,9] L-NMMA and not D-NMMA (the inactive isomer), gave a dose-dependent reduction in resting gastric mucosal blood flow.[8] These effects could be reversed with L-arginine but not D-arginine. Using laser Doppler flowmetry, both L-NMMA and N^G-nitro-L-arginine methyl ester (L-NAME) were shown to reduce mucosal blood flow,[10] suggesting a role for endogenous NO in the maintenance of basal mucosal blood flow.

Several investigators have demonstrated that a disruption in gastric mucosal blood flow can lead to the formation of mucosal lesions. For example, endothelin (ET)-1, a potent vasoconstrictor peptide formed by the endothelium, induces extensive mucosal injury after local intra-arterial infusion to the rat stomach.[11] Similar results have been shown with thromboxane A_2 in stressed rats.[12] The gastric haemorrhagic injury induced by intra-arterial infusion of ET-1 was dose dependently reduced by concurrent local arterial infusion of nitrovasodilators glyceryl trinitrate or *S*-nitroso-N-acetylpenicillamine (SNAP).[13] These findings suggest that exogenous NO can protect the gastric mucosa from damage induced by the vasoconstrictor ET-1. Moreover, L-NAME exacerbates damage associated with ET-1 infusion,[11] raising the possibility that endogenous NO counters the effect of ET-1. These types of experiments demonstrate a correlation between vasoconstriction and injury versus vasodilatation and protection. Analogous to this are studies where compromised mucosal blood flow potentiates the damaging effects of lumen-based insults. For example, ethanol-induced macroscopic and microscopic mucosal injury is accompanied by a decrease in mucosal blood flow and haemoglobin oxygen saturation. Inhibition of endogenous NO production by N^G-nitro-L-arginine (L-NNA) aggravated the damage and exacerbated the decrease in mucosal blood flow and haemoglobin oxygen saturation.[13] Similarly, gastric lesions induced by 0.45 N HCl are enhanced by inhibitors of endothelium-dependent vasodilatation such as methylene blue, collagenase, haemoglobin and ascorbic acid.[14] In this same study, increasing mucosal haemodynamics with oral nitrates (sodium nitrite and isoamyl nitrite) prevented the formation of HCl-induced gastric lesions.

Other agents that are acknowledged gastric damaging compounds are non-steroidal anti-inflammatory drugs (NSAIDs). These drugs, whilst being prescribed for their anti-inflammatory, analgesic and antipyretic effects, cause significant gastrointestinal damage (stomach and small intestine) in the form of acute or chronic ulceration and bleeding. Administration of NSAIDs to rats results in a significant reduction in gastric blood flow. It is thought that this reduction in blood flow contributes significantly to the increased susceptibility of the gastric mucosa to injury following NSAID administration.[15] NSAIDs also stimulate leucocyte adherence to the vascular endothelium and this has also been postulated as part of the mechanism by which they induce their deleterious side effects.[16] Recently, a novel series of NSAIDs has been generated which has been modified to incorporate a NO-releasing moiety.[17,18] These drugs have been found to possess anti-inflammatory effects equipotent with those of their parent compounds, yet do not have the undesirable side-effects. With the administration of NO-NSAIDs, gastric mucosal blood flow is maintained and leucocyte adhesion is not induced.[19] Presumably, the modified NSAIDs release NO, reducing the injury to the gastrointestinal tissue by an effect on gastric blood flow and leucocyte adhesion. It is unclear whether the reduction in blood flow is the primary mechanism of gastric damage or a secondary effect due to leucocyte adhesion in the microvasculature. However, after aspirin administration, Kitahora and Guth[20] described the appearance of 'white thrombi' in the gastric microcirculation prior to the reduction in gastric blood flow, suggesting that these aggregates may cause the reduction in flow.

The aforementioned studies suggest that in the gastric microcirculation, NO plays an important role in maintaining gastric integrity under basal conditions and in protecting the mucosa against luminal insult, by increasing gastric mucosal blood flow and in part by inhibiting leucocyte adhesion.

Nitric Oxide and the Small Bowel

Elsewhere in the gastrointestinal tract, the role of NO in protecting against mucosal dysfunction by improving organ perfusion is not established. MacKendrick *et al.*[21] in 1993, demonstrated that small bowel injury induced by platelet-activating factor (PAF) is markedly exacerbated by L-NAME pretreatment. Although this damage was accompanied by a pronounced mesenteric vasoconstriction, a similar vasoconstriction by phenylephrine failed to exacerbate the PAF-induced injury. Similarly, it has been shown that L-NMMA pretreatment enhanced endotoxin-induced macroscopic and histological intestinal damage and increased intestinal permeability, measured by leakage of ^{125}I-labelled human serum albumin into the lumen of the bowel.[22] Although L-NMMA increased mean systemic arterial blood pressure, similar systemic pressor effects induced by phenylephrine did not exacerbate endotoxin-induced injury. From these studies, it is tempting to conclude that endogenous NO protects by blood flow-independent mechanisms, although the possibility that NO synthesis inhibitors and phenylephrine may affect blood flow distribution to layers of the gastrointestinal tract differently cannot be discounted and therefore similar reductions in total flow may not directly reflect microcirculatory alterations.

Data accumulated from experiments involving mucosal injury induced by ischaemia-reperfusion (I-R) suggest that the beneficial effects of NO donors, as well as L-arginine, are not as a result of improved perfusion. Reperfusion of the ischaemic intestine associated with haemorrhage and other shock states is characterized by a number of microvascular and mucosal alterations, including endothelial cell swelling, capillary plugging, a prolonged reduction in intestinal blood flow and mucosal barrier dysfunction.[23–25] The mucosal damage can be exploited by commensal bacteria in the lumen of the intestine and can lead to bacterial translocation and ultimately sepsis. Clearly, restoration of an intact mucosal barrier after a period of ischaemia is a requirement for tissue function and viability. Exogenous sources of NO, including SIN-1, CAS-754, and nitroprusside, as well as L-arginine, all reduced reperfusion-induced mucosal barrier dysfunction as assessed by ^{51}Cr-labelled ethylenediaminetetraacetic acid (EDTA) clearance from blood to lumen, at concentrations that did not improve intestinal blood flow.[26]

In addition, administration of 8-bromo-cGMP, the permeable analogue of guanosine 3',5'-cyclic monophosphate (cGMP), improved reperfusion-induced intestinal blood flow without protecting against the barrier dysfunction. Moreover, no apparent correlation between the reduction in blood flow and increased mucosal permeability was observed. Andrews *et al.*[27] demonstrated that neither nitroprusside nor acetylcholine (which releases NO from endothelial cells) significantly affected gastric blood flow, yet both agents protected the stomach against reperfusion injury. A non-NO vasodilator (papaverine) did not provide protection against I-R-induced gastric injury. Taken together, these data suggest the beneficial effect of NO in the post-ischaemic gastrointestinal tissue is not as a result of improved perfusion. How then does NO protect the gut from injury?

Role of NO in Leucocyte Adhesion

It is clear from studies using inhibitors of NO production that the role of constitutive NO production in the microvasculature goes beyond its role as a vasodilator. The use of intravital microscopy, which permits visualization of post-capillary venules in feline mesentery, demonstrated that inhibition of constitutive NO production induced a profound increase in the number of adherent leucocytes in post-capillary venules. This effect was reversible with the substrate for NO synthase, L-arginine,[28] NO donors,[29] and cGMP analogues.[30] This initial observation that inhibition of NO synthesis increases leucocyte adhesion has been reproduced in a variety of organ systems and in a range of species.[31–35]

The leucocyte adhesion observed after L-NAME administration coincided closely with a reduction in blood flow and shear forces within the venules. As these hydrodynamic dispersal forces can affect leucocyte adhesion (lower shear force = increased adhesion), it was conceivable that NO synthesis inhibition was reducing shear and thereby promoting adhesion. However, in additional experiments, when blood flow was controlled mechanically over a wide range of shear forces, leucocyte adhesion was of much greater magnitude in the presence of L-NAME than without L-NAME at any given shear, suggesting that the mechanism of adhesion was, at least in part, independent of changes in shear forces within the vessel.[28]

In inflammatory conditions such as I-R[36,37] or indomethacin administration[38–40] there is adequate evidence that neutrophils are a major contributor to injury. This contention is based on the following observations: (1) neutrophils infiltrate inflamed tissue, (2) neutropaenia reduces tissue injury, and (3) treatment with monoclonal antibodies directed against adhesion molecules reduces various indices of inflammation. However, pretreatment with anti-CD18 antibody, which prevents leucocyte adhesion to postcapillary venules, significantly reduced the increase in microvascular but not mucosal permeability induced in the first hour after reperfusion.[41] These data suggest that circulating neutrophils which infiltrate into the intestine are not responsible for the early damage to the mucosal barrier. Taken together with the data that illustrated that NO donors can reduce both mucosal and microvascular permeability in the intestine,[26,42] this suggests that NO must have other beneficial effects in addition to its anti-adhesive properties.

It should be noted that although infiltrating neutrophils can be discounted as participants of the early mucosal dysfunction induced by reperfusion of ischaemic tissue, recent evidence suggests a role for the resident population of neutrophils in the intestinal tissue.[41] Clancy *et al.*[43] have demonstrated that NO can inhibit superoxide production from neutrophils by directly inhibiting NADPH oxidase, the enzyme system responsible for the oxidative burst. Therefore, NO may also affect the potential of resident neutrophils to be cytotoxic by inhibiting their production of oxidants and may explain the adhesive-independent effect of NO donors in I-R-induced mucosal dysfunction.

Neutrophil adhesion is generally thought to be dependent on CD11/CD18 (β_2 integrin). Using a monoclonal antibody (IB_4) directed against the CD11/CD18 glycoprotein complex in the cat mesentery, essentially all of the L-NAME-induced adhesion was inhibited.[28] However, the use of flow cytometry revealed that, unlike PAF or leukotriene (LT)B_4, L-NAME did not directly increase the expression or adhesiveness of CD11/CD18 *in vitro*.[28,31] Clearly, L-NAME did not directly activate CD11/CD18 on leucocytes and therefore changes in CD11/CD18 adhesiveness were not responsible for the increase in leucocyte adhesion observed *in vivo*. Arndt *et al.*[44] reported that a PAF-receptor antagonist (WEB 2086) and an LTB_4-receptor antagonist (SC 41930) reduced the increased leucocyte adhesion associated with L-NAME. These data suggest that the leucocyte adherence involved PAF and LTB_4. Although the source of these chemoattractants remained unknown, based on the *in vitro* data it was unlikely to originate from neutrophils, suggesting that inhibition of NO synthesis activated some other cell type, perhaps endothelium, to recruit adhering leucocytes. However, co-incubation of neutrophils, endothelium and a NO synthesis inhibitor also was insufficient to induce adhesion, suggesting the absence of some factor and/or cell that modulates the L-NAME-induced leucocyte adhesion *in vivo*.[31]

Mast Cells and NO

Studies from our laboratory have demonstrated that connective tissue mast cells are located in close proximity to the mesenteric vasculature. These cells, which are capable of releasing a multitude of pro-inflammatory mediators, such as histamine, proteases, chemokines, PAF, and tumour necrosis factor (TNF)-α became activated following inhibition of NO synthesis.[31] In another study, we demonstrated that mast cell-derived protease II levels from mucosal mast cells were elevated following systemic administration of L-NAME.[45] These two studies suggest that both connective tissue mast cells (from mesentery) and mucosal mast cells (intestinal mucosa) may be under physiological control from the continuous release of endogenous NO. Moreover, the L-NAME-induced leucocyte recruitment could be prevented by treatment with various mast cell stabilizers, suggesting that mast cells played an important role in leucocyte adhesion following NO synthesis inhibition. One can speculate on the relationship between PAF and NO and mast cells. As mentioned earlier, the PAF antagonist WEB 2086 blocked the adhesion induced by NO synthesis inhibition[44] and as mast cells synthesize the majority of PAF in the intestine,[46] it is conceivable that PAF is released from activated mast cells following NO synthesis inhibition to induce leucocyte adhesion.

NO donors have been shown to decrease release of histamine and PAF from mast cells,[47,48] and to dramatically reduce leucocyte rolling and adhesion in a model of mast cell degranulator-induced leucocyte recruitment in the rat mesentery, using compound 48/80.[49] Similar anti-adhesive effects of NO were noted *in vitro* in an endothelial cell-neutrophil culture system wherein mast cells were added.[50] If NO was given before (but not after) mast cell activation, leucocyte adhesion was dramatically reduced. These studies provide evidence that NO is capable of reducing mast cell-induced leucocyte recruitment primarily by preventing release of mast cell-derived mediators.

Antioxidant Properties of NO

Reactive oxygen metabolites such as superoxide and hydrogen peroxide (H_2O_2) are produced by enzymes (xanthine oxidase, NADPH oxidase, amine oxidase) which are present in the gastrointestinal tract in cells of the microvascular endothelium and mucosal epithelium as well as neutrophils and macrophages.[51] All tissues contain antioxidant enzymes (superoxide dismutase (SOD), catalase) and free radical scavengers (ascorbate, glutathione) and disease may develop when there is an overload of oxidants with respect to the ability of the antioxidants to absorb the radicals. *In vitro*, NO has been shown to react rapidly with superoxide derived from an enzymatic source or from activated neutrophils to render it biologically inactive.[52] *In vivo*, intra-arterial infusion of hypoxanthine/xanthine oxidase (a radical generating system) in healthy intestine induced an increase in microvascular permeability.[53] Also *in vivo*, NO donors were shown to double the antioxidant capacity of plasma and could directly inhibit neutrophil influx induced by a superoxide-generating system.[54] Wink *et al.*[55] have suggested that NO donors have a protective role in cellular injury induced by H_2O_2. Superoxide and H_2O_2 are known to induce a P-selectin-dependent leucocyte rolling in post-capillary venules which can be inhibited by NO donors.[54,56] Indeed, administration of NO prior to ischaemia prevented leucocyte rolling and adhesion during reperfusion in rat mesentery,[57] whereas administration of NO at the onset of reperfusion prevented only the adhesive response in cat mesentery.[58] It is not clear whether NO is inactivating the signalling pathways for these oxidants to induce P-selectin expression or acting as an antioxidant to inactivate them before they can stimulate the endothelium.

There is a growing body of evidence to suggest that NO acts as an antioxidant and in certain inflamed states may be anti-inflammatory, in part, due to this activity. Both superoxide and H_2O_2 are implicated in I-R injury of the gastric mucosa[59] and small intestine[60] as well as indomethacin-induced inflammation[61] in the gastrointestinal tract. Therefore it is possible that NO may be acting as an antioxidant in these conditions to reduce oxidant-induced tissue damage. However, direct evidence is lacking and designing experiments to test this hypothesis may be very difficult if not impossible. Indirect evidence suggests that antioxidants such as superoxide scavenging with SOD[53] or detoxification of H_2O_2 with catalase[23] significantly protects against I-R-induced microvascular permeability to a similar degree as NO donors, but whether the two events are related is unknown. Interestingly, inhibition of NO by L-NAME exacerbated the tissue damage after PAF administration;[21] however, pretreatment with SOD or catalase could not reverse the L-NAME effect, suggesting that endogenous NO was unlikely to be acting as an antioxidant. Of course, the likelihood that SOD or catalase could gain access to the same intracellular compartments as NO is highly unlikely and leaves the data by MacKendrick *et al.*[21] open to interpretation.

In addition to acting as an antioxidant, NO has a role in reducing oxidant production. *In vitro* treatment of endothelial cells with L-NAME induced oxidative stress within 30 minutes, which continued to increase for up to 3 hours.[62] In the presence of mast cells, this increased oxidative stress induced rapid endothelial/neutrophil adhesion.[50] SOD was capable of reducing the L-NAME-induced rapid increase in leucocyte adhesion both *in vitro* and *in vivo*. These results were extended by two groups, demonstrating both *in vivo*[34,63] and *in vitro*[50] that exposure of tissue to L-NAME caused a detectable increase in oxidative stress within mast cells. These results are suggestive of a role for endothelial release of extracellular oxidants, leading to activation of mast cells, subsequently inducing leucocyte adhesion.[50] It also suggests an ongoing interaction between constitutive NO and constitutive superoxide, which is continuously produced in small amounts from all cells as part of mitochondrial respiration.[64] Disrupting the balance by inhibiting NO may lead to mast cell activation by oxidants and increased leucocyte/endothelial cell adhesion.

NO and Microvascular Permeability

There is abundant evidence now to state that NO plays a homeostatic regulatory role in fluid and protein movement across the vasculature. Inhibition of NO synthesis with L-NAME in the cat intestine results in a very rapid and persistent increase in microvascular permeability.[29] D-NAME, the biologically inert enantiomer, did not have any effect on the microvasculature, suggesting that the increased transvascular and protein flux associated with L-NAME was not related to a non-specific effect of the NO synthesis inhibitor. L-arginine given at five times the dose of L-NAME could overcome the effects of the inhibitor. Compatible with a role for NO in microvascular permeability was the ability of exogenous NO in the form of nitroprusside or SIN-1 to reverse the L-NAME-induced effects.[29,65] The increase in fluid and protein efflux from the microcirculation was not as a result of increased hydrostatic pressure and therefore may be due to overt endothelial retraction.

Furthermore, there is evidence that NO regulates endothelial cell relaxation and contraction. NO synthesis inhibition diminishes the cGMP content in endothelial cells[4] and it is well known that increasing intracellular cGMP induces endothelial cell relaxation and thereby reduces the size of the intraendothelial cell junctions.[66] It follows that a reduction in cGMP, after L-NAME treatment for example, would cause endothelial cell contraction and increase the leakiness of the microvascular barrier.[67] In the cat mesentery, the L-NAME-induced increase in microvascular permeability could be prevented by treatment with 8 bromo-cGMP, which bypasses the guanylate cyclase system and replenishes the intracellular cGMP content.[65] Moreover, an inhibitor of cGMP, methylene blue, moderately increased the transvascular fluid and protein flux. Subsequently, it was demonstrated that 8-bromo-cGMP could reverse the L-NAME-induced increase in fluorescein isothiocyanate (FITC)-labelled albumin out of postcapillary venules.[30] Collectively, these data support the view that depletion of intracellular cGMP plays an integral role in the increase in microvascular permeability induced by L-NAME.

As discussed previously, the ability of NO to reduce damage induced by I-R in the intestine was dependent on its anti-adhesive effects on leucocytes and not on increased blood flow. The reduction in leucocyte adhesion observed in the rat mesentery after application of three NO donors correlated well with microvascular dysfunction,[68] suggesting that the ability of NO donors to reduce leucocyte adhesion was related to their ability to prevent the increase in microvascular permeability. Inhibition of endogenous NO with L-NAME has been reported to exacerbate damage to the intestine during partial[69] or complete[26] I-R. D-NAME did not enhance intestinal dysfunction, whereas L-arginine reversed the effects of L-NAME. These findings support the view that NO may be an endogenous regulator with beneficial properties during reperfusion.

Based on the aforementioned data, NO from both cNOS and from exogenous NO-generating reagents appears to reduce the sequelae of acute inflammation. However, as the inflammatory response progresses, the role played by excess NO produced from the inducible NOS becomes an important factor for consideration (see below).

The Role of Nitric Oxide in Sepsis

Injection of animals with LPS causes various systemic disorders, including severe hypotension, blood flow maldistribution, thrombosis, neutrophil activation and organ dysfunction. Associated with the progression of sepsis is the stimulation of iNOS, which produces much larger amounts of NO. iNOS is activated *via* cytokines and microbial biproducts. An elevated level of iNOS, known to occur in macrophages, hepatocytes, epithelium and other cells, serves an essential role in host-defence as a cytostatic or cytotoxic mediator against tumours or invading microorganisms (helminths, protozoa, bacteria, viruses). However, it is now also thought that overproduction of NO can: (1) induce profound hypotension, and (2) impact on host cells and cause inflammatory damage to tissues, perhaps by forming potent reactive nitrogen intermediates.

Sepsis is a very complex condition, with many cascades of mediators being produced and released to invoke a multiplicity of biological alterations. It is therefore not surprising that the role of NO in sepsis is presently so confusing. For example, in the first hour of sepsis, tissue injury can be greatly enhanced if NO synthesis is inhibited,[22] whereas administration of the same inhibitors at 4 hours can reduce the injury.[70] However, in other studies, inhibition of NO synthesis in sepsis (4–8 hours) worsened outcome.[71,72] Some investigators have suggested that perhaps inhibition of iNOS, without affecting eNOS, may improve the septic condition. Although some data support this view, there is one clear inconsistency. There is evidence that NO production from cNOS may be lost: down-regulation of cNOS gene expression 4 hours after LPS treatment has been reported in a rat model.[73] If indeed constitutively generated NO is lost, then any source of NO, whether it be from NO donors, inhaled NO or from iNOS, could conceivably replace some of the essential functions of cNOS, including limiting inappropriate cell adhesion.

It may be tempting to suggest that only partial inhibition of NO might improve blood pressure without decreasing adequate perfusion of organs, causing blood flow maldistribution and stasis, and increasing leucocyte adhesion. However, the results from partial inhibition of NO are also not entirely clear. It has been shown that systemic treatment with low doses of bacterial endotoxin (or LPS) that cause only a slight increase in leucocyte recruitment in the murine liver[35] could be significantly exacerbated if NO production is inhibited by L-NMMA at a concentration that did not affect baseline leucocyte adhesion. Moreover, work by Mulder *et al.*[74] has demonstrated that even partial inhibition of NO synthesis was detrimental. On the other hand, various other investigators have not observed increased leucocyte recruitment in tissues in septic animals treated with small amounts of NO synthesis inhibitors. In an ovine model of sepsis, low doses of the NOS inhibitor L-NMMA did not elevate leucocyte accumulation.[75] Preliminary work from our own laboratory has observed that LPS, at a concentration that only subtly affects leucocyte recruitment in normal mice, induces a marked increase in leucocyte recruitment in iNOS-deficient mice (Hickey & Kubes, unpublished).

In addition, *in vitro* studies of murine macrophages using antisense techniques to inhibit iNOS have shown that LPS causes lesser adhesion to cytokine-stimulated murine endothelial cells.[76] These studies provide some *in vitro* evidence to suggest that NO from iNOS may also have anti-adhesive effects by acting on monocytes in an autocrine fashion. However, the mechanism remains entirely unclear. Further work in this regard is still needed.

Nitric Oxide and the Large Bowel

Attention has recently been focused on the overproduction of NO as a key player in the pathogenesis of inflammatory bowel disease (IBD). This view is based on the fact that NO levels have been shown to be elevated in the rectal dialysates of patients with ulcerative colitis.[77] Increased activity of NOS in the colonic mucosa of ulcerative colitis patients has been shown to be largely Ca^{2+}-independent, suggesting the presence of iNOS.[78] In animal models of colitis there is evidence of: (1) increased luminal nitrate/nitrite levels, (2) increased iNOS immunoreactivity, and (3) increased enzymatic conversion of L-arginine to L-citrulline.[79–83]

When inhibitors of NO synthase are used in various animal models of IBD, however, ambiguous results have been achieved. For example, the non-specific NO synthase inhibitor L-NAME has been effective in reducing intestinal inflammation associated with acetic acid,[84] trinitrobenzene sulphonic acid,[79,84] and peptidoglycan/polysaccharide.[85] On the other hand, the same inhibitor, in the same or different models of intestinal inflammation, ameliorated some of the sequelae[86] or provided no protection[87] or further exacerbated the inflammation.[88] Similarly, dichotomous results have been obtained with purported selective inhibitors of iNOS.[80,85,89] These results could, in part, be attributed to different drug regimens achieving partial or complete inhibition of cNOS, iNOS or both types of enzyme. Therefore, using these different inhibitors over prolonged periods of time continues to provide equivocal results.

A new approach to the role of NO in inflammation has become available with recent advances in recombinant DNA technology, which has permitted the development of mice that specifically lack the capacity to express the iNOS gene.[90] Therefore, these animals produce NO constitutively but are unable to overproduce NO *via* the inducible NOS enzyme. The development of these mice gave us the opportunity to study the role of NO in intestinal inflammation without the complications of specificity related to NOS inhibitors. Using iNOS-deficient mice, colitis was induced by intra-rectal acetic acid and, as early as 24 hours, the macroscopic damage (haemorrhage and ulceration compared with mild erythema) and neutrophil infiltration (as measured by myeloperoxidase activity) was worse in the iNOS deficient mice.[91] Consistent with this was an increase in epithelial permeability, as determined by ^{51}Cr-EDTA leakage from the

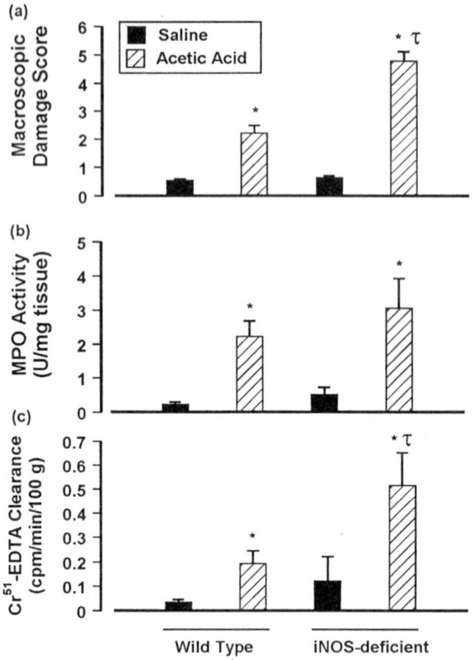

Figure 1. Illustrates (a) macroscopic damage score, (b) neutrophil infiltration measured by myeloperoxidase (MPO) activity and (c) mucosal dysfunction measured by blood-lumen-clearance of radiolabelled EDTA, 24 h after saline or acetic acid (3%) treatment in iNOS-deficient mice or their wild type controls. *, significance from saline control group, τ, significance from wild type inflamed group.

bloodstream to the lumen of the colon (Figure 1). Moreover, with time (1 week) the intestine was completely healed in wild-type mice, consistent with previously reported data.[92] In contrast, the iNOS deficient mice, which lack the ability to overproduce NO, had persistent macroscopic damage over the 7 days (Figure 2). The wild-type animals had evidence of induction of iNOS message at both 24 and 72 hours of inflammation. These findings suggest that the iNOS deficient mice have an inability to resolve the intestinal inflammation associated with simple damage. These data are consistent with our unpublished observations that such mice have difficulty resolving skin lesions following scratches and bites. Therefore the presence of iNOS and its induction in the wild type mice ameliorates the inflammatory process induced by an acute insult to the colon. Although we have not measured NO levels in our model, increased NO levels have been demonstrated in the acetic acid-induced colitis model.[84]

The increase in NO produced in this acute inflammation may exert an anti-inflammatory effect through one or more of the mechanisms discussed previously in this chapter. The presence of NO and its induction may improve the perfusion

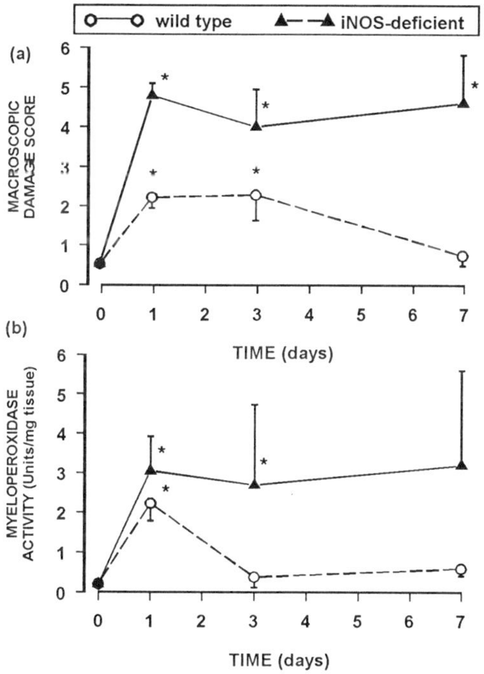

Figure 2. Illustrates (a) macroscopic damage score and (b) neutrophil infiltration measured by myeloperoxidase activity over 7 days, in acetic acid (3%)-treated iNOS-deficient mice (▲) or their wild type controls (○). *, significance from zero time point group.

of the colon, helping to dissipate the noxious stimuli. Leucocytes have been demonstrated to play a role in IBD. For example, inhibition of leucocyte infiltration using an antibody directed against intercellular adhesion molecule (ICAM)-1 on endothelial cells has been demonstrated to reduce acetic acid-induced colitis in rats.[93] Shiratora *et al.*[94] observed that the oxygen-derived free radical generating capacity of polymorphonuclear leucocytes in patients with active ulcerative colitis was markedly enhanced compared with controls. NO may or may not play a role in mast cell reactivity. Chin and Barrett, in 1994, showed that colitis could be induced in both mast cell deficient mice and their wild type littermates in a similar fashion,[95] demonstrating that, in at least one model of colitis, mast cells were not involved in the development of inflammation. However, if NO does have a role in inhibiting leucocyte infiltration and oxidant activity in IBD, it is not surprising that endogenous NO from iNOS contributes to the resolution of an acute inflammatory response.

It is likely, based on our data, that iNOS is a normal component of the bowel's inflammatory response and perhaps only in particular situations does iNOS actually contribute to an inappropriate inflammatory situation. Indeed, most inflammatory responses resolve without incident in the presence of iNOS induction. However, it cannot be excluded that in the inappropriate, complex IBD condition which the acetic acid model almost certainly does not reflect, iNOS may be a contributing factor to the inflammation.

Summary

From the evidence presented in this chapter, NO clearly has many roles in the haemodynamic regulation of the gastrointestinal system, both physiologically and in pathophysiological conditions. NO can alter blood flow not only by a vasodilatory effect through relaxation of smooth muscle in blood vessels but by also affecting the behaviour of leucocytes and platelets within the microvasculature. Together with the other roles of NO as an antioxidant and a regulator of mast cell activation, the interaction of the various roles of NO in the intestinal circulation leads to a multifaceted attack to dampen inflammatory processes. The extent to which each of these roles of NO is important depends on the pathophysiological state being studied. Whether the overproduction of NO from the inducible NOS is beneficial or deleterious in inflammatory conditions remains the subject of active research.

References

1. Ignarro LJ, Buga GM, Wood KS *et al.* Endothelium-derived relaxing factor produced and released from artery and vein is nitric oxide. *Proc Natl Acad Sci USA* 1987; **84**: 9265–9269.
2. Furchgott RF, Zawadzki JV. The obligatory role of endothelial cells in the relaxation of arterial smooth muscle by acetylcholine. *Nature* 1980; **288**: 373–376.
3. Nathan C, Xie Q-W. Nitric oxide synthases: roles, tolls and controls. *Cell* 1994; **78**: 915–918.
4. Moncada S. The L-arginine: nitric oxide pathway. *Acta Physiol Scand* 1992; **145**: 201–227.
5. Radomski MW, Palmer RMJ, Moncada S. Glucocorticoids inhibit the expression of an inducible, but not the constitutive, nitric oxide synthase in vascular endothelial cells. *Proc Natl Acad Sci USA* 1990; **87**: 10043–10047.
6. Tepperman BL, Brown JF, Korolkiewicz R, Whittle BJR. Nitric oxide synthase activity, viability and cyclic GMP levels in rat colonic epithelial cells: effect of endotoxin challenge. *J Pharmacol Exp Ther* 1994; **271**: 1477–1482.
7. Goldring CEP, Narayanan R, Lagadec P, Jeannin J-F. Transcriptional inhibition of the inducible nitric oxide synthase gene by competitive binding of NF-κB/Rel proteins. *Biochem Biophys Res Commun* 1995; **209**: 73–79.

8. Piqué JM, Whittle BJR, Esplugues JV. The vasodilator role of endogenous nitric oxide in the rat gastric microcirculation. *Eur J Pharmacol* 1989; **174**: 293–296.
9. Pizcueta P, Piqué JM, Bosch J *et al.* Involvement of endogenous nitric oxide in the regulation of the rat splanchnic circulation [Abstract]. *Gastroenterology* 1991; **100**: A240.
10. Tepperman BA, Whittle BJR. Endogenous nitric oxide and sensory neuropeptides interact in the modulation of the rat gastric microcirculation. *Br J Pharmacol* 1992; **105**: 171–175.
11. Filep JG, Földes-Filep E, Rousseau A *et al.* Vascular responses to endothelin-1 following inhibition of nitric oxide synthesis in the conscious rat. *Br J Pharmacol* 1993; **110**: 1213–1221.
12. Kitagawa H, Kurahashi K, Fujiwara M. Gastric mucosal erosion due to a mucosal ischemia produced by thromboxane A_2-like substance in rats under water-immersion stress. *J Pharmacol Exp Ther* 1986; **237**: 300–304.
13. Lopez-Belmonte J, Whittle BJR, Moncada S. The actions of nitric oxide donors in the prevention or induction of injury to the rat gastric mucosa. *Pharmacology* 1993; **108**: 73–78.
14. Kitagawa H, Takeda F, Kohei H. Effect of endothelium-derived relaxing factor on the gastric lesion induced by HCl in rats. *J Pharmacol Exp Ther* 1990; **253**: 1133–1137.
15. Wallace JL, McKnight GW. The mucoid cap over superficial gastric damage in the rat. A high-pH environment dissipated by non-steroidal anti-inflammatory drugs and endothelin. *Gastroenterology* 1990; **99**: 295–304.
16. Asako H, Kubes P, Wallace JL *et al.* Modulation of leukocyte adhesion in rat mesenteric venules by aspirin and salicylate. *Gastroenterology* 1992; **103**: 146–152.
17. Conforti A, Donini M, Brocco G *et al.* Acute anti-inflammatory activity and gastrointestinal tolerability of diclofenac and nitrofenac. *Agents Actions* 1993; **40**: 176–180.
18. Wallace JL, Reuter BK, Cirino G. Nitric oxide-releasing non-steroidal anti-inflammatory drugs: a novel approach for reducing gastrointestinal toxicity. *J Gastroenterol Hepatol* 1994; **9**: S40–S44.
19. Reuter B, Wallace JL. A therapeutic application of nitric oxide: GI-sparing NSAIDs. In: *Nitric Oxide: A Modulator of Cell-Cell Interactions in the Microcirculation*, Kubes P (Ed). RG Landes Company, Austin 1995: 157–168.
20. Kitahora T, Guth PH. Effects of asprin plus hydrochloric acid on the gastric mucosal microcirculation. *Gastroenterology* 1987; **93**: 810–817.
21. MacKendrick W, Caplan M, Hsueh W. Endogenous nitric oxide protects against platelet-activating factor-induced bowel injury in the rat. *Pediatr Res* 1993; **34**: 222–228.
22. Hutcheson IR, Whittle BJR, Boughton-Smith NK. Role of nitric oxide in maintaining vascular integrity in endotoxin-induced acute intestinal damage in the rat. *Br J Pharmacol* 1990; **101**: 815–820.

23. Granger DN, McCord JM, Parks DA, Höllwarth ME. Xanthine oxidase inhibitors attenuate ischemia-induced vascular permeability changes in the cat intestine. *Gastroenterology* 1986; **90**: 80–84.
24. Haglund U, Bulkley GB, Granger DN. On the pathophysiology of intestinal ischemic injury. Clinical review. *Acta Chir Scand* 1987; **153**: 321–324.
25. Schoenberg MH, Muhl E, Sellin D *et al.* Posthypotensive generation of superoxide free radicals — possible role in the pathogenesis of the intestinal mucosal damage. *Acta Chir Scand* 1984; **150**: 301–309.
26. Payne D, Kubes P. Nitric oxide donors reduce the rise in reperfusion-induced intestinal mucosal permeability. *Am J Physiol* 1993; **265**: G189–G195.
27. Andrews FJ, Malcontenti-Wilson C, O'Brien PE. Protection against gastric ischemia-reperfusion injury by nitric oxide generators. *Dig Dis Sci* 1994; **39**: 366–373.
28. Kubes P, Suzuki M, Granger DN. Nitric oxide: an endogenous modulator of leukocyte adhesion. *Proc Natl Acad Sci USA* 1991; **88**: 4651–4655.
29. Kubes P, Granger DN. Nitric oxide modulates microvascular permeability. *Am J Physiol* 1992; **262**: H611–H615.
30. Kurose I, Kubes P, Wolf R *et al.* Inhibition of nitric oxide production: mechanisms of vascular albumin leakage. *Circ Res* 1993; **73**: 164–171.
31. Kubes P, Kanwar S, Niu X-F, Gaboury J. Nitric oxide synthesis inhibition induces leukocyte adhesion via superoxide and mast cells. *FASEB J* 1993; **7**: 1293–1299.
32. May GR, Crook P, Moore PK, Page CP. The role of nitric oxide as an endogenous regulator of platelet and neutrophil activation within the pulmonary circulation of the rabbit. *Br J Pharmacol* 1991; **102**: 759–763.
33. Ma X, Weyrich AS, Lefer DJ, Lefer AM. Diminished basal nitric oxide release after myocardial ischemia and reperfusion promotes neutrophil adherence to coronary endothelium. *Circ Res* 1993; **72**: 403–412.
34. Suematsu M, Tamatani T, Delano FA *et al.* Microvascular oxidative stress preceding leukocyte activation elicited by *in vivo* nitric oxide suppression. *Am J Physiol* 1994; **266**: H2410–H2415.
35. Nishida J, McCuskey RS, McDonnell D, Fox ES. Protective role of nitric oxide in hepatic microcirculatory dysfunction during endotoxemia. *Am J Physiol* 1994; **267**: G1135–G1141.
36. Hernandez LA, Grisham MB, Twohig B *et al.* Role of neutrophils in ischemia-reperfusion-induced microvascular injury. *Am J Physiol* 1987; **253**: H699–H703.
37. Grisham MB, Granger DN. Neutrophil-mediated mucosal injury. Role of reactive oxygen metabolites. *Dig Dis Sci* 1988; **33**: 6S–15S.
38. Wallace JL, Keenan CM, Granger DN. Gastric ulceration induced by nonsteroidal anti-inflammatory drugs is a neutrophil-dependent process. *Am J Physiol* 1990; **259**: G462–G467.

39. Wallace JL, Arfors KE, McKnight GW. A monoclonal antibody against the CD18 leukocyte adhesion molecule prevents indomethacin-induced gastric damage in the rabbit. *Gastroenterology* 1991; **100**: 878–883.
40. Alican I, Coskun T, Corak A *et al.* Role of neutrophils in indomethacin-induced gastric mucosal lesions in rats. *Inflamm Res* 1995; **44**: 164–168.
41. Kubes P, Hunter JA, Granger DN. Ischemia/reperfusion-induced feline intestinal dysfunction: importance of granulocyte recruitment. *Gastroenterology* 1992; **103**: 807–812.
42. Kurose I, Wolf R, Grisham MB, Granger DN. Modulation of ischemia/reperfusion-induced microvascular dysfunction by nitric oxide. *Circ Res* 1994; **74**: 376–382.
43. Clancy RM, Leszczynska-Piziak J, Abramson SB. Nitric oxide, and endothelial cell relaxation factor, inhibits neutrophil superoxide production *via* a direct action on the NADPH oxidase. *J Clin Invest* 1992; **90**: 1116–1121.
44. Arndt H, Russell JM, Kurose I *et al.* Mediators of leukocyte adhesion in rat mesenteric venules elicited by inhibition of nitric oxide synthesis. *Gastroenterology* 1993; **105**: 675–680.
45. Kanwar S, Wallace JL, Befus D, Kubes P. Nitric oxide synthesis inhibition increases epithelial permeability *via* mast cells. *Am J Physiol* 1994; **266**: G222–G229.
46. Wallace JL. PAF as a mediator of gastrointestinal damage. In: *Platelet-Activating Factor and Diseases*. Saito K, Hanahan DJ (Eds). International Medical Publishers 1989: 153–186.
47. Salvemini D, Masini E, Pistelli A *et al.* Nitric oxide: a regulatory mediator of mast cell reactivity. *J Cardiovasc Pharmacol* 1991; **17 (Suppl 3)**: S258–S264.
48. Hogaboam CM, Befus AD, Wallace JL. Modulation of rat mast cell reactivity by IL-1b. Divergent effects on nitric oxide and platelet-activating factor release. *J Immunol* 1993; **151**: 3767–3774.
49. Gaboury JP, Niu X-F, Kubes P. Nitric oxide inhibits numerous features of mast cell-induced inflammation. *Circulation* 1996; **93**: 318–326.
50. Niu X-F, Ibbotson G, Kubes P. A balance between nitric oxide and oxidants regulates mast cell-dependent neutrophil-endothelial cell interactions. *Circ Res* 1996; **79**: 992–999.
51. Klebanoff SJ. Oxygen metabolism and the toxic properties of phagocytes. *Ann Int Med* 1980; **93**: 480–489
52. Rubanyi GM, Ho EH, Cantor EH *et al.* Cytoprotective function of nitric oxide: inactivation of superoxide radicals produced by human leukocytes. *Biochem Biophys Res Commun* 1991; **181**: 1392–1397.
53. Parks DA, Shah AK, Granger DN. Oxygen radicals: effects on intestinal vascular permeability. *Am J Physiol* 1984; **247**: G167–G170.
54. Gaboury J, Woodman RC, Granger DN *et al.* Nitric oxide prevents leukocyte adherence: role of superoxide. *Am J Physiol* 1993; **265**: H862–H867.

55. Wink DA, Hanbauer I, Krishna MC *et al.* Nitric oxide protects against cellular damage and cytotoxicity from reactive oxygen species. *Proc Natl Acad Sci USA* 1993; **90**: 9813–9817.
56. Johnston B, Kanwar S, Kubes P. Hydrogen peroxide induces leukocyte rolling: modulation by endogenous antioxidant mechanisms including NO. *Am J Physiol* 1996; **271**: H614–H621.
57. Gauthier TW, Davenpeck KL, Lefer AM. Nitric oxide attenuates leukocyte-endothelial interaction *via* P-selectin in splanchnic ischemia-reperfusion. *Am J Physiol* 1994; **267**: G562–G568.
58. Kubes P, Kurose I, Granger DN. NO donors prevent integrin-induced leukocyte adhesion, but not P-selectin-dependent rolling in postischemic venules. *Am J Physiol* 1994; **267**: H931–H937.
59. Itoh M, Guth PH. Role of oxygen-derived free radicals in ischemic gastric injury in the rat. *Gastroenterology* 1985; **88**: 1162–1167.
60. Granger DN, Höllwarth ME, Parks DA. Ischemia-reperfusion injury: role of oxygen-derived free radicals. *Acta Physiol Scand* 1986; **Suppl 548**: 47–63.
61. Vaananen PM, Meddings JB, Wallace JL. Role of oxygen-derived free radicals in indomethacin-induced gastric injury. *Am J Physiol* 1991; **261**: G470–G475.
62. Niu X-F, Smith CW, Kubes P. Intracellular oxidative stress induced by nitric oxide synthesis inhibition increases endothelial cell adhesion to neutrophils. *Circ Res* 1994; **74**: 1133–1140.
63. Kurose I, Wolf R, Grisham MB *et al.* Microvascular responses to inhibition of nitric oxide production: role of active oxidants. *Circ Res* 1995; **76**: 30–39.
64. Cross CE. Oxygen radicals and human disease. *Ann Int Med* 1987; **107**: 526–545.
65. Kubes P. Nitric oxide-induced microvascular permeability alterations: a regulatory role for cGMP. *Am J Physiol* 1993; **265**: H1909–H1915.
66. Rapoport RM, Draznin MB, Murad F. Endothelium-dependent relaxation in rat aorta may be mediated through cyclic GMP-dependent protein phosphorylation. *Nature* 1983; **306**: 174–176.
67. Miller FN, Sims DE. Contractile elements in the regulation of macromolecular permeability. *Fed Proc* 1986; **45**: 84–88.
68. Kurose I, Wolf R, Grisham MB *et al.* Modulation of ischemia/reperfusion-induced microvascular dysfunction by nitric oxide. *Circ Res* 1994; **74**: 376–382.
69. Kubes P. Ischemia/reperfusion in the feline small intestine: a role for nitric oxide. *Am J Physiol* 1993; **264**: G143–G149.
70. Boughton-Smith NK, Evans SM, Laszlo F *et al.* The induction of nitric oxide synthase and intestinal vascular permeability of endotoxin in the rat. *Br J Pharmacol* 1993; **110**: 1189–1195.
71. Billiar TR, Curran RD, Harbrecht BG *et al.* Modulation of nitrogen oxide synthesis *in vivo*: N^G-monomethyl-L-arginine inhibits endotoxin-induced nitrite-nitrate

biosynthesis while promoting hepatic damage. *J Leukocyte Biol* 1990; **48**: 565–569.

72. Tiao G, Rafferty J, Ogle C *et al.* Detrimental effect of nitric oxide synthase inhibition during endotoxemia may be caused by high levels of tumor necrosis factor and interleukin-6. *Surgery* 1994; **116**: 332–338.
73. Liu SF, Adcock IM, Old RW *et al.* Differential regulation of the constitutive and inducible nitric oxide synthase mRNA by lipopolysaccharide treatment *in vivo* in the rat. *Crit Care Med* 1996; **24**: 1219–1225.
74. Mulder MF, Van Lambalgen AA, Huisman E *et al.* Protective role of NO in the regional hemodynamic changes during acute endotoxemia in rats. *Am J Physiol* 1994; **266**: H1558–H1564.
75. Lingnau W, McGuire R, Dehing DJ *et al.* Changes in regional hemodynamics after nitric oxide inhibition during ovine bacteremia. *Am J Physiol* 1996; **270**: R207–R216.
76. Cartwright JE, Johnstone AP, Whitley GSJ. Inhibition of nitric oxide synthase by antisense techniques: investigation of the roles of NO produced by murine macrophages. *Br J Pharmacol* 1997; **120**: 146–152.
77. Roediger WEW, Lawson MJ, Nance SH, Radcliffe BC. Detectable colonic nitrite levels in inflammatory bowel disease — mucosal or bacterial malfunction? *Digestion* 1986; **35**: 199–204.
78. Boughton-Smith NK, Evans SM, Hawkey CJ *et al.* Nitric oxide synthase activity in ulcerative colitis and Crohn's disease. *Lancet* 1993; **342**: 338–340.
79. Miller MJS, Sadowska-Krowicka H, Chotinaruemol S *et al.* Amelioration of chronic ileitis by nitric oxide synthase inhibition. *J Pharmacol Exp Ther* 1993; **264**: 11–16.
80. Thompson JA, Sadowska-Krowicka H, Rossi J *et al.* Inducible nitric oxide synthase gene expression in guinea pig ileitis: a model of IBD prevented by aminoguanidine [Abstract]. *Gastroenterology* 1994; **106**: A782.
81. Florquin S, Amraoui Z, Dubois C *et al.* The protective role of endogenously synthesized nitric oxide in staphylococcal enterotoxin B-induced shock in mice. *J Exp Med* 1994; **180**: 1153–1158.
82. Yamada T, Sartor RB, Marshall S *et al.* Mucosal injury and inflammation in a model of chronic granulomatous colitis in rats. *Gastroenterology* 1993; **104**: 759–771.
83. Ribbons KA, Zhang X-J, Thompson JH *et al.* Potential role of nitric oxide in a chronic model of colitis in rhesus macaques. *Gastroenterology* 1995; **108**: 705–711.
84. Rachmilewitz D, Karmeli F, Okon E, Bursztyn M. Experimental colitis is ameliorated by inhibition of nitric oxide synthase activity. *Gut* 1995; **37**: 247–255.
85. Grisham MB, Specian RD, Zimmerman TE. Effects of nitric oxide synthase inhibition on the pathophysiology observed in a model of chronic granulomatous colitis. *J Pharmacol Exp Ther* 1994; **271**: 1114–1121.

86. Hogaboam CM, Jacobson K, Collins SM, Blennerhassett MG. The selective beneficial effects of nitric oxide inhibition in experimental colitis. *Am J Physiol* 1995; **268**: G673–G684.
87. Conner EM, Chen Y, Grisham MB. Effect of nitric oxide synthase (NOS) inhibition on dextran sulfate sodium (DSS)-induced colitis in rats and mice [Abstract]. *Gastroenterology* 1995; **108**: A725.
88. Pfeiffer CJ, Qiu BS. Effects of chronic nitric oxide synthase inhibition on TNB-induced colitis in rats. *J Pharm Pharmacol* 1995; **47**: 827–832.
89. Ribbons KA, Clark DA, Currie MG *et al.* Inducible nitric oxide synthase and idiopathic colitis in rhesus macaques [Abstract]. *Gastroenterology* 1995; **108**: A903.
90. MacMicking JD, Nathan C, Hom G *et al.* Altered responses to bacterial infection and endotoxic shock in mice lacking inducible nitric oxide synthase. *Cell* 1995; **81**: 641–650.
91. McCafferty D-M, Mudgett JS, Swain MG, Kubes P. Inducible nitric oxide synthase plays a critical role in resolving intestinal inflammation. *Gastroenterology* 1997; **112**: 1022–1027.
92. McCafferty D-M. Studies on a short chain fatty acid-induced colitis in mice. *PhD thesis*. Strathclyde University, Glasgow 1991.
93. Wong PY, Yue G, Yin K *et al.* Antibodies to intracellular adhesion molecule-1 ameliorate the inflammatory response in acetic acid-induced inflammatory bowel disease. *J Pharmacol Exp Ther* 1995; **274**: 475–480.
94. Shiratora Y, Aoki S, Takada H *et al.* Oxygen-derived free radical generating capacity of polymorphonuclear cells in patients with ulcerative colitis. *Digestion* 1989; **44**: 163–171.
95. Chin KW, Barrett KE. Mast cells are not essential to inflammation in murine model of colitis. *Dig Dis Sci* 1994; **39**: 513–525.

Chapter 16

The Pathophysiological and Physiological Effects of Nitric Oxide on the Renal Circulation

Edgar A Jaimes, Hiroshi Hayakawa and Leopoldo Raij

Introduction

In the kidney, constitutive NO synthase (cNOS), identified by immunohistochemical staining or by reverse transcription and polymerase chain reaction, has been found in glomeruli, macula densa, the collecting ducts and the inner medullary thin limb.[1] Endothelial cells of the renal arteries and arterioles also express cNOS; however, efferent arterioles (Ef-art) show stronger cNOS expression than afferent arterioles (Af-art), particularly in juxtamedullary glomeruli.[1] The inducible NOS (iNOS), which is expressed in response to certain cytokines and bacterial products such as endotoxin, occurs in vascular smooth muscle within the terminal portion of the glomerular Af-art and medullary thick ascending limb and it has also been reported in cultured cells from the proximal tubule, inner medullary collecting duct, and mesangium.[1]

Nitric Oxide in Renal Physiology

Glomerular Haemodynamics

The glomerulus is situated between two sets of resistance vessels arranged in series. The preglomerular or Af-art lies between the systemic circulation and the glomerular capillaries, while the postglomerular or Ef-art forms the terminus of the glomerular capillary network. Inspection of this arrangement reveals that glomerular capillary pressure (P_{gc}) and glomerular capillary flow can be modulated by independent changes in afferent and efferent resistances. Regulation of these two parameters, as well as the colloid osmotic pressure in the glomerular capillary and the effective hydraulic ultrafiltration coefficient (K_f), which is modulated by the state of mesangial contraction and relaxation, determines the single-nephron glomerular filtration rate (GFR_{sn}).[1]

Changes in renal haemodynamics induced by NO synthesis inhibition

Systemic administration of NO synthesis inhibitors produces dose-dependent and prolonged vasoconstriction, thereby increasing systemic blood pressure,[2] which is accompanied by a fall in renal plasma flow (RPF), an increase in renal vascular resistance, and a smaller decline in glomerular filtration rate (GFR).[2] Micropuncture studies have shown that intravenous pressor doses of NO synthesis inhibitors decrease RPF without substantially modifying GFR_{sn}.[3] In these studies, glomerular Af-art and Ef-art resistances are markedly increased by NO inhibition but with a predominant effect on the Ef-art. As a consequence of this imbalance and the concurrent systemic arterial hypertension, the glomerular hydraulic pressure is elevated and may explain the preserved GFR_{sn} despite the reduced K_f and RPF.[3] It should be recognized, however, that such studies do not exclude the influence of systemic haemodynamics, hormones, and the sympathetic nervous system. Infusion of pressor doses of NO synthesis inhibitors into the renal artery of the isolated perfused rat kidney causes a less significant reduction in RPF, GFR, and urinary flow rate than occurs during systemic NO synthesis inhibition, suggesting that a substantial part of the vasoconstriction observed in response to systemic NO inhibition is due to secondary phenomena perhaps mediated by renal nerves.[4]

Sub-pressor intravenous doses of NO synthesis inhibitors decrease urine flow, sodium excretion, and RPF,[5] suggesting that the renal vasculature is particularly sensitive to NOS inhibition. In micropuncture studies, sub-pressor doses of NO inhibitors produce a small renal vasoconstriction with an increase in Af-art resistances but without any change in Ef-art resistances and no change in P_{gc}; K_f is reduced and a small reduction in GFR_{sn} is also seen,[6] implying that in the euvolaemic rat, the cortical Af-art is tonically under the control of locally produced NO whereas Ef-art is not.[6] Similar findings have been reported in microperfused isolated rabbit arterioles,[7] indicating that local release of NO in the Af-art may play an important role in the regulation of renal haemodynamics. By contrast, in juxtamedullary nephrons NO synthesis inhibition produces constriction of both Af-art and Ef-art,[8] suggesting that in this region both Af-art and Ef-art are under the tonic control of NO.

Neuronal NOS (nNOS) is abundantly expressed in the macula densa cells of the juxtaglomerular apparatus, indicating that NO generated from macula densa may influence the vascular tone of the Af-art and/or Ef-art.[9] The macula densa monitors the NaCl concentration of the tubular fluid, serving as a sensor that regulates tubuloglomerular feedback (TGF). In this system, an increase in NaCl concentration at the macula densa results in constriction of the Af-Art and a decrease in GFR_{sn}.[10] The macula densa also controls renin release and therefore the circulating and probably intrarenal levels of angiotensin II (Ang-II).[11] Studies *in vitro* have shown that NO synthesis inhibition at the macula densa induces vasoconstriction of Af-art when concentration of NaCl at the macula densa is high but not when it is low, indicating that NO synthesis in the macula densa regulates TGF-induced Af-art constriction in response to high tubular NaCl concentration.[11]

Interaction with the renin-angiotensin system

In vitro studies using isolated rabbit arterioles have shown that sensitivity to Ang-II is higher in the isolated microperfused Ef-art than in the Af-art.[12] NO synthesis inhibition causes stronger vasoconstriction in the Af-art than in the Ef-art and augments the vasoconstrictor action of Ang-II in the Af-art,[12] implying that NO antagonizes the vasoconstrictor action of Ang-II in Af-art and may therefore contribute to the differential Ang-II sensitivity between pre-glomerular and post-glomerular arterioles. In dogs, Ang-II infusion has a predominant effect on post-glomerular resistance;[13] however, when NO synthesis is inhibited, Ang-II has a profound effect on pre-glomerular resistance, indicating that NO protects pre-glomerular vessels from Ang-II-induced vasoconstriction.[13] In contrast, studies in microvessels of rat juxtamedullary nephrons have shown that Ang-II-induced vasoconstriction of Ef-art and Af-art is not exaggerated by NO synthesis inhibition.[8] Taken together, these findings suggest that tonic release of NO is confined to the Af-art, possibly accounting for lesser Ang-II sensitivity of the superficial Af-art; however, in the juxtamedullary nephron preparation, both the Af-art and Ef-art are under tonic NO control.

Several *in vitro* studies have suggested that the renal vascular endothelium may directly modulate renin release. However, the cellular mechanisms controlling renin secretion are not completely understood and conflicting data have been reported regarding the role of NO on renin release and/or synthesis. Early studies have shown that NO donors inhibit renin release in renal cortical slices.[14] In isolated Af-art, however, inhibition of NO synthesis abolishes renin mRNA levels, total renin content, and renin secretion.[15] Inhibition of NO synthesis increases plasma renin activity in conscious dogs and anaesthetized rats.[16,17] Other studies, however, have shown that acute inhibition of NO synthesis decreases plasma renin activity, accompanied by increased systemic blood pressure.[18] Hypertension induced by chronic NOS inhibition seems to be partially Ang-II-dependent, indicating heightened function of the renin-angiotensin system (RAS).[19] Moreover, chronic NO synthesis inhibition induces coronary vascular remodelling and myocardial hypertrophy that is accompanied by increased angiotensin-converting enzyme (ACE) expression and activity.[20] These growth responses occur independently of changes in systemic blood pressure, and appear to be linked to increased production of superoxide anion in vascular smooth muscle cells[21] and mesangial cells[22] due to Ang-II activation of NADH/NADPH oxidase in those cells. The reasons for these discrepancies remain unclear; however, there is evidence that NO may have both a stimulatory and an inhibitory effect on renin production, perhaps related to the duration of exposure to NO.[23]

Finally, in conditions such as sodium depletion in which the RAS is stimulated, blockade of Ang-II type 1 (AT_1) receptors results in increased synthesis of intrarenal NO and cGMP due to activation of renal Ang-II type 2 (AT_2) receptors.[24]

Interaction with endothelin

Endothelin (ET) is a potent vasoconstrictor peptide that comprises a family of three distinct isoforms: ET-1, ET-2 and ET-3.[25] The two receptors for ET, ET_A and ET_B, are both present in the kidney.[26] ET-1 stimulation of the ET_A receptor, which is widely distributed in vascular smooth muscle cells, mediates most of the vasoconstrictor response to ET[27] while stimulation of the ET_B receptor (largely present on endothelial cells) mediates endothelium-dependent vasodilatation by stimulating NO production.[28] ET-1 administration produces renal vasoconstriction in the rabbit, dog, and rat and is potentiated by simultaneous NO synthesis inhibition.[29–31] In the renal microcirculation of the rabbit, ET-1 causes vasoconstriction of afferent arterioles that is potentiated by NO synthesis inhibition.[32]

Recent evidence also suggests that NO regulates the interaction of ET with its receptor by displacing bound ET-1 from its ET_A receptor and interfering with post-receptor pathways involved in calcium mobilization.[33] Furthermore, blockade of the ET_A receptor ameliorates systemic and renal vasoconstriction induced by NO synthesis inhibition,[34] indicating that ET acting through the ET_A receptor may contribute to the systemic and renal vasoconstriction induced by NO inhibition.[34]

In porcine aorta, NO synthesis inhibition augments thrombin-stimulated release of ET, while 8-bromo cGMP has an inhibitory effect, therefore suggesting that NO inhibits ET production *via* a cGMP-dependent mechanism.[35] On the other hand, it has been proposed that Ang-II upregulates ET-1 synthesis.[36] Increased vascular ET-1 expression and plasma levels have been observed in severely hypertensive spontaneously hypertensive rats (SHR) treated with NO synthesis inhibitors.[37] In this model of severe hypertension, treatment with ACE inhibitors produces a significant reduction in blood pressure and plasma ET levels, suggesting that this model has an angiotensin-dependent component and that ACE inhibitors may reduce ET spillover into the bloodstream.[37]

In hypertensive Dahl salt-sensitive (DS) rats, NO synthase is down-regulated in the aorta and kidney.[38] In these rats, contraction of aorta to ET-1 is attenuated, suggesting that vascular ET-1 receptors are down-regulated, probably due to increased local ET-1 synthesis linked to decreased NO production. Moreover, urinary excretion of ET-1 is markedly increased in hypertensive DS rats given high dietary salt, but not in Dahl salt-resistant (DR) rats.[39] This finding supports the notion that hypertension, when accompanied by decreased NO synthesis, promotes up-regulation of vascular and renal ET-1 synthesis.

Medullary Blood Flow

Several studies have investigated the effect of selective alterations of renal medullary flow on sodium and water excretion and the role of NO on medullary circulation. Infusion of N^G-nitro-L-arginine methyl ester (L-NAME) into the medullary interstitium has been shown to produce a reduction in interstitial pressure of 24% and a concomitant decrease in water and sodium excretion of approximately 35%, as well as an increase in systemic blood pressure. These changes occur without alteration in cortical blood flow or GFR.[40] These effects of intrarenal NO synthesis inhibition indicate that NO may be tonically active in the medullary circulation and that changes in medullary flow alter sodium and water excretion. Furthermore, intravenous infusion of L-NAME at a dose that has no effect on cortical blood flow also reduces medullary blood flow by 22% and is accompanied by sodium retention and hypertension.[41] Measurement of cGMP or NO trapping by oxyhaemoglobin[42,43] indicates that NO production in the renal medulla is greater than in the renal cortex. Western blot analysis as well as direct measurement of NOS activity has confirmed a greater expression of NOS in the renal medulla.[44,45]

All these studies suggest that NO plays an important role in the regulation of medullary blood flow. There is vast literature supporting the view that changes in medullary flow are sufficient to reset the pressure-natriuresis relationship, promote sodium retention, and contribute to the development of hypertension. Interestingly, however, treatment with amiloride and furosemide does not reduce blood pressure in N^G-monomethyl-L-arginine (L-NMMA)-induced hypertension despite increases in urinary sodium excretion, indicating that changes in sodium excretion alone are not enough to modify L-NMMA-induced hypertension.[46]

Pressure Natriuresis

Acetylcholine (ACh), a well-known stimulator of NO release from the endothelium, induces renal vasodilatation as well as natriuresis.[40] Lameire *et al.*[47] reported that ACh and secretin induce renal vasodilatation but only ACh increases urinary sodium excretion (UNaV). In addition, they reported that ACh-induced increases in UNaV were inhibited by meclofenamate, an inhibitor of prostaglandin synthesis. They concluded that the renal haemodynamic and natriuretic effects of ACh are probably mediated by prostaglandin release. On the other hand, it has been reported that ACh-induced natriuresis is mediated by NO because intrarenal infusion of ACh in meclofenamate-treated dogs increased renal blood flow, diuresis, and natriuresis.[48] However, since NO stimulates cyclo-oxygenase (COX) activity, leading to augmented production of prostaglandins, both NO and prostaglandins appear to collaborate in the regulation of natriuresis.[49]

Endothelium-derived factors such as ET, Ang-II, and NO have been reported to affect renal haemodynamics and function (e.g. sodium reabsorption and excretion).[50,51] Chronic inhibition of NOS with L-arginine analogues promotes systemic hypertension and renal dysfunction; however, the exact mechanism of NOS inhibitors in the induction of hypertension is not clear. Several investigators have reported that sodium retention contributes to the development of hypertension in this model. Salazar *et al.*[52] showed that a sub-pressor dose of L-NAME decreases diuresis and natriuresis and that these effects are abolished by intravenous infusion of L-arginine.

Hayakawa *et al.*[45] recently found that NOS activity in renal medulla is decreased in hypertensive DS rats compared with normotensive DS rats. DS rats have been reported to have impaired pressure natriuresis and medullary blood flow.[53] These data further support the notion that NO may not only participate in renal haemodynamics but may also contribute to the regulation of pressure natriuresis, which plays an important role in the development and/or maintenance of hypertension.

Nitric Oxide in Renal Physiopathology

Diabetes Mellitus

In experimental models of diabetes, it has been proposed that glomerular hyperfiltration associated with increased intra-glomerular pressures and flows may play a role in the development and progression of renal injury.[54] Micropuncture studies have revealed renal vasodilatation, predominantly of the afferent resistance vessels,[54] rendering the glomerular capillaries vulnerable to hypertensive injury. Studies have demonstrated an increased NO production early in the course of experimental diabetes mellitus, and this may contribute to hyperfiltration.[55] Other studies, however, have shown impaired vasodilatation in response to exogenous and endogenous NO in diabetes,[56] indicating either inactivation of NO or decreased reactivity of the vascular smooth muscle to NO. Advanced glycosylation products can quench NO activity *in vivo* and *in vitro*, which results in impaired NO-dependent vasodilatation.[57] Furthermore, in response to carabamylcholine and the calcium ionophore A23187, glomeruli from diabetic rats have an impaired NO generation and/or stability that appears to be mediated by increased protein kinase C activation.[58] Acute L-arginine administration has been shown to normalize defective relaxation to ACh in streptozotocin-induced diabetic rat aortic rings.[59] However, the mechanisms for these effects of L-arginine remain to be established.

Pregnancy

Marked vasodilatation of maternal peripheral and renal vasculature occurs during normal pregnancy in humans and rats.[60,61] This vasodilatation is associated with a decrease in blood pressure, increased renal blood flow (RBF), and increased GFR. Both GFR and RBF increase in early pregnancy and peak during mid-gestation at 20–40% above preconception levels.[62] This rise in GFR and RBF is secondary to a fall in renal vascular resistance.[62] Although renal vasodilatation is probably under hormonal control, the mechanisms involved remain unclear. Gestational increases in RBF and GFR are unaffected by administration of COX inhibitors to conscious pregnant rabbits[63] and rats,[64] suggesting that vasodilatory prostaglandins are not involved in the haemodynamic changes of pregnancy.

Studies in pregnant rats have shown increased production of NO and cGMP, and in pregnant women urinary excretion[65] as well as plasma levels of cGMP[66] have been reported to be increased. These findings indicate that NO may be involved in the vasodilatation of normal pregnancy.[67] In midterm pregnant rats, acute NO inhibition induces a greater decline in GFR and RPF and rise in renal vascular resistance when compared with virgin rats, which suggests that NO mediates gestational hyperfiltration and renal vasodilatation.[68] In the same studies, pregnant rats were shown to be less responsive to the vasoconstrictor effects of Ang-II, implying that the increased vasoconstrictor response to NO inhibition is not secondary to a non-specific increased sensitivity to vasoconstrictor agents.[68] Chronic NO blockade has been shown to prevent pregnancy-induced renal and peripheral vasodilatation and increased GFR, to promote proteinuria, and to compromise maternal and fetal health.[69] These studies support the view that increased NO production is necessary to accommodate the haemodynamic status of pregnancy.[67,69]

Animals[70] and humans[71] are known to be more susceptible in late pregnancy to development of glomerular thrombosis and renal failure due to pathological processes that result in endothelial injury and increased blood coagulability. Plasma levels of certain amino acids, including L-arginine, are decreased in humans and animals in late pregnancy.[72] Administration of gram-negative endotoxin in a dose that has no effect in virgin rats induces glomerular thrombosis in rats in late pregnancy. This is accompanied by a significant decrease in L-arginine levels and is prevented by L-arginine supplementation,[73] suggesting that in late pregnancy, maternal reserve capability to increase NO synthesis may be limited due to inadequate substrate.

Hepatorenal Syndrome

Portal hypertension and arteriolar vasodilatation leading to hyperdynamic circulation characterized by arterial hypotension, hypervolaemia, and high cardiac output are the most significant haemodynamic alterations in patients and experimental animals with cirrhosis and ascites.[74] Secondary to the arterial hypotension there is also hyperactivity of the RAS and the sympathetic nervous system and increased levels of antidiuretic hormone.[74] These changes are associated with sodium retention and impaired free water excretion. The pathogenesis of sodium retention in cirrhosis has been the subject of numerous studies, but the mechanisms involved are not all completely established.

It has been proposed that overproduction of NO by the vascular endothelium may play a critical role in the hyperdynamic circulation and in the secondary neurohumoral abnormalities and renal dysfunction of cirrhosis accompanied by ascites.[75] Cirrhotic rats with ascites have increased pressor sensitivity to NO synthesis inhibition, together with significant increases in GFR and urinary sodium excretion.[76] In addition, the aorta of cirrhotic rats has been shown to possess increased cGMP — which was reduced to control levels by chronic inhibition of NO synthesis — together with increased endothelial cNOS protein expression.[77]

On the other hand, increased endothelial cNOS expression and NO synthesis in cirrhosis or hepatorenal syndrome (HRS) may be secondary to increased flow and shear stress.[78] In the latter case, increased NO synthesis would be a consequence rather than a cause of the hyperdynamic state of cirrhosis and HRS.

Patients with advanced cirrhosis exhibit a greater forearm vasoconstrictor response to locally infused L-NMMA and a decreased response to locally infused noradrenaline, suggesting that NO may contribute to the peripheral vasodilatation and hyperdynamic circulatory state in advanced liver cirrhosis in humans.[79] It has been postulated that endotoxin and/or cytokines may trigger increased NO production in cirrhosis by inducing iNOS expression in the vascular wall.[75] Despite evidence that iNOS mRNA expression is increased in arteries from cirrhotic rats, however, this increase does not appear to be accompanied by iNOS protein expression.[80] The role of NO in the haemodynamics of the hepatic and splanchnic circulations in cirrhosis is discussed in Chapter 14.

Current information concerning ET in cirrhosis is conflicting. Early studies reported normal or decreased levels;[81] more recent studies have found ET levels consistently increased in cirrhosis.[82] The source of ET in cirrhosis appears to be the liver itself, as tissue levels of preproET-1 mRNA and immunoreactive ET-1 are increased in the liver but not in other organs from cirrhotic rats.[83] ET may have a role as counter-regulatory hormone to the vascular vasodilatation of cirrhosis. However, current evidence indicates that blockade of the ET_A receptor in cirrhosis does not have any effect on arterial blood pressure or portal pressure in cirrhotic rats, suggesting that ET is not involved in the homeostasis of arterial or portal pressure.[83]

F_2-isoprostanes are a series of prostaglandin F_2-like compounds produced *in vivo* by a COX-independent, free radical-catalyzed lipid peroxidation mechanism.[84] Chronic endotoxaemia and tissue hypoxia may be present in HRS[74], an environment that could lead to the generation of free radicals and generation of F_2-isoprostanes. Indeed, plasma levels of F_2-isoprostanes have been reported to be increased in humans with HRS[85] and in rats with carbon tetrachloride hepatotoxicity.[86] One of these isoprostanes in particular, 8-epi-$PGF_{2\alpha}$, has been found to be an extremely potent renal vasoconstrictor and therefore is a potential mediator for the renal vasoconstriction of HRS.[87]

HRS is a unique form of acute renal failure occurring in patients with liver disease for which a specific cause cannot be elucidated. A strong body of evidence supports the concept that the renal failure in HRS is functional in nature. Despite severe derangement in renal function, pathological abnormalities are minimal and inconsistent.[74] In addition, tubular functional integrity is maintained during HRS, as manifested by unimpaired sodium reabsorptive capacity and concentrating ability.[74] Several studies have demonstrated a significant reduction in renal perfusion;[88,89] however it is unlikely that a decrease in mean blood flow *per se* is responsible for the oliguria of HRS.[74] HRS may constitute an extreme condition of underfilling of the arterial circulation with the most extreme elevations of vasoactive hormones including renin, noradrenaline, and vasopressin and the most extreme degree of renal vasoconstriction.[90] It has also been proposed that renal ischaemia in cirrhosis is not an expression of the normal neural and hormonal response to liver damage but rather represents an alteration in the synthesis, degradation, or potency of a vasoactive substance or a malfunction of the normal feedback regulation of its release.[74] Currently, however, the effectors that promote renal ischaemia and a decrease in GFR in HRS remain incompletely defined.

Acute Renal Failure

Ischaemic acute renal failure (ARF) in the rat is associated with a decreased vasodilatory response to bradykinin and acetylcholine, despite the fact that cNOS protein expression in renal resistance vessels is increased.[91] Increased cGMP generation has been reported in gentamicin-induced ARF,[92] and NO synthesis inhibition aggravates gentamicin-induced renal failure, suggesting that endogenously released NO may be protective in ARF.[93] Furthermore, L-arginine administration improves GFR in glycerol-, uranyl nitrate-, and cyclosporine-induced ARF.[94–96] The protective mechanism of L-arginine, however, remains to be determined. Radiocontrast administration has been shown to increase outer medullary blood flow.[97] This effect can be prevented by pretreatment with indomethacin and NO synthesis inhibitors, implying that NO and prostanoids have an important role in the renal response to radiocontrast.[97]

Renal Haemodynamics in Sepsis

Lipopolysaccharide (LPS) from gram-negative bacteria has been shown to be the cause of the clinical syndrome of septic shock, characterized by severe vasodilatation, hypotension, vascular damage, and disseminated intravascular coagulation.[98] LPS and certain cytokines increase the expression of iNOS in glomerular mesangium[99] and vascular smooth muscle cells,[100] suggesting that NO may play an important role in the hypotension observed during endotoxaemia or sepsis. In fact, NOS inhibition restores blood pressure to normal in animals[101] and humans[102] with endotoxaemia. However, NOS inhibition in rats and dogs with endotoxaemia results in increased mortality,[103] showing that vasodilatation induced by NO may also help to maintain critical organ perfusion during periods of systemic hypotension. In rats with endotoxaemia, NO synthesis inhibition leads to glomerular thrombosis[103] that can be prevented with an exogenous NO donor,[104] indicating that NO may have a reno-protective role during endotoxaemia by maintaining renal vasodilatation and inhibiting platelet adhesion and aggregation. Following injection of LPS to rats, there was an initial increase in NO production in all measured tissues, which was sustained with protracted endotoxaemia in the glomerulus and aorta but not in the liver or spleen. Raised glomerular NO synthesis may maintain perfusion and prevent thrombosis in the kidney, while upregulated NO production in major arteries may explain the persistence of hyporesponsiveness to vasopressors during chronic endotoxaemia.[105]

During glomerular inflammation, NO, reactive oxygen species (ROS) and prostanoids are produced in increased amounts within the glomerulus.[1] These autacoids, which are released in large quantities during inflammatory processes, can have complex interactions with one another. Indeed, NO upregulates COX-2 activity,[49] while ROS such as superoxide anion promote peroxynitrite generation.[106] On the other hand, hydrogen peroxide has been shown to down-regulate interleukin-1 stimulated iNOS activity in mesangial cells[107] and therefore to mitigate NO production in response to this inflammatory cytokine.

Hypertension

Salt-sensitive hypertension

Because populations of patients with essential hypertension show blood pressure responses to high dietary salt intake, hypertensive individuals are usually classified as having salt-sensitive or salt-resistant hypertension.[108] Salt-sensitive hypertensives more often have abnormal renal function, greater sodium retention, and frequent target organ diseases.[108]

Abnormal endothelial function has been reported in several experimental models of hypertension. The severity and type of dysfunction varies in different experimental models of hypertension. Impaired endothelium-dependent relaxation has been reported in aortas from SHR,[109] DS rats,[110] and deoxycorticosterone acetate (DOCA)-salt hypertensive rats.[111] In SHR, impaired endothelium-dependent relaxation to ACh in the aorta was reported to be due to increased release of endothelium-derived constricting factor(s)[112] which could be inhibited by a COX inhibitor or a thromboxane A_2 blocker. On the other hand, in DOCA-salt or DS hypertensive rats, impaired response to ACh was not improved by treatment with COX inhibitors.[111] These results suggest that in DS or DOCA-salt rats, the response to ACh is attenuated because of less production of NO from vascular endothelium. In general, clinical and experimental data indicate that the physiological response to an increase in cyclic strain, such as that imposed by hypertension, is to upregulate vascular cNOS activity.[78] Hayakawa and Raij,[38] however, recently studied the cNOS activity (conversion of ^{14}C L-arginine to ^{14}C L-citrulline) in aortas of SHR and DS rats matched for blood pressure (Figure 1). Compared with their normotensive counterparts, aortic cNOS activity was increased 106% in SHR but reduced by 73% in DS rats. These results explain why endothelium-dependent relaxation mediated by NO is impaired in salt-sensitive DS rats but not in SHR.

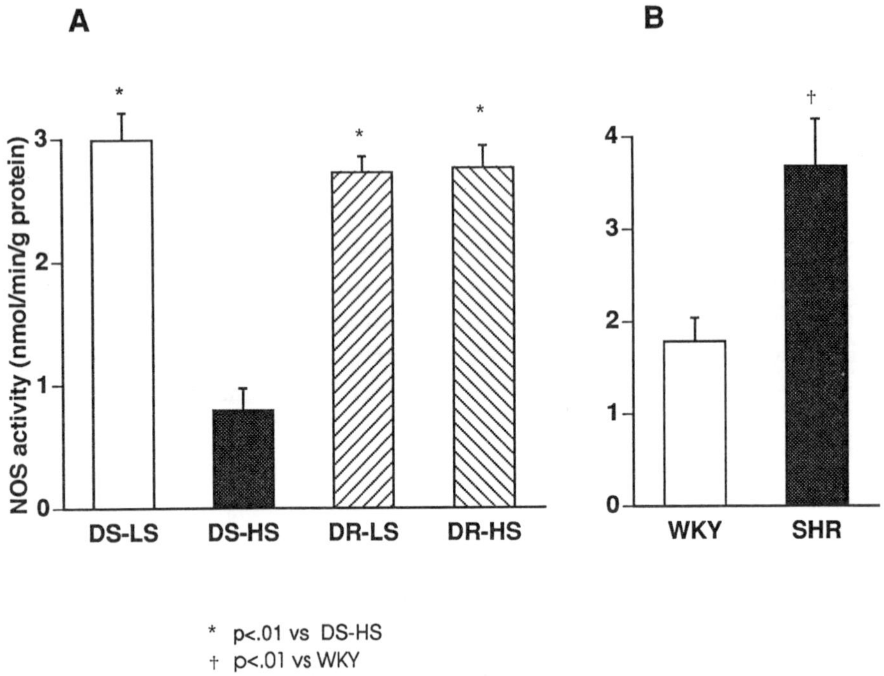

Figure 1. cNOS activity in the aortae of (A) Dahl rats and (B) WKY rats and SHR. NOS activity was determined by conversion of [^{14}C]-L-arginine to [^{14}C]-L-citrulline. DS, Dahl salt-sensitive rats; DR, Dahl salt-resistant rats; LS (low salt), fed 0.5% NaCl; HS (high salt), fed 4% NaCl. *$P < 0.01$ vs DS-HS, †$P < 0.01$ vs WKY. Values are mean, S.E.M. (Reproduced, with permission, from Ref. 38).

As observed in aortas, impaired endothelium-dependent relaxation was also demonstrated in renal vessels in these three models of hypertension. Hayakawa *et al.*[113] reported the effects of ACh on renal vascular resistance and the release rate of NO in kidneys isolated from SHR, DOCA-salt, and DS rats (Figure 2). Endothelium-dependent renal vasorelaxation mediated by ACh was attenuated in all hypertensive rats compared with their normotensive controls. However, the release rate of NO into the perfusate was not decreased in SHR compared with normotensive Wistar-Kyoto rats, whereas the release rate of NO was markedly decreased in both DOCA-salt rats and DS rats.[114, 115] These studies indicate that in salt-sensitive hypertension, endothelial function in conduit or resistance vessels is impaired because of a failure to increase NO bioactivity in response to hypertension. This failure may be genetically conditioned.

In humans, the relationship among endothelial function, NO production, and dietary sodium intake is still unclear. In salt-sensitive and salt-resistant hypertensive patients with mild/moderate hypertension, the plasma levels of NO_2^- and NO_3^- (stable metabolites of NO) are increased.[116] These findings agree

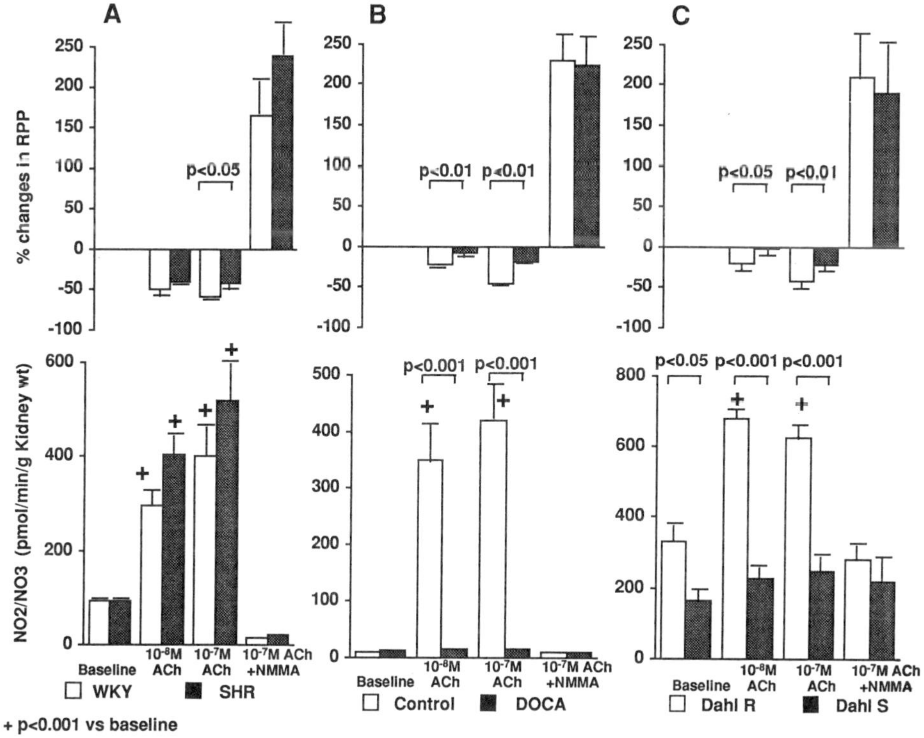

Figure 2. Effects of acetylcholine (ACh) and N^G-monomethyl-L-arginine (NMMA) on renal perfusion pressure (RPP) and release rates of NO_2/NO_3 into perfusate from isolated perfused kidneys of: (A) WKY rats and SHR, (B) DOCA-salt hypertensive rats and their controls, (C) Dahl salt-sensitive (Dahl S) rats and salt-resistant (Dahl R) rats. $+P < 0.001$ vs baseline. Values are mean, S.E.M. (Reproduced, with permission, from Ref. 114).

with experimental data which indicated that the 'physiological' response to an increase in blood pressure is upregulation of endothelial NOS activity.[38] If the experimental data apply to humans, they suggest that salt-sensitive hypertensives with untreated moderate/severe hypertension would, over time, suffer a decrease in endothelial NOS, accompanied by more severe end organ disease (renal injury, left ventricular hypertrophy, hypertension) than similarly hypertensive salt-resistant patients.[38,116]

NOS in hypertensive renal injury

NOS inhibition results not only in hypertension but proteinuria and glomerular injury.[117] DS rats develop hypertension and renal injury when given high dietary salt.[118] Hypertension results in remarkably more severe proteinuria and renal injury in DS rats than SHR matched for blood pressure.[118]

Hayakawa and Raij[45] determined NOS activity in renal medulla and cortex, proteinuria and glomerular injury in DS rats and SHR with hypertension of similar duration and severity. DS rats showed decreased cNOS activity in renal medulla, severe proteinuria, and glomerular injury compared with normotensive controls. On the other hand, despite similar severity of hypertension, SHR had increased NOS activity, minimal glomerular injury, and proteinuria. These studies further support the notion that increased cNOS activity in SHR may play a beneficial role by mitigating hypertensive renal injury.

The comparative studies in SHR and DS hypertensive rats thus imply that upregulation of vascular cNOS activity has a protective cardiovascular homeostatic role in hypertension and may contribute to forestalling renal injury, left ventricular hypertrophy and stroke.

Summary

NO produced by cNOS, nNOS and iNOS participates in a variety of renal physiological and pathophysiological processes. NO plays an important role in the regulation of glomerular microcirculation by modulating afferent arteriolar tone and relaxation of the mesangium. NO may also contribute to the regulation of medullary circulation, renal sodium excretion and renin release. NO can also interact with other vasoactive agents such as ET and Ang-II, antagonizing Ang-II-induced vasoconstriction in glomerular arterioles and modulating ET-1 synthesis and activity. In addition, NO is a powerful endogenous inhibitor of growth-related responses in vascular cells and thereby participates in vascular remodelling in reaction to injury. Alterations in the L-arginine/NO pathway have been incriminated in the pathophysiology of diabetic nephropathy, the vasodilatation of normal pregnancy and the hyperdynamic circulation of cirrhosis and hepatorenal syndrome. Large quantities of NO produced by iNOS may participate in the hypotension of endotoxaemia or sepsis. In the kidney, however, increased NO during sepsis may be renoprotective by maintaining renal vasodilatation and inhibiting platelet adhesion and aggregation. Endothelial dysfunction due to decreased NO production occurs in several experimental models of hypertension. Recent studies indicate that the physiological adaptation to the increased haemodynamic workload of hypertension is to upregulate vascular cNOS activity. Increased cNOS activity may have a cardiovascular homeostatic role in hypertension that contributes to forestall end-organ complications of hypertension, such as renal failure, left ventricular hypertrophy, coronary artery disease and stroke.

References

1. Raij L, Baylis C. Glomerular actions of nitric oxide. *Kidney Int* 1995; **48**: 20–32.
2. Tolins JP, Palmer RMJ, Moncada S, Raij L. Role of endothelium-derived relaxing factor in regulation of renal hemodynamic responses. *Am J Physiol* 1990; **258**: H655–H662.
3. Zatz R, de Nucci G. Effects of acute nitric oxide inhibition on rat glomerular microcirculation. *Am J Physiol* 1991; **261**: F360–F363.
4. Sakuma I, Togashi H, Yoshioka M *et al.* N^G-methyl-L-arginine, an inhibitor of L-arginine-derived nitric oxide synthesis, stimulates sympathetic renal nerve activity *in vivo*. A role for nitric oxide in the central regulation of sympathetic tone? *Circ Res* 1992; **70**: 607–611.
5. Lahera V, Salom MG, Miranda-Guardiola F *et al.* Effects of N^G-nitro-L-arginine methyl ester on renal function and blood pressure. *Am J Physiol* 1991; **261**: F1033–F1037.
6. Deng A, Baylis C. Locally produced EDRF controls preglomerular resistance and ultrafiltration coefficient. *Am J Physiol* 1993; **264**: F212–F215.
7. Ito SC, Johnson CS, Carretero OA. Modulation of angiotensin II-induced vasoconstriction by endothelium-derived relaxing factor in the isolated microperfused rabbit afferent arteriole. *J Clin Invest* 1991; **87**: 1656–1663.
8. Ohishi K, Carmines PK, Inscho EW, Navar LG. EDRF-angiotensin II interactions in rat juxtamedullary afferent and efferent arterioles. *Am J Physiol* 1992; **263**: F900–F906.
9. Mundel P, Bachmann S, Bader M *et al.* Expression of nitric oxide synthase in kidney macula densa cells. *Kidney Int* 1992; **42**: 1017–1019.
10. Wilcox CS, Baylis C, Wingo CS. Glomerular-tubular balance and proximal regulation. In: *The Kidney: Physiology and Pathophysiology*, Seldin DW, Giebisch G (Eds). Raven Press, New York 1992: 1807–1842.
11. Wilcox CS, Welch WJ, Murad F *et al.* Nitric oxide synthase in macula densa regulates glomerular capillary pressure. *Proc Natl Acad Sci USA* 1992; **89**: 11993–11997.
12. Ito S, Arima S, Ren YL *et al.* Endothelium-derived relaxing factor/nitric oxide modulates angiotensin II action in the isolated microperfused rabbit afferent but not efferent arteriole. *J Clin Invest* 1993; **91**: 2012–2019.
13. Schnackenberg CG, Wilkins FC, Granger JP. Role of nitric oxide in modulating the vasoconstrictor actions of angiotensin II in preglomerular and postglomerular vessels in dogs. *Hypertension* 1995; **26**: 1024–1029.
14. Vidal MJ, Romero JC, Vanhoutte PM. Endothelium-derived relaxing factor inhibits renin release. *Eur J Pharmacol* 1988; **149**: 401–402.
15. Tharaux PL, Dussaule JC, Pauti MD *et al.* Activation of renin synthesis is dependent on intact nitric oxide production. *Kidney Int* 1997; **51**: 1780–1787.

16. Sigmon DH, Carretero OA, Beierwaltes WH. Endothelium-derived relaxing factor regulates renin release *in vivo*. *Am J Physiol* 1992; **32**: F256–F261.
17. Beierwaltes WH, Carretero OA. Nonprostanoid endothelium-derived factors inhibit renin release. *Hypertension* 1992; **19 (Suppl II)**: II-68–II-73.
18. Persson PB, Baumann JE, Ehmke H *et al*. Endothelium-derived NO stimulates pressure-dependent renin release in conscious dogs. *Am J Physiol* 1993; **264**: F943–F947.
19. Pollock DM, Polakowski JS, Divish BJ, Opgenorth TJ. Angiotensin blockade reverses hypertension during long-term nitric oxide synthase inhibition. *Hypertension* 1993; **21**: 660–666.
20. Takemoto M, Egashira K, Usui M *et al*. Important role of tissue angiotensin-converting enzyme activity in the pathogenesis of coronary vascular and myocardial structural changes induced by long-term blockade of nitric oxide synthase in rats. *J Clin Invest* 1997; **99**: 278–287.
21. Griendling KK, Minieri CA, Ollerenshaw JD, Alexander RW. Angiotensin II stimulates NADH and NADPH oxidase activity in cultured vascular smooth muscle cells. *Circ Res* 1994; **74**: 1141–1148.
22. Jaimes EA, Galceran JM, Raij L. Angiotensin II induces superoxide anion production by mesangial cells. *Kidney Int* 1998; **54**: 775–784.
23. Greenberg SG, He X-R, Schnermann JB, Briggs JP. Effect of nitric oxide on renin secretion. I. Studies in isolated juxtaglomerular granular cells. *Am J Physiol* 1995; **268**: F948–F952.
24. Sigmon DH, Carretero OA, Beierwaltes WH. Plasma renin activity and the renal response to nitric oxide. *J Am Soc Nephrol* 1992; **3**: 1288-1294.
25. Inoue A, Yanagisawa M, Kimura S *et al*. The human endothelin family: three structurally and pharmacologically distinct isopeptides predicted by three separate genes. *Proc Natl Acad Sci USA* 1989; **86**: 2863–2867.
26. Kohan DE. Endothelins in the kidney: physiology and pathophysiology. *Am J Kidney Dis* 1993; **22**: 493–510.
27. Télémaque S, Gratton J-P, Claing A, D'Orléans-Juste P. Endothelin-1 induces vasoconstriction and prostacyclin release *via* the activation of endothelin ET_A receptors in the perfused rabbit kidney. *Eur J Pharmacol* 1993; **237**: 275–281.
28. Brooks DP, Depalma PD, Pullen M, Nambi P. Characterization of canine renal endothelin receptor subtypes and their function. *J Pharmacol Exp Ther* 1994; **268**: 1091–1097.
29. Madeddu P, Anania V, Parpaglia PP *et al*. Endothelin-1-induced renal vasoconstriction is blunted by enalaprilat and enhanced by EDRF antagonist in awake normotensive rats. *Clin Invest Med* 1991; **14**: 600–606.
30. Rogerson ME, Cairns HS, Fairbanks LD *et al*. Endothelin-1 in the rabbit: interactions with cyclo-oxygenase and NO-synthase products. *Br J Pharmacol* 1993; **108**: 838–843.

31. Lerman A, Sandok EK, Hildebrand FL Jr, Burnett JC Jr. Inhibition of endothelium-derived relaxing factor enhances endothelium-mediated vasoconstriction. *Circulation* 1992; **85**: 1894–1898.
32. Ito S, Juncos LA, Nushiro N *et al*. Endothelium-derived relaxing factor modulates endothelin action in afferent arterioles. *Hypertension* 1991; **17**: 1052–1056.
33. Goligorsky MS, Tsukahara H, Magazine H *et al*. Termination of endothelin signaling: role of nitric oxide. *J Cell Physiol* 1994; **158**: 485–494.
34. Thompson A, Valeri CR, Lieberthal W. Endothelin receptor A blockade alters hemodynamic response to nitric oxide inhibition in rats. *Am J Physiol* 1995; **269**: H743–748.
35. Boulanger CM, Lüscher TF. Differential effect of cyclic GMP on the release of endothelin-1 from cultured endothelial cells and porcine aorta. *J Cardiovasc Pharmacol* 1991; **17**: S264–S266.
36. Sung CP, Arleth AJ, Storer BL, Ohlstein BH. Angiotensin type 1 receptors mediate smooth muscle proliferation and endothelin biosynthesis in rat vascular smooth muscle. *J Pharmacol Exp Ther* 1994; **271**: 429–437.
37. Li JS, Deng LY, Grove K *et al*. Comparison of effect of endothelin antagonism and angiotensin-converting enzyme inhibition on blood pressure and vascular structure in spontaneously hypertensive rats treated with N^{ω}-nitro-L-arginine methyl ester. Correlation with topography of vascular endothelin-1 gene expression. *Hypertension* 1996; **28**: 188–195.
38. Hayakawa H, Raij, L. The link among nitric oxide synthase activity, endothelial function, and aortic and ventricular hypertrophy in hypertension. *Hypertension* 1997; **29**: 235–241.
39. Hayakawa H, Coffee K, Raij, L. Endothelial dysfunction and cardio-renal injury in experimental salt-sensitive hypertension: effects of antihypertensive therapy. *Circulation* 1997; **96**: 2407–2413.
40. Mattson DL, Roman RJ, Cowley AW Jr. Role of nitric oxide in renal papillary blood flow and sodium excretion. *Hypertension* 1992; **19**: 766–769.
41. Nakanishi K, Mattson DL, Cowley AW, Jr. Role of renal medullary blood flow in the development of L-NAME hypertension in rats. *Am J Physiol* 1995; **268**: R310–316.
42. Zou AP, Cowley AW Jr. Nitric oxide in renal cortex and medulla. An *in vivo* microdialysis study. *Hypertension* 1997; **29**: 194–198.
43. Biondi ML, Dousa TP, Vanhoutte PM, Romero JC. Evidence for the existence of endothelium-derived relaxing factor in the renal medulla. *Am J Hypertens* 1990; **3**: 876–878.
44. Mattson DL, Higgins DJ. Influence of dietary sodium intake on renal medullary nitric oxide synthase. *Hypertension* 1996; **27**: 688–692.
45. Hayakawa H, Coffee K, Raij L. Does vascular nitric oxide synthase activity mitigate end-organ injury in hypertension? [Abstract]. *J Hypertens* 1997; **10**: 160.

46. Uzuner K, Banks RO. Endothelin-induced natriuresis and diuresis are pressure-dependent events in rats. *Am J Physiol* 1993; **265**: R90–R96.
47. Lameire N, Vanholder R, Ringoir S, Leusen I. Role of medullary hemodynamics in the natriuresis of drug-induced renal vasodilation in the rat. *Circ Res* 1980; **47**: 839–844.
48. Lahera V, Salom MG, Fiksen-Olsen MJ *et al.* Effects of N^G-monomethyl-L-arginine and L-arginine on acetylcholine renal response. *Hypertension* 1990; **15**: 659–663.
49. Salvemini D, Misko TP, Masferrer JL *et al.* Nitric oxide activates cyclooxygenase enzymes. *Proc Natl Acad Sci USA* 1993; **90**: 7240–7244.
50. Hirata Y, Matsuoka H, Kimura K *et al.* Renal vasoconstriction by the endothelial cell-derived peptide endothelin in spontaneously hypertensive rats. *Circ Res* 1989; **65**: 1370–1379.
51. Llinas MT, Gonzalez JD, Salazar FJ. Interactions between angiotensin and nitric oxide in the renal response to volume expansion. *Am J Physiol* 1995; **269**: R504–R510.
52. Salazar FJ, Pinilla JM, Lopez F *et al.* Renal effects of prolonged synthesis inhibition of endothelium-derived nitric oxide. *Hypertension* 1992; **20**: 113–117.
53. Roman RJ. Abnormal renal hemodynamics and pressure-natriuresis relationship in Dahl salt-sensitive rats. *Am J Physiol* 1986; **251**: F57–F65.
54. Hostetter TH, Troy JL, Brenner BM. Glomerular hemodynamics in experimental diabetes mellitus. *Kidney Int* 1981; **19**: 410–415.
55. Tolins JP, Shultz PJ, Raij L *et al.* Abnormal renal hemodynamic response to reduced renal perfusion pressure in diabetic rats: role of NO. *Am J Physiol* 1993; **265**: F886–F895.
56. Calver A, Collier J, Vallance P. Inhibition and stimulation of nitric oxide synthesis in the human forearm arterial bed of patients with insulin-dependent diabetes. *J Clin Invest* 1992; **90**: 2548–2554.
57. Bucala R, Tracey KJ, Cerami A. Advanced glycosylation products quench nitric oxide and mediate defective endothelium-dependent vasodilation in experimental diabetes. *J Clin Invest* 1991; **87**: 432–438.
58. Craven PA, Studer RK, DeRubertis FR. Impaired nitric oxide-dependent cyclic guanosine monophosphate generation in glomeruli from diabetic rats. Evidence for protein kinase C-mediated suppression of the cholinergic response. *J Clin Invest* 1994; **93**: 311–320.
59. Pieper GM, Peltier BA. Amelioration by L-arginine of a dysfunctional arginine/nitric oxide pathway in diabetic endothelium. *J Cardiovasc Pharmacol* 1995; **25**: 397–403.
60. Robson SC, Hunter S, Boys RJ, Dunlop W. Serial studies of factors influencing changes in cardiac output during human pregnancy. *Am J Physiol* 1989; **256**: H1060–H1065.

61. Gilson GJ, Mosher MD, Conrad KP. Systemic hemodynamics and oxygen transport during pregnancy in chronically instrumented, conscious rats. *Am J Physiol* 1992; **263**: H1911–H1918.
62. Conrad KP. Renal hemodynamics during pregnancy in chronically catheterized, conscious rats. *Kidney Int* 1986; **26**: 24–29.
63. Venuto RC, Donker AJM. Prostaglandin E2, plasma renin activity, and renal function throughout rabbit pregnancy. *J Lab Clin Med* 1982; **99**: 239–246.
64. Conrad KP, Colpoys MC. Evidence against the hypothesis that prostaglandins are the vasodepressor agents of pregnancy. *J Clin Invest* 1986; **77**: 236–245.
65. Kopp L, Paradiz G, Tucci JR. Urinary excretion of cyclic 3',5'-guanosine monophosphate during and after pregnancy. *J Clin Endocrinol Metab* 1977; **44**: 590–594.
66. Boccardo P, Soregaroli M, Aiello S *et al.* Systemic and fetal-maternal nitric oxide synthesis in normal pregnancy and pre-eclampsia. *Br J Obstet Gynaecol* 1996; **103**: 879–886.
67. Conrad KP, Joffe GM, Kruszyna H *et al.* Identification of increased nitric oxide biosynthesis during pregnancy in rats. *FASEB J* 1993; **7**: 566–571.
68. Danielson LA, Conrad KP. Acute blockade of nitric oxide synthase inhibits renal vasodilation and hyperfiltration during pregnancy in chronically instrumented conscious rats. *J Clin Invest* 1995; **96**: 482–490.
69. Baylis C, Engels K. Adverse interactions between pregnancy and a new model of systemic hypertension produced by chronic blockade of EDRF in rat. *Clin Exp Hypertens B* 1992; **11**: 117–129.
70. Bergstein JM, Hoyer JR, Michael AFJ. Glomerular fibrinolytic activity following endotoxin-induced glomerular fibrin deposition in the pregnant rat. *Am J Pathol* 1974; **75**: 195–202.
71. Lindheimer MD, Katz AI. Renal physiology and diseases in the kidney. In: *The Kidney: Physiology and Pathophysiology*, D. W. Seldin and G. Giebisch (Ed). Raven Press, New York, 1992: 3371–3431.
72. Domenech M, Gruppuso PA, Nishino VT *et al.* Preserved fetal plasma amino acid concentrations in the presence of maternal hypoaminoacidemia. *Pediatr Res* 1986; **20**: 1071–1076.
73. Raij L. Glomerular thrombosis in pregnancy: role of the L-arginine-nitric oxide pathway. *Kidney Int* 1994; **45**: 775–781.
74. Epstein M. Hepatorenal syndrome: emerging perspectives of pathophysiology and therapy. *J Am Soc Nephrol* 1994; **4**: 1735–1753.
75. Vallance P, Moncada S. Hyperdynamic circulation in cirrhosis: a role for nitric oxide? *Lancet* 1991; **337**: 776–778.
76. Clària J, Jiménez W, Ros J *et al.* Pathogenesis of arterial hypotension in cirrhotic rats with ascites: role of endogenous nitric oxide. *Hepatology* 1992; **15**: 343–349.

77. Niederberger M, Ginès P, Martin P-Y *et al.* Comparison of vascular nitric oxide production and systemic hemodynamics in cirrhosis versus prehepatic portal hypertension in rats. *Hepatology* 1996; **24**: 947–951.
78. Awolesi MA, Sessa WC, Sumpio BE. Cyclic strain upregulates nitric oxide synthase in cultured bovine aortic endothelial cells. *J Clin Invest* 1995; **96**: 1449–1454.
79. Campillo B, Chabrier PE, Pelle G *et al.* Inhibition of nitric oxide synthesis in the forearm arterial bed of patients with advanced cirrhosis. *Hepatology* 1995; **22**: 1423–1429.
80. Martin P-Y, Xu DL, Niederberger M *et al.* Upregulation of the endothelial constitutive NOS: a major role in the increased NO production in cirrhotic rats. *Am J Physiol* 1996; **270**: F494–F499.
81. Veglio F, Pinna G, Melchio R *et al.* Plasma endothelin levels in cirrhotic subjects. *J Hepatol* 1992; **15**: 85–87.
82. Asbert M, Ginès A, Ginès P *et al.* Circulating levels of endothelin in cirrhosis. *Gastroenterology* 1993; **104**: 1485–1491.
83. Leivas A, Jiménez W, Lamas S *et al.* Endothelin 1 does not play a major role in the homeostasis of arterial pressure in cirrhotic rats with ascites. *Gastroenterology* 1995; **108**: 1842–1848.
84. Morrow JD, Hill KE, Burk RF *et al.* A series of prostaglandin F2-like compounds are produced *in vivo* in humans by a non-cyclooxygenase, free radical-catalyzed mechanism. *Proc Natl Acad Sci USA* 1990; **87**: 9383–9387.
85. Morrow JD, Moore KP, Awad JA *et al.* Marked overproduction of non-cyclooxygenase derived prostanoids (F2-isoprostanes) in the hepatorenal syndrome. *J Lipid Mediat* 1993; **6**: 417–420.
86. Morrow JD, Awad JA, Kato T *et al.* Formation of novel non-cyclooxygenase-derived prostanoids (F2-isoprostanes) in carbon tetrachloride hepatotoxicity. An animal model of lipid peroxidation. *J Clin Invest* 1992; **90**: 2502–2507.
87. Takahashi K, Nammour TM, Fukunaga M *et al.* Glomerular actions of a free radical-generated novel prostaglandin, 8-epi-prostaglandin $F_{2\alpha}$, in the rat. Evidence for interaction with thromboxane A_2 receptors. *J Clin Invest* 1992; **90**: 136–141.
88. Schroeder ET, Shear L, Sancetta SM, Gabuzda GJ. Renal failure in patients with cirrhosis of the liver. 3. Evaluation of intrarenal blood flow by para-aminohippurate extraction and response to angiotensin. *Am J Med* 1967; **43**: 887–896.
89. Epstein M, Schneider N, Befeler B. Relationship of systemic and intrarenal hemodynamics in cirrhosis. *J Lab Clin Med* 1977; **89**: 1175–1187.
90. Schrier RW, Arroyo V, Bernardi M *et al.* Peripheral arterial vasodilation hypothesis: a proposal for the initiation of renal sodium and water retention in cirrhosis. *Hepatology* 1988; **8**: 1151–1157.
91. Conger J, Robinette J, Villar A *et al.* Increased nitric oxide synthase activity despite lack of response to endothelium-dependent vasodilators in postischemic acute renal failure in rats. *J Clin Invest* 1995; **96**: 631–638.

92. Rivas-Cabanero L, Montero A, Lopez-Novoa JM. Increased glomerular nitric oxide synthesis in gentamicin-induced renal failure. *Eur J Pharmacol* 1994; **270**: 119–121.
93. Rivas-Cabanero L, Rodriguez-Barbero A, Arevalo M, Lopez-Novoa JM. Effect of N^G-nitro- L-arginine methyl ester on nephrotoxicity induced by gentamicin in rats. Nephron 1995; **71**: 203–207.
94. Maree A, Peer G, Schwartz D *et al.* Role of nitric oxide in glycerol-induced acute renal failure in rats. *Nephrol Dial Transplant* 1994; **9 (Suppl 4)**: 78–81.
95. Schramm L, Heidbreder E, Lopau K *et al.* Influence of nitric oxide on renal function in toxic acute renal failure in the rat. *Miner Electrolyte Metab* 1996; **22**: 168–177.
96. De Nicola L, Thomson SC, Wead LM *et al.* Arginine feeding modifies cyclosporine nephrotoxicity in rats. *J Clin Invest* 1993; **92**: 1859–1865.
97. Chintala MS, Chiu PJ, Vemulapalli S *et al.* Inhibition of endothelial derived relaxing factor (EDRF) aggravates ischemic acute renal failure in anesthetized rats. *Naunyn-Schmiedebergs Arch Pharmacol* 1993; **348**: 305–310.
98. Bone RC. The pathogenesis of sepsis. *Ann Intern Med* 1991; **115**: 457–469.
99. Shultz PJ, Tayeh MA, Marletta MA, Raij L. Synthesis and action of nitric oxide in rat glomerular mesangial cells. *Am J Physiol* 1991; **261**: F600–F606.
100. Beasley DJ, Schwartz JH, Brenner BM. Interleukin-1 induces prolonged L-arginine dependent cyclic guanosine monophosphate and nitric production in rat vascular smooth muscle cells. *J Clin Invest* 1991; **87**: 602–608.
101. Kilbourn RG, Jubran A, Gross SS *et al.* Reversal of endotoxin-mediated shock by N^G-methyl-L-arginine, an inhibitor of nitric oxide synthesis. *Biochem Biophys Res Commun* 1990; **172**: 1132–1138.
102. Petros A, Bennett D, Vallance P. Effect of nitric oxide synthase inhibitors on hypotension in patients with septic shock. *Lancet* 1991; **338**: 1557–1558.
103. Shultz PJ, Raij L. Endogenously synthesized nitric oxide prevents endotoxin-induced glomerular thrombosis. *J Clin Invest* 1992; **90**: 1718–1725.
104. Westberg G, Shultz PJ, Raij L. Exogenous nitric oxide prevents endotoxin-induced glomerular thrombosis in rats. *Kidney Int* 1994; **46**: 711–716.
105. del Castillo D, Agarwal A, Jaimes EA, Raij L. Glomerular and vascular tissues do not down-regulate nitric oxide synthesis during protracted endotoxemia. *Kidney Int* 1997; **52**: 460–467.
106. Beckman JS, Beckman TW, Chen J *et al.* Apparent hydroxyl radical production by peroxynitrite: implications for endothelial injury from nitric oxide and superoxide. *Proc Natl Acad Sci USA* 1990; **87**: 1620–1624.
107. Jaimes EA, Nath K, Raij L. Hydrogen peroxide downregulates IL-1 stimulated iNOS activity in mesangial cells: implications for glomerulonephritis. *Am J Physiol* 1997; **272**: F721–F728.

108. Campese VM. Salt sensitivity in hypertension. Renal and cardiovascular implications. *Hypertension* 1994; **23**: 531–550.
109. Konishi M, Su C. Role of endothelium in dilator responses of spontaneously hypertensive rat arteries. *Hypertension* 1983; **5**: 881–886.
110. Raij L, Lüscher TF, Vanhoutte PM. High potassium diet augments endothelium-dependent relaxations in the Dahl rat. *Hypertension* 1988; **12**: 562–567.
111. Lockette W, Otsuka Y, Carretero O. The loss of endothelium-dependent vascular relaxation in hypertension. *Hypertension* 1986; **8**: II61–II66.
112. Lüscher TF, Vanhoutte PM. Endothelium-dependent contractions to acetylcholine in the aorta of the spontaneously hypertensive rat. *Hypertension* 1986; **8**: 344–348.
113. Hayakawa H, Hirata Y, Suzuki E *et al.* Mechanisms for altered endothelium-dependent vasorelaxation in isolated kidneys from experimental hypertensive rats. *Am J Physiol* 1993; **264**: H1535–H1541.
114. Hayakawa H, Hirata Y, Suzuki E *et al.* Long-term administration of L-arginine improves nitric oxide release from kidney in deoxycorticosterone acetate-salt hypertensive rats. *Hypertension* 1994; **23**: 752–756.
115. Hayakawa H, Hirata Y, Suzuki E *et al.* Endothelium-derived relaxing factors in the kidney of spontaneously hypertensive rats. *Life Sci* 1995; **56**: PL401–PL408.
116. Campese VM, Tawadrous M, Bigazzi R *et al.* Salt intake and plasma atrial natriuretic peptide and nitric oxide in hypertension. *Hypertension* 1996; **28**: 335–340.
117. Baylis C, Mitruka B, Deng A. Chronic blockade of nitric oxide synthesis in the rat produces systemic hypertension and glomerular damage. *J Clin Invest* 1992; **90**: 278–281.
118. Raij L, Azar S, Keane WF. Role of hypertension in progressive glomerular immune injury. *Hypertension* 1985; **7**: 398–404.

Part 3

Clinical Implications of Nitric Oxide

Chapter 17

Organ Ischaemia-reperfusion Injury: The Role and Therapeutic Potential of Nitric Oxide

Alastair G Stewart and Jane E Barker

Introduction

Myocardial and cerebral infarction, organ transplantation and replantation of amputated digits represent some of the more significant forms of clinically encountered ischaemia-reperfusion (I-R) injury. In addition, when a large mass of tissue is affected by ischaemia, multiple organs may be secondarily affected upon reperfusion of the ischaemic mass, resulting in multiple organ failure and possibly death. Other more chronic forms of I-R injury, such as peripheral vascular disease, have a different pathogenesis and will not be considered further in this review.

Acute I-R injury occurs when blood flow to a tissue is compromised for a critical period of time, which varies according to the organ and the prevailing conditions. Inadequate blood flow can result from vascular thrombosis or vasospasm, or systemic problems such as arterial hypotension. In surgery, occlusion of vessels may result from external mechanical compression of blood vessels by kinking, haematoma, excessively tight wound closure or pressure from a wound drain.[1] Irrespective of the cause, there are three theoretical approaches to the treatment of the I-R injury. The first is the use of prophylactic drugs or preconditioning to diminish the impact of either the ischaemia or the reperfusion. The second is early detection and physical restoration of blood flow prior to the development of irreversible damage by replantation surgery of severed body parts or thrombolysis to remove an occlusive blood clot. The third approach is to treat the incipient reperfusion injury with pharmacological agents when blood flow to the affected tissue is restored. It is generally accepted that following ischaemia of limited duration there is a component of the ensuing inflammatory response that, if successfully treated, will diminish the ultimate extent of necrosis or degree of permanent functional impairment. Over the last 5–6 years interest in the therapeutic potential of manipulation of nitric oxide (NO) levels has escalated, but as yet, practical applications are restricted to the use of inhaled NO in lung injury.[2]

The beneficial effects of NO as a vasodilator or anti-thrombotic agent have been recognised in conditions such as angina and atherosclerosis.[3] More recently, it has become equally evident that excessive production of NO may be deleterious in many pathological processes and that this overproduction may even play a dominant role in some conditions (e.g. septic shock). The use of selective inhibitors of NO synthesis, NO donors and transgenic mice deficient in one of the NOS isoforms is providing substantial evidence for a complex role of NO in I-R injury. However, as the specific role and enzymatic derivation (constitutively expressed NOS, inducible NOS or non-enzymatic) of NO in various forms of I-R injury appear to be dependent on factors such as species, organ, duration and origin of the injury, the development of therapeutic strategies based on manipulation of NO levels awaits the further characterization of the precise role(s) played by NO.

This chapter summarizes the role(s) of NO in I-R injury by describing the sequence of events in the injury and the associated perturbations in endogenous NO production. The influence of NO on the pathogenesis of I-R injury has been simplified, when possible, to indicate beneficial or deleterious effects. Organ-specific aspects of the influence of NO or NOS inhibitors have been highlighted. Finally, the therapeutic potential of manipulating NO levels in I-R injury is discussed.

Ischaemia-reperfusion Injury

Sequence of Events in I-R Injury

Ischaemia

Ischaemia is characterized by a reduction in blood flow which results in the failure of metabolic homeostasis in the affected tissue (Figures 1 and 2). Decreased delivery of oxygen and glucose leads to a conversion of cellular metabolism to anaerobic pathways utilising intrinsic substrates. More prolonged oxygen deficiency

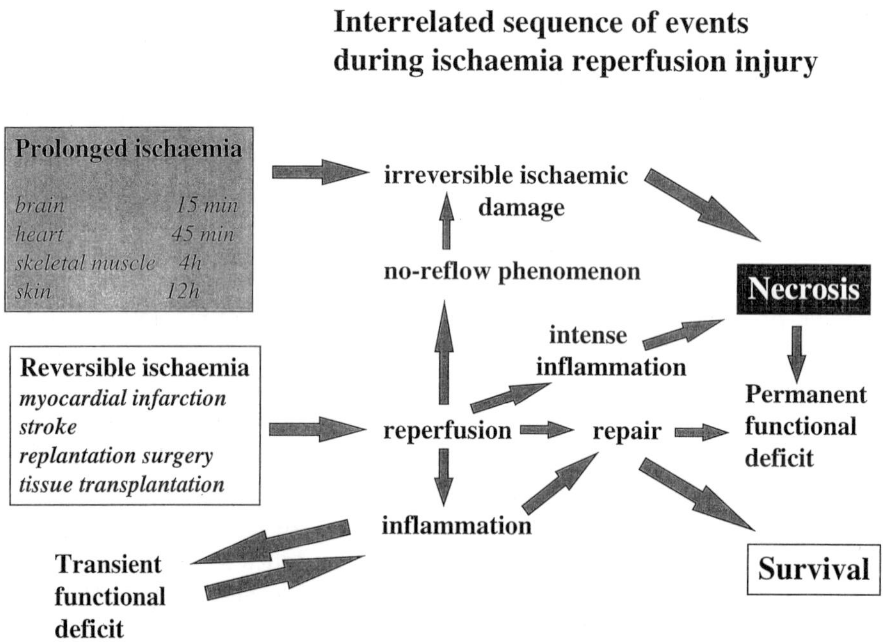

Figure 1. The outcome of ischaemia and reperfusion is related to the duration of the initial ischaemia, with irreversible damage occurring in the brain and heart after much shorter periods than in tissues such as skin and skeletal muscle. Reversible ischaemia is followed by inflammation during the reperfusion phase which, if severe, may lead to necrosis. Reperfusion is important for the repair process (fibrosis) which may nevertheless leave a permanent functional deficit.

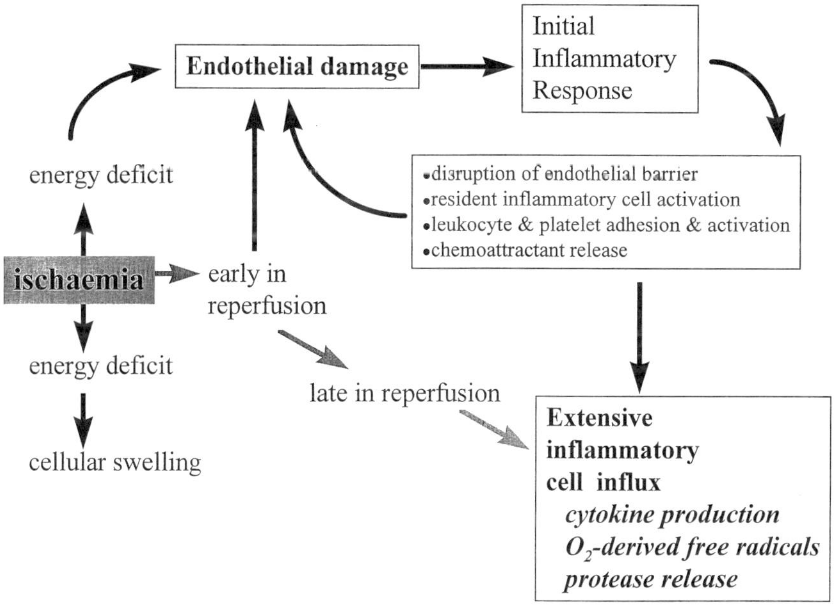

Figure 2. Ischaemia causes an energy deficit, which is sufficient to induce endothelial damage, initiating an inflammatory response that occurs later in reperfusion. Ischaemia itself causes cell swelling and may be cytotoxic without tissue reperfusion. Early in reperfusion, the endothelial damage is increased by production of reactive oxygen intermediates and by an initial inflammatory response caused by release of other mediators from resident and rapidly recruited inflammatory cells. Later in reperfusion, the leucocyte influx greatly amplifies the inflammatory response in what is likely to constitute a positive feedback.

leads to an increased local concentration of lactic acid, acidosis, and exhaustion of high energy phosphates (ATP, phosphocreatine). Consequently, cytoplasmic metabolites accumulate and energy- and pH-dependent membrane transport systems malfunction, resulting in a marked increase in intracellular Ca^{2+} concentration. Ca^{2+} overload may, if prolonged, lead to mitochondrial membrane dysfunction and irreversible damage.[4] Elevated intracellular Ca^{2+} also activates various enzymes, including Ca^{2+}-dependent isoforms of NOS and proteases.

Reperfusion

The restoration of blood flow following ischaemia replenishes energy levels and allows the removal of accumulated toxic metabolites. Upon reperfusion, however, there is paradoxically a continued, and often intensified tissue injury.[5] Many studies have implicated oxygen-derived free radicals, such as the superoxide anion ($O_2^{\bullet -}$), hydrogen peroxide and the hydroxyl radical in the pathophysiology of the reperfusion injury.[1] Reactive oxygen intermediates (ROI) lead to lipid peroxidation, resulting in structural and functional cell damage. ROI have also been implicated in the early complement activation in I-R injury.[6] A variety of microvascular alterations occurs, including endothelial cell swelling and increased capillary permeability. In addition to these direct physical responses of the endothelium to I-R injury, further increases in intracellular Ca^{2+} concentration activate non-specific proteases and phospholipases, leading to synthesis of pro-inflammatory mediators, including platelet-activating factor (PAF), leukotriene B_4 (LTB_4) and thromboxane A_2 (TxA_2).[1] Resident inflammatory cells, such as mast cells, and the parenchymal cells are an additional source of pro-inflammatory mediators, whereas later in reperfusion, infiltrating leucocytes may constitute a more significant source.

Inflammation

In response to chemotactic inflammatory mediators, neutrophils adhere to the local venular endothelium, transmigrate to the extravascular space, accumulate in the affected area, degranulate and promote the development of oedema.[7] The mechanisms of neutrophil-induced tissue damage include microvascular plugging, production of free radicals and the release of vasoactive mediators. In addition to the acute chemoattractant and vasoactive effects of PAF and LTB_4, these mediators amplify the inflammatory response by augmenting the more delayed production of peptide mediators such as tumour necrosis factor (TNF)-α and interleukin (IL)-1, the primary stimulus for which may be endothelial hypoxia and reoxygenation.[8] However, the exact mechanisms underlying production and the cellular sources of cytokines in I-R injury have yet to be fully elucidated.

No-reflow phenomenon

Restoration of an arterial blood supply does not necessarily ensure that the entire mass of the ischaemic tissue will be reperfused. In some regions flow may not return at all; in others it may be suboptimal. This no-reflow phenomenon[9] has been observed in many organs and in different species and may be a major contributor to the ultimate tissue damage when flow to a particular zone resumes. The principal mechanisms associated with the phenomenon include endothelial and tissue cell swelling, narrowing of the blood vessel lumen, vasospasm and plugging of capillaries with thrombi.[9] The no-reflow phenomenon extends the ischaemia in the affected tissue to a duration which is itself sufficient to cause cell death or to intensify inflammatory damage when blood flow does eventually resume.

Organ-specific Factors Determining the Severity of I-R Injury

Parenchymal cells

Following I-R injury, it is generally assumed that tissue capacity to survive will depend on the functional properties of the parenchymal cells. Thus, the metabolically more active cells expend energy reserves faster and therefore have a lower tolerance to ischaemia. This is clearly evident from the marked differences in organ sensitivity to I-R injury (Figure 1). For example, skeletal muscle, because of its more active metabolism, tolerates a shorter period of ischaemia compared with skin. The corollary of this is the use of hypothermia to retard cellular metabolism and increase tissue tolerance to ischaemia. Other aspects of cellular biochemistry, such as antioxidant status, enzyme substrate and cofactor availability in the parenchymal cell are also likely to be crucial determinants of the intracellular events after I-R.

Endothelial function

In addition to organ-specific properties of parenchymal cells, the physiological response of the endothelium lining the local capillary and post-capillary venular vessels is an important determinant of the severity of I-R injury.[10] Dysregulation of the endothelial influence on vascular tone induced by I-R injury may contribute to adverse clinical sequelae. Post-ischaemic reduction of endothelium-dependent vasorelaxation has been shown in small vessels, larger vessel ring segments,[11] intact coronary[10] and renal vasculature and in the rat hind limb.[12] Oxidative changes resulting in $O_2^{\cdot-}$ generation are implicated in hypoxia-induced decreases in the thromboresistance of the endothelium,[13] which may make a particular contribution to stasis of venous blood flow following I-R injury.

Early in reperfusion there is an increase in the stickiness of endothelium for neutrophils that is mediated by mobilisation of P-selectin from Weibel-Palade bodies and by PAF expression on the endothelial cell surface.[14] There is also a delayed increase in expression of intercellular adhesion molecule (ICAM)-1 and E-selectin which can be simulated by subjecting cultured endothelial cells to hypoxia/reoxygenation.[15] Nevertheless, whilst there has been a general acceptance of a central role for neutrophil infiltration in the pathogenesis of I-R injury, several studies now question the importance of neutrophil adherence and tissue infiltration in more severe I-R injury.[16,17] Indeed, neutrophil infiltration is likely to play a key role in the removal of cellular debris accumulating from partial necrosis under such circumstances.

Measuring the Severity of I-R Injury

Studies of I-R injury use a wide range of assays and parameters to determine the severity and investigate the protective potential of drugs and other interventions.

These parameters include oedema, increased vascular permeability, infiltration of inflammatory cells, parenchymal cell function (e.g. muscle contraction) and cell survival measured by membrane integrity, mitochondrial enzyme activity or morphometrically. The choice and timing of these measurements may well be important in the conclusions reached and their relevance to the ultimate level of functional deficit. For example, studies using early oedema responses as an index have implicated neutrophils in injury, whereas measurements of tissue viability do not support a dominant role for neutrophils.[16,17]

Nitric Oxide Production During Ischaemia-reperfusion Injury

The biosynthesis and fate of NO are covered in detail in Chapters 3 & 6. In brief, three distinct genes for NOS producing 3 forms of the enzyme have been identified: neuronal (nNOS) and endothelial (eNOS), both of which are constitutively expressed and Ca^{2+}-dependent and are collectively referred to as cNOS; and the inducible isoform (iNOS), which is Ca^{2+}-independent and is widely distributed in inflammatory and non-inflammatory cell types following exposure to highly specific stimuli.[18] iNOS produces NO at levels 2–3 orders of magnitude greater than cNOS, which may in part explain many of the differences in biological activity of NO produced by iNOS compared with cNOS. The stimuli that induce iNOS expression, including IL-1 and TNF-α, also activate inflammatory cell production of $O_2^{\cdot -}$. The simultaneous production of NO and $O_2^{\cdot -}$ in I-R injury may lead to peroxynitrite formation.[19]

Increased NO Production

Activation of Ca^{2+}-dependent NOS

The mechanism of NO production during ischaemia and its role(s) are not clear, especially in view of the requirement for oxygen and other co-factors that are lowered in concentration during ischaemia (Table 1). Early in reperfusion, increased intracellular Ca^{2+} activates eNOS, but this is short-lived, and in 2–5 minutes NO-mediated endothelium-dependent dilatation is impaired.[20] In addition to damage-induced cellular Ca^{2+} entry, a wide range of mediators implicated in I-R injury, including bradykinin, tachykinins, serotonin, PAF and histamine have the potential to stimulate eNOS *via* receptor-mediated increases in intracellular Ca^{2+}. cNOS has been described in mast cells,[21] which are activated during reperfusion[22] and may therefore be an additional source of NO.

During cerebral ischaemia, NO reaches micromolar concentrations. N-methyl-D-aspartate (NMDA) receptor-mediated increases in intracellular Ca^{2+} are implicated as the major stimulus to NO production, since levels of excitatory amino acids, glutamate and aspartate increase during I-R,[23] suggesting that NO

Table 1. Influence of ischaemia on cofactors for NO production/availability.

Factor	Change during IR injury	Predicted effect on NO levels	Observed effect on NO in IR injury
Ca^{2+}	Increase	↑ NO *via* ↑ cNOS activity	Ca^{2+} channel blockers ↓ NOS activity protect against cerebral IRI
tetrahydrobiopterin (BH4)	↓ availability	↓ NO	synthetic BH_4 restores eNOS activity.[112]
L-Arginine	Decrease (inferred)	↓ NO	L-arginine supplementation beneficial (pre-ischaemically)
Glutathione	Decrease	↓ NO	N-acetyl cysteine ↑ glutathione protective in variety of IRI models
pH	Decrease	↓ NO, eNOS, optimum pH 7.5[113] ↑ NO, nNOS, optimum pH 6.7 ↑ NO *via* nitrite/nitrate reduction	deleterious to IR injury ? deleterious to IR injuty
Oxygen	Decrease	↓ NO *via* NOS activity[114]	variable

is involved in glutamate-mediated neurotoxicity.[24] Conversely, the selective nNOS inhibitor, 7-nitroindazole, decreases the release of excitatory neurotransmitters and gamma-aminobutyric acid (GABA),[25] as does local, but not systemic, treatment with the cNOS inhibitor, N^G-nitro-L-arginine methyl ester (L-NAME).[26] These observations suggest a positive feedback between NO and amino acid transmitter release in cerebral I-R injury. NO may also ameliorate cerebral ischaemia by inhibiting NMDA receptors. The nitrosonium (NO^+) redox form of NO reacts with the thiol group of the redox modulatory site of the NMDA receptor, blocking receptor-mediated increases in intracellular Ca^{2+} and NOS activation.[27]

NO, or a closely related factor, is the neurotransmitter of the nitrergic nervous system supplying the intestine, retina, adrenal medulla and blood vessels.[28] It is not known whether excessive production of NO leads to neurotoxicity in the periphery, but nNOS activity is associated with protection against ventricular fibrillation during reperfusion of rat isolated ischaemic hearts.[29]

Activation of Ca^{2+}-independent NOS

A wide range of cell-type specific stimuli for iNOS induction have been characterized, but the most active of these implicated in I-R damage are TNF-α and IL-1. Similarly, of the many cell types known to express iNOS following appropriate cytokine activation, neutrophils, macrophages, mast cells, endothelium and vascular smooth muscle cells have been documented as expressing iNOS in the reperfusion phase. Inducible NOS has been demonstrated following I-R injury to skeletal muscle,[30] brain,[31] retina and heart.[32] It may be of significance that glutathione, which is generally accepted as being cytoprotective in I-R injury, is required for optimal induction of iNOS in liver.[33]

It is likely that iNOS rather than cNOS plays an important part in the oedema and inflammation in I-R injury, since iNOS induction is delayed (hours — days), and the activity persists, whereas cNOS activity decreases during the early stages of reperfusion after ischaemia.[30] Although the oedema and leucocyte infiltration commence during the early stages of reperfusion, in most tissues these responses do not reach a maximal level until some hours later, coinciding with the synthesis of iNOS.

Non-enzymatic production of NO in I-R injury

Until recently it was accepted that NO was produced enzymatically from L-arginine in biological systems, although it has long been known that acidification of nitrite solutions could generate NO. However, studies on myocardial I-R injury using direct measurement of NO by spin trapping techniques provide clear evidence of NO production that is not inhibited by L-arginine analogue inhibitors of NOS.[34] Moreover, the low pH environment in post-ischaemic tissue is conducive for production of NO from nitrite. The importance of this pathway of NO formation for the pathogenesis of I-R injury has yet to be established. We have recently obtained evidence that the non-enzymatic NO production in post-ischaemic skeletal muscle is inhibited by the anti-inflammatory steroid dexamethasone[35] which has a more extended period of efficacy than NOS inhibition in the treatment of I-R injury when administered after reperfusion has commenced.[30]

Decreased NO Production

Early in reperfusion of the ischaemic myocardium, vasodilator responses to acetycholine are decreased, but the vascular responses to endothelium-independent vasodilators such as $NaNO_2$ are preserved,[11,36,37] suggesting that the impairment is more likely to be due to diminished release or availability of NO than to reduced capacity of vascular smooth muscle to relax. This observation, which has also been observed in other microcirculatory beds, such as the post-ischaemic hindlimb[38] and the brain[39], may be due to inactivation of NO by $O_2^{\cdot -}$. since both endothelial cells and infiltrating leucocytes produce $O_2^{\cdot -}$.[40] which is known to scavenge NO.[41] Furthermore, superoxide dismutase (SOD), which inactivates $O_2^{\cdot -}$. preserves vascular reactivity to acetylcholine,[36] although this may be explained by the preservation of endothelial integrity rather than by diminished neutralization of NO. Neutrophils release a factor spontaneously or upon stimulation with TNF-α which contracts isolated endothelium-intact coronary arteries by decreasing NO-dependent relaxation.[42] Release of this factor from activated neutrophils during I-R injury may contribute to the compromised dilator responses early in reperfusion. Serum nitrite + nitrate concentrations, measured as an index of NO release, decrease by 98% in the venous effluent of the microcirculation of the ischaemic small intestine after 30 minutes of reperfusion[43] providing evidence that diminished production of NO is the major cause of loss of endothelium-dependent responses. In the skeletal musculature of the rat, cNOS activity levels in cell-free homogenates decrease to less than 20% of the control level 8 hours after reperfusion,[30] consistent with diminished production as the cause of the impaired vasodilator responses in severe ischaemia. In rabbit skeletal muscle, such injury may occur after only 10 minutes of reperfusion.[44] Immunohistochemical studies on reperfused rat skeletal muscle also indicate that eNOS immunoreactivity is lost from all but the largest vessels after 8 hours of reperfusion.[45] TNF-α decreases the stability of eNOS mRNA and may therefore play a role in the observed decrease in eNOS levels.[46]

Studies of vascular responsiveness that have focused on short-term changes to endothelial activation have generally shown that directly acting vasodilators retain activity. Endothelial production of peroxynitrite ($ONOO^-$)[47] has acute, albeit rapidly tachyphylactic vasodilator activity[48] and may subsequently non-specifically impair vasodilator mechanisms and predispose to vasospasm,[48,49] resulting in exacerbation of the no-reflow phenomenon. In the gut, these actions appear to be outweighed by the ability of $ONOO^-$ to reduce leucocyte/endothelial adhesion *in vivo* which protects from I-R injury.[50]

Regulation of NOS activity in I-R injury

There are two direct negative feedback mechanisms for NOS activity: the ability of NO to reduce intracellular Ca^{2+} levels attenuates activity of the cNOS isoforms; NO competes with molecular oxygen for attachment to the haem moiety of all isoforms of NOS.[51] In addition, NO may limit the induction of iNOS by interfering with the interaction of the nuclear transcription factor NFκB with the promotor region of the iNOS gene.[52] The cofactors involved in NOS activity, particularly the availability of $\cdot O_2$, are also influenced by ischaemia (Table 1).

Factors Determining the Availability of NO

NO persists for approximately 10 seconds in aqueous solution, depending on the pH and oxygen concentration. It may persist for much longer periods in tissue, especially in the absence of O_2. Under normal physiological conditions, it is rapidly bound to oxyhaemoglobin in the vascular compartment,[53] although the nature of this binding may itself be influenced by local O_2 concentrations (see Chapters 2 and 6). As discussed above, in tissue rendered ischaemic for prolonged periods, NO may also be produced non-enzymatically and persist for much longer periods than currently appreciated.[34,35] The fate of NO produced during I-R injury is therefore highly dependent on the cellular milieu and its level of production. A large proportion of NO is oxidized to nitrite and nitrate. The loss of an electron from NO results in the formation of the nitrosonium ion (NO^+). In the presence of $O_2^{\cdot -}$ and low pH, conditions predominating in I-R injury, NO is likely to combine with $O_2^{\cdot -}$ to form $ONOO^-$,[19] which undergoes protonation to species with the chemical characteristics of the hydroxyl radical.[54] Peroxynitrite directly nitrosates proteins to form nitrotyrosine residues, providing a 'footprint' that can be detected immunohistochemically;[55] it may also combine with glucose and other sugars with an alcohol functional group or with thiols such as glutathione.[56] Nitrosothiols, particularly nitrosoglutathione, share some of the biological activities of NO, but are more stable, allowing diffusion over greater distances and more persistent effects, and are cardioprotective when added to cardioplegic solutions.[57] In contrast, glutathione depletion enhances both cerebral ischaemic damage *in vivo* and the toxic effects of $ONOO^-$ *in vitro*, suggesting that the reaction of NO or $ONOO^-$ with glutathione limits toxicity.[58]

Table 2. Influence of NO on key cellular targets in IR injury.

Target	Influence of NO		
	Beneficial	**Detrimental**	**Uncertain/Variable**
Endothelium	preservation of endothelium-dependent dilator response ↓ adhesion molecule expression ↓ inter-endothelial gap formation ↓ endothelin production	lipid peroxidation ↑	↑ cyclo-oxygenase activity
Platelet	↓ adhesion and activation		
Leukocyte	↓ adhesion and activation	↑ formation of $ONOO^-$	
Vascular smooth muscle	relaxation	↑ blood flow to facilitate reperfusion ↑ blood flow enhances plasma leak tissue rewarming & provision of O_2 for formation of ROIs	redistribution of blood flow
Skeletal muscle	inhibition of contractility → energy conservation		
Cardiac muscle		inhibition of contractility hypotension → exacerbation of ischaemia	
Neurones	↓ release of excitatory neurotransmitters		

Influence of Nitric Oxide on Ischaemia-reperfusion Injury

The influence of NO on I-R injury is complicated, time-dependent and in some aspects, variable between different tissues (Table 2).

Potentially Beneficial Effects

NO stimulates the activity of soluble guanylate cyclase (sGC), causing accumulation of guanosine 3',5'-cyclic monophosphate (cGMP), resulting in relaxation of smooth muscle (see Chapter 4), decreased cardiac contractility and reduced platelet and leucocyte activation.[59] The effects of transient increases in NO on sGC activity are short-lived, with vascular relaxant responses lasting only a few minutes in response to bolus exposure to NO.

Vasodilatation

Endothelium-dependent relaxation of subjacent vascular smooth muscle is largely due to blood flow-induced shear stress causing continuous release of NO.[60] Such shear stress is diminished or absent during ischaemia. Interestingly, shear stress increases the amount of eNOS protein as well as its activity.[61] The vasodilator response to cNOS-dependent NO releasing agents (histamine, serotonin, etc.) early in reperfusion may be important in restoring blood flow and opposing the vasospastic component of the no-reflow phenomenon. There are a large number of studies documenting the protective effects of NO donors on subsequent I-R injury, possibly as a result of their dilator activity.[62] Early in reperfusion, activation of cNOS may increase the calibre of flow-limiting resistance vessels, facilitating blood flow.[63] In rat hindlimb I-R injury, pre-ischaemic administration of the NOS inhibitor L-NAME exacerbated I-R injury, perhaps due to vasospasm in the absence of NO dilator tone. In contrast, muscle survival was improved by iNOS inhibition using dexamethasone administered after the start of ischaemia.[30]

Anti-inflammatory effects

NO alone decreases both platelet aggregation and adhesion and shows synergy in these effects with the anti-platelet, vasorelaxant endothelial mediator, prostacyclin.[64] Peroxynitrite also inhibits platelet activation.[56] Neutrophil adhesion to endothelium is decreased by NO[65] and $ONOO^-$,[50] possibly by reducing expression of adhesion molecules such as P-selectin and ICAM-1.[20] High concentrations of NO decrease neutrophil release of $O_2^{\bullet-}$ both through neutralization of $O_2^{\bullet-}$ and by elevation of cGMP-regulating intracellular signalling pathways for nicotinamide adenine dinucleotide phosphate (NADPH) oxidase activation on the neutrophil membrane surface.[66] The contribution of thrombosis and oedema to the no-reflow phenomenon is likely to be decreased by the anti-platelet and endothelial effects of NO in I-R injury, but direct evidence is as yet unavailable. Nevertheless, in skeletal muscle, inhibition of NOS by local infusion of N^5-(1-iminoethyl)-L-ornithine (L-NIO) reduced subsequent necrosis, despite producing a marked decrease in post-ischaemic blood flow.[67]

Mast cell activation in post-ischaemic mesentery is inhibited by NO donors.[22] Mast cells produce NO through cNOS, and inhibition of degranulation by both endogenous and exogenous NO is well documented.[21] We have recently obtained immunohistochemical evidence of iNOS in rodent mast cells.[45] The functional significance of the mast cell iNOS is not yet clear, but the high and persistent level of NO production by these cells may well have different effects on the degranulation response compared with those of NO produced *via* the cNOS pathway. The effects of $ONOO^-$ on mast cell function have yet to be investigated. Further discussion of mast cells and NO in the intestine is given in Chapter 15.

NO may also have a general anti-inflammatory effect through inhibition of the activation of NFκB,[68] which regulates expression of a wide range of pro-inflammatory cytokines and adhesion molecules implicated in I-R injury.

Glucose utilization

Simulation of brain ischaemia by inhibition of mitochondrial ATP synthesis produces a marked increase in lactate production by astrocytes, but not by neurones[69] and the resulting increase in glycolysis maintains ATP levels. In cultured astrocytes, inhibition of the mitochondrial respiratory chain by iNOS-derived NO causes a switch to glycolysis to maintain cellular ATP levels.[70] Suppression of 2-deoxyglucose uptake into skeletal muscle by treatment with NOS inhibitors suggests that NO promotes glucose transport and utilization.[71] These observations are compatible with the suggestion that during ischaemia and early in reperfusion NO may facilitate maintenance and recovery of ATP levels by enhancing glucose utilization (Figure 3).

Potentially Detrimental Effects

There are multiple mechanisms for the cytotoxic effects of NO. Many of the toxic reactions have long lasting effects as they involve covalent modification of macromolecules requiring *de novo* synthesis for recovery of function. The direct actions of NO and its products include nitrosylation of iron-sulphur groups and tyrosine residues, inactivation of haem centres in proteins, lipid peroxidation[72] and the formation of toxic products, including $ONOO^-$ and nitrogen dioxide.[54]

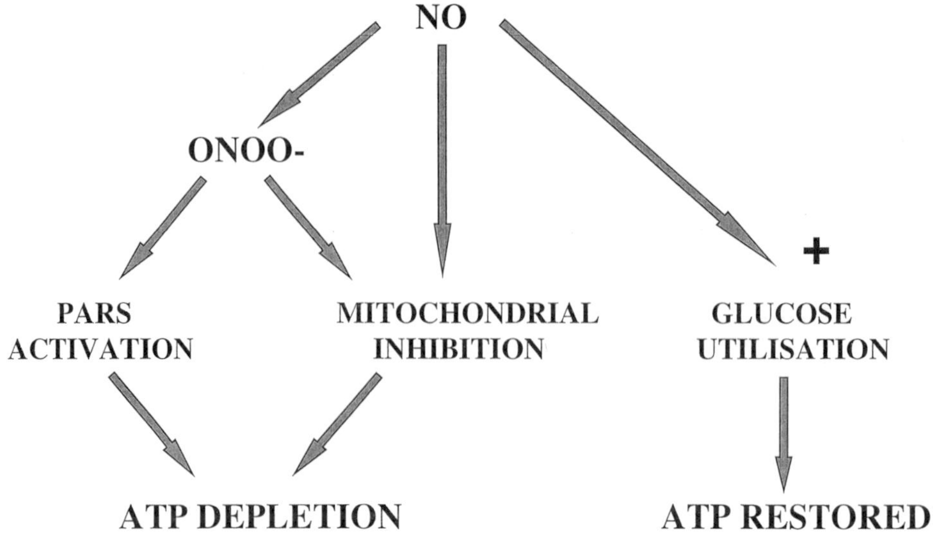

Figure 3. NO acts reversibly to inhibit mitochondrial oxidative metabolism, decreasing ATP levels, whereas promotion of glucose uptake tends to maintain ATP levels. Higher levels of NO production in proximity to the mitochondria result in $ONOO^-$ production which irreversibly inhibits mitochondrial function and activates the enzyme poly (ADPribose) synthetase (PARS) that also depletes ATP.

Peroxynitrite formation

The reaction between NO and $O_2^{\bullet -}$ has been considered as an anti-oxidant effect of NO on $O_2^{\bullet -}$ levels. However, the peroxynitrite ($ONOO^-$) formed from this reaction is a powerful oxidant and responsible for many of the toxic effects of NO, including persistent inhibition of mitochondrial function.[53] Peroxynitrite-mediated inhibition of glutathione reductase[70,73] or direct interaction with glutathione depletes glutathione, rendering post-ischaemic tissue more susceptible to oxidative stress. In I-R injury, proteolytic tissue destruction by $ONOO^-$ may be further increased by inactivation of tissue inhibitor of metalloproteinase (TIMP), removing the restraint on metalloproteinase activity.[74] Endothelial cell activation by bradykinin or elevation of intracellular Ca^{2+} by A23187 elicits $ONOO^-$ formation.[75] Endothelial generation of NO may therefore have adverse effects[47] in I-R injury, in which endothelial $O_2^{\bullet -}$ generation is well documented.[1] Sustained production of NO by iNOS has been most commonly implicated in $ONOO^-$ formation. This is of particular relevance to I-R injury, as the stimuli for iNOS induction also induce NADPH oxidase activity in inflammatory cells, resulting in simultaneous production of large amounts of NO and $O_2^{\bullet -}$. Peroxynitrite has established cytotoxic effects in I-R injury to isolated perfused rat heart[76,77] and ischaemic lung.[78]

Impairment of vital cell functions

DNA damage

Exposure of DNA to NO leads to damage by deamination and direct inhibition of ribonucleotide reductase, the rate-limiting enzyme for DNA synthesis.[79] Such damage is known to activate p53, a transcription factor implicated in triggering apoptosis. There is now evidence for limited amounts of apoptosis in several forms of I-R injury,[80] especially in the CNS. In macrophage and pancreatic cell lines, NO-induced apoptosis has been linked to induction of p53[81].

NO-mediated damage also activates poly(ADP-ribose) synthetase (PARS), a nuclear enzyme identified in pancreatic islet cells and neurones.[82] Activation of the PARS pathway results in massive depletion of NAD and ATP and eventually induces irreversible cytotoxicity and cell death. In vascular smooth muscle cells, this process leads to suppression of mitochondrial respiration and suppression of contractility.[83] PARS inhibitors reduce the extent of I-R injury to rabbit myocardium and skeletal muscle.[84]

Energy depletion

NO and/or its derivatives depress cellular energy production by inhibiting the glycolytic enzyme glyceraldehyde-3-phosphate dehydrogenase (GAPDH),[85] the Krebs' cycle enzyme aconitase,[86] and the mitochondrial respiratory chain enzymes[73]. NO may act as a physiological regulator of mitochondrial respiration by direct and reversible inhibition of cytochrome c oxidase through an interaction of NO with the haem moiety.[73] However, exposure of mitochondria to sustained high amounts of NO can cause irreversible damage to the respiratory chain enzymes cytochrome c oxidase and succinate-cytochrome c, possibly by an indirect mechanism involving $ONOO^-$.[70] In bovine myocardial slices, hypoxia/reoxygenation-induced inhibition of mitochondrial respiration was found to be enhanced by L-arginine and attenuated by the NOS inhibitor, nitro-L-arginine, or the $ONOO^-$ scavenger urate, implicating NO/$ONOO^-$.[87]

Exacerbation of inflammation

Endothelial integrity

Many mediators of I-R injury are capable of inducing inter-endothelial gaps and thereby eliciting plasma leakage.[1] Local production of NO would counteract the endothelial contraction, but the dominant influence of NO may be to exacerbate the leakage response through vasodilatation, promoting enhanced post-capillary blood flow. There are contradictory reports on the influence of NO on microvascular leakage. In the gut, NO reduces leak while NOS inhibitors promote the response.[88]

However, in the rat skin NOS inhibitors decrease oedema formation in response to inflammogens[89] and local oedema is elicited by NO and the NO donor, sodium nitroprusside.[90] Furthermore, in the CNS, NO may enhance permeability and compromise the integrity of the blood brain barrier.

Inflammatory cell survival

Leucocytes normally undergo apoptosis in the absence of survival factors such as granulocyte-macrophage colony-stimulating factor (GM-CSF). Neutrophil and eosinophil survival is increased following activation of sGC.[91] By contrast, high levels of NO production induce apoptosis in macrophages.[92] There have been no studies on the influence of NO on neutrophil survival in I-R injury, but it is clear that the survival of inflammatory cells may be an important determinant of the ultimate amount of necrosis in post-ischaemic tissue.

Influence of Nitric Oxide in Ischaemia-reperfusion Injury to Specific Organs

The preceding general overview of the influence of NO in I-R injury indicates the potential beneficial or detrimental effects of endogenous and exogenous NO. The data available from studies of particular organs suggest that generalizations are difficult to make about the net effect of NO. Some recent observations begin to provide a more consistent picture of the dynamics of NO involvement in I-R injury.

Cerebral I-R Injury

The role of NO in cerebral I-R injury is potentially more complex than in peripheral tissues, since there is evidence that all 3 isoforms of NOS are involved, albeit at different stages in the I-R injury.[93] NOS knockout (KO) mice have been examined to isolate the influence of the different NOS isoforms (see also Chapter 10). Larger infarcts are observed in eNOS KO mice following middle cerebral artery occlusion than in wild-types.[94] On the other hand, nNOS KO results in smaller lesions in the same model.[95] iNOS has also been implicated in cerebral I-R injury in a rat model of middle cerebral artery occlusion, since the iNOS selective inhibitor aminoguanidine reduces infarct volume.[31] These complexities are further exemplified by the dose- and time-dependence of the protective effects of L-NAME in rats subjected to bilateral common carotid artery and left middle cerebral occlusion.[96] L-NAME at doses of 1 and 3 mg/kg, but not at 10 mg/kg, reduces infarct volume when administered between 5 min and 9 h after ischaemia, but not thereafter, contrasting with previous studies showing that pre-ischaemic administration of L-NAME could increase infarct volume.[97]

Cardiac I-R Injury

The majority of studies have suggested that NO supplementation has protective effects in myocardial I-R injury, possibly as a result of decreasing leucocyte-mediated increases in vascular permeability.[98] However, elevation of NO levels using NO donors during the reperfusion phase may increase the duration of sustained arrythmias,[99] and L-arginine supplementation during reperfusion decreased the recovery of contractile function in isolated, buffer-perfused rat hearts.[100] These findings are in agreement with several studies documenting cardioprotective effects of NOS inhibitors, indicating that, at least during the reperfusion phase, endogenous NO is detrimental to myocardial function in I-R injury, possibly due to inhibition of mitochondrial function.[87]

Splanchnic I-R Injury

Kubes and co-workers have carried out extensive studies on I-R injury to the gut (see also Chapter 15) which suggest that supplementation of NO levels protects the microcirculation and limits injury.[101] However, in a model of splanchnic artery occlusion-induced shock, a combination of NOS inhibition and infusion of an NO donor throughout the reperfusion phase resulted in greater protection (increased survival, less impairment of endothelium-dependent vascular function) than either agent administered alone.[102] The role of NO in I-R injury of the liver is discussed in Chapter 14.

Therapeutic Prospects for Manipulation of Nitric Oxide Levels in Ischaemia-reperfusion Injury

Ischaemic Preconditioning

Ischaemic preconditioning results in two distinct periods of protection from subsequent, more prolonged ischaemia. The first period of protection occurs within minutes, has a relatively short duration, and is reported to be dependent on the release of adenosine and/or NO. The second period of protection occurs with a delay of up to 24 hours and may be associated with the formation of heat shock proteins (HSP). Preconditioning has been described in several organs, including heart,[103] skeletal muscle[104] and liver.[105] The existence of a relationship between NO and HSP induction is emerging from recent studies in which NO is reported to induce HSP *in vitro*,[106] whereas induction of HSP appears to protect from NO-mediated toxicity.[107] These observations are compatible with the idea that peri-ischaemic increases in NO are protective[108] but higher levels of NO produced during reperfusion contribute to the injury. The short-term protective effects of preconditioning through NO could result from vasodilatation, decreased leucocyte activation/adhesion, decreased platelet activation and/or increased glucose uptake.

Oestrogen as Stimulant of NO Production

Pre-menopausal women have a lower incidence of coronary artery disease and myocardial infarction than men, and epidemiological data suggest that this is probably due to different levels of oestrogens. Recently, it has been suggested that oestrogen has vasoprotective actions as a result of stimulation of cNOS as well as its well-established effects on lipid metabolism.[109] In human endothelial cells, 17β-oestradiol increases eNOS protein.[110] Oestrogen has recently been shown to reduce myocardial necrosis following I-R injury,[111] but the relationship of this protection to any change in NO levels remains to be investigated.

A Rationale for the Use of NO Supplementation Followed by NOS Inhibition in I-R Injury

The therapeutic exploitation of the ever-increasing knowledge of the role(s) of NO in I-R injury is a considerable challenge, given the complexity of the effects of this molecule, which are further increased in the post-ischaemic environment. A number of theoretical strategies are available to influence NO levels, including the use of L-arginine, nitrovasodilators and NO donors, NOS isoform-selective inhibitors, and combinations of these. The primary considerations are the *timing* of commencement and *duration* of the treatment, and whether combination treatments are feasible. To date, studies in experimental animals and isolated tissues/cells have generally used a single strategy to influence NO levels and have then generalized a conclusion as to the importance of NO. However, we suggest that the influence of NO on I-R injury is highly dependent on the location of its production, the context (cellular pH, concurrent $O_2^{\cdot -}$ generation, anti-oxidant status), the amount produced, the NOS enzyme isoform involved and whether the production is during ischaemia, early reperfusion or late reperfusion.

There is a single, albeit complex, pattern of data to have emerged which supports supplementation of endogenous levels of NO production immediately prior to ischaemia, during ischaemia and in the first few minutes of reperfusion, but later in reperfusion inhibitors of NOS would provide improved outcome. Selective iNOS inhibitors may provide better protection than non-selective compounds, but there are insufficient data to reach a firm conclusion. Results of recent studies suggest that combinations of NO donors and NOS inhibitors are superior to these agents used singly. In peripheral tissues, the matter of selective NOS isoform inhibition has not been addressed in a systematic fashion. Furthermore, the actions of NOS inhibitors administered during reperfusion may well be altered by peri-ischaemic

NO supplementation or by the use of anti-oxidant agents that decrease $O_2^{\cdot -}$ generation and therefore limit production of $ONOO^-$. Non-enzymatic production of NO may contribute to the pathology of I-R injury, warranting the use of biocompatible NO scavenging agents. Further studies are required to define treatment regimens for specific organs. However, it is important that such studies examine the long-term function and tissue survival outcomes rather than early aspects of the pathogenesis of the injury (e.g. oedema) that may not be closely related to the ultimate therapeutic goal of minimising permanent functional deficit.

Summary

The physiological roles of NO derived from vascular endothelium are lost early during the reperfusion of ischaemic tissue. Strategies to replace this NO, including L-arginine supplementation and the use of NO donors, reduce the ischaemia-reperfusion injury. Later in reperfusion, however, increased NO production (in part from iNOS) contributes to the injury, and inhibitors of NOS are then protective. These somewhat conflicting observations may be resolved by considering the impact of the microenvironment on the actions and/or the amounts of NO produced at different stages of the I-R injury. The therapeutic application of this understanding of the role of NO will require carefully timed administration of NO donors and NOS inhibitors to achieve optimal tissue protection.

Acknowledgements

The authors' work cited in this review is supported by grants from the NHMRC (Australia), the HM Schutt Trust and the Transport Accident Commission.

References

1. Grace PA. Ischaemia-reperfusion injury. *Br J Surg* 1994; **81**: 637–647.
2. Fullerton DA, McIntyre RC Jr. Inhaled nitric oxide: therapeutic applications in cardiothoracic surgery. *Ann Thorac Surg* 1996; **61**: 1856–1864.
3. Cooke JP, Dzau VJ. Nitric oxide synthase: role in the genesis of vascular disease. *Ann Rev Med* 1997; **48**: 489–509.
4. Ferrari R, Pedersini P, Bongrazio M *et al.* Mitochondrial energy production and cation control in myocardial ischaemia and reperfusion. *Basic Res Cardiol* 1993; **88**: 495–512.
5. Manche A, Edmondson SJ, Hearse DJ. Dynamics of early postischemic myocardial functional recovery. Evidence of reperfusion-induced injury? *Circulation* 1995; **92**: 526–534.

6. Weiser MR, Williams JP, Moore FD *et al.* Reperfusion injury of ischemic skeletal muscle is mediated by natural antibody and complement. *J Exp Med* 1996; **183**: 2343–2348.
7. Williams FM. Neutrophils and myocardial reperfusion injury. *Pharmacol Ther* 1996; **72**: 1–12.
8. Fandrey J. Hypoxia-inducible gene expression. *Resp Physiol* 1995; **101**: 1–10.
9. Ames A, Wright RL, Kowada M *et al.* Cerebral ischaemia. II. The no-reflow phenomenon. *Am J Pathol* 1968; **52**: 437–453.
10. Lefer AM, Lefer DJ. Pharmacology of the endothelium in ischemia-reperfusion and circulatory shock. *Ann Rev Pharmacol Toxicol* 1993; **33**: 71–90.
11. Sobey CG, Dalipram RA, Dusting GJ, Woodman OL. Impaired endothelium-dependent relaxation of dog coronary arteries after myocardial ischaemia and reperfusion: prevention by amlodipine, propranolol and allopurinol. *Br J Pharmacol* 1992; **105**: 557–562.
12. Summers ST, Zinner MJ, Freischlag JA. Production of endothelium-derived relaxing factor (EDRF) is compromised after ischemia and reperfusion. *Am J Surg* 1993; **166**: 216–220.
13. Verrier E. The microvascular cell and ischemia-reperfusion injury. *J Cardiovasc Pharmacol* 1996; **27 (Suppl 1)**: S26–S30.
14. Breidahl A, Hickey M, Stewart A *et al.* The role of adhesion molecules in surgery. *Aust NZ J Surg* 1995; **65**: 838–847.
15. Clark ET, Desai TR, Hynes KL, Gewertz BL. Endothelial cell response to hypoxia-reoxygenation is mediated by IL-1. *J Surg Res* 1995; **58**: 675–681.
16. Williams FM, Kus M, Tanda K *et al.* Effect of duration of ischaemia on reduction of myocardial infarct size by inhibition of neutrophil accumulation using an anti-CD18 monoclonal antibody. *Br J Pharmacol* 1994; **111**: 1123–1128.
17. Breidahl AF, Hickey MJ, Stewart AG *et al.* Effects of low dose intra-arterial monoclonal antibodies to ICAM-1 and CD11/CD18 on ischaemia reperfusion injury in the rabbit rectus femoris muscle. *Br J Plast Surg* 1996; **49**: 202–209.
18. Förstermann U, Kleinert H. Nitric oxide synthase: expression and expressional control of the three isoforms. *Naunyn Schmiedebergs Arch Pharmacol* 1995; **352**: 351–364.
19. Szabó C. The pathophysiological role of peroxynitrite in shock, inflammation, and ischemia-reperfusion injury. *Shock* 1996; **6**: 79–88.
20. Lefer AM, Lefer DJ. The role of nitric oxide and cell adhesion molecules on the microcirculation in ischaemia-reperfusion. *Cardiovasc Res* 1996; **32**: 743–751.
21. Salvemini D, Masini E, Pistelli A *et al.* Nitric oxide: a regulatory mediator of mast cell reactivity. *J Cardiovasc Pharmacol* 1991; **17**: S258–S264.

22. Kanwar S, Kubes P. Mast cells contribute to ischemia-reperfusion-induced granulocyte infiltration and intestinal dysfunction. *Am J Physiol* 1994; **267**: G316–G321.
23. Paakkari I, Lindsberg P. Nitric oxide in the central nervous system. *Ann Med* 1995; **27**: 369–377.
24. Schulz JB, Matthews RT, Jenkins BG *et al.* Blockade of neuronal nitric oxide synthase protects against excitotoxicity *in vivo*. *J Neurosci* 1995; **15**: 8419–8429.
25. Kahn RA, Panah M, Weinberger J. Modulation of ischemic excitatory neurotransmitter and γ-aminobutyric acid release during global temporary cerebral ischemia by selective neuronal nitric oxide synthase inhibition. *Anesth Analg* 1997; **84**: 997–1003.
26. Kahn RA, Panah M, Kiffel S, Weinberger J. Modulation of ischemic excitatory neurotransmitter and gamma-aminobutyric acid release during global temporary cerebral ischemia by local nitric oxide synthase inhibition. *Anesth Analg* 1997; **84**: 1004–1010.
27. Lipton SA, Choi Y, Pan Z *et al.* A redox-based mechanism for the neuro-protective and neurodestructive effects of nitric oxide and related nitroso-compounds. *Nature* 1993; **364**: 626–631.
28. Gibson A, Brave SR, McFadzean I *et al.* The nitrergic transmitter of the anococcygeus — NO or not? *Arch Int Pharmacodyn Ther* 1995; **329**: 39–51.
29. Pabla R, Curtis MJ. Endogenous protection against reperfusion-induced ventricular fibrillation: role of neuronal versus non-neuronal sources of nitric oxide and species dependence in the rat versus rabbit isolated heart. *J Mol Cell Cardiol* 1996; **28**: 2097–2110.
30. Zhang B, Knight KR, Dowsing B *et al.* Timing of administration of dexamethasone or the nitric oxide synthase inhibitor, nitro-L-arginine methyl ester, is critical for effective treatment of ischaemia-reperfusion injury to rat skeletal muscle. *Clin Sci* 1997; **93**: 167–174.
31. Iadecola C, Zhang F, Casey R *et al.* Inducible nitric oxide synthase gene expression in vascular cells after transient focal cerebral ischemia. *Stroke* 1996; **27**: 1373–1380.
32. Wildhirt SM, Dudek RR, Suzuki H, Bing RJ. Involvement of inducible nitric oxide synthase in the inflammatory process of myocardial infarction. *Int J Cardiol* 1995; **50**: 253–261.
33. Harbrecht BG, Di Silvio M, Chough V *et al.* Glutathione regulates nitric oxide synthase in cultured hepatocytes. *Ann Surg* 1997; **225**: 76–87.
34. Zweier JL, Wang P, Samouilov A, Kuppusamy P. Enzyme-independent formation of nitric oxide in biological tissues. *Nat Med* 1995; **1**: 804–809.
35. Lepore DA, Koslov A, Stewart AG *et al.* Electron spin resonance measurement of nitric oxide production in the rat skeletal muscle subjected to ischaemia and reperfusion. In: *Progress in Microcirculation Research — Proceedings of the Ninth Australian and New Zealand Symposium*, Rogers PAW, Cooke BM,

Andrews FJ, Knight KR (Eds.), The Australian and New Zealand Microcirculation Society 1997: 28.

36. Tsao PS, Aoki N, Lefer DJ *et al.* Time course of endothelial dysfunction and myocardial injury during myocardial ischemia and reperfusion in the cat. *Circulation* 1990; **82**: 1402–1412.
37. Ma XL, Weyrich AS, Lefer DJ, Lefer AM. Diminished basal nitric oxide release after myocardial ischemia and reperfusion promotes neutrophil adherence to coronary endothelium. *Circ Res* 1993; **72**: 403–412.
38. Sternbergh WC, Makhoul RG, Adelman B. Nitric oxide-mediated, endothelium-dependent vasodilation is selectively attenuated in the postischemic extremity. *Surgery* 1993; **114**: 960–967.
39. Mayhan WG, Amundesen SM, Faraci FM, Heistad DD. Responses of cerebral arteries after ischemia and reperfusion in cats. *Am J Physiol* 1988; **255**: H879–H884.
40. Suematsu M, Tamatani T, Delano FA *et al.* Microvascular oxidative stress preceding leukocyte activation elicited by *in vivo* nitric oxide suppression. *Am J Physiol* 1994; **226**: H2410–H2415.
41. Rubanyi GM, Vanhoutte PM. Superoxide anions and hyperoxia inactivate endothelium-derived relaxing factor. *Am J Physiol* 1986; **250**: H822–H827.
42. Sobey CG, Dusting GJ, Stewart AG. Tumour necrosis factor α augments the release of an endothelium-dependent vasoconstrictor from human polymorphonuclear leukocytes. *J Cardiovasc Pharmacol* 1992; **20**: 813–819.
43. Kurose I, Wolf R, Grisham MB, Granger DN. Modulation of ischemia/reperfusion-induced microvascular dysfunction by nitric oxide. *Circ Res* 1994; **74**: 376–382.
44. Hickey MJ, Hurley JV, Angel MF, O'Brien BM. The response of the rabbit *rectus femoris* muscle to ischemia and reperfusion. *J Surg Res* 1992; **53**: 369–377.
45. Stewart AG, Phan LH, Zhang BM *et al.* Nitric oxide contributes to necrosis of skeletal muscle following ischaemia reperfusion injury. In: *Progress in Microcirculation Research — Proceedings of the Ninth Australian and New Zealand Symposium.* Rogers PAW, Cooke BM, Andrews FJ, Knight KR (Eds.), The Australian and New Zealand Microcirculation Society 1997: 25–27.
46. Yoshizumi M, Perrella MA, Burnett JC Jr, Lee M-E. Tumor necrosis factor downregulates an endothelial nitric oxide synthase mRNA by shortening its half-life. *Circ Res* 1993; **73**: 205–209.
47. Beckman JS, Beckman TW, Chen J *et al.* Apparent hydroxyl radical production by peroxynitrite: implications for endothelial injury from nitric oxide and superoxide. *Proc Natl Acad Sci USA* 1990; **87**: 1620–1624.
48. Villa LM, Salas E, Darley-Usmar VM *et al.* Peroxynitrite induces both vasodilatation and impaired vascular relaxation in the isolated perfused rat heart. *Proc Natl Acad Sci USA* 1994; **91**: 12383–12387.

49. Zou MH, Ullrich V. Peroxynitrite formed by simultaneous generation of nitric oxide and superoxide selectively inhibits bovine aortic prostacyclin synthase. *FEBS Lett* 1996; **382**: 101–104.
50. Lefer DJ, Scalia R, Campbell B *et al.* Peroxynitrite inhibits leukocyte-endothelial cell interactions and protects against ischemia-reperfusion injury in rats. *J Clin Invest* 1997; **99**: 684–691.
51. Griscavage JM, Fukuto JM, Komori Y, Ignarro LJ. Nitric oxide inhibits neuronal nitric oxide synthase by interacting with the heme prosthetic group. Role of tetrahydrobiopterin in modulating the inhibitory action of nitric oxide. *J Biol Chem* 1994; **269**: 21644–21649.
52. Lander HM, Jacovina AT, Davis RJ, Tauras JM. Differential activation of mitogen-activated protein kinases by nitric oxide-related species. *J Biol Chem* 1996; **271**: 19705–19709.
53. Beckman JS, Koppenol WH. Nitric oxide, superoxide, and peroxynitrite: the good, the bad, and ugly. *Am J Physiol* 1996; **271**: C1424–C1437.
54. Pryor WA, Squadrito GL. The chemistry of peroxynitrite: a product from the reaction of nitric oxide with superoxide. *Am J Physiol* 1995; **268**: L699–L722.
55. Ischiropoulos H, Zhu L, Chen J *et al.* Peroxynitrite-mediated tyrosine nitration catalyzed by superoxide dismutase. *Arch Biochem Biophys* 1992; **298**: 431–437.
56. Moro MA, Darley-Usmar VM, Lizasoain I *et al.* The formation of nitric oxide donors from peroxynitrite. *Br J Pharmacol* 1995; **116**: 1999–2004.
57. Konorev EA, Joseph J, Tarpey MM, Kalyanaraman B. The mechanism of cardioprotection by *S*-nitrosoglutathione monoethyl ester in rat isolated heart during cardioplegic ischaemic arrest. *Br J Pharmacol* 1996; **119**: 511–518.
58. Barker JE, Bolanos JP, Land JM *et al.* Glutathione protects astrocytes from peroxynitrite-mediated mitochondrial damage: implications for neuronal/astrocytic trafficking and neurodegeneration. *Dev Neurosci* 1996; **18**: 391–396.
59. Ånggård E. Nitric oxide: mediator, murderer and medicine. *Lancet* 1994; **343**: 1199–1206.
60. Shen W, Zhang X, Zhao G *et al.* Nitric oxide production and NO synthase gene expression contribute to vascular regulation during exercise. *Med Sci Sports Exerc* 1995; **27**: 1125–1134.
61. Topper JN, Cai J, Falb D, Gimbrone MA Jr. Identification of vascular endothelial genes differentially responsive to fluid mechanical stimuli: cyclooxygenase-2, manganese superoxide dismutase, and endothelial cell nitric oxide synthase are selectively up-regulated by steady laminar shear stress. *Proc Natl Acad Sci USA* 1996; **93**: 10417–10422.
62. Nonami Y. The role of nitric oxide in cardiac ischemia-reperfusion injury. *Jpn Circ J* 1997; **61**: 119–132.

63. Pabla R, Curtis MJ. Effects of NO modulation on cardiac arrhythmias in the rat isolated heart. *Circ Res* 1995; **77**: 984–992.
64. Radomski MW, Palmer RMJ, Moncada S. Endogenous nitric oxide inhibits human platelet adhesion to vascular endothelium. *Lancet* 1987; **ii**: 1057–1058.
65. Kubes P, Suzuki M, Granger DN. Nitric oxide: an endogenous modulator of leukocyte adhesion. *Proc Natl Acad Sci USA* 1991; **88**: 4651–4655.
66. Iha S, Orita K, Kanno T *et al.* Oxygen-independent inhibition of neutrophil respiratory burst by nitric oxide. *Free Radic Res* 1996; **25**: 489–498.
67. Phan LP, Hickey MJ, Niazi ZBM, Stewart AG. Nitric oxide synthase inhibitor, nitro-iminoethyl-L-ornithine, reduces ischemia-reperfusion injury in rabbit skeletal muscle. *Microsurgery* 1994; **15**: 703–707.
68. Peng H-B, Libby P, Liao JK. Induction and stabilization of IκBα by nitric oxide mediates inhibition of NF-κB. *J Biol Chem* 1995; **270**: 14214–14219.
69. Pauwels PJ, Opperdoes FR, Tiouet A. Effects of antimycin, glucose deprivation, and serum on cultures of neurons, astrocytes and neuroblastoma cells. *J Neurochem* 1985; **44**: 143–148.
70. Bolanos JP, Peuchen S, Heales SJR *et al.* Nitric oxide-mediated inhibition of the mitochondrial respiratory chain in cultured astrocytes. *J Neurochem* 1994; **63**: 910–916.
71. Wang D, McMillin JB, Bick R, Buja M. Response of the neonatal rat cardiomycyte in culture to energy depletion: effects of cytokines, nitric oxide, and heat shock proteins. *Lab Invest* 1996; **75**: 809–818.
72. Rubbo H, Radi R, Trujillo M *et al.* Nitric oxide regulation of superoxide and peroxynitrite-dependent lipid peroxidation. *J Biol Chem* 1994; **269**: 26066–26075.
73. Bolanos JP, Almeida A, Stewart V *et al.* Nitric oxide-mediated mitochondrial damage in the brain: mechanisms and implications for neurodegenerative diseases. *J Neurochem* 1997; **68**: 2227–2240.
74. Frears ER, Zhang Z, Blake DR *et al.* Inactivation of tissue inhibitor of metalloproteinase-1 by peroxynitrite. *FEBS Lett* 1996; **381**: 21–24.
75. Kooy NW, Royall JA. Agonist-induced peroxynitrite production from endothelial cells. *Arch Biochem Biophys* 1994; **310**: 352–359.
76. Yasmin W, Strynadka KD, Schulz R. Generation of peroxynitrite contributes to ischemia-reperfusion injury in isolated rat hearts. *Cardiovasc Res* 1997; **33**: 422–432.
77. Wang P, Zweier JL. Measurement of nitric oxide and peroxynitrite generation in the postischemic heart. Evidence for peroxynitrite-mediated reperfusion injury. *J Biol Chem* 1996; **271**: 29223–29230.
78. Ischiropoulos H, al-Mehdi AB, Fisher AB. Reactive species in ischemic rat lung injury: contribution of peroxynitrite. *Am J Physiol* 1995; **269**: L158–L164.
79. Lepoivre M, Fieschi F, Coves J *et al.* Inactivation of ribonucleotide reductase by nitric oxide. *Biochem Biophys Res Commun* 1991; **179**: 442–448.

80. Knight KR, Messina A, Hickey MJ *et al.* Evidence of apoptosis in oxidatively stressed skeletal muscle following ischaemia-reperfusion injury. *Proceedings of the Sixth World Congress for Microcirculation* 1996; 443–447.
81. Messmer UK, Ankarcrona M, Nicotera P, Brune B. p53 expression in nitric oxide-induced apoptosis. *FEBS Lett* 1994; **355**: 23–26.
82. Zhang J, Dawson VL, Dawson TM, Snyder SH. Nitric oxide activation of poly (ADP-ribose) synthetase in neurotoxicity. *Science* 1994; **26**: 3687–3689.
83. Szabó C, Zingarelli B, Salzman AL. Role of poly-ADP ribosyltransferase activation in the vascular contractile and energetic failure elicited by exogenous and endogenous nitric oxide and peroxynitrite. *Circ Res* 1996; **78**: 1051–1063.
84. Bowes J, Thiemermann C. Inhibition of the activity of poly(ADP-ribose) synthetase reduces infarct size in a rabbit model of myocardial ischaemia and reperfusion [Abstract]. *Br J Pharmacol* 1997; **120**: 87P.
85. Dimmler S, Brune B. Characterisation of nitric oxide catalysed ADP-ribosylation of glyceraldehyde-3-phosphate dehydrogenase. *Eur J Biochem* 1992; **210**: 305–310.
86. Drapier JC, Hibbs JB. Murine cytotoxic activated macrophages inhibit aconitase in tumor cells. *J Clin Invest* 1986; **78**: 700–797.
87. Xie YW, Wolin MS. Role of nitric oxide and its interaction with superoxide in the suppression of cardiac muscle mitochondrial respiration. Involvement in response to hypoxia/reoxygenation. *Circulation* 1996; **94**: 2580–2586.
88. Kubes P. Nitric oxide-induced microvascular permeability alterations: a regulatory role for cGMP. *Am J Physiol* 1993; **265**: H1909–H1915.
89. Ialenti A, Ianaro A, Moncada S, Di Rosa M. Modulation of acute inflammation by endogenous nitric oxide. *Eur J Pharmacol* 1992; **211**: 177–182.
90. Chander CL, Moore AR, Desa FM *et al.* The local modulation of vascular permeability by endothelial cell derived products. *J Pharm Pharmacol* 1988; **40**: 745–746.
91. Beauvais F, Michel L, Dubertret L. The nitric oxide donors, azide and hydroxylamine, inhibit the programmed cell death of cytokine-deprived human eosinophils. *FEBS Lett* 1995; **361**: 229–232.
92. Sarih M, Souvannavong V, Adam A. Nitric oxide synthase induces macrophage death by apoptosis. *Biochem Biophys Res Commun* 1993; **191**: 503–508.
93. Dawson VL, Dawson TM. Nitric oxide neurotoxicity. *J Chem Neuroanat* 1996; **10**: 179–190.
94. Huang Z, Huang PL, Ma J *et al.* Enlarged infarcts in endothelial nitric oxide synthase knockout mice are attenuated by nitro-L-arginine. *J Cereb Blood Flow Metab* 1996; **16**: 981–987.
95. Zaharchuk G, Hara H, Huang PL *et al.* Neuronal nitric oxide synthase mutant mice show smaller infarcts and attenuated apparent diffusion coefficient changes in the peri-infarct zone during focal cerebral ischemia. *Mag Res Med* 1997; **37**: 170–175.

96. Margaill I, Allix M, Boulu RG, Plotkine M. Dose- and time-dependence of L-NAME neuroprotection in transient focal cerebral ischaemia in rats. *Br J Pharmacol* 1997; **120**: 160–163.
97. Kuluz JW, Prado RJ, Dietrich WD *et al.* The effect of nitric oxide synthase inhibition on infarct volume after reversible focal cerebral ischemia in conscious rats. *Stroke* 1993; **24**: 2023–2029.
98. Kupatt C, Zahler S, Seligmann C *et al.* Nitric oxide mitigates leukocyte adhesion and vascular leak after myocardial ischemia. *J Mol Cell Cardiol* 1996; **28**: 643–654.
99. Bilinska M, Maczewski M, Beresewicz A. Donors of nitric oxide mimic effects of ischaemic preconditioning on reperfusion induced arrhythmias in isolated rat heart. *Mol Cell Biochem* 1996; **161**: 265–271.
100. Engelman DT, Watanabe M, Maulik N *et al.* Critical timing of nitric oxide supplementation in cardioplegic arrest and reperfusion. *Circulation* 1996; **94**: 407–411.
101. Alican I, Kubes P. A critical role for nitric oxide in intestinal barrier function and dysfunction. *Am J Physiol* 1996; **270**: G225–G237.
102. Squadrito F, Altavilla D, Squadrito G *et al.* Effects of *S*-ethylisothiourea, a potent inhibitor of nitric oxide synthase, alone or in combination with a nitric oxide donor in splanchnic artery occlusion shock. *Br J Pharmacol* 1996; **119**: 23–28.
103. Sumeray MS, Yellon DM. Myocardial preconditioning. What have we learned? *Eur Heart J* 1997; **18 Suppl A**: A8–A14.
104. Carroll CM, Carroll SM, Overgoor ML *et al.* Acute ischemic preconditioning of skeletal muscle prior to flap elevation augments muscle-flap survival. *Plast Reconstr Surg* 1997; **100**: 58–65.
105. Peralta C, Hotter G, Closa D *et al.* Protective effect of preconditioning on the injury associated to hepatic ischemia-reperfusion in the rat: role of nitric oxide and adenosine. *Hepatology* 1997; **25**: 934–937.
106. Malyshev IY, Malugin AV, Golubeva LU *et al.* Nitric oxide donor induces HSP70 accumulation in the heart and in cultured cells. *FEBS Lett* 1996; **391**: 21–23.
107. Bellmann K, Jaatela M, Wissing D *et al.* Heat shock protein hsp70 overexpression confers resistance against nitric oxide. *FEBS Lett* 1996; **391**: 185–188.
108. Parratt JR, Vegh A, Papp JG. Bradykinin as an endogenous myocardial protective substance with particular reference to ischemic preconditioning — a brief review of the evidence. *Can J Physiol Pharmacol* 1995; **73**: 837–842.
109. Ruehlmann DO, Mann GE. Actions of oestrogen on vascular endothelial and smooth-muscle cells. *Biochem Soc Trans* 1997; **25**: 40–45.
110. Hayashi T, Yamada K, Esaki T *et al.* Estrogen increases endothelial nitric oxide by a receptor-mediated system. *Biochem Biophys Res Commun* 1995; **214**: 847–855.

111. Hale SL, Birnbaum Y, Kloner RA. β-Estradiol, but not α-estradiol, reduces myocardial necrosis in rabbits after ischemia and reperfusion. *Am Heart J* 1996; **132**: 258–262.
112. Tiefenbacher CP, Chilian WM, Mitchell M, DeFily DV. Restoration of endothelium-dependent vasodilation after reperfusion injury by tetrahydrobiopterin. *Circulation* 1996; **94**: 1423–1429.
113. Chen D, Hurst WJ, Ding JM *et al.* Localization and characterization of nitric oxide synthase in the rat suprachiasmatic nucleus: evidence for a nitrergic plexus in the biological clock. *J Neurochem* 1997; **68**: 855–861.
114. Abu-Soud HM, Rousseau DL, Stuehr DJ. Nitric oxide binding to the heme of neuronal nitric oxide synthase links its activity to changes in oxygen tension. *J Biol Chem* 1996; **271**: 32515–32518.

Chapter 18

Oxidative Tissue Injury, Nitric Oxide and Atherosclerosis

Rakesh P Patel, Joanne McAndrew, Nicole Purcell, Matthew C Maland, Hanjoong Jo, C Roger White and Victor M Darley-Usmar

Introduction

Atherosclerosis is a disease of the artery wall associated with progressive loss of endothelial cell function and subendothelial accumulation of lipid-enriched "foam cells" derived largely from macrophages.[1–7] Elevated plasma concentrations of low density lipoprotein (LDL) and decreased high density lipoprotein are recognized risk factors for the disease.[1] In order to form foam cells, modification of LDL must occur in the artery wall, and there is compelling evidence that oxidation of lipids and proteins is a key event in this process. For example, antibodies raised against LDL modified by lipid oxidation products such as malondialdehyde (MDA) or 4-hydroxynonenal[6–8] cross-react with atherosclerotic tissues,[5] suggesting that aldehyde-dependent modification of LDL occurs *in vivo*. Furthermore, oxidized LDL (oxLDL) is itself immunogenic and can be used to raise antibodies which cross-react with material in human atherosclerotic lesions.[9] Indeed, an autoimmune component to atherosclerosis is suggested by the finding that human subjects have auto-antibodies to oxLDL and the titre shows a positive correlation with the progression of atherosclerotic lesions.[10]

A corollary of the oxidative hypothesis of atherosclerosis is that antioxidants, natural or synthetic, should inhibit the atherosclerotic process. While it is important to be aware that "antioxidants" may possess pharmacological properties other than scavenging free radicals, a diverse range of such compounds have been shown to be anti-atherosclerotic. These include chain-breaking antioxidants such as probucol or butylated hydroxytoluene, which retard the development of atherosclerosis in animal models of the disease such as the Watanabe rabbit or the apolipoprotein E (apo-E) deficient mouse.[11–13] Intervention trials in humans designed to test whether antioxidants prevent atherosclerosis have not yet been reported.

The biological properties of oxLDL extend beyond the ability to stimulate foam cell formation, and include cytotoxicity, induction of smooth muscle cell proliferation, increased monocyte binding and disruption of essential endothelial cell signalling pathways (including the synthesis of NO).[14–20] It appears, therefore, that oxidative modification of LDL is a key event in atherogenesis. In the present chapter we shall review the role of NO in vascular disease, focusing specifically on atherosclerosis, and develop the theme that NO can function as both a pro- and anti-atherogenic agent, depending on its interactions with other biological molecules.

Nitric Oxide, Superoxide, Peroxynitrite and Atherosclerosis

Endothelial cells synthesize NO *via* a constitutive Ca^{2+}-calmodulin dependent NO synthase (eNOS), using L-arginine, O_2 and NADPH as substrates (see Chapters 1 and 3).[21] This enzyme may be involved in the anti-atherogenic effects of shear stress. An inducible isoform (iNOS) is found in macrophages and smooth muscle cells and is Ca^{2+}-calmodulin independent.[21] This latter enzyme appears to play a role in the inflammatory response and has been implicated in the development of atherosclerotic lesions.[22] Endothelium-dependent responses mediated by NO are functionally attenuated early in hypercholesterolaemia in a process associated with the stimulation of both NO and $O_2^{\bullet -}$ (superoxide) formation.[23–27] Since NO production may be increased in atherosclerosis, it is reasonable to ask whether NO *per se* can cause oxidative damage. It is frequently assumed that NO must be extremely reactive because it is a free radical. This is not the case and, in fact, NO is relatively stable and unable to oxidize proteins, lipids or DNA. Indeed, recent reports suggest that NO has anti-atherogenic actions that include scavenging of the lipid-centred radicals that lead to LDL oxidation.[28–30] This will be discussed in more detail in a later section.

The mechanisms that underlie loss of NO-dependent relaxation also play a central role in the genesis of atherosclerosis (Figure 1). The reaction between NO and $O_2^{\bullet -}$ is particularly important since it occurs at almost diffusion-limiting rates ($k = 6.7 \times 10^9\ M^{-1}cm^{-1}$).[31] The product of this reaction is peroxynitrite ($ONOO^-$) which is now thought to be formed at most sites of chronic inflammation. The rapid rate of reaction between NO and $O_2^{\bullet -}$ means that $ONOO^-$ formation will be favoured if the two radicals are formed in close proximity to each other. Peroxynitrite can oxidize sulphydryl groups,[32] and nitrate phenolic compounds[33]

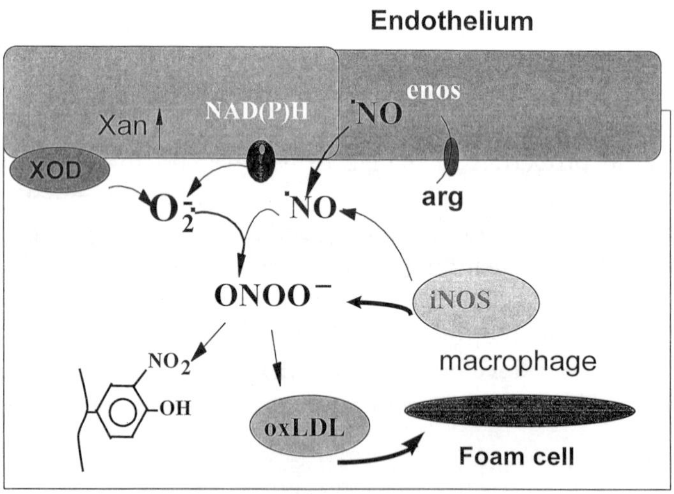

Figure 1. Conversion of NO to a pro-oxidant by reaction with $O_2^{\bullet -}$ in atherosclerosis. Several factors conspire at different stages of the atherosclerotic lesion to diminish the consequences of NO synthesis from L-arginine (arg) by the endothelial NO synthase (eNOS). These include: (1) formation of $O_2^{\bullet -}$ from xanthine (Xan) by xanthine oxidase (XOD) or NAD(P)H oxidases in the endothelium; (2) formation of oxidized LDL (oxLDL) which inhibits eNOS and promotes infiltration with macrophage-derived foam cells. Atherogenesis is also associated with induction of iNOS and nitrotyrosine formation.

and has been shown to decay, *via* peroxynitrous acid, to yield an oxidant with 'hydroxyl radical like' activity.[34,35] This reaction of NO with $O_2^{\bullet-}$ is critical since, in addition to preventing vasorelaxation, it channels the metabolic fate of NO towards nitration, nitrosation or oxidation reactions.

In further support of $ONOO^-$ formation during the disease process, a stable residue of its reactivity, nitrotyrosine, has been detected in atherosclerotic lesions with the distribution of staining being almost identical to that for iNOS.[22,36] An alternative, and possibly biologically relevant mechanism of nitrotyrosine formation by neutrophils has also recently been described which involves reactions between hypochlorous acid and nitrite that are dependent on the enzyme myeloperoxidase.[37] While neutrophils are not normally associated with human atherosclerotic plaques, the detection of myeloperoxidase in such lesions raises the possibility that this pathway could also contribute to oxidative damage in the arterial wall.[37,38]

In summary, evidence to support the formation of $ONOO^-$ *in vivo* derives from the $O_2^{\bullet-}$-dependent loss of endothelium-derived NO in the earliest stages of atherosclerotic lesions, the stimulation of $ONOO^-$ formation in endothelial cells by agonists such as bradykinin, the detection of tyrosine nitration in atherosclerotic lesions, and the identification of iNOS in lesions associated with nitrotyrosine formation.[22,36,39,40] In addition to the loss of NO-dependent vasorelaxation through its reaction with $O_2^{\bullet-}$, $ONOO^-$ will neutralize other "anti-atherogenic" effects of NO. These include the prevention of monocyte or platelet adhesion to endothelium and inhibition of lipid oxidation.[30,41] Recent studies have shown that oxLDL and $ONOO^-$ can induce apoptosis, and that both necrotic and apoptotic mechanisms of cell death occur in atherosclerotic lesions.[42–44] Although insights into the pro-atherogenic effects of $ONOO^-$ have been derived largely from biochemical approaches, it has become important to understand how the relative amounts of NO and $O_2^{\bullet-}$ formed in the vasculature are controlled physiologically, as the balance between these radicals is likely to be a key factor in the initiation of atherosclerosis.

Shear Stress and Formation of NO and Other Antioxidants

The endothelium is continuously subjected to haemodynamic shear forces that exert a regulatory effect on a broad range of functions including vascular tone, vessel wall remodelling, the interaction of blood cells with endothelium, coagulation and fibrinolysis.[45] The prevailing view is that high shear is beneficial to the maintenance of a normal vasculature and prevents the development of atherosclerotic lesions.[46–50] Shear stress is a key regulator of NO production in endothelial cells through signalling pathways that include G-proteins[51] and tyrosine kinases, and since these are discussed in detail in Chapter 2, the present review will focus on aspects related to oxidative stress.

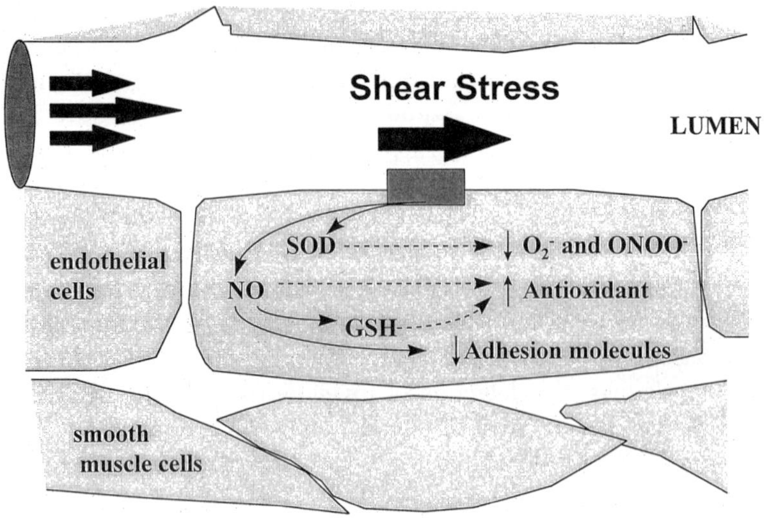

Figure 2. Anti-atherogenic effects of shear stress. Shear stress, *via* an as yet unidentified mechanosensor, increases intracellular concentrations of NO which in turn downregulates adhesion molecule expression. Superoxide dismutase (SOD) is also upregulated by shear stress. Taken together with an NO-dependent stimulation of glutathione (GSH) synthesis, shear stress increases the antioxidant concentration, thereby protecting the cell against oxidative damage and subsequent atherosclerotic lesion development.

The full spectrum of changes in endothelial function initiated by shear stress that could retard atherogenesis includes inhibition of monocyte binding to the endothelium by an NO-dependent mechanism.[52] This effect appears to be mediated by inhibition of nuclear factor-κB (NF-κB), which in turn results in the downregulation of vascular cell adhesion molecule-1 (VCAM-1) gene transcription.[28] Shear stress also induces the synthesis of the antioxidant enzyme manganese superoxide dismutase (MnSOD) and the intracellular antioxidant glutathione.[53,54] Both these effects are likely to inhibit generation of $O_2^{\cdot -}$ by the endothelium (Figure 2). Although the signalling pathways regulating these antioxidants are not known, an NO-dependent mechanism is a possibility. Other pathways involved in the response to shear forces include activation of mitogen-activated protein (MAP) kinases. For example, shear differentially regulates the activities of two MAP kinase family members, ERK and JNK, in bovine aortic endothelial cells.[55] These proteins play a key role as intermediate signalling molecules linking cell surface signals to nuclear events and form a natural focus for dissecting the molecular control of shear-dependent events. Interestingly, these proteins also respond in cells subjected to oxidative stress.[56]

Ultimately, the control mechanisms that transduce the cellular response to shear stress may be overwhelmed by the atherosclerotic process itself, since flow-dependent dilatation is lost in patients with hypercholesterolaemia.[57] Understanding the molecular mechanisms involved in shear-dependent modulation of antioxidants may therefore lead to strategies that could help sustain endogenous antioxidant defences in those regions of the arterial system that are subjected to low or unstable shear stresses and are therefore prone to develop atherosclerosis.

Anti-atherogenic Effects of NO

The extent to which beneficial shear-dependent effects can be ascribed specifically to stimulation of NO synthesis is not clear at the present time. Nevertheless, there is accumulating evidence for NO as an anti-atherogenic agent. For example, *in vitro* macrophage- or endothelial cell-dependent oxidative modification of LDL is inhibited by cytokine-stimulated NO production or administration of exogenous NO donors.[58–60] Furthermore, activation of NO synthase decreases the formation of oxLDL peroxides and the subsequent uptake of LDL by macrophages.[61] Conversely, inhibition of NO synthase or depletion of L-arginine increases LDL oxidation and the uptake of LDL by macrophages.[58,61] Such observations have been extended to animal models of hypercholesterolaemia in which dietary administration of the NO synthase inhibitor L-NAME enhances neointima formation in the aorta of rabbits,[62,63] whereas L-arginine retards development of atherosclerotic lesions.[64]

Recently, the idea that NO can exhibit antioxidant properties has gained considerable support. Of particular relevance to the oxidative hypothesis for atherosclerosis is the finding that NO reacts with the class of oxidants, peroxyl radicals, that cause lipid oxidation at almost diffusion limiting rates.[65] This is a necessary, but not sufficient condition, to make NO an antioxidant. An additional requirement is that the products of the reaction of NO with lipid-centred radicals are stable and do not lead to further propagation of the oxidation reaction. This condition also appears to be satisfied since NO has been shown to inhibit lipid oxidation in LDL and liposomes initiated by different oxidation systems.[30,66] The potential significance of these effects are highlighted by comparing the peroxyl radical scavenging efficiency of NO with that of α-tocopherol, the "classical" endogenous lipophilic antioxidant. Calculations demonstrate that on a molar basis NO is approximately 3 times more efficient, and indeed spares consumption of α-tocopherol during oxidation of liposomes.[66]

Sources of Superoxide in Atherosclerosis

Unlike NO, the origin of $O_2^{\bullet-}$ radicals that react with NO to form $ONOO^-$ is not yet clearly defined. As illustrated in Figure 1, potential sources of $O_2^{\bullet-}$ which have been recently identified include a plasma membrane NAD(P)H oxidase,[67,68] the eNOS enzyme itself when deprived of its substrate L-arginine,[69] and the enzyme xanthine oxidase (XO).[70] The NAD(P)H oxidase system appears to be induced in a variety of disease states, including those involving mediators of hypertension such as the renin-angiotensin system. Acute infusion of angiotensin II (Ang II) in rats thus results in a hypertensive response, characterized by defective acetylcholine-mediated, NO-dependent vascular relaxation.[68] Decreased relaxation in this model can be reversed by treatment with superoxide dismutase and is associated with activation of the $O_2^{\bullet-}$ generating membrane-bound enzyme NAD(P)H oxidase.[68,71] eNOS may itself be an important source of $O_2^{\bullet-}$ under conditions of arginine depletion and in hypercholesterolaemia. Pre-treatment of endothelial cells with native LDL stimulates $O_2^{\bullet-}$ formation. [69] Furthermore, exposure to LDL is associated with quantitative increases in immunoreactive nitrotyrosine, a marker for $ONOO^-$ formation.[69] Thus, eNOS may be an important source of $O_2^{\bullet-}$ under conditions of hypercholesterolaemia and arginine depletion.[67,69]

Xanthine oxidase (XO) is produced via sulphydryl oxidation or limited proteolysis of xanthine dehydrogenase (XDH).[70] In the presence of O_2 and purine substrate, XO yields uric acid and the oxidants $O_2^{\bullet-}$ and H_2O_2[72] and has been widely implicated in pathological conditions such as ischaemia-reperfusion injury[73,74] and respiratory distress syndrome.[75] The vascular endothelium is a common site of injury associated with these conditions, and oxidative damage at this locus has been linked to the enhanced production of $O_2^{\bullet-}$ by XO.[73–75] The importance of XO in cardiovascular injury has become the subject of debate since the appearance of reports suggesting that human tissues exhibit low or undetectable levels of XO.[76–78] Recently, however, the concept that XO may be released from one tissue and bind to the endothelium at distant sites has gained experimental support. Regional ischaemia in organs such as the liver and intestine results in a significant release of XO into the circulation after aortic cross-clamping procedures.[79] Remote cardiac and pulmonary injury are characteristic responses to ischaemia-reperfusion, as is elevation in the plasma concentration of XO.[79,80] Adachi and co-workers have shown that XO binds electrostatically to glycosaminoglycans (GAGs) on the surface of cultured endothelial cells, and that this binding phenomenon can be reversed by treatment of cells with heparin or by enzymatically cleaving GAGs with heparinase and heparitinase.[81] Recent evidence suggests that numerous oxidizing agents, including xanthine/XO, stimulate the synthesis of GAGs[82] and reduce expression of endothelial cell superoxide dismutase and antioxidant activity.[83]

High levels of XO have also been found in atherosclerotic human vascular tissue removed surgically during aneurysm repair or endarterectomy.[84] It has been suggested that in atherosclerosis, localized hypoxia in the vessel wall may favour the conversion of XDH to XO, thus promoting oxidative injury and providing substrate for the enzyme.[85] The potential for this interaction to cause vascular dysfunction is illustrated by a rabbit model of atherosclerosis in which cholesterol feeding promotes an increase in the circulating levels of XO and its binding to the endothelium.[86] In vessels from hypercholesterolaemic rabbits, but not in controls, local hypoxia results in sufficient substrate for XO to sustain the production of

$O_2^{\bullet -}$, thereby impairing NO-dependent vasorelaxation and ultimately leading to $ONOO^-$ formation.[86]

Interactions of Peroxynitrite with Cells and Lipoproteins

Since oxLDL and $ONOO^-$ are likely to be formed in atherosclerotic vessels, and play major pathogenic roles, it is reasonable to postulate that the maintenance of cellular antioxidants is important to retard the development of atherosclerosis. This aspect is perhaps best understood from the perspective of the intracellular antioxidant glutathione (GSH) and its associated antioxidant enzymes. It is thought that the toxic mediators produced in the oxidation of LDL contribute to the pathogenesis of atherosclerosis, and this will be illustrated at a mechanistic level using the $ONOO^-$-dependent oxidation of LDL as an example.

Glutathione and Atherosclerosis

The tripeptide GSH (γ-glu-cys-gly) is found in cells predominantly (90–95%) in its reduced form (GSH) at a concentration that varies between 1 and 5 mM.[87,88] It is essential in the detoxification of inorganic peroxides (e.g. $ONOO^-$ and H_2O_2) and lipid peroxides and their decomposition products (e.g. aldehydes and epoxides).[89–91] In support of an important role for GSH-dependent detoxification pathways in atherosclerosis, glutathione *S*-transferases have been localized in vascular smooth muscle cells. The enzyme is induced under conditions of oxidative stress, and this response is found in models of atherosclerosis suggesting that it is involved in protecting cells from oxidized LDL or $ONOO^-$.[92]

Glutathione can activate purified neuronal NO synthase (nNOS) up to 4-fold and is essential for the activation of NOS in both endothelial cells and macrophages.[93–95] Short term exposure of endothelial cells to oxLDL is capable of depleting GSH, and depressed levels are associated with inhibition of eNOS.[20,91] Furthermore, NO induces GSH synthesis in endothelial and smooth muscle cells at low concentrations through a mechanism that is not cytotoxic.[96] Formation of $ONOO^-$ may therefore lower the net antioxidant concentration present in the cell through direct consumption of NO, which may in turn lead to a decrease in GSH levels.

The two forms of glutathione peroxidase (GPX) metabolize lipid or organic peroxides.[97] They are normally undetectable in the human arterial wall but have been localized to macrophages and degenerated smooth muscle cells in atherosclerotic human lesions.[98] In the rabbit, GPX can be found in the normal endothelium and is present in a greater proportion of the artery wall in hypertensive and hyperlipidaemic animals.[99] Glutathione levels also appear to be important in reducing lipid peroxidation. Macrophages have been shown to respond to

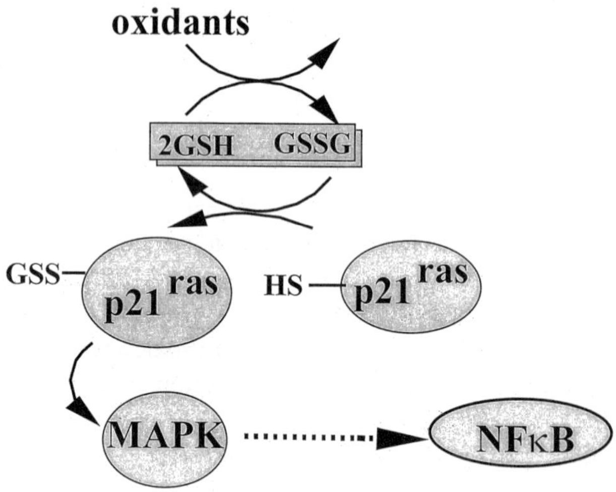

Figure 3. Mechanism for the modulation of the inflammatory response through glutathione (GSH) oxidation. The free thiol on p21^{ras} is shown (SH) forming a mixed disulphide with GSH and, through subsequent steps, activation of mitogen-activated protein kinases (MAPK) and translocation of the transcription factor NFκB to the nucleus. There is also evidence for multiple interactions between oxidants and thiols in the modulation of NFκB activity.

increased levels of oxLDL by increasing the synthesis of GSH which prevents cytotoxicity and modulates foam cell formation.[89,90]

Glutathione also has the capacity to transduce a pro-oxidant stimulus to the cell to modulate protein function, including key elements of cell signalling pathways. Although the majority of protein thiols are reduced under physiological conditions, many proteins have thiols in specific domains which are in equilibrium with the low concentration of oxidized glutathione (GSSG) present in the cell, thus modulating enzymatic activity.[100] Since many of these proteins play an essential role in ion transport, expression of inflammatory mediators, metabolism or signal transduction, it has been proposed that *S*-thiolation by GSH is an important cellular regulatory mechanism.[101–103] A specific example, relevant to the inflammatory process, has been reported in rat phaeochromocytoma (PC12) cells, in which it has been shown that the protein p21^{ras} is activated by inhibition of GSH synthesis and thiol reactive agents (including NO) which are likely to result in formation of a protein mixed disulphide with GSH (Figure 3). This results in increased activation of p21^{ras} and downstream elements of the signalling pathway, including the MAP kinases.[103] It has been shown that the pro-oxidant events which are known to occur in the atherosclerotic lesion lead to a decrease in the GSH:GSSG ratio.[89,91] Recently, this has been extended to include the onset of pathological conditions. Indeed, thiol oxidation and reduction probably underlies the observation that endothelial cells sense changes in oxidative stress through activation of "redox sensitive signal transduction pathways".[101] Peroxynitrite directly oxidizes both free and protein-bound thiols.[32,104] Taken together with its ability to generate lipid peroxides, intracellular $ONOO^-$ production may therefore influence protein function by modulating the redox state of cellular thiols.

These mechanisms link the response of cells to peroxides with the concentration and oxidation state of GSH. Some of these effects are mediated through changes in cellular Ca^{2+} homeostasis and could be relevant to vascular disease. For example, addition of the model peroxide t-butyl hydroperoxide to endothelial cells results in three temporally-separated phases of response. Initially, agonist-stimulated Ca^{2+} influx is inhibited; later, agonist-stimulated release of Ca^{2+} from intracellular stores is inhibited due to decreased inositol trisphosphate (IP_3) production. Finally, basal intracellular Ca^{2+} becomes elevated and cells lose their responsiveness to agonists. Depletion of cellular GSH has been found to increase the rate and extent of oxidant induced changes in Ca^{2+} signalling.[105] The level of GSSG may thus be the important factor controlling Ca^{2+} signalling since increased levels of GSSG reduce Ca^{2+}mobilization from IP_3-sensitive stores and capacitive Ca^{2+} influx.[106]

NO Donor Formation Due to the Reactions of $ONOO^-$

A possible protective pathway against $ONOO^-$ is its conversion to an *S*-nitrosothiol. This has been shown to occur with GSH where it results in the formation of *S*-nitrosoglutathione.[107,108] The biological advantage, and hence protective effect of such reactions, may derive from the fact that GS-NO can be metabolized by cells, release NO, and thereby increase cGMP levels.[108] In this way, NO may be able to avoid neutralization by $O_2^{\cdot-}$. $ONOO^-$ can also form NO-donating molecules through reactions with polyhydroxylated compounds such as glucose or glycerol, whose biological half-lives are much longer than those of NO or $ONOO^-$, yet whose action can still be inhibited by the NO scavenger oxyhaemoglobin, suggesting conversion to NO in the vasculature.[109] $ONOO^-$-derived NO donating species can thus inhibit platelet aggregation and mediate smooth muscle relaxation.[107,110,111] Indeed, the reaction of $ONOO^-$ with glycerol yields glyceryl mononitrite and glyceryl mononitrate, which are metabolites of nitroglycerine (glyceryl trinitrate), a vasodilator used for decades in the treatment of acute angina.

These mechanisms may represent a cytoprotective pathway through conversion of NO to products that allow expression of the antioxidant properties of NO- and cGMP-mediated regulation of intracellular [Ca^{2+}]. However, long-term exposure to organic nitrates and nitrites is known to induce tolerance, resulting in a diminished response to nitrovasodilators.[112] Therefore, it is possible that prolonged exposure of the vasculature to $ONOO^-$ may also contribute to the induction of tolerance to NO and ultimately to cardiovascular pathophysiology.

$ONOO^-$-mediated Oxidative Modification of LDL

The detection of nitrotyrosine in LDL isolated from atherosclerotic lesions suggests that reactive nitrogen species such as $ONOO^-$ are involved in the atherogenic process through direct chemical reactions. It is through the decomposition of a "$\cdot OH/NO_2$"-like species, that $ONOO^-$ may directly initiate lipid peroxidation by abstraction of an allylic hydrogen from a polyunsaturated fatty acid.[113,114] The resultant lipid radicals can then self-propagate the peroxidation process. This is the postulated mechanism by which $ONOO^-$ causes LDL oxidation, and has been substantiated by studies demonstrating inhibitory effects of lipophilic antioxidants *in vitro*.[115,116] The reaction leading to lipid peroxidation does not require either metal ions or pre-formed lipid hydroperoxide.[113]

The principal antioxidant component of LDL is α-tocopherol, the concentration of which decreases after addition of $ONOO^-$ to the lipoprotein.[117] A direct two electron oxidation reaction between $ONOO^-$ and α-tocopherol has been shown to occur.[118] This would compromise the ability of other antioxidants such as ascorbic acid to regenerate α-tocopherol and thus significantly decrease the antioxidant defences present to prevent oxidative damage to LDL. Furthermore, as $ONOO^-$ addition to LDL results in the formation of lipid hydroperoxide, this would render the LDL particle more susceptible to subsequent oxidative modification by other oxidants such as metal ions.[119] In agreement, LDL pre-exposed to varying concentrations of $ONOO^-$ is less resistant to oxidation by Cu^{2+} ions. Recent studies suggest that γ-tocopherol, which forms a stable nitrated product on reaction with $ONOO^-$, may also play an important role in protecting LDL against this oxidant.[120]

In vitro studies have shown that $ONOO^-$ can transform LDL into a particle taken up by macrophage scavenger receptors and so lead to the foam cell formation characteristic of atherosclerotic lesions.[121] Peroxynitrite may mediate this reaction *via* formation of aldehydic lipid peroxidation products which then react with apolipoprotein B (apoB), or through a direct reaction with its protein component.[113,116] Other lipid oxidation products formed by the reactions of $ONOO^-$ with LDL possess diverse biological activities which may then contribute to atherogenesis. For example, LDL-associated cholesterol is oxidized to a variety of cytotoxic oxysterols.[116] A novel class of compounds formed during free radical mediated oxidation of arachidonic acid called F_2-isoprostanes is also produced during $ONOO^-$-dependent LDL oxidation. One of these, 8-epi-$PGF2_{\alpha}$, has been shown to be a potent vasoconstrictor.[115] $ONOO^-$ can result in the formation of this compound when added to LDL or plasma, thus providing another pathway by which NO-derived pro-oxidant species can affect vascular tone (Figure 4).

Addition of SIN-1, a synthetic compound which decomposes to form NO, $O_2^{\cdot -}$ and subsequently $ONOO^-$, to LDL produces greater yields of lipid oxidation products than does addition of a pre-formed solution of $ONOO^-$ itself.[115,116] NO

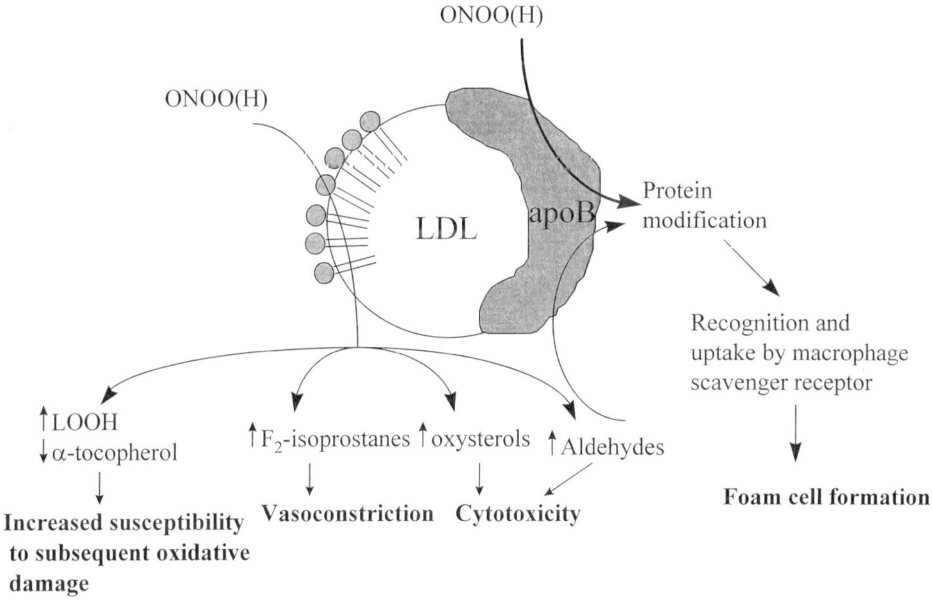

Figure 4. Peroxynitrite-mediated lipid peroxidation reactions in the LDL particle. This process consumes endogenous antioxidants such as α-tocopherol, and causes the formation of lipid hydroperoxide (LOOH), resulting in a lipoprotein particle primed for subsequent oxidative damage. Vasoreactive (F_2-isoprostanes) and cytotoxic (oxysterols and aldehydes) lipid oxidation products may also be formed. A direct reaction between aldehydes and/or $ONOO^-$ with apoB can result in a particle which is recognized by macrophage scavenger receptors and thus lead to foam cell formation.

is a lipophilic molecule and thus in cells where NO and $O_2^{\cdot -}$ are generated simultaneously it is likely that $ONOO^-$ formation occurs close to lipid domains. *In vivo*, therefore, it would be expected that the oxidative modification of membrane lipids would be a dominant reaction pathway for $ONOO^-$.

Summary

In this short overview we have discussed how the physiological actions of NO may be modified in atherosclerosis through interactions with $.O_2^-$ to form $ONOO^-$. This reaction explains the loss of endothelium-derived relaxing factor-like activity as an early response to hypercholesterolaemia, since NO is no longer available to activate guanylate cyclase. Furthermore, the secondary reactions of $ONOO^-$ with lipids leads to the formation of oxidized lipid molecules, which mimic the properties of lipid messengers such as oxysterols or prostanoids. The balance between antioxidant and pro-oxidant pathways in the vasculature are critical in determining the outcome of the atherosclerotic process.

Acknowledgements

The authors gratefully acknowledge support from the National Institutes of Health Grants #HL 48676-04, HL 54815, HL 53601, from the American Heart Association. Alabama Aff. #AL-G960035, and from the United Soybean Board Project #7312.

References

1. Witztum JL, Steinberg D. Role of oxidized low density lipoprotein in atherogenesis. *J Clin Invest* 1991; **88**: 1785–1792.
2. Esterbauer H, Gebicki J, Puhl H, Jurgens G. The role of lipid peroxidation and antioxidants in the oxidative modification of LDL. *Free Radic Biol Med* 1992; **13**: 341–390.
3. Navab M, Fogelman AM, Berliner JA *et al.* Pathogenesis of atherosclerosis. *Am J Cardiol* 1995; **76**: 18C–23C.
4. Darley-Usmar VM, Hassall DG. The role of macrophages and the modification of LDL in the pathogenesis of atherosclerosis. In: *Immunopharmacology of the Heart*, Curtis M (Ed). Academic Press, London 1993: 43–51.
5. Ylä-Herttuala S, Palinski W, Rosenfeld ME *et al.* Evidence for the presence of oxidatively modified low density lipoprotein in atherosclerotic lesions of rabbit and man. *J Clin Invest* 1989; **84**: 1086–1095.
6. Palinski W, Rosenfeld ME, Ylä-Herttuala S *et al.* Low density lipoprotein undergoes oxidative modification *in vivo*. *Proc Natl Acad Sci USA* 1989; **86**: 1372–1376.
7. Parthasarathy S, Printz DJ, Boyd D *et al.* Macrophage oxidation of low density lipoprotein generates a modified form recognized by the scavenger receptor. *Arteriosclerosis* 1986; **6**: 505–510.
8. Jürgens G, Ashy A, Esterbauer H. Detection of new epitopes formed upon oxidation of low density lipoprotein, lipoprotein (a) and very low density lipoprotein. Use of an antiserum against 4-hydroxynonenal-modified low-density lipoprotein. *Biochem J* 1990; **265**: 605–608.
9. Mowri H, Ohkuma S, Takano T. Monoclonal DLR1a/104G antibody recognizing peroxidized lipoproteins in atherosclerotic lesions. *Biochim Biophys Acta* 1988; **963**: 208–214.
10. Salonen JT, Ylä-Herttuala S, Yamamoto R *et al.* Autoantibody against oxidised LDL and progression of carotid atherosclerosis. *Lancet* 1992; **339**: 883–887.
11. Carew TE, Schwenke DC, Steinberg D. Antiatherogenic effect of probucol unrelated to its hypocholesterolemic effect: evidence that antioxidants *in vivo* can selectively inhibit low density lipoprotein degradation in macrophage-rich fatty streaks and slow the progression of atherosclerosis

in the Watanabe heritable hyperlipidemic rabbit. *Proc Natl Acad Sci USA* 1987; **84**: 7725–7729.

12. Björkhem I, Henriksson-Freyschuss A, Breuer O *et al.* The antioxidant butylated hydroxytoluene protects against atherosclerosis. *Arterioscler Thromb* 1991; **11**: 15–22.
13. Tangirala RK, Casanada F, Miller E *et al.* Effect of the antioxidant N,N'-diphenyl 1,4-phenylenediamine (DPPD) on atherosclerosis in apoE-deficient mice. *Arterioscler Thromb Vasc Biol* 1995; **15**: 1625–1630.
14. Hamilton TA, Guoping MA, Chisolm GM. Oxidized low density lipoprotein suppresses the expression of tumor necrosis factor-α mRNA in stimulated murine peritoneal macrophages. *J Immunol* 1990; **144**: 2343–2350.
15. Autio I, Jaakkola O, Solakivi T, Nikkari T. Oxidized low-density lipoprotein is chemotactic for arterial smooth muscle cells in culture. *FEBS Lett* 1990; **277**: 247–249.
16. Chin JH, Azhar S, Hoffman BB. Inactivation of endothelial derived relaxing factor by oxidized lipoproteins. *J Clin Invest* 1992; **89**: 10–18.
17. Bernheimer AW, Robinson WG, Linder R *et al.* Toxicity of enzymically-oxidized low density lipoprotein. *Biochem Biophys Res Commun* 1987; **148**: 260–266.
18. Boulanger CM, Tanner FC, Béa M-L *et al.* Oxidized low density lipoproteins induce mRNA expression and release of endothelin from human and porcine endothelium. *Circ Res* 1992; **70**: 1191–1197.
19. Malden LT, Ross R, Chait, A. Oxidatively modified LDL inhibits expression of platelet derived growth factor by human monocyte derived macrophages [Abstract]. *Clin Res* 1990; **38**: 291A.
20. Liao JK, Shin WS, Lee WY, Clark SL. Oxidized low-density lipoprotein decreases the expression of endothelial nitric oxide synthase. *J Biol Chem* 1995; **270**: 319–324.
21. Moncada S, Palmer RMJ, Higgs EA. Nitric oxide: physiology, pathophysiology and pharmacology. *Pharmacol Rev* 1991; **43**: 109–142.
22. Buttery LDK, Springall DR, Chester AH *et al.* Inducible nitric oxide synthase is present within human atherosclerotic lesions and promotes the formation and activity of peroxynitrite. *Lab Invest* 1996; **75**: 77–85.
23. Mugge A, Elwell JH, Peterson TE *et al.* Chronic treatment with polyethylene-glycolated superoxide dismutase partially restores endothelium-dependent vascular relaxations in cholesterol-fed rabbits. *Circ Res* 1991; **69**: 1293–1300.
24. White CR, Brock TA, Chang LY *et al.* Superoxide and peroxynitrite in atherosclerosis. *Proc Natl Acad Sci USA* 1994; **91**: 1044–1048.
25. Ohara Y, Peterson TE, Harrison DG. Hypercholesterolemia increases endothelial superoxide anion production. *J Clin Invest* 1993; **91**: 2546–2551.
26. Naruse K, Shimizu K, Muramatsu M *et al.* Long-term inhibition of NO synthesis promotes atherosclerosis in the hypercholesterolemic rabbit thoracic aorta.

PGH_2 does not contribute to impaired endothelium-dependent relaxation. *Arterioscler Thromb* 1994; **14**: 746–752.

27. Cooke JP, Singer AH, Tsao P *et al.* Antiatherogenic effects of L-arginine in the hypercholesterolemic rabbit. *J Clin Invest* 1992; **90**: 1168–1172.
28. De Caterina R, Libby P, Peng HB *et al.* Nitric oxide decreases cytokine-induced endothelial activation. Nitric oxide selectively reduces endothelial expression of adhesion molecules and proinflammatory cytokines. *J Clin Invest* 1995; **96**: 60–68.
29. Hogg N, Kalyanaraman B, Joseph J *et al.* Inhibition of low-density lipoprotein oxidation by nitric oxide. Potential role in atherogenesis. *FEBS Lett* 1993; **334**: 170–174.
30. Rubbo H, Parthasarathy S, Barnes S *et al.* Nitric oxide inhibition of lipoxygenase-dependent liposome and low-density lipoprotein oxidation: termination of radical chain propagation reactions and formation of nitrogen-containing oxidized lipid derivatives. *Arch Biochem Biophys* 1995; **324**: 15–25.
31. Huie RE Padmaja S. The reaction of NO with superoxide. *Free Radic Res Commun* 1993; **18**: 195–199.
32. Radi R, Beckman JS, Bush KM, Freeman BA. Peroxynitrite oxidation of sulfhydryls. The cytotoxic potential of superoxide and nitric oxide. *J Biol Chem* 1991; **266**: 4244–4250.
33. Beckman JS, Koppenol WH. Nitric oxide, superoxide and peroxynitrite: the good, the bad, and ugly. *Am J Physiol* 1996; **271**: C1424–C1437.
34. Beckman JS, Beckman TW, Chen J *et al.* Apparent hydroxyl radical production by peroxynitrite: implications for endothelial injury from nitric oxide and superoxide. *Proc Natl Acad Sci USA* 1990; **87**: 1620–1624.
35. Hogg N, Darley-Usmar VM, Wilson MT, Moncada S. Production of hydroxyl radicals from the simultaneous generation of superoxide and nitric oxide. *Biochem J* 1992; **281**: 419–424
36. Beckman JS, Ye YZ, Anderson PG *et al.* Extensive nitration of protein tyrosines in human atherosclerosis detected by immunohistochemistry. *Biol Chem Hoppe Seylers* 1994; **375**: 81–88.
37. Eiserich JP, Hristova M, Cross CE *et al.* Formation of nitric oxide-derived inflammatory oxidants by myeloperoxidase in neutrophils. *Nature* 1998; **391**: 393–397.
38. Daugherty A, Dunn JL, Rateri DL, Heinecke JW. Myeloperoxidase, a catalyst for lipoprotein oxidation, is expressed in human atherosclerotic lesions. *J Clin Invest* 1994; **94**: 437–444.
39. Kooy NW, Royall JA. Agonist-induced peroxynitrite production from endothelial cells. *Arch Biochem Biophys* 1994; **310**: 352–359.
40. Leeuwenburgh C, Hardy MM, Hazen SL *et al.* Reactive nitrogen intermediates promote low density lipoprotein oxidation in human atherosclerotic intima. *J Biol Chem* 1997; **272**: 1433–1436.

41. Flavahan NA. Atherosclerosis or lipoprotein-induced endothelial dysfunction. Potential mechanisms underlying reduction in EDRF/nitric oxide activity. *Circulation* 1992; **85**: 1927–1938.
42. Shin JT, Barbeito L, MacMillan-Crow LA *et al*. Acidic fibroblast growth factor enhances peroxynitrite-induced apoptosis in primary murine fibroblasts. *Arch Biochem Biophys* 1996; **335**: 32–41.
43. Lin KT, Xue JY, Nomen M *et al*. Peroxynitrite-induced apoptosis in HL-60 cells. *J Biol Chem* 1995; **270**: 16487–16490.
44. Hegyi L, Skepper JN, Cary NR, Mitchinson MJ. Foam cell apoptosis and the development of the lipid core of human atherosclerosis. *J Pathol* 1996; **180**: 423–429.
45. Davies PF, Tripathi SC. Mechanical stress mechanisms and the cell. An endothelial paradigm. *Circ Res* 1993; **72**: 239–245.
46. Ku DN, Giddens DP, Zarins CK, Glagov S. Pulsatile flow and atherosclerosis in the human carotid bifurcation. Positive correlation between plaque location and low and oscillation shear stress. *Arteriosclerosis* 1985; **5**: 293–302.
47. Zarins CK, Giddens DP, Bharadvaj BK *et al*. Carotid bifurcation in atherosclerosis. Quantitative correlation of plaque localization with flow velocity profiles and wall shear stress. *Circ Res* 1983; **53**: 502–514.
48. Kraiss LW, Kirkman TR, Kohler TR *et al*. Shear stress regulates smooth muscle proliferation and neointimal thickening in porous polytetrafluoroethylene grafts. *Arterioscler Thromb* 1991; **11**: 1844–1952.
49. Langille BL, O'Donnell F. Reductions in arterial diameter produced by chronic decreases in blood flow are endothelium-dependent. *Science* 1986; **231**: 405–407.
50. Jo H, Dull RO, Hollis TM, Tarbell JM. Endothelial albumin permeability is shear dependent, time dependent, and reversible. *Am J Physiol* 1991; **260**: H1992–H1996.
51. Kuchan MJ, Jo H, Frangos JA. Role of G proteins in shear stress-mediated nitric oxide production by endothelial cells. *Am J Physiol* 1994; **267**: C753–C758.
52. Tsao PS, Buitrago R, Chan JR, Cooke JP. Fluid flow inhibits endothelial adhesiveness. Nitric oxide and transcriptional regulation of VCAM-1. *Circulation* 1996; **94**: 1682–1689.
53. Topper JN, Cai J, Falb D, Gimbrone MA Jr. Identification of vascular endothelial genes differentially responsive to fluid mechanical stimuli: cyclooxygenase-2, manganese superoxide dismutase, and endothelial cell nitric oxide synthase are selectively up-regulated by steady laminar shear stress. *Proc Natl Acad Sci USA* 1996; **93**: 10417–10422.
54. Ohno M, Gibbons GH, Dzau VJ, Cooke JP. Shear stress elevates endothelial cGMP. Role of a potassium channel and G protein coupling. *Circulation* 1993; **88**: 193–197.

55. Jo H, Sipos K, Go YM *et al.* Differential effect of shear stress on extracellular signal regulated kinase and N-terminal Jun kinase in endothelial cells. G_i2- and $G\beta/\gamma$-dependent signaling pathways. *J Biol Chem* 1997; **272**: 1395–1401.
56. Suzuki YJ, Forman HJ, Sevanian A. Oxidants as stimulators of signal transduction. *Free Radic Biol Med* 1997; **22**: 269–285.
57. Cox DA, Vita JA, Treasure CB *et al.* Atherosclerosis impairs flow-mediated dilation of coronary arteries in humans. *Circulation* 1989; **80**: 458–465.
58. Yates MT, Lambert LE, Whitten JP *et al.* A protective role for nitric oxide in the oxidative modification of low density lipoproteins by mouse macrophages. *FEBS Lett* 1992; **309**: 135–138.
59. Jessup W, Mohr D, Gieseg SP *et al.* The participation of nitric oxide in cell free- and its restriction of macrophage-mediated oxidation of low-density lipoprotein. *Biochim Biophys Acta* 1992; **1180**: 73–82.
60. Struck AT, Hogg N, Thomas JP, Kalyanaraman B. Nitric oxide donor compounds inhibit the toxicity of oxidized low-density lipoprotein to endothelial cells. *FEBS Lett* 1995; **361**: 291–294.
61. Jessup W, Dean RT. Autoinhibition of murine macrophage-mediated oxidation of low-density lipoprotein by nitric oxide synthesis. *Atherosclerosis* 1993; **101**: 145–155.
62. Cayatte AJ, Palacino JJ, Horten K, Cohen RA. Chronic inhibition of nitric oxide production accelerates neointima formation and impairs endothelial function in hypercholesterolemic rabbits. *Arterioscler Thromb* 1994; **14**: 753–759.
63. Naruse K, Shimizu K, Muramatsu M *et al.* Long-tern inhibition of NO synthesis promotes atherosclerosis in the hypercholesterolemic rabbit thoracic aorta. PGH_2 does not contribute to impaired endothelium-dependent relaxation. *Arterioscler Thromb* 1994; **14**: 746–752.
64. Candipan RC, Wang BY, Buitrago R *et al.* Regression or progression. Dependency on vascular nitric oxide. *Arterioscler Thromb Vasc Biol* 1996; **16**: 44–50.
65. Padmaja S, Huie RE. The reaction of nitric oxide with organic peroxyl radicals. *Biochem Biophys Res Commun* 1993; **195**: 539–544.
66. Hayashi K, Noguchi N, Niki E. Action of nitric oxide as an antioxidant against oxidation of soybean phosphatidylcholine liposomal membranes. *FEBS Lett* 1995; **370**: 37–40.
67. Mohązzab KM, Wolin MS. Properties of a superoxide anion-generating microsomal NADH oxidoreductase, a potential pulmonary artery PO_2 sensor. *Am J Physiol* 1994; **267**: L823–L831.
68. Rajagopalan S, Kurz S, Munzel T *et al.* Angiotensin II-mediated hypertension in the rat increases vascular superoxide production via membrane NADH/NADPH oxidase activation. Contribution to alterations of vasomotor tone. *J Clin Invest* 1996; **97**: 1916–1923.

69. Pritchard KA Jr, Groszek L, Smalley DM *et al.* Native low-density lipoprotein increases endothelial cell nitric oxide synthase generation of superoxide anion. *Circ Res* 1995; **77**: 510–518.
70. Amaya Y, Yamazaki K, Sato M *et al.* Proteolytic conversion of xanthine dehydrogenase from the NAD-dependent type to the O_2-dependent type. Amino acid sequence of rat liver xanthine dehydrogenase and identification of the cleavage sites of the enzyme protein during irreversible conversion by trypsin. *J Biol Chem* 1990; **265**: 14170–14175.
71. Griendling KK, Minieri CA, Ollerenshaw JD, Alexander RW. Angiotensin II stimulates NADH and NADPH oxidase activity in cultured vascular smooth muscle cells. *Circ Res* 1994; **74**: 1141–1148.
72. Dupont GP, Huecksteadt TP, Marshall BC *et al.* Regulation of xanthine dehydrogenase and xanthine oxidase activity and gene expression in cultured rat pulmonary endothelial cells. *J Clin Invest* 1992; **89**: 197–202.
73. Linas SL, Whittenburg D, Repine JE. Role of xanthine oxidase in ischemia/reperfusion injury. *Am J Physiol* 1990; **258**: F711–F716.
74. Terada LS, Dormish JJ, Shanley PF *et al.* Circulating xanthine oxidase mediates lung neutrophil sequestration after intestinal ischemia-reperfusion. *Am J Physiol* 1992; **263**: L394–L401.
75. Quinlan GJ, Lamb NJ, Tilley R *et al.* Plasma hypoxanthine levels in ARDS: implications for oxidative stress, morbidity, and mortality. *Am J Respir Crit Care Med* 1997;**155**:479–484.
76. Hopson SB, Lust RM, Sun YS *et al.* Allopurinol improves myocardial reperfusion injury in a xanthine oxidase-free model. *J Natl Med Assoc* 1995; **87**: 480–484.
77. Toivonen HJ, Ahotupa M. Free radical reaction products and antioxidant capacity in arterial plasma during coronary artery bypass grafting. *J Thorac Cardiovasc Surg* 1994; **108**: 140–147.
78. Dorion D, Zhong A, Chiu C *et al.* Role of xanthine oxidase in reperfusion injury of ischemic skeletal muscles in the pig and human. *J Appl Physiol* 1993; **75**: 246–255.
79. Tan S, Gelman S, Wheat JK, Parks DA. Circulating xanthine oxidase in human ischemia reperfusion. *South Med J* 1995; **88**: 479–482.
80. Till GO, Friedl HP, Ward PA. Lung injury and complement activation: role of neutrophils and xanthine oxidase. *Free Radic Biol Med* 1991; **10**: 379–386.
81. Adachi T, Fukushima T, Usami Y, Hirano K. Binding of human xanthine oxidase to sulphated glycosaminoglycans on the endothelial cell surface. *Biochem J* 1993; **289**: 523–527.
82. Tanaka H, Okada T, Konishi H, Tsuji T. The effect of reactive oxygen species on the biosynthesis of collagen and glycosaminoglycans in cultured human dermal fibroblasts. *Arch Dermatol Res* 1993; **285**: 352–355.

83. Stralin P, Marklund SL. Effects of oxidative stress on expression of extracellular superoxide dismutase, CuZn-superoxide dismutase and Mn-superoxide dismutase in human dermal fibroblasts. *Biochem J* 1994; **298**: 347–352.
84. Swain J, Gutteridge JM. Prooxidant iron and copper, with ferroxidase and xanthine oxidase activities in human atherosclerotic material. *FEBS Lett* 1995; **368**: 513–515.
85. Crawford DW, Blankenhorn DH. Arterial wall oxygenation, oxyradicals and atherosclerosis. *Atherosclerosis* 1991; **89**: 97–108.
86. White CR, Darley-Usmar V, Berrington WR *et al.* Circulating plasma xanthine oxidase contributes to vascular dysfunction in hypercholesterolemic rabbits. *Proc Natl Acad Sci USA* 1996; **93**: 8745–8749.
87. Meister A. On the antioxidant effects of ascorbic acid and glutathione. *Biochem Pharmacol* 1992; **44**: 1905–1915.
88. Deneke SM, Fanburg BL. Regulation of cellular glutathione. *Am J Physiol* 1989; **257**: L163–L173.
89. Darley-Usmar VM, Severn A, O'Leary VJ, Rogers M. Treatment of macrophages with oxidised low-density lipoprotein increases their intracellular glutathione content. *Biochem J* 1991; **278**: 429–434.
90. Gotoh N, Graham A, Nikl E, Darley-Usmar VM. Inhibition of glutathione synthesis increases the toxicity of oxidised low-density lipoprotein to human monocytes and macrophages. *Biochem J* 1993; **296**: 151–154.
91. Kuzuya M, Naito M, Funaki C *et al.* Protective role of intracellular glutathione against oxidized low-density lipoprotein in cultured endothelial cells. *Biochem Biophys Res Commun* 1989; **163**: 1466–1472.
92. Misra P, Srivastava SK, Singhal SS *et al.* Glutathione *S*-transferase 8-8 is localized in smooth muscle cells of rat aorta and is induced in an experimental model of atherosclerosis. *Toxicol Appl Pharmacol* 1995; **133**: 27–33.
93. Ghigo D, Alessio P, Foco A *et al.* Nitric oxide synthesis is impaired in glutathione-depleted human umbilical vein endothelial cells. *Am J Physiol* 1993; **265**: C728–C732.
94. Hofmann H, Schmidt HHHW. Thiol dependence of nitric oxide synthase. *Biochemistry* 1995; **34**: 13443–13452.
95. Buchmuller-Rouiller Y, Corrandin SB, Smith J *et al.* Role of glutathione in macrophage activation: effect of cellular glutathione depletion on nitrite production and leishmanicidal activity. *Cell Immunol* 1995; **164**: 73–80.
96. Moellering D, McAndrew J, Patel RP *et al.* Nitric oxide dependent induction of glutathione synthesis through increased expression of γ-glutamylcysteine synthetase. *Arch Biochem Biophys* 1998; **358**: 74–82.
97. Thomas JP, Maiorino M, Ursini F, Girotti AW. Protective action of phospholipid hydroperoxide glutathione peroxidase against membrane-damaging lipid peroxidation. *In situ* reduction of phospholipid and cholesterol hydroperoxides. *J Biol Chem* 1990; **265**: 454–461.

98. Mori I, Yoshimura S, Watanabe K. Immunohistochemical localization of glutathione peroxidase, a lipid peroxidase scavenger, in atherosclerotic lesions of human arteries. *Pathol Int* 1995; **45**: 343–351.
99. Sharma RC, Crawford DW, Kramsch DM *et al.* Immunolocalization of native antioxidant scavenger enzymes in early hypertensive and atherosclerotic arteries. Role of oxygen free radicals. *Arterioscler Thromb* 1992; **12**: 403–415.
100. Gilbert HF. Molecular and cellular aspects of thiol-disulfide exchange. *Adv Enzymol Relat Areas Mol Biol* 1990; **63**: 69–172.
101. Alexander RW. Hypertension and the pathogenesis of atherosclerosis. Oxidative stress and the mediation of arterial inflammatory response: a new perspective. *Hypertension* 1995; **25**: 155–161.
102. Cappel RE, Gilbert HF. Oxidative inactivation of 3-hydroxy-3-methylglutaryl-coenzyme A reductase and subunit cross-linking involve different dithiol/disulfide centers. *J Biol Chem* 1993; **268**: 342–348.
103. Lander HM, Ogiste JS, Teng KK, Novogrodsky A. p21RAS as a common signaling target of reactive free radicals and cellular redox stress. *J Biol Chem* 1995; **270**: 21195–21198.
104. Vasquez-Vivar J, Santos AM, Junqueira VB, Augusto O. Peroxynitrite-mediated formation of free radicals in human plasma; EPR detection of ascorbyl, albumin-thiyl and uric acid derived free radicals. *Biochem J* 1996; **314**: 869–876.
105. Elliott SJ, Doan TH, Henschke PN. Reductant substrate for glutathione peroxidase modulates oxidant inhibition of Ca^{2+} signaling in endothelial cells. *Am J Physiol* 1995; **268**: H278–H287.
106. Henschke PN, Elliott SJ. Oxidized glutathione decreases luminal Ca^{2+} content of the endothelial cell Ins(1,4,5)P_3-sensitive Ca^{2+} store. *Biochem J* 1995; **312**: 485–489.
107. Moro MA, Darley-Usmar VM, Goodwin DA *et al.* Paradoxical fate and biological action of peroxynitrite on human platelets. *Proc Natl Acad Sci USA* 1994; **91**: 6702–6706.
108. Mayer B, Schrammel A, Klatt P *et al.* Peroxynitrite-induced accumulation of cyclic GMP in endothelial cells and stimulation of purified guanylyl cyclase. Dependence on glutathione and possible role of 5-nitrosation. *J Biol Chem* 1995; **270**: 17355–17360.
109. Moro MA, Darley-Usmar VM, Lizasoain I *et al.* The formation of nitric oxide donors from peroxynitrite. *Br J Pharmacol* 1995; **116**: 1999–2004.
110. Liu S, Beckman JS, Ku DD. Peroxynitrite, a product of superoxide and nitric oxide, produces coronary vasorelaxation in dogs. *J Pharmacol Exp Ther* 1994; **268**: 1114–1121.
111. Villa LM, Salas E, Darley-Usmar VM *et al.* Peroxynitrite induces both vasodilatation and impaired vascular relaxation in the isolated perfused heart. *Proc Natl Acad Sci USA* 1994; **91**: 12383–12387.

112. Fung HL. Clinical pharmacology of organic nitrates. *Am J Cardiol* 1993; **72**: 9C–15C.
113. Darley-Usmar VM, Hogg N, O'Leary VJ. The simultaneous generation of superoxide and nitric oxide can initiate lipid peroxidation in human low density lipoprotein. *Free Radic Res Commun* 1992; **17**: 9–20.
114. Radi R, Beckman JS, Bush KM, Freeman BA. Peroxynitrite-induced membrane lipid peroxidation: the cytotoxic potential of superoxide and nitric oxide. *Arch Biochem Biophys* 1991; **288**: 481–487.
115. Moore KP, Darley-Usmar V, Morrow J, Roberts LJ. Formation of F_2 isoprostanes during oxidation of human low density lipoprotein and plasma by peroxynitrite. *Circ Res* 1995; **77**: 335–341.
116. Patel RP, Diczfalusy U, Dzeletovic S *et al.* Formation of oxysterols during oxidation of low density lipoprotein by peroxynitrite, myoglobin, and copper. *J Lipid Res* 1996; **37**: 2361–2371.
117. Hogg N, Darley-Usmar VM, Wilson MT, Moncada S. The oxidation of α-tocopherol in human low-density lipoprotein by the simultaneous generation of superoxide and nitric oxide. *FEBS Lett* 1993; **326**: 199–203.
118. Hogg N, Joseph J, Kalyanaraman B. The oxidation of α-tocopherol and trolox by peroxynitrite. *Arch Biochem Biophys* 1994; **314**: 153–158.
119. Hogg N, Darley-Usmar VM, Graham A, Moncada S. Peroxynitrite and atherosclerosis. *Biochem Soc Trans* 1993; **21**: 358–361.
120. Christen S, Woodall AA, Shigenaga MK *et al.* γ-tocopherol traps mutagenic electrophiles such as NO_x and complements α-tocopherol: physiological implications. *Proc Natl Acad Sci USA* 1997; **94**: 3217–3222.
121. Graham A, Hogg N, Kalyanaraman B *et al.* Peroxynitrite modification of low-density lipoprotein leads to recognition by the macrophage scavenger receptor. *FEBS Lett* 1993; **330**: 181–185.

Chapter 19

Endothelial Dysfunction Associated with Cardiovascular Disease and Transplantation

Gregory J Dusting and Anthony M Dart

Introduction

For more than a century, living blood vessels, specifically their endothelial cell linings, have been thought to play a vital role in maintaining the fluidity of the blood passing through them. However, it was not until the 1970s that chemical candidates for this endothelial 'protective' function were identified. Prostacyclin (PGI_2) and later EDRF emerged as short-acting contenders for this role, although it is clear that the endothelium possesses other metabolic pathways that produce bioactive peptides and lipids which can also participate in thromboresistance and anti-adhesive mechanisms. As discussed elsewhere in this book, NO is responsible for most of the vascular functions formerly ascribed to EDRF, one notable exception being the endothelium-dependent hyperpolarization triggered by vasoactive agents and transmitters in some arteries (see Chapters 1 and 2). The present review focuses on the disruption of NO mechanisms in the endothelial cell and the arterial intima in hyperlipidaemia, atherosclerosis and ischaemia-reperfusion injury, as well as other disturbances in NO function associated with atherosclerosis and transplantation, including induction of the iNOS isoform. Pathophysiological defects in other endothelial mechanisms associated with vascular disease, including disturbances in PGI_2 and endothelin production, have been reviewed elsewhere[1,2] and are beyond the scope of this review.

Principal Physiological Functions of Nitric Oxide

The primary target of NO in both vascular smooth muscle and platelets is the haem centre of the soluble guanylate cyclase (sGC) enzyme, which is activated and thereby leads to an elevation in intracellular levels of guanosine 3',5'-cyclic monophosphate (cGMP).[1–6] This mechanism provides an explanation for the action of the nitrovasodilator drugs which have been used for more than 100 years and are metabolized to NO in smooth muscle[3] (see Chapters 1 and 4). Some of the most important physiological functions of NO become impaired in vascular disease.[4] The release of NO from the endothelium may be compromised, with consequences for blood flow regulation and cell adhesion. In addition, synthesis of iNOS may be stimulated in vascular disease, thus contributing to cytotoxicity associated with inflammatory reactions. The physiological functions of NO are detailed in preceding chapters of this book, but those most relevant to vascular disease will be summarized here.

Inhibition of Platelet and Leucocyte Functions

The maintenance of the fluidity of the blood is an important function of the vascular endothelium, and NO is one of the factors involved. NO is capable of regulating haemostasis by inhibiting cell-cell contact, both adhesion and aggregation, at the interface between platelets and endothelial cells.[5] Elevations in cGMP inhibit the expression on the platelet and endothelial cell membranes of crucial glycoproteins involved in adhesion, including IIb/IIIa and P-selectin.[4,7] In addition, NO appears capable of regulating leucocyte adhesion and activation on endothelial cells[8]. The thromboresistance of the vascular system may thus depend upon the combined effects of NO, PGI_2 (see below) and possibly other endothelial factors, whilst NO may have an additional role in modulating the activation and migration of inflammatory leucocytes into the arterial wall. Disturbances in NO production by the endothelium are therefore likely to have profound pathophysiological consequences.

Physiological Vasodilatation

The vasodilator actions of many hormones and autacoids are mediated by release of NO from the endothelium, and the direct vasoconstrictor action of other hormones and transmitters is modulated by activation of the eNOS isoform in these cells.[2,4,5] Agents acting through NO include bradykinin and substance P, as well as acetylcholine (ACh), the classical stimulus for EDRF release, which acts *via* endothelial muscarinic receptors. Teleologically, the existence of muscarinic receptors for ACh on endothelial cells remains an enigma, as ACh is unlikely to be blood-borne under physiological circumstances. However, neurogenic vasodilatation induced by stimulation of cholinergic nerves in the rat and dog hindquarters[9,10] does involve the release of NO, most likely of endothelial origin,[10] suggesting that ACh may indeed mediate vasodilatation under physiological conditions by interacting with the endothelium.[4] Nonetheless, it is the vasodilatation elicited by increased shear stress that may be the most important physiological function of endothelium-derived NO. By this means, NO exerts a flow-related dilator feedback mechanism that opposes the myogenic response to raised intraluminal pressure[11] and, as discussed in Chapters 2 and 9, flow-dependent dilatation may be of central importance during physical exercise when there is increased demand for oxygen in the myocardium and skeletal muscle.[12] Interestingly, chronic exercise training upregulates eNOS in coronary vessels, thereby enhancing the potential for NO release when the metabolic demands of the myocardium increase.[13] Clearly, pathological disruption of NO synthesis may cause inappropriate distribution of blood flow and increased propensity for vasoconstriction, as well as increased cell adhesion.

Synergy between NO and Prostacyclin in Blood Vessels and Platelets

NO, like PGI_2, is a potent inhibitor of platelet aggregation, and promotes disaggregation of clumped platelets.[1,5,7] In addition, there is a strong synergy between PGI_2 and NO in promoting these effects in the platelet, which is probably played out at the level of cyclic nucleotides and their respective protein kinases, since PGI_2 acts through stimulation of adenylate cyclase and adenosine 3',5'-cyclic monophosphate (cAMP).[1,5] Thus NO and PGI_2 work together to prevent intraluminal aggregation of platelets but allow extraluminal adhesion and haemostatic function to proceed with minimal hindrance.

Synergy between PGI_2 and NO may also be important in the regulation of vascular tone. For example, the smooth muscle of rat aorta and porcine coronary artery is itself insensitive to low concentrations of PGI_2, but a vasorelaxant action of PGI_2 and other activators of cAMP — including isoprenaline and calcitonin gene-related peptide (CGRP) — is revealed if endothelium is present.[14–16] This can be attributed to the background release of NO from the endothelium.[14,15] Therefore, the synergy between the cyclic nucleotides explains the amplification by the endothelium of vasodilatation induced by cAMP activators in these tissues.[1] This synergy may be particularly relevant *in vivo*, where changing shear stresses on the endothelial cells promote release of both NO and PGI_2[17] (see also Chapter 2). These mediators thus act in concert to ensure appropriate distribution of blood flow to areas of increased oxygen demand, such as may occur during functional hyperaemia and at sites of stenoses. It is yet to be determined how widely this synergy occurs across species, particularly in the microvasculature.

Inflammation and Cytotoxicity

NO-induced Vasodilatation and Sensory Nerves in Inflammation

Acute inflammation, which accompanies some cardiac and vascular diseases, depends upon the release of chemical mediators that increase the permeability of the venular endothelium. This leads to the extravasation of leucocytes, fluid and proteins, thus producing oedema. It is well known that the release of endogenous vasodilators that open up microvascular arterioles and increase local blood flow to the site of inflammation potentiate inflammatory oedema. NO and PGI_2 derived from the endothelium are vasodilator mediators that have prominent roles in inflammation.[18]

Another component of inflammation involves the activation of sensory nerves which brings about pain, and also initiates local vasodilatation and protein extravasation. At their peripheral ends, these sensory nerves release neuropeptides, particularly substance P and CGRP which, in turn, recruit mast cells, nerves and the endothelium to participate in so-called 'neurogenic' inflammation. In the skin of rats and rabbits, substance P releases NO which causes vasodilatation and promotes extravasation of protein from the microvasculature.[4,19] In contrast, NO does not contribute substantially to the vasodilator action of CGRP.[4,20] However, direct stimulation of the sensory nerves and the ensuing inflammatory oedema do appear to involve the production of NO from constitutively expressed enzymes, localized mainly in the nerves.[21] Interestingly, the vasodilator response to activation of sensory nerves in rats increases with advancing age, and this appears to be attributable to an age-related increase in the contribution from endogenous NO.[22]

Cytotoxicity of NO

Constitutive Ca^{2+}-dependent NO synthase isoforms (eNOS and nNOS) produce relatively small amounts of NO upon cell activation by appropriate stimuli, producing rapid effects on neighbouring cells through the cGMP system.[5] Much larger amounts of NO are released over longer periods of time by the Ca^{2+}-independent, inducible NO synthase (iNOS), which is expressed in many cell types following their exposure to endotoxin or cytokines. The primary function of NO produced in this way is probably to kill invading microorganisms and cancer cells, although excessive NO may incidentally cause local vasodilatation, inhibit mitogenesis, or cause damage to surrounding cells of the host, or even to the cells that are producing it.[5] The molecular targets identified in such cytostatic or cytotoxic actions are the iron-centred enzymes involved in mitochondrial respiration, aconitase activity and DNA synthesis.[23] However, another potential route of NO toxicity is through its interaction with superoxide anion ($O_2^{\bullet -}$). Originally it was demonstrated that $O_2^{\bullet -}$ inactivated or 'neutralized' NO derived from endothelial cells,[4,24] but the reaction between these two free radicals can also produce the peroxynitrite anion ($ONOO^-$).[25] This powerful oxidant is generated by activated macrophages *in vitro*,[25] and at appropriate pH it decomposes to tissue damaging $\cdot OH$ and $\cdot NO_2$ radical species (see also Chapters 17 and 18). Although the conditions under which peroxynitrite may be toxic *in vivo* remain to be fully defined, the presence of nitrotyrosine (3-nitro-L-tyrosine), a stable product of the reaction between peroxynitrite and tyrosine residues present in proteins, has been taken as an indicator of iNOS activity and peroxynitrite-induced tissue damage.[26]

Hypercholesterolaemia and Atherosclerosis

The pathophysiological consequences of abnormalities in endothelial NO production are nowhere more prominent than in atherosclerosis. A striking feature of atherosclerotic coronary artery disease in man is that ACh can cause a constrictor response, whereas in normal coronary arteries it produces endothelium-dependent vasodilatation.[27] Serotonin also causes exaggerated coronary vasoconstriction in individuals susceptible to vasospasm assumed to be in the early stages of atherogenesis, and after coronary angioplasty.[28] Hypercholesterolaemia is associated with more subtle changes in vascular reactivity that precede the development of lesions in conduit arteries.[29] Adhesion, followed by migration of inflammatory cells and smooth muscle cells within the vessel wall are factors common to these three conditions, and some of the mechanisms involved in the vascular remodelling that occurs after angioplasty are probably shared with the much slower process of atherogenesis. Various animal models of atherosclerosis have been studied to investigate the mechanisms of these changes in endothelial NO function, and its potential consequences.

Experimental Studies

The development of intimal thickening in hypercholesterolaemic rabbits is associated with progressive impairment of endothelium-dependent vasodilatation,[30] and atheroma may compromise the apparent ability of ACh to release NO from the endothelium.[31] In other models of atherosclerosis, such as a normocholesterolaemic rabbit model that we have studied,[32–35] impairment of endothelium-dependent vasodilatation actually precedes intimal thickening. In this model, a neo-intima develops rapidly in the carotid artery after application of a peri-arterial collar and, despite maintaining a cholesterol-free diet, these neo-intimal lesions accumulate cholesteryl esters.[35] The sensitivity of the smooth muscle sGC to stimulation by NO is unchanged during lesion development.[32,34] Most animal studies suggest that the bioavailability of NO from the endothelium is reduced in hypercholesterolaemia or atherosclerosis; this defect could lie at the level of the endothelial receptors (particularly the muscarinic receptor for ACh) or in the coupling of the receptor (*via* intracellular transducers) to activation of eNOS. Alternative explanations of the dysfunction include the possibility that substrate availability might be compromised, since the provision of exogenous L-arginine partially restores endothelium-dependent vasodilatation.[36,37] Impaired function can also be attributed to accelerated inactivation of NO by superoxide anions, for these radicals are also produced in atherosclerotic lesions.[38] Different models provide evidence for each of these mechanisms.

An additional factor that needs to be considered in this context is that the total production of nitrogen oxides may paradoxically be *increased* in hypercholesterolaemic arteries.[39] Indeed, we have recently shown immunohistochemically[40] that lesions induced by peri-arterial collars express iNOS in the smooth muscle and endothelial cells of the neo-intima (Figure 1), and similar observations have been made in advanced human atheromatous lesions. In some studies,[26,41] but not all,[42,43] iNOS and nitrotyrosine have been found together in atheromatous arteries, thus suggesting a direct role for peroxynitrite in the development of lesions. The expression of iNOS probably results from increased cytokine and growth factor activity, and a second consequence may be the downregulation of eNOS in the endothelium, for tumour necrosis factor (TNF)-α is known to promote the degradation of mRNA for eNOS.[44] Moreover, iNOS expression may lead to depletion of substrate for the endothelial enzyme, thus contributing to endothelial dysfunction. Clearly we need a better understanding

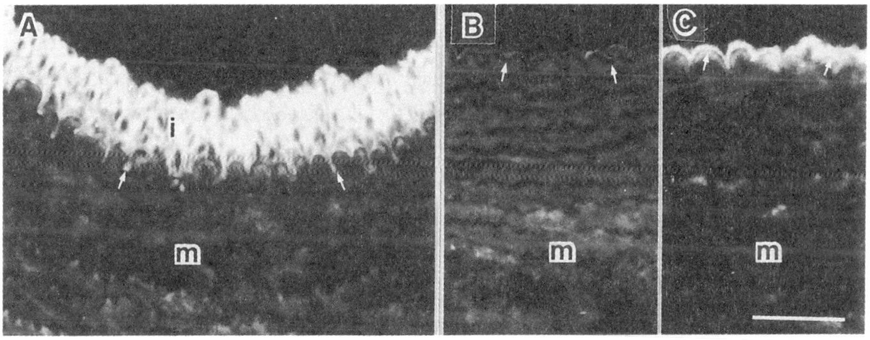

Figure 1. Fluorescence micrographs demonstrating localization of different isoforms of NO synthase (NOS) in atheroma-like neo-intima induced in rabbit carotid arteries.[32,40] A: Inducible NOS immunoreactivity is expressed in the neo-intima induced by a peri-arterial collar, but not in the medial layer of the artery. B and C: Control carotid artery sections processed for iNOS (B) or eNOS (C) immunohistochemistry. Control sections do not express iNOS in any part of the artery wall, and eNOS is restricted to the endothelial cells. Arrows indicate internal elastic lamina; i, neo-intima; and m, media. Scale bar = 50μm.

of the NO biosynthetic pathways and their precursor requirements in the participating cells to clarify the role of NO in atherogenesis.

Interactions with other enzyme systems may also be of major importance. As in the case of iNOS induction, inflammatory cytokines generated in atheroma can induce the expression of cyclooxygenase (COX)-2.[45] Indeed, in rabbit models of atherosclerosis and patients with atheroma there is excess production of prostanoids,[46,47] and oxidized low density lipoprotein (LDL) may exert effects *via* increased COX-2 expression.[48,49] Under certain conditions, NO may itself enhance COX activity, and COX can generate $O_2^{\bullet -}$ radicals, thereby providing the potential for enhanced peroxynitrite formation.[50–52] The likelihood of cross-talk between the iNOS and COX-2 enzymes is strengthened by their co-localization in animal models of inflammation.[50,53]

As in man, the impairment of endothelium-dependent vasodilatation in hypercholesterolaemic animal models extends to the coronary resistance vessels that, unlike the large arteries, do not develop intimal thickenings or sequester macrophages.[54] This also occurs in the resistance circulation downstream of collar-induced neo-intimal lesions in rabbit carotid arteries,[55] suggesting it arises at least partly from chemical mediators or from cells activated in the conduit artery lesions, such as neutrophil-derived vasoconstrictor factors (see below).

Human Studies

Probably the first direct evidence that endothelial NO function is compromised in human atherosclerosis was obtained by Förstermann *et al.*,[56] who reported that endothelium-dependent relaxation in human, isolated coronary arteries was reduced in atherosclerotic segments. Endothelial dysfunction, determined as an impaired vasodilator response to administration of an endothelium-dependent agonist or an abrupt increase in flow, has been demonstrated in human experiments both in the presence of established vascular disease as well as at earlier stages during disease development.[27,29,57] As mentioned above, the generalized nature of endothelial dysfunction is evidenced by its demonstration in vessels and vascular beds which are not, *per se*, sites at which the vascular pathology is apparent.[58–60] Endothelial dysfunction may have direct haemodynamic consequences but additionally it serves as a marker for other processes affected by endothelial NO production, such as cell adhesion.

Abnormalities in Conduit and Resistance Arteries

Confirming the animal studies, evidence of an impaired vasodilator response in humans has been obtained both in conduit and resistance vessels *in vivo* as well as in small arteries *ex vivo*.[61] In some circumstances, such as the impairment of dilatation elicited by ischaemia in the brachial artery of patients with hypercholesterolaemia, there is no direct pathophysiological correlate but the impairment may indicate a systemic problem (such as platelet aggregation and adhesion of leucocytes and platelets to the endothelium) likely to be of importance at more critical sites. Studies of platelet function have confirmed alterations consistent with compromised endothelial NO production.[62] Alterations in circulating levels of endothelium-derived protein S and von Willebrand Factor, as well as evidence of changes in plasminogen-activator inhibitor function, have been found in patients with coronary artery disease and with acute coronary syndromes.[63] In conduit arteries, such as the coronary and carotid arteries, impairment of endothelial NO function may be directly responsible for pathophysiological consequences by promoting vasospasm.[64]

Evidence of an impaired vasodilator response at the level of resistance vessels has been obtained both *in vivo* and with *ex vivo* biopsy material and may affect the availability of nutritive flow. *Ex vivo* studies of subcutaneous vessels obtained from gluteal biopsies or explanted human hearts confirmed that endothelium-dependent vasorelaxation is compromised in hypercholesterolaemic vessels.[65,66] *In vivo* studies have been conducted on the circulation in the forearm, which is accessible for flow measurement (by plethysmography or Doppler ultrasound), for the infusion of drugs, and for the simultaneous collection of arterial and venous samples for determination of nitrate and nitrite. A noteworthy feature of the *in vivo* studies has been the reliance in many cases of only a limited number of endothelial stimuli, usually ACh, raising the possibility that some of the demonstrated effects could be receptor specific, although endothelial dysfunction has recently been reported with both ACh and substance P in coronary arteries *in vivo*.[67] Future studies should address these questions using a broader range of agonists.

Endothelial Dysfunction Linked to Other Risk Factors for Atherosclerosis

The risk factors known to increase the susceptibility to atherosclerotic disease in humans (apart from hypercholesterolaemia) include age, hypertension, cigarette smoking, diabetes mellitus, the post-menopausal state, hyperhomocystinaemia, as well as family history. Some of these factors have been clearly linked to a diminished response to endothelium-dependent vasodilators, whereas for others, such as hypertension, the evidence for such a link is more controversial.[68–71] These putative links between endothelial function and early disease have been established mainly in conduit vessels such as the epicardial coronary arteries. Data from a small number of patients have objectively confirmed myocardial ischaemia in the absence of demonstrable coronary artery disease ('cardiological syndrome X'). The mechanism of the ischaemia in these cases remains to be clarified, but here the endothelium in resistance vessels is implicated.[72] An association between the 'metabolic syndrome' (which includes hyperinsulinaemia) and cardiological syndrome X is of particular interest in view of evidence that the vasodilator effects of insulin appear to be attenuated by endothelial dysfunction.[73,74]

Effects on Outcome of Atherosclerosis

The clinical consequences of established atherosclerotic plaques can be subdivided into chronic/stable and acute/unstable events. In addition, ischaemic heart disease is a major cause of end-stage heart failure, in which endothelial dysfunction has been well documented. While it might seem unlikely that endothelial dysfunction could contribute to *stable* angina pectoris, in which metabolic vasodilatation might be expected to achieve maximal flow, recent studies have shown that cholesterol reduction significantly reduces the prevalence of 'silent' ischaemia as reflected by episodes of ST segment depression during ambulatory monitoring.[75] Endothelial dysfunction may therefore contribute to ischaemia in this setting. Furthermore, myocardial infarction and unstable angina are both characterized by thrombosis at the site of a significant but non-occlusive atherosclerotic plaque (which in the case of myocardial infarction has generally been subject to tear or rupture). Clearly, there are a number of ways in which the endothelial production of NO could influence the outcome of these events, although at the present time direct evidence is lacking. The early benefit of cholesterol reduction in terms of patient survival may reflect modulation of these acute endothelial events.

Restoration of Endothelial Function

Several studies have found it possible to restore endothelial function using a number of different approaches. In subjects in whom a clearly identified abnormality, such as hypercholesterolaemia, is responsible for the endothelial dysfunction, treatment rapidly reverses the endothelial defect, even within one month of therapy or after one treatment with LDL apheresis.[76] Interestingly, dietary supplementation with n-3 fatty acids (marine oils) restores the impaired dilator responses to ACh without lowering LDL cholesterol.[60,77] The mechanism of this effect has not been established, but could relate to the synergy between NO and PGI_2, discussed above. Some, but not all, studies have demonstrated improvement in endothelial function following treatment with the NO precursor, L-arginine.[29,78]

Treatment with direct NO donor drugs has achieved limited success in preventing restenosis after coronary angioplasty.[79] In rabbits, NO donors have been shown to prevent neo-intima formation and iNOS expression, whilst preserving endothelial NO function in the peri-arterial collar model, probably by suppressing smooth muscle migration and proliferation.[80,81] The demonstration that the endothelial defect caused by LDL cholesterol may depend on the presence of lipoprotein oxidation products has also led to the investigation of the effects of antioxidant treatment.[82]

Lastly, beneficial effects on endothelial dysfunction have also been found in recent trials with angiotensin converting enzyme (ACE) inhibitors.[83] These drugs are effective in limiting neo-intima development in balloon-denuded arteries of rats,[84] and reducing atheroma-like neo-intimal thickening in hypercholesterolaemia- and collar-induced lesions in rabbits[85–88] and hypercholesterolaemic cynomolgus monkeys.[89] These anti-atherosclerotic effects were exerted without significant changes in plasma cholesterol levels or blood pressure. As in humans, in both the collar and the hyperlipidaemic rabbit models, chronic treatment with ACE inhibitors reverses defective endothelial NO function.[85,87,88] Other studies have linked antiproliferative effects of ACE inhibitors, and restoration of endothelial function, to augmentation of endogenous kinins,[85,90] which in turn may derive from potentiating NO release.[90] How chronic treatment with ACE inhibitors restores endothelial NO function requires further study, and the potential involvement of PGI_2, which is also released from the endothelium by kinins and synergizes with NO, should not be overlooked (see above).

Reperfusion Injury of the Vasculature in Ischaemic Myocardium Produces Vasoconstrictor Factors

Following acute coronary events, the reperfusion of the ischaemic myocardium causes an inflammatory reaction associated with the migration of cells through the vessel wall and their sequestration in the myocardium and even with the

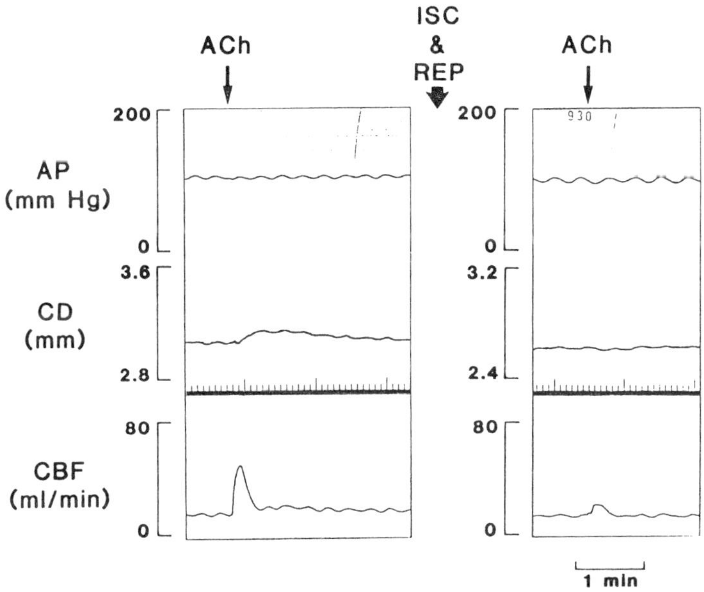

Figure 2. Impairment of endothelium-dependent vasodilatation in the coronary circulation after a period of myocardial ischaemia (60 min) and myocardial injury following reperfusion (30 min). Records are of arterial pressure (AP), circumflex artery diameter (CD) and circumflex coronary blood flow (CBF) in an anaesthetised dog. After ischaemia and reperfusion (ISC & REP) the acetylcholine-induced dilatation of the circumflex artery and the increase in blood flow are virtually abolished. Note also that the baseline circumflex CD has reduced, indicating marked vasoconstriction, possibly due to neutrophil-derived constrictor factors (see text). (Reproduced, with permission, from Sobey *et al.*[92])

vessel wall.[91] NO produced by invading inflammatory cells (associated with cytokine-induced expression of iNOS) could damage myocytes, particularly in the infarct border. This would be exacerbated by the simultaneous production of superoxide and hence peroxynitrite by these cells (see above), and the role of NO in this context is discussed in detail in Chapter 17 of this volume. However, there are endogenous inhibitors of NO, demonstrated to exert effects in the coronary vasculature, which have the potential to modulate these toxic reactions in the myocardium and coronary vasculature. Coronary occlusion followed by reperfusion of the heart injured the endothelial cells in a canine model, thus compromising coronary vasodilatation (Figure 2) and augmenting vasoconstrictor mechanisms.[92,93] We found that although the endothelial cells remain attached in the epicardial coronary arteries, leucocytes adhere to them and endothelium-dependent vasodilatation is abolished. Moreover, the coronary arteries constrict *in vivo* and the action of direct vasodilators is compromised, suggesting that the arteries are under the influence of local vasoconstrictor factors.[92,93] Indeed, invading neutrophils release several factors known to cause vasoconstriction. These include superoxide

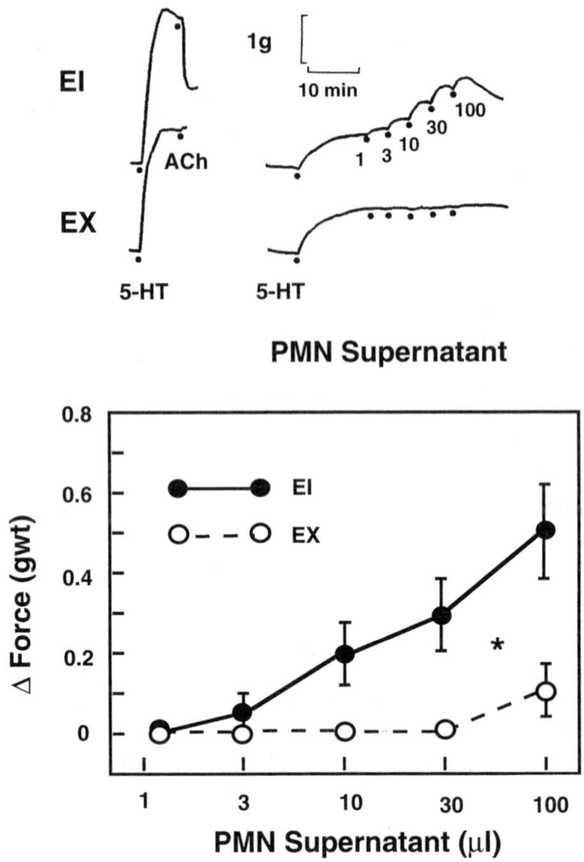

Figure 3. A neutrophil-derived inhibitor of NO activity detected by bioassay on rabbit aortic rings. The supernatant of rabbit polymorphonuclear leucocytes (PMN) causes concentration-dependent contractions of rings partially constricted with serotonin (5-HT), but only if they contain intact endothelium (EI), and not if they are endothelium denuded (EX).[96,97] (Reproduced, with permission, from Dusting[4])

(that could neutralize endothelium-derived NO)[94] as well as stable factors, including a peptide of molecular weight less than 5 kDa[95] and another peptide of greater than 5 kDa. We found the latter[96] to be released from rabbit neutrophils *in vitro* and this causes vasoconstriction by inactivating endothelium-derived NO (Figure 3). The release of a similar factor (above 30 kDa) from human neutrophils was augmented by priming them with TNF-α.[97] The identities of these endogenous peptides that are potentially important regulators of NO action have yet to be determined, but these factors might well contribute to abnormalities of myocardial reperfusion, including the 'no-reflow' phenomenon, and have roles in other forms of inflammation. Treatment with a variety of drugs, including β-adrenoceptor antagonists, Ca^{2+} antagonists and free radical scavengers, protects against the loss of NO function through diverse mechanisms.[98] Thus, pharmacological intervention to preserve endothelial function could significantly improve coronary blood flow and cardiac function after ischaemic episodes, and this might be a useful approach for increasing myocardial salvage after thrombolytic therapy.

Roles of Nitric Oxide in Transplantation and Rejection

Endothelial dysfunction does not only occur during classical atherogenesis, but it is an early and marked feature of the coronary circulation in transplanted hearts.[99] Indeed, endothelial dysfunction in cardiac transplant recipients has been found to be predictive of the subsequent development of coronary artery disease, which usually takes the form of a widespread concentric arterial narrowing rather than the classical focal disease of non-transplant atheroma. This might reflect the initiation of processes that could lead to allograft rejection were it not for immunosuppressant therapy.

In 1982 it was reported that L-arginine was able to accelerate rejection of skin allografts in mice,[100] although neither the metabolic fate of arginine nor the existence of endogenous NO was known at that time. There is now substantial evidence both from *in vitro* and *in vivo* models that NO production is increased during acute allograft rejection. Langrehr *et al.*[101] observed that there was a marked rise in plasma nitrate levels in association with both host-versus-graft and graft-versus-host allogeneic responses in a variety of transplant models in the rat. No increase in plasma nitrate levels was observed in animals who received syngeneic grafts. In a similar rat cardiac allograft model, Winlaw *et al.*[102] reported a rise in urinary nitrate excretion in association with acute rejection, which was blocked by the NO synthase inhibitor N-monomethyl-L-arginine (L-NMMA). The time course of the increase in urinary nitrate excretion mirrored the increase in plasma nitrate levels observed in the study of Langrehr *et al.*[101] and confirmed that the increase in plasma levels was due to increased endogenous NO production rather than a decrease in nitrate excretion. In both studies, the rise in plasma and urinary nitrate preceded the onset of rejection as detected by daily graft palpation. The increased plasma and urinary nitrate levels during allograft rejection were suppressed by treatment with L-NMMA, the macrolide tacrolimus (FK 506),[101] cyclosporin and dexamethasone[102] in doses that prevented or prolonged the time to rejection.

The increased NO production during acute cardiac allograft rejection in the rat correlates closely with activation of iNOS within the allograft.[103] Increased iNOS activity was observed in macrophages, vascular endothelial cells and cardiac myocytes, the peak enzyme activity occurring on day 5 corresponding to the peak in plasma and urinary nitrate levels[101,102] and the onset of graft dysfunction. It is thus most likely that iNOS expression is induced by cytokines released by activated lymphocytes and monocytes, and the immunosuppressive agents inhibit NO production by suppressing iNOS expression,[104,105] or indirectly *via* their inhibitory effects on cytokine activation.

The functional consequences of increased NO production during acute allograft rejection are yet to be clarified, but they appear to be quite complex. Chan *et al.*[106] demonstrated that total coronary blood flow was increased during acute rejection in heart transplant patients. The cause of this increased coronary flow remains uncertain, but cytokine-induced expression of iNOS in the heart or coronary vasculature is probably important. Increased coronary blood flow exacerbates myocardial oedema formation and augments the influx of inflammatory cells, resulting in reduced diastolic compliance, which is the earliest functional abnormality observed in acute cardiac rejection.[107] On the other hand, NO improves ventricular relaxation and distensibility[108] and in this context may reduce diastolic dysfunction during rejection. In addition, NO production may account for the negative inotropic actions of a number of pro-inflammatory cytokines, including interleukin (IL)-1,[109] IL-2 and TNF-α. Thus NO may contribute to the systolic dysfunction seen in more advanced cardiac allograft rejection through its negative inotropic actions.

NO may also exacerbate graft dysfunction through its cytotoxic actions, either directly or by generating peroxynitrite (see above). Electron paramagnetic resonance studies of the rat heart during rejection demonstrate the formation of iron-nitrosyl complexes not present in native immunocompetent tissues, consistent with NO-mediated destruction of iron-containing respiratory enzymes within the allograft.[110]

The immune sytem may also be modulated indirectly by NO. Some actions could be seen as immunostimulatory: for example, L-arginine enhances surgical wound healing,[111] has a trophic effect on the thymus and stimulates release of a number of pituitary hormones, including growth hormone and prolactin.[112] Inhibition of prolactin production by bromocryptine has been shown to potentiate the immunosuppressive efficacy of cyclosporin.[113] On the other hand, NO may have immunomodulatory actions, including inhibition of lymphocyte proliferation *in vitro*[114] and curbing the expansion of alloreactive lymphocyte colonies during rejection.[115] The net effect of NO on allograft function may thus depend on the balance between these immunostimulatory and immunomodulatory actions. Conceivably, this balance may vary between transplanted organs at different times and between species. Recently, Winlaw *et al.*[116] have shown that treatment of rats with L-NMMA produced a small but significant increase in rat cardiac allograft survival, although this effect was weaker than that of conventional immunosuppressive agents. This finding indicates that acute cardiac allograft rejection can proceed to complete graft loss despite total suppression of NO production, indicating that NO is not a key effector molecule in the rejection process. However, this finding does not exclude the possibility that selective inhibition of NO production may potentiate the action of other immunosuppressive agents or that NO may exacerbate graft dysfunction (*via* its negative inotropic action in the heart, for example), even in the absence of any direct cytotoxic effect.

Inhibition of NOS Induction by Immunosuppressant Drugs

iNOS can be expressed with adverse consequences for the circulation during the activation of the immune system following transplantation of organs, including heart, lung, liver or kidney. Rejection of allografts is associated with increased production of NO, which may be cytotoxic to the transplanted tissue in this setting (see above). Immunosuppressant drugs such as cyclosporin and FK 506 suppressed serum nitrite and nitrate to control levels in rats that were rejecting allografts of liver, heart or small bowel.[101] It has been assumed that the suppressant action of these drugs on NO production stemmed from inhibition of the cytokines required to activate iNOS gene expression. However, we found in monocultures of vascular smooth muscle and a murine macrophage cell line (J774 cells), that cyclosporin A and FK 506 suppress iNOS induction stimulated by individual cytokines or by lipopolysaccharide.[104] Cyclosporin, which is some 30–100-fold *less* potent than FK 506 in suppressing IL-2 production, was active in suppressing iNOS at therapeutic concentrations and was *more potent* than the macrolide in these cell lines.[105] These data raise the possibility that cyclosporin has a specific action on iNOS gene expression or activity that is not shared by other immunosuppressants and operates independently of its well-defined action on cytokines.[117] Whether this property is useful in the setting of transplant rejection is yet to be determined.

It has also been reported recently that cyclosporin and FK 506 block the activity (rather than the expression) of nNOS. The intracellular mechanism of the suppressant action appears to involve inhibition of the dephosphorylation of the NOS enzyme, which is a post-translational modification of the nNOS protein required for full expression of its activity.[118] We found no evidence that this occurs for iNOS in vascular smooth muscle cells or J774 macrophages, although cyclosporin inhibited the transcription of iNOS (unpublished).

Some years ago, immunosuppressant drugs were suggested to have potential for managing recent onset, insulin-dependent diabetes. Such diabetes appears to be caused by the attack of cytotoxic T-cells and macrophages on β-islet cells of the pancreas, producing NO and other radical species that increase toxicity.[119] Cyclosporin initially showed promise in preserving β-islet cell function in such patients,[120] although the serious side effects of hypertension and nephrotoxicity associated with this drug have limited its application in this condition. Nevertheless, specific suppression of iNOS expression may be a property of cyclosporin and other immunosuppressants that could be developed for this potential application in preventing diabetes, and selective iNOS inhibitors might prove to be of value in this context.

Summary

NO has complex roles in cardiac and vascular function, and under certain conditions may be cytotoxic. There are several pathological circumstances where

NO is produced in excess by the iNOS isoform in blood vessels. This is often accompanied by defects in the operation of eNOS, which in turn compromises the moment-to-moment regulation of blood flow and promotes adhesion of platelets and leucocytes. A more complete understanding of the cross-regulation of the NOS isoforms will be necessary to develop strategies for manipulating NO production therapeutically. The advantages of selectively suppressing iNOS activity must be analysed carefully, for this might suggest new approaches to improving cardiovascular function in atherosclerosis and ischaemia-reperfusion injury, and for preserving the function of transplanted organs in the longer term. There is incontrovertible evidence of a link between dysfunction in the endothelial production of NO and these common vascular problems in humans. In atherosclerosis, endothelial NO dysfunction can be reversed or improved by a variety of therapeutic interventions, such as NO donor compounds and ACE inhibitors. Questions remain regarding the exact mechanisms underlying these endothelial defects in man, but studies are continuing to assess the extent to which improvement in clinical outcomes can be attributed to restoration of endothelial NO production.

References

1. Dusting GJ, Macdonald PS. Prostacyclin and vascular function: implications for hypertension and atherosclerosis. *Pharmacol Ther* 1990; **48**: 323–344.
2. Lüscher TF, Noll G. The pathogenesis of cardiovascular disease: role of the endothelium as a target and mediator. *Atherosclerosis* 1995; **118 (Suppl)**: S81–S90.
3. Ignarro LJ, Lippton H, Edwards JC *et al.* Mechanism of vascular smooth muscle relaxation by organic nitrates, nitrites, nitroprusside and nitric oxide: evidence for the involvement of *S*-nitrosothiols as active intermediates. *J Pharmacol Exp Ther* 1981: **218**: 739–749.
4. Dusting GJ. Nitric oxide in cardiovascular disorders. *J Vasc Res* 1995; **32**: 143–161.
5. Moncada S, Palmer RMJ, Higgs EA. Nitric oxide: physiology, pathophysiology, and pharmacology. *Pharmacol Rev* 1991; **43**: 109–142.
6. Lincoln TM, Cornwell TL. Towards an understanding of the mechanism of action of cyclic AMP and cyclic GMP in smooth muscle relaxation. *Blood Vessels* 1991; **28**: 129–137.
7. deBelder AJ, Radomski MW. Nitric oxide in the clinical arena. *J Hypertens* 1994; **12**: 617–624.
8. Kubes P, Suzuki M, Granger DN. Nitric oxide: an endogenous modulator of leukocyte adhesion. *Proc Natl Acad Sci USA* 1991; **88**: 4651–4655.
9. Loke KE, Sobey CG, Dusting GJ, Woodman OL. Cholinergic neurogenic vasodilatation is mediated by nitric oxide in the dog hindlimb. *Cardiovasc Res* 1994: **28**: 542–547.

10. Loke KE, Sobey CG, Dusting GJ, Woodman OL. Requirement for endothelium-derived nitric oxide in vasodilation produced by stimulation of cholinergic nerves in rat hindquarters. *Br J Pharmacol* 1994; **112**: 630–634.
11. Griffith TM, Edwards DH. Myogenic autoregulation of flow may be inversely related to endothelium-derived relaxing factor activity. *Am J Physiol* 1990; **258**: H1171–H1180.
12. Wilson JR, Kapoor S. Contribution of endothelium-derived relaxing factor to exercise-induced vasodilation in humans. *J Appl Physiol* 1993; **75**: 2740–2744.
13. Sessa WC, Pritchard K, Seyedi N *et al.* Chronic exercise in dogs increases coronary vascular nitric oxide production and endothelial cell nitric oxide synthase gene expression. *Circ Res* 1994; **74**: 349–353.
14. Grace GC, Dusting GJ, Kemp BE, Martin TJ. Endothelium and the vasodilator action of rat calcitonin gene-related peptide (CGRP). *Br J Pharmacol* 1987; **91**: 729–733.
15. Grace GC, Macdonald PS, Dusting GJ. Cyclic nucleotide interactions involved in endothelium-dependent dilatation in rat aortic rings. *Eur J Pharmacol* 1988; **148**: 17–24.
16. Shimokawa H, Flavahan NA, Lorenz RR, Vanhoutte PM. Prostacyclin releases endothelium-derived relaxing factor and potentiates its action in coronary arteries of the pig. *Br J Pharmacol* 1988; **95**: 1197–1203.
17. Koller A, Kaley G. Prostaglandins mediate arteriolar dilation to increased blood flow velocity in skeletal muscle microcirculation. *Circ Res* 1990; **67**: 529–534.
18. Ialenti A, Ianaro A, Moncada S, Di Rosa M. Modulation of acute inflammation by endogenous nitric oxide. *Eur J Pharmacol* 1992; **211**: 177–182.
19. Ralevic V, Khalil Z, Helme RD, Dusting GJ. Role of nitric oxide in the actions of substance P and other mediators of inflammation in rat skin microvasculature. *Eur J Pharmacol* 1995; **284**: 231–239.
20. Ralevic V, Khalil Z, Dusting GJ, Helme RD. Nitric oxide and sensory nerves are involved in the vasodilator response to acetylcholine but not calcitonin gene-related peptide in rat skin microvasculature. *Br J Pharmacol* 1992; **106**: 650–655.
21. Merhri M, Dusting GJ, Khalil Z. CGRP and nitric oxide of neuronal origin mediate neurogenic vasodilatation in rat skin microvasculature. *Br J Pharmacol* 1998; **123**: 863–868.
22. Khalil Z, Ralevic V, Bassirat M *et al.* Effects of ageing on sensory nerve function in rat skin. *Brain Res* 1994; **641**: 265–272.
23. Hibbs JB Jr, Taintor RR, Vavrin Z *et al.* Synthesis of nitric oxide from a terminal guanidino nitrogen atom of L-arginine: a molecular mechanism regulating cellular proliferation that targets intracellular iron. In: *Nitric Oxide from L-Arginine: A Bioregulatory System*, Moncada S, Higgs E (Eds). Elsevier, Amsterdam 1990: 189–223.

24. Rubanyi GM, Vanhoutte PM. Superoxide anions and hyperoxia inactivate endothelium-derived relaxing factor. *Am J Physiol* 1986; **250**: H822–H827.
25. Beckman JS, Beckman TW, Chen J *et al*. Apparent hydroxyl radical production by peroxynitrite: implications for endothelial injury from nitric oxide and superoxide. *Proc Natl Acad Sci USA* 1990; **87**: 1620–1624.
26. Beckmann JS, Ye YZ, Anderson PG *et al*. Extensive nitration of protein tyrosines in human atherosclerosis detected by immunohistochemistry. *Biol Chem Hoppe Seyler* 1994; **375**: 81–88.
27. Ludmer PL, Selwyn AP, Shook TL *et al*. Paradoxical vasoconstriction induced by acetylcholine in atherosclerotic coronary arteries. *N Engl J Med* 1986; **315**: 1046–1051.
28. McFadden EP, Bauters C, Lablanche J *et al*. Response of human coronary arteries to serotonin after injury by coronary angioplasty. *Circulation* 1993; **88**: 2076–2085.
29. Drexler H, Zeiher AM, Meinzer K, Just H. Correction of endothelial dysfunction in coronary microcirculation of hypercholesterolaemic patients by L-arginine. *Lancet* 1991; **338**: 1546–1550.
30. Verbeuren TJ, Jordaens FH, Zonnekeyn LL *et al*. Effect of hypercholesterolemia on vascular reactivity in the rabbit. I. Endothelium-dependent and endothelium-independent contractions and relaxations in isolated arteries of control and hypercholesterolemic rabbits. *Circ Res* 1986; **58**: 552–564.
31. Verbeuren TJ, Jordaens FH, Van Hove CE *et al*. Release and vascular activity of endothelium-derived relaxing factor in atherosclerotic rabbit aorta. *Eur J Pharmacol* 1990; **191**: 173–178.
32. Dusting GJ, Curcio A, Harris PJ *et al*. Supersensitivity to vasoconstrictor action of serotonin precedes the development of atheroma-like lesions in the rabbit. *J Cardiovasc Pharmacol* 1990; **16**: 667–674.
33. Arthur JF, Dusting GJ. Selective endothelial dysfunction in early atheroma-like lesions in the rabbit. *Coron Artery Dis* 1992; **3**: 623–629.
34. De Meyer GR, Bult H, Van Hoydonck A-E *et al*. Neointima formation impairs endothelial muscarinic receptors while enhancing prostacyclin-mediated responses in the rabbit carotid artery. *Circ Res* 1991; **68**: 1669–1680.
35. Booth RF, Martin JF, Honey AC *et al*. Rapid development of atherosclerotic lesions in the rabbit carotid artery induced by perivascular manipulation. *Atherosclerosis* 1989; **76**: 257–268.
36. Girerd XJ, Hirsch AT, Cooke JP *et al*. L-arginine augments endothelium-dependent vasodilation in cholesterol-fed rabbits. *Circ Res* 1990; **67**: 1301–1308.
37. Rossitch E Jr, Alexander E, Black PM, Cooke JP. L-arginine normalizes endothelial function in cerebral vessels from hypercholesterolemic rabbits. *J Clin Invest* 1991; **87**: 1295–1299.
38. Ohara Y, Peterson TE, Harrison DG. Hypercholesterolemia increases endothelial superoxide anion production. *J Clin Invest* 1993; **91**: 2546–2551.

39. Minor RL Jr, Myers PR, Guerra R Jr *et al.* Diet-induced atherosclerosis increases the release of nitrogen oxides from rabbit aorta. *J Clin Invest* 1990; **86**: 2109–2116.
40. Arthur JF, Yin ZL, Young HM, Dusting GJ. Induction of nitric oxide synthase in the neointima induced by a periarterial collar in rabbits. *Arterioscler Thromb Vasc Biol* 1997; **17**: 737–740.
41. Buttery LDK, Springall DR, Chester AH *et al.* Inducible nitric oxide synthase is present within human atherosclerotic lesions and promotes the formation and activity of peroxynitrite. *Lab Invest* 1996; **75**: 77–85.
42. LaFond-Walker A, Chen CL, Augustine S *et al.* Inducible nitric oxide synthase expression in coronary arteries of transplanted human hearts with accelerated graft atherosclerosis. *Am J Pathol* 1997; **151**: 919–925.
43. Evans P, Kaur H, Mitchinson MJ, Halliwell B. Do human atherosclerotic lesions contain nitrotyrosine? *Biochem Biophys Res Commun* 1996; **226**: 346–351.
44. Yoshizumi M, Perrella MA, Burnett JC Jr, Lee ME. Tumor necrosis factor downregulates an endothelial nitric oxide synthase mRNA by shortening its half-life. *Circ Res* 1993; **73**: 205–209.
45. Wu KK. Inducible cyclooxygenase and nitric oxide synthase. *Adv Pharmacol* 1995; **33**: 179–207.
46. FitzGerald GA, Smith B, Pedersen AK, Brash AR. Increased prostacyclin biosynthesis in patients with severe atherosclerosis and platelet activation. *N Engl J Med* 1984; **310**: 1065–1068.
47. Tremoli E, Socini A, Petroni A, Galli C. Increased platelet aggregability is associated with increased prostacyclin production by vessel walls in hypercholesterolemic rabbits. *Prostaglandins* 1982; **24**: 397–404.
48. Maier JA, Barenghi L, Bradamante S, Pagani F. Modulators of oxidized LDL-induced hyperadhesiveness in human endothelial cells. *Biochem Biophys Res Commun* 1994; **204**: 673–677.
49. Wohlfeil ER, Campbell WB. 25-Hydroxycholesterol enhances eicosanoid production in cultured bovine coronary artery endothelial cells by increasing prostaglandin G/H synthase-2. *Biochim Biophys Acta* 1997; **1345**: 109–120.
50. Salvemini D, Masferrer JL. Interactions of nitric oxide with cyclooxygenase: *in vitro, ex vivo* and *in vivo* studies. *Methods Enzymol* 1996; **269**: 12–25.
51. Salvemini D, Misko TP, Masferrer JL *et al.* Nitric oxide activates cyclooxygenase enzymes. *Proc Natl Acad Sci USA* 1993; **90**: 7240–7244.
52. Inoue T, Fukuo K, Morimoto S *et al.* Nitric oxide mediates interleukin-1-induced prostaglandin E2 production by vascular smooth muscle cells. *Biochem Biophys Res Commun* 1993; **194**: 420–424.
53. Mitchell JA, Larkin S, Williams TJ. Cyclooxygenase-2: regulation and relevance in inflammation. *Biochem Pharmacol* 1995; **50**: 1535–1542.

54. Sellke FW, Armstrong ML, Harrison DG. Endothelium-dependent vascular relaxation is abnormal in the coronary microcirculation of atherosclerotic primates. *Circulation* 1990; **81**: 1586–1593.
55. Arthur JF, Dusting GJ, Woodman OL. Impaired vasodilator function of nitric oxide associated with developing neo-intima in conscious rabbits. *J Vasc Res* 1994; **31**: 187–194.
56. Förstermann U, Mügge A, Alheid U *et al.* Selective attenuation of endothelium-mediated vasodilation in atherosclerotic human coronary arteries. *Circ Res* 1988; **62**: 185–190.
57. Zeiher AM, Drexler H, Saurbier B, Just H. Endothelium-mediated coronary blood flow modulation in humans. Effects of age, atherosclerosis, hypercholesterolemia and hypertension. *J Clin Invest* 1993; **92**: 652–662.
58. Lieberman EH, Gerhard MD, Uehata A *et al.* Flow-induced dilation of the human brachial artery is impaired in patients < 40 years of age with coronary artery disease. *Am J Cardiol* 1996; **78**: 1210–1214.
59. Celermajer DS, Sorensen KE, Gooch VM *et al.* Non-invasive detection of endothelial dysfunction in children and adults at risk of atherosclerosis. *Lancet* 1992; **340**: 1111–1115.
60. Chin JPF, Dart AM. Therapeutic restoration of endothelial function in hypercholesterolaemic subjects: effect of fish oils. *Clin Exp Pharmacol Physiol* 1994; **21**: 749–755.
61. Calver A, Collier J, Vallance P. Nitric oxide and cardiovascular control. *Exp Physiol* 1993; **78**: 303–326.
62. Langford EJ, Wainwright RJ, Martin JF. Platelet activation in acute myocardial infarction and unstable angina is inhibited by nitric oxide donors. *Arterioscler Thromb Vasc Biol* 1996; **16**: 51–55.
63. Dart A, Sherrard B, Salem H. Protein S, von Willebrand factor, and euglobulin lysis time as markers of coronary heart disease [Letter]. *Ann Intern Med* 1997; **126**: 89.
64. Kugiyama K, Yasue H, Okumura K *et al.* Nitric oxide activity is deficient in spasm arteries of patients with coronary spastic angina. *Circulation* 1996; **94**: 266–271.
65. Kichuk MR, Seyedi N, Zhang X *et al.* Regulation of nitric oxide production in human coronary microvessels and the contribution of local kinin formation. *Circulation* 1996; **94**: 44–51.
66. Goode GK, Heagerty AM. *In vitro* responses of human peripheral small arteries in hypercholesterolemia and effects of therapy. *Circulation* 1995; **91**: 2898–2903.
67. Quyyumi AA, Mulcahy D, Andrews NP *et al.* Coronary vascular nitric oxide activity in hypertension and hypercholesterolemia. Comparison of acetylcholine and substance P. *Circulation* 1997; **95**: 104–110.
68. Celermajer DS, Sorensen KE, Georgakopoulos D *et al.* Cigarette smoking is associated with dose-related and potentially reversible impairment of

endothelium-dependent dilation in healthy young adults. *Circulation* 1993; **88**: 2149–2155.

69. Cockcroft JR, Chowienczyk PJ, Benjamin N, Ritter JM. Preserved endothelium dependent vasodilatation in patients with essential hypertension. *N Engl J Med* 1994; **330**: 1036–1040.
70. Clarkson P, Celermajer DS, Donald AE *et al.* Impaired vascular reactivity in insulin-dependent diabetes mellitus is related to disease duration and low density lipoprotein cholesterol levels. *J Am Coll Cardiol* 1996; **28**: 573–579.
71. Zeiher AM, Drexler H, Wollschlager H, Just H. Modulation of coronary vasomotor tone in humans. Progressive endothelial dysfunction with different early stages of coronary atherosclerosis. *Circulation* 1991; **83**: 391–401.
72. Egashira K, Inou T, Hirooka Y *et al.* Evidence of impaired endothelium-dependent coronary vasodilatation in patients with angina pectoris and normal coronary angiograms. *N Engl J Med* 1993; **328**: 1659–1664.
73. Goodfellow J, Owens D, Henderson A. Cardiovascular syndromes X, endothelial dysfunction and insulin resistance. *Diabetes Res Clin Pract* 1996; **31 (suppl)**: S163–S171.
74. Petrie JR, Ueda S, Webb DJ *et al.* Endothelial nitric oxide production and insulin sensitivity. A physiological link with implications for pathogenesis of cardiovascular disease. *Circulation* 1996; **93**: 1331–1333.
75. Andrews TC, Raby K, Barry J *et al.* Effect of cholesterol reduction on myocardial ischemia in patients with coronary disease. *Circulation* 1997; **95**: 324–328.
76. Tamai O, Matsuoka H, Itabe H *et al.* Single LDL apheresis improves endothelium-dependent vasodilatation in hypercholesterolemic humans. *Circulation* 1997; **95**: 76–82.
77. Chin JPF, Dart AM. How do fish oils affect vascular function? *Clin Exp Pharmacol Physiol* 1995; **22**: 71–81.
78. Bode-Böger SM, Böger RH, Alfke H *et al.* L-arginine induces nitric oxide-dependent vasodilation in patients with critical limb ischemia. A randomized, controlled study. *Circulation* 1996; **93**: 85–90.
79. Lablanche JM, Grollier G, Lusson J-R *et al.* Effect of the direct nitric oxide donors linsidomine and molsidomine on angiographic restenosis after coronary balloon angioplasy. The ACCORD study. *Circulation* 1997; **95**: 83–89.
80. Assender JW, Southgate KM, Newby AC. Does nitric oxide inhibit smooth muscle proliferation? *J Cardiovasc Pharmacol* 1991; **17 (Suppl 3)**: S104–S107.
81. Sarkar R, Meinberg EG, Stanley JC *et al.* Nitric oxide reversibly inhibits the migration of cultured vascular smooth muscle cells. *Circ Res* 1996; **78**: 225–230.
82. Anderson TJ, Meredith IT, Charbonneau F *et al.* Endothelium-dependent coronary vasomotion relates to the susceptibility of LDL to oxidation in humans. *Circulation* 1996; **93**: 1647–1650.

83. Mancini GBJ, Henry GC, Macaya C *et al.* Angiotensin-converting enzyme inhibition with quinapril improves endothelial vasomotor dysfunction in patients with coronary artery disease. The TREND (Trial on Reversing Endothelial Dysfunction) study. *Circulation* 1996; **94**: 258–265.
84. Powell JS, Clozel J-P, Muller RK *et al.* Inhibitors of angiotensin-converting enzyme prevent myointimal proliferation after vascular injury. *Science* 1989; **245**: 186–188.
85. Becker RH, Wiemer G, Linz W. Preservation of endothelial function by ramipril in rabbits on a long-term atherogenic diet. *J Cardiovasc Pharmacol* 1991; **18 (Suppl 2)**: S110–S115.
86. Fennessy PA, Campbell JH, Campbell GR. Perindopril inhibits both the development of atherosclerosis in the cholesterol-fed rabbit and lipoprotein binding to smooth muscle cells in culture. *Atherosclerosis* 1994; **106**: 29–41.
87. Hickey H, Makdissi M, Hyland R *et al.* Perindopril treatment prevents the loss of endothelial nitric oxide function and development of neointima in rabbits. *J Mol Cell Cardiol* 1996; **28**: 1985–1994.
88. Dusting GJ, Hyland R, Hickey H, Makdissi M. Angiotensin-converting enzyme inhibitors reduce neointimal thickening and maintain endothelial nitric oxide function in rabbit carotid arteries. *Am J Cardiol* 1995; **76**: 24E–27E.
89. Aberg G, Ferrer P. Effects of captopril on atherosclerosis in cynomolgus monkeys. *J Cardiovasc Pharmacol* 1990; **15 (Suppl 5)**: S65–S72.
90. Farhy RD, Ho KL, Carretero OA, Scicli AG. Kinins mediate the antiproliferative effect of ramipril in rat carotid artery. *Biochem Biophys Res Commun* 1992; **182**: 283–288.
91. Lucchesi BR, Mullane KM. Leukocytes and ischaemia-induced myocardial injury. *Annu Rev Pharmacol Toxicol* 1986; **26**: 201–224.
92. Sobey CG, Dusting GJ, Grossman H, Woodman OL. Impaired vasodilatation of epicardial coronary arteries and resistance vessels following myocardial ischaemia and reperfusion in anaesthetized dogs. *Coron Artery Dis* 1990; **1**: 363–373.
93. Sobey CG, Woodman OL. Myocardial ischaemia: what happens to the coronary arteries? *Trends Pharmacol Sci* 1993; **14**: 448–453.
94. Ohlstein E, Nichols AJ. Rabbit polymorphonuclear neutrophils elicit endothelium-dependent contraction in vascular smooth muscle. *Circ Res* 1989; **65**: 917–924.
95. Sessa WC, Mullane KM. Release of a neutrophil-derived vasoconstrictor agent which augments platelet-induced contractions of blood vessels *in vitro*. *Br J Pharmacol* 1990; **99**: 553–559.
96. Sobey CG, Dusting GJ, Stewart AG, Woodman OL. Rabbit polymorphonuclear leucocytes cause endothelium-dependent contraction of rabbit aorta. *J Vasc Med Biol* 1990; **2**: 107–115.

97. Sobey CG, Dusting GJ, Stewart AG. Tumour necrosis factor α augments the release of an endothelium-dependent vasoconstrictor from human polymorphonuclear leukocytes. *J Cardiovasc Pharmacol* 1992; **20**: 813–819.
98. Sobey CG, Dalipram RA, Dusting GJ, Woodman OL. Impaired endothelium-dependent relaxation of dog coronary arteries after myocardial ischaemia and reperfusion: prevention by amlodipine, propranolol and allopurinol. *Br J Pharmacol* 1992; **105**: 557–562.
99. Davis SF, Yeung AC, Meredith IT *et al.* Early endothelial dysfunction predicts the development of transplant coronary artery disease at 1 year posttransplant. *Circulation* 1996; **93**: 457–462.
100. Seifter E, Rettura G, Levenson S, Barbul A. Supplemental arginine and ornithine promote allograft rejection [Abstract]. *Am Chem Soc* 1982; **183**: Agfd–11.
101. Langrehr JM, Hoffman RA, Lancaster JR Jr, Simmons RL. Nitric oxide — a new endogenous immunomodulator. *Transplantation* 1993; **55**: 1205–1212.
102. Winlaw D, Schyvens CG, Smythe GA *et al.* Urinary nitrate excretion is a noninvasive indicator of acute cardiac allograft rejection and nitric oxide production in the rat. *Transplantation* 1994; **58**: 1031–1036.
103. Yang X, Chowdhury N, Cai B *et al.* Induction of myocardial nitric oxide synthase by cardiac allograft rejection. *J Clin Invest* 1994; **94**: 714–721.
104. Akita K, Dusting GJ, Hickey H. Cyclosporin A and FK 506 suppress nitric oxide production by a murine macrophage cell line J774 and rat vascular smooth muscle cells. In: *Biology of Nitric Oxide Vol. 4*, Moncada S, Feelisch M, Busse R, Higgs E (Eds). Portland Press, London 1994: 479–484.
105. Akita K, Dusting GJ, Hickey H. Suppression of nitric oxide production by cyclosporin A and FK506A in rat vascular smooth muscle cells. *Clin Exp Pharmacol Physiol* 1994; **21**: 231–233.
106. Chan SY, Kobashigawa J, Stevenson LW *et al.* Myocardial blood flow at rest and during pharmacological vasodilation in cardiac transplants during and after successful treatment of rejection. *Circulation* 1994; **90**: 204–212.
107. Valantine HA, Yeoh TK, Gibbons R *et al.* Sensitivity and specificity of diastolic indexes for rejection surveillance: temporal correlation with endomyocardial biopsy. *J Heart Lung Transplant* 1991; **10**: 757–765.
108. Paulus WJ, Vantrimpont PJ, Shah AM. Acute effects of nitric oxide on left ventricular relaxation and diastolic distensibility in humans. Assessment by bicoronary sodium nitroprusside infusion. *Circulation* 1994; **89**: 2070–2078.
109. Evans HG, Lewis MJ, Shah AM. Interleukin-1β modulates myocardial contraction *via* dexamethasone sensitive production of nitric oxide. *Cardiovasc Res* 1993; **27**: 1486–1490.
110. Lancaster JR Jr, Langrehr JM, Bergonia HA *et al.* EPR detection of heme and nonheme iron-containing protein nitrosylation by nitric oxide during rejection of rat heart allograft. *J Biol Chem* 1992; **267**: 10994–10998.
111. Bulgrin JP, Shabani M, Smith DJ. Arginine-free diet suppresses nitric oxide production in wounds. *J Nutr Biochem* 1993; **4**: 588–593.

112. Barbul A. Physiology and pharmacology of arginine. In: *Nitric Oxide from L-Arginine: A Bioregulatory System,* Moncada S, Higgs EA (Eds). Elsevier, Amsterdam 1990: 317–329.
113. Wilner ML, Ettenger RB, Koyle MA, Rosenthal JT. The effect of hypoprolactinemia alone and in combination with cyclosporine on allograft rejection. *Transplantation* 1990; **49**: 264–267.
114. Hibbs JB Jr, Taintor RR, Vavrin Z, Rachlin EM. Nitric oxide: a cytotoxic activated macrophage effector molecule. *Biochem Biophys Res Commun* 1988; **157**: 87–94.
115. Langrehr JM, Dull KE, Ochoa JB *et al.* Evidence that nitric oxide production by *in vivo* allosensitized cells inhibits the development of allospecific CTL. *Transplantation* 1992; **53**: 632–640.
116. Winlaw DS, Schyvens CG, Smythe GA *et al.* Selective inhibition of nitric oxide production during cardiac allograft rejection causes a small increase in graft survival. *Transplantation* 1995; **60**: 77–82.
117. Ryffel B. Pharmacology of cyclosporine. VI. Cellular activation: regulation of intracellular events by cyclosporine. *Pharmacol Rev* 1990; **41**: 407–422.
118. Dawson TM, Steiner JP, Dawson VL *et al.* Immunosuppressant FK506 enhances phosphorylation of nitric oxide synthase and protects against glutamate neurotoxicity. *Proc Natl Acad Sci USA* 1993; **90**: 9808–9812.
119. Kroncke KD, Kolb-Bachofen V, Berschick B *et al.* Activated macrophages kill pancreatic syngeneic islet cells *via* arginine-dependent nitric oxide generation. *Biochem Biophys Res Commun* 1991; **175**: 752–758.
120. The Canadian-European Randomized Control Trial Group. Cyclosporin-induced remission of IDDM after early intervention. Association of 1 yr of cyclosporin treatment with enhanced insulin secretion. *Diabetes* 1988; **37**: 1574–1582.

Chapter 20

The Effects of Nitric Oxide in the Pulmonary Circulation

Tom Siddons and Tim Higenbottam

Introduction

The unique anatomical arrangement of the three major cell types present in the alveoli of the lung has important implications for the fate of NO released by the pulmonary vascular endothelium (Figure 1). Partitioning between the liquid and gaseous phases of the aqueous layer of surfactant that lines the epithelial cells is such that under conditions of normal atmospheric pressure only 3% of any NO present stays in solution, with the remaining 97% entering the airspace.[1] The large surface area for gas exchange provided by the alveoli thus permits the detection of NO released from the pulmonary endothelium in exhaled air. Conversely, inhaled NO is taken up from within the alveolus and can be used to assess lung diffusing capacity in the same way as carbon monoxide, both gases having a high affinity for the haemoglobin present in circulating red cells.[2] NO released by the endothelium is also scavenged by red cells in the intravascular compartment (Figure 1).

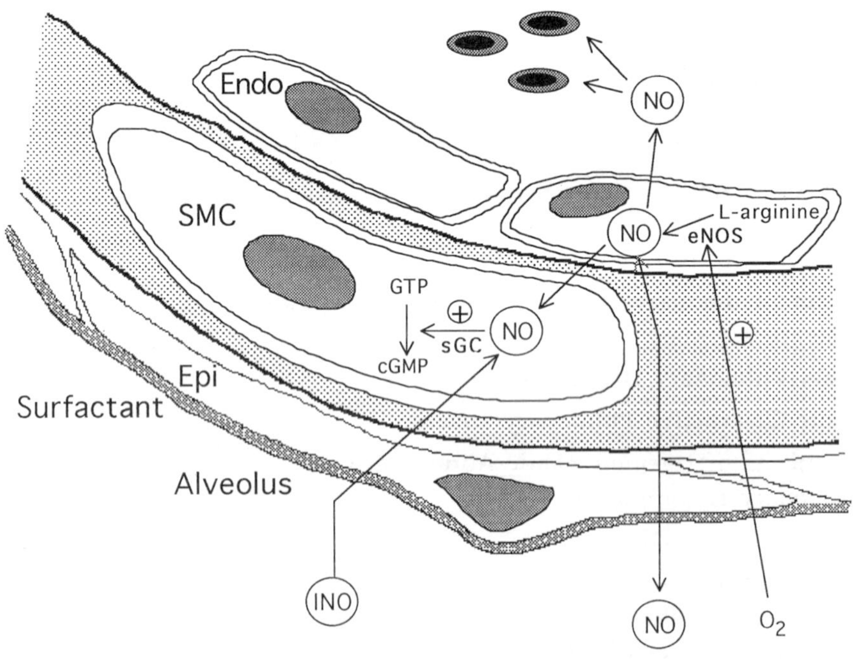

Figure 1. Schematic representation of the wall of an alveolus. NO is synthesized by the constitutive endothelial nitric oxide synthase (eNOS) and diffuses to adjacent smooth muscle cells (SMC) in which it promotes dilatation by stimulating the soluble guanylate cyclase (sGC). NO can diffuse across the alveolar epithelium (Epi) and its superficial layer of surfactant to enter the air space but is also scavenged by haemoglobin present within red cells in the intravascular compartment. Inhaled NO (INO) may also diffuse into smooth muscle cells and thereby contribute to the optimization of ventilation/perfusion matching. NO synthesis by the pulmonary endothelium (Endo) declines when alveolar O_2 concentrations are low, thereby enhancing hypoxic vasoconstriction.

Experiments with isolated perfused ventilated lungs from many species, including rabbit, guinea pig and man, have established that endothelial NO is an important regulator of basal pulmonary vascular resistance (PVR) and introduction of red cells into the perfusate reduces the exhaled NO concentration, thus confirming that at least some of the NO present in exhaled air originates from the vascular endothelium.[3,4] The amount of endothelium-derived NO released from isolated lung preparations can be manipulated by intravascular administration of NO-releasing agonists such as acetylcholine (ACh) which increase exhaled NO, or L-arginine analogues such as N^G-nitro-L-arginine methyl ester (L-NAME) which cause a decrease.[3] Inhibition of endothelial NO synthase (eNOS) in porcine lungs also increases PVR, whereas the thromboxane analogue U46689 increases PVR but, unlike L-NAME, fails to alter exhaled NO.[3] Subsequent administration of ACh markedly increases exhaled NO with an expected drop in PVR. In isolated preparations, there is thus a close correlation between NO production by endothelial cells and its concentration in exhaled air. It is also apparent that changes in PVR do not modulate NO release to any great extent *per se*.

L-NAME also increases PVR in sheep and man, but has no effect in the rat or the dog in which the relationship between pulmonary blood flow and perfusion pressure is unchanged.[1] Of the species studied, the dog has the lowest initial PVR, which suggests that the response to L-NAME may be masked. This is also thought to be the case in the rat, in which Barer *et al.* have suggested that NO is not released under normoxic conditions unless vascular tone is raised.[5] It is possible that species differences reflect differences in the need to adapt to increased blood flow.[1]

NO concentrations in exhaled gas from man and many other species are within the range 20–30 ppb, but substantially lower (2–5 ppb) in air sampled from the lower airways of intubated or tracheotomized subjects.[6–8] There is now evidence that a major component of the NO detectable in exhaled air *in vivo* derives from the paranasal sinuses rather than the lung itself. NO concentrations in air sampled directly from the maxillary sinus of healthy human volunteers may thus be 1000-fold higher than in air from the distal airways, and L-NAME instilled locally into the maxillary antrum markedly reduces sinus air NO concentrations, whereas instillation into the nose does not affect nasal NO concentrations.[6] Notably, NO levels (1 ppb) measured in the exhaled air of baboons (the only mammal known to lack paranasal sinuses) correspond closely with the levels found in the lower airways of other species.[9] The source of non-pulmonary NO appears to be a NOS which is expressed constitutively by the epithelial cells that line the paranasal sinuses, but nevertheless possesses the characteristics of an inducible isoform.[10] Chapter 8 discusses some of these issues further, and describes techniques available for measuring NO in exhaled air.

Many disease states affecting the lung are associated with pulmonary hypertension, which can be primary or secondary in aetiology, and is characterized by an increase in PVR. Given that endothelial cells produce vasodilator substances such as prostacyclin (PGI_2) and NO, as well as vasoconstrictors such as thromboxane A_2 and endothelin, endothelial dysfunction may contribute to the pathophysiology of this condition, either through loss of endothelium-dependent vasodilator activity or through enhanced endothelium-dependent vasoconstriction. As will be discussed below, there is accumulating evidence for abnormalities in endothelial NO production in pulmonary hypertension, and altered NO production has also been proposed as a contributing factor in systemic hypertension (see Chapter 21).[11]

While the present chapter aims to summarize the role of endogenous and exogenous NO in the control of pulmonary haemodynamics in health and disease, it should be noted that NO is also thought to mediate the nonadrenergic, noncholinergic neural bronchodilator responses of human airway smooth muscle.[12]

Nitric Oxide and Hypoxia

Ventilation of isolated perfused lungs with a low fractional inspired O_2 (F_iO_2) of 10% results in a fall in exhaled NO and a rise in PVR,[3] thus suggesting that NO synthesis by the pulmonary endothelium is under the control of the ambient alveolar O_2 concentration. Shaul & Wells[13] demonstrated directly that hypoxia depresses NO release from fetal pulmonary endothelial cells by measuring vascular smooth muscle cyclic guanosine monophosphate (cGMP) accumulation in co-cultures of endothelial/smooth muscle cells. In marked contrast, there was no change in smooth muscle cGMP accumulation when the endothelial cells were obtained from systemic vessels, thus suggesting that there may be specific effects of O_2 on pulmonary endothelial cell Ca^{2+} homeostasis.[13] In rabbit pulmonary artery rings, ambient O_2 concentrations do not modulate the activity of the soluble guanylate cyclase enzyme itself in the acute situation, as the induction of moderate hypoxia depresses cGMP accumulation in response to endothelium-dependent stimuli, but not elevations in cGMP induced by sodium nitroprusside (SNP), an exogenous source of NO.[14] By contrast, chronic hypoxia impairs cGMP accumulation and relaxation to both ACh and SNP in rat pulmonary artery without affecting these markers of NO synthesis in aortic rings.[15] The potential difficulties in interpreting the effects of hypoxia on NO-dependent responses, and the apparently discrepant results that may be obtained with different assay techniques, are discussed in detail in Chapter 8.

Arterial, venous and double occlusion techniques have shown that the role of NO within the lung exhibits regional differences and that reduced NO synthesis, such as may be observed during hypoxia, results in non-uniform changes in pulmonary resistance.[16] L-NAME causes constriction only in muscular arterial and precapillary arteries which can be reversed by inhalation of a gas mixture containing 40 ppm NO, thus confirming that NO can be taken up not only in the alveolar capillaries but also by resistance arteries.[1] Occlusion techniques also permit the measurement of compliance in different segments of the pulmonary bed, and again the effects of L-NAME show marked inhomogeneity in that the compliance of the alveolar capillaries, precapillary arteries and postcapillary veins falls to a greater extent than that of the main arterial and venous compartments.[1]

It follows that in hypoxic regions of the lung, reductions in NO synthesis will increase the resistance of the distributive pulmonary arteries and precapillary arteries and reduce the compliance of capillaries. These effects will serve to divert blood flow from underventilated regions and complement hypoxic vasoconstriction by contributing to the optimal matching of ventilation and perfusion. Residual NO production may nevertheless attenuate the ability of pulmonary vascular smooth muscle to constrict when O_2 concentrations fall,[17] a mechanism thought to involve *inter alia* inhibition of Ca^{2+}-independent K^+ channels.[18]

Inhaled Nitric Oxide

Pulmonary Effects

The therapeutic potential afforded by local delivery of NO to the lung by inhalation has been explored in a number of disorders, including various types of pulmonary hypertension and acute respiratory distress syndrome (ARDS). These specific applications of inhaled NO (INO) therapy are discussed in detail in the corresponding sections below. The concept that INO produces selective pulmonary vasodilatation is an attractive one, and relies on recruiting blood to functional lung units to which the inspired gas has access. Treatment with INO thus leads to a simultaneous improvement in ventilation/perfusion ($V/\dot{Q}$) matching, a reduction in shunt fraction and a fall in PVR.[19] Its effects thus contrast fundamentally with intravenous vasodilators which tend to worsen mismatching by dilating the pulmonary vasculature in a nonselective fashion. Anatomically, the alveoli are closely adjacent to small resistance arteries and veins (see Figure 1)[20] and *in vitro* experiments have demonstrated that arteries and veins are both responsive to NO.[21] Under both normal conditions and following acute injury, however, INO has been shown primarily to reduce arterial rather than venular resistance in the isolated perfused rabbit lung.[22] Increased concentrations of INO exhibit successively larger effects on oxygenation, PVR and bronchial smooth muscle tone: INO may thus also decrease alveolar dead space through a mechanism that has been attributed to local bronchodilatation.[23]

Inhibitors of phosphodiesterase that reduce the catabolism of cGMP may potentiate and prolong the effects of INO without changing its pulmonary selectivity, thus increasing the response rate and allowing the use of lower concentrations of INO.[24,25] The effects of INO can also be enhanced by administration of vasoconstrictor agents which augment hypoxic vasoconstriction.[26] The combination of a gravitational strategy (prone position) and INO may also result in an increase in the beneficial effects of both therapies on oxygenation.[27] NO may also attenuate sympathetically-mediated increases in pulmonary artery tone,[28] but it is unknown whether this action contributes to the fall in PVR observed during therapy with INO.

Convincing evidence suggests that NO originating from the paranasal sinuses is inhaled and resorbed in the lower respiratory tract,[8] an 'aerocrine' role of NO that may normally contribute to the optimization of $V/\dot{Q}$ matching in the pulmonary bed. Indeed, in ventilated patients, autoinhalation of NO may increase PaO_2 by as much as 20%, implying that low dose INO may then be regarded as NO replacement.[29]

Systemic Effects

In principle, INO should be devoid of major deleterious systemic effects,[30] as it is rapidly bound to haemoglobin and oxidized to nitrate (see Chapter 6). However, 25% of inhaled radiolabelled ^{15}NO appears in the urine as urea, suggesting that there may indeed be a systemic action.[31] INO (at 40 ppm) thus increases renal blood flow, glomerular filtration rate and diuresis in pigs, independently of its pulmonary effects.[32] These effects may be secondary to the reversible binding of NO to albumin or haemoglobin to generate a circulating reservoir of biologically active NO carried as *S*-nitrosoalbumin and *S*-nitrosohaemoglobin.[33,34] Haemoglobin has a 6-fold higher affinity for NO than for O_2,[35] but NO release from *S*-nitrosohaemoglobin is promoted by low O_2 levels and therefore regulated by the ambient O_2 concentration.[34] NO may thus be taken up by oxyhaemoglobin in the lung and act at sites distant, both spatially and temporally, from its original site of production or uptake when O_2 levels fall. Recent data have also shown that INO may decrease platelet aggregation in patients with acute respiratory distress syndrome (ARDS).[36]

Adverse Effects

In air, NO reacts with oxygen to form nitrogen dioxide ($\cdot NO_2$) (see also Chapters 3, 6 and 8). This is a second-order reaction with respect to NO, so that the time taken to yield the recommended upper limit for $.NO_2$ inhalation of 5 ppm depends upon the initial concentration of NO. In order to reduce the formation of $\cdot NO_2$, it is advisable that the amount of time NO is in contact with oxygen is minimized as $\cdot NO_2$ can cause pneumonitis, pulmonary oedema and emphysema.[37] When inhaled in clinically-effective concentrations of 40–80 ppm, however, virtually no NO is taken up until it reaches the small airways,[2] and exposure studies have revealed little evidence for injurious effects at these low concentrations.[38] Pulmonary artery pressure (PAP) and oxygenation levels may return to their pre-treatment level when INO therapy is discontinued.[39,40] In some patients, however, discontinuation of therapy may promote marked elevations in PAP and decreases in O_2 saturation that can be life-threatening.[40] The onset of these adverse effects can be as early as 5 minutes after stopping INO therapy and dependency may build up over a 24 hour period.[40] The exact mechanism of this 'rebound' phenomenon is currently unknown. Since NO release from cultured endothelial cells has been shown to function as a negative feedback modulator by inhibiting NO synthase, it is possible that rebound is secondary to downregulation of pulmonary eNOS activity.[41,42]

Acute Respiratory Distress Syndrome

ARDS is characterized by diffuse changes in the pulmonary interstitium, with the alveoli becoming inundated with exudate, leading to reduced compliance and, in turn, a reduction in total lung capacity.[43] ARDS can occur as a result of trauma or inhaled irritants such as wood smoke,[44] infectious agents[45] or remote ischaemia-reperfusion injury.[46] Initially there is pulmonary oedema, followed by an increase in PAP.[43] The damage to the capillary networks surrounding the alveoli causes them to become highly permeable, allowing the entry of proteins, water and electrolytes. Fluid entering the alveoli causes the dilution of pulmonary surfactant, which normally decreases surface tension and the work of breathing. As a consequence, some regions of the lung become unventilated or only partially ventilated, leading to an imbalance in $V/\dot{Q}$ matching and increased shunting, i.e. an increased fraction of total blood volume passing through the lungs without taking part in alveolar gas exchange.

Patient Assistance

Patients with ARDS require increased F_iO_2. In addition, ventilation needs to be mechanically assisted, partly to overcome the increased work of breathing, but also to enable the administration of continuous positive airway pressure (CPAP) and reverse airway collapse. On breathing pure oxygen, however, only a modest increase in arterial saturation is observed,[47] this being indicative of a large shunt fraction. The $V/\dot{Q}$ mismatch, loss of vascular bed, and hypoxic pulmonary vasoconstriction in underventilated regions culminate in modest pulmonary hypertension.[47]

If insufficient ventilated lung units are available, INO-induced vasodilatation may not have a significantly measurable effect. In such cases, the use of CPAP has been shown to improve $V/\dot{Q}$ mismatch by recruiting additional lung units and dilating previously unexposed blood vessels.[48] By increasing the PaO_2/FiO_2 ratio, INO may reduce the FiO_2 level required to achieve acceptable PaO_2 concentrations.[49] Thus, INO therapy can help to avoid the possible deleterious effects of excess O_2.

The selective nature of INO causes vasodilatation only in ventilated alveoli, so that hypoxic pulmonary vasoconstriction persists in regions affected by the ARDS disease process.[47] This reduces PAP and improves the $V/\dot{Q}$ mismatch.[48] The degree of improvement in arterial oxygenation induced by INO has been demonstrated to be related to the level of PVR before treatment: the higher the PVR, the greater the response.[50] In contrast, intravenously administered NO donors and PGI_2 increase perfusion to the inundated regions, which causes a worsening of the $V/\dot{Q}$ mismatch.[51] The extrapulmonary effect of INO against platelet activation and preserving renal function could also be helpful against the systemic dysfunction associated with ARDS.[32,36]

Inhibition of NO Synthesis

The development of NOS inhibitors has led to the investigation of novel approaches to enhance the efficacy of INO, but such attempts have so far generated conflicting findings. In a sheep model of ARDS (bilateral bronchial lavage), INO has been demonstrated to decrease PAP and PVR and increase PaO_2.[43] In this model, the administration of L-NAME enhanced hypoxic pulmonary vasoconstriction in all lung regions. It was hoped that INO would overcome this vasoconstrictor response only in the ventilated regions, thereby improving its overall effect, but the improvement in haemodynamic parameters was similar in the presence or absence of NOS inhibition.[43] In contrast, in an oleic acid-induced acute lung injury model in the dog, INO improved gas exchange to a greater extent in the presence of the NOS inhibitor N^G-monomethyl-L-arginine (L-NMMA) than when administered alone.[52] The impact of INO in combination with the use of selective inhibitors of inducible NOS (iNOS) in the treatment of the inflammatory processes associated with ARDS has not been reported.

Persistent Pulmonary Hypertension of the Neonate

During gestation, the fetus receives O_2 and loses CO_2 *via* placental gas exchange mechanisms. Only 10% of the fetal cardiac output is directed towards the pulmonary circulation, as the lungs play no role in gas exchange before birth. At this stage of development, the PVR is greater than systemic vascular resistance (SVR)[53] so that blood from the right heart bypasses the lungs *via* the patent ductus arteriosus and foramen ovale.[54] At birth, PVR falls, thereby allowing the lungs to take over the role of gas exchanger.[53] In cases where SVR does not exceed the level of PVR, shunting of blood persists and can cause severe hypoxaemia.[53,55]

Persistent pulmonary hypertension of the neonate (PPHN) manifests during the first hours or days after birth. It can be of unknown cause or secondary to conditions such as respiratory distress syndrome, pneumonia, sepsis, meconium aspiration or congenital diaphragmatic hernia.[53] It tends to be further complicated by multi-system dysfunction.[53] For instance, decreased cardiac performance along with a reduction in left ventricular output can lead to a decrease in SVR, thereby causing a relative increase in the PVR/SVR ratio which exacerbates the shunting of blood.

Therapeutic Management

PPHN is associated with high morbidity and mortality, which is related to the severity and duration of systemic hypoxaemia,[54] so that management of this condition is directed towards improving oxygenation. Extracorporeal membrane oxygenation (ECMO) reduces mortality,[56] and before it became available, an 'oxygenation index' exceeding 40 carried a risk of mortality of around 80%.[57] Conventional therapeutic regimens such as oxygen support, mechanical ventilation and induction of alkalosis have been shown not to reduce mortality or the need for ECMO.[58]

Due to the invasive nature of ECMO, alternative strategies have been sought. The selective action of INO, its lack of major systemic vascular effects, and its relative simplicity of application have led to investigations into whether INO could provide a more effective therapeutic strategy. Such studies have so far provided conflicting results. INO therapy of 30 minutes duration, delivered to infants with persistent hypoxaemia, produced a mean increase in PaO_2 from 49 mmHg to 75 mmHg, while the oxygenation index fell from 37 to 20.[59] Others have found INO to cause a lasting improvement in oxygenation and a reduced requirement for ECMO.[60,61] These findings have been supported by data from a recently published clinical trial.[62] On the other hand, the Neonatal Inhaled Nitric Oxide Study Group at Yale University reported that INO did not reduce mortality overall and could not replace ECMO in the majority of patients.[58] Possible reasons for such discrepant findings could relate to the magnitude of the respiratory tidal volume (RTV). Responders to INO possessed a mean RTV of 8.9 ml/kg, whereas non-responders had a mean RTV of 4.2 ml/kg,[55] suggesting that low alveolar ventilation could reduce the efficacy of INO. As discussed earlier with reference to the use of INO in ARDS, lung recruitment may prove beneficial, along with high frequency oscillatory ventilation.[63]

Primary Pulmonary Hypertension

Primary pulmonary hypertension (PPH) differs from other conditions in which there is elevated pulmonary artery pressure in that there is no specific pulmonary or cardiac disease to explain its aetiology. It is characterized by a PAP greater than 25 mmHg during rest and increasing to 30 mmHg or more on exertion, with a cardiac output below 5 L/min.[64] The disease is more common in women than in men, by a ratio of nearly 2:1.[65] There are also familial forms of PPH.[66] The prognosis is poor, with median life expectancy being only 2–3 years from the onset of symptoms,[67,68] and correlating with the severity of the haemodynamic changes.[69] Spontaneous recovery has been reported, though very rarely.[70] The condition, unlike ARDS, is slowly-developing and there is consequently often a considerable delay from the onset of the symptoms to definitive diagnosis.[65]

Right heart catheterization is essential to confirm the diagnosis,[64] and provides the opportunity to assess the degree to which functional pulmonary vasoconstriction contributes to the increased PVR.[71] At present, PGI_2 is the vasodilator most commonly used but, due to its non-selective characteristics and high cost, alternatives have been sought. With the advantage of selectivity and relatively low cost, INO has been under investigation as a suitable alternative and in some patients, NO induces pulmonary vasodilatation comparable to that achieved with PGI_2.[64] The response to INO varies widely, however, presumably because chronic pulmonary hypertension leads to varying degrees of remodelling and medial hypertrophy in muscular pulmonary arteries. Due to the nature of PPH, patients often lead very restricted lifestyles and, with this in mind, delivery systems for INO are being developed. Pulsing the delivery at the onset of inspiration allows the volume of NO wasted to the atmosphere to be reduced. This means that the size of the gas cylinder can also be reduced, which is clearly advantageous when designing an ambulatory delivery system. Such a device has been designed and tested and in certain cases has successfully reduced mean PAP.[72]

Chronic Obstructive Pulmonary Disease

The ability of INO to decrease secondary pulmonary hypertension in chronic obstructive pulmonary disease (COPD) is established, although its effect on gas exchange is somewhat disappointing.[73] In patients with COPD, INO may inappropriately dilate pulmonary vessels in chronically underventilated alveoli. This will exert a detrimental effect on gas exchange by reversing hypoxic vasoconstriction in regions of the lung where this mechanism is important for maintaining $V/\dot{Q}$ matching[73], resulting in a decrease in PaO_2. If the concentration of NO is high and the delivery continuous, NO will diffuse into the partially ventilated regions more readily than if the delivery was given as a 'spike' at the onset of inspiration.

Mechanisms of Haemodynamic Abnormality in Pulmonary Hypertension

In addition to abnormal pulmonary artery thickening, PPH is also characterized by functional vasoconstriction that contributes to increased PVR. The pathophysiology of the process(es) underlying these changes remains incompletely understood, although it has been proposed that the vascular remodelling and an associated increase in PVR are linked to the production or effectiveness of vasoactive molecules and/or vascular growth factors.[74] Interestingly, diminished expression of eNOS (correlating inversely with the severity of the condition) has been observed in the pulmonary arteries of patients with PPH,[75] suggesting that impaired release of endothelial NO contributes to the functional haemodynamic abnormality. It may be relevant to this hypothesis that patients with PPH failed to elicit a normal increase in the rate of NO production in expired air during standard exercise testing.[76] More recently, it has been demonstrated that eNOS is selectively downregulated in small arterioles but 'paradoxically' upregulated in the plexiform lesions associated with the most severe and irreversible forms of the condition.[77] However, the functional importance of such lesions currently remains unknown, though potentially these findings indicate the possibility of selective drug targeting. Pulmonary hypertension associated with other lung diseases similarly reveals reduced expression of eNOS in small arteries and increased expression in the plexiform lesions that are also found in secondary forms of the disease.[77] Immunocytochemical studies, while demonstrating eNOS expression, do not reflect absolute enzyme activity (see Chapter 8). Functional studies, with explanted lungs from patients undergoing heart-lung transplant, nevertheless confirm the above conclusions. In isolated ventilated perfused lungs in which the pulmonary bed is pharmacologically constricted, maximal dilatation to ACh is impaired in both PPH and COPD.[1] Whether changes in endothelial function and NOS expression lead directly to the development of pulmonary hypertension or are secondary to a sustained increase in PAP remains an unresolved question.

Congenital Heart Disease

Intractable pulmonary hypertension is the major cause of death in the immediate post-operative period following the repair of congenital heart defects.[78] In the most severe cases, the response to alkalosis-induced vasodilatation is blunted.[79] However, it has been shown that alkalosis-induced mechanisms of vasodilatation are independent of NO,[79] thus permitting the use of NO in the treatment of such severe cases. The acute increase in PAP puts the right ventricle (RV) under increased strain, and this may result in RV failure.[80] Intervention at this stage with INO therapy may improve perfusion efficiency and reduce the amount of RV work needed to sustain a given blood flow.[49]

In children, low dose INO has been shown to cause a reduction in severe post-operative pulmonary hypertension after the successful surgical correction of a congenital defect.[78] In such cases, much of the increase in PVR is thought to be due to endothelial cell dysfunction and therefore sensitive to the dilatory effects of NO,[81] with arteries being particularly responsive.[42] The response to INO correlates positively with the pulmonary/systemic pressure ratio, i.e. the higher the ratio the greater the response to INO.[82] At present, the administration of vasodilators plus high levels of F_iO_2, along with lung volume recruitment and hyperventilation, is the standard therapy in such cases.[82]

Transplantation

Patient Suitability

Preoperative assessment of the likely outcome of heart/lung transplantation is essential before surgery is undertaken, as early post-operative right heart failure is associated with substantial morbidity and mortality[83–85] and there is limited availability of donor organs. Prior to the discovery of NO, intravenously-delivered vasodilator agents such as SNP and prostaglandin E_1 were used to identify patients at acceptable risk levels, but systemic effects limited their effectiveness.[86] The vasodilatory effects of NO have thus been found to be useful in assessing patient suitability. If it is found that decreased endogenous production of NO, rather than irreversible vascular damage, is responsible for vasoconstriction in patients with high pulmonary artery pressure, then NO is likely to be an effective treatment. The integrity of endothelial function can be investigated by examining the vascular response to ACh.[87] Successful reduction of an increased PVR may postpone the need for transplantation or permit exchange of the heart alone, as opposed to both the heart and the lungs.[88] The dose of NO considered relevant in this assessment is higher (approx 80 ppm)[85] than during attempts to treat $V/\dot{Q}$ abnormalities, as the maximum potential acute decrease in PVR is the measurement of physiological importance. Patients who do not respond to vasodilator treatment are excluded from becoming transplant recipients[85] and generally have a PVR greater than 200 $dyne.s.cm^{-5}$, either at baseline or during vasodilator administration.[89]

Transplant Options in Pulmonary Hypertension

In the case of end-stage pulmonary hypertension, three types of lung transplant are possible: single, double, and heart + lung. Transplantation carries a two year survival rate of 70–80% but each technique has its advantages and disadvantages.[90–92] In the case of single lung transplant, the poor functional reserve, coupled with the increased $V/\dot{Q}$ inequality during transient graft dysfunction, is notable in 20% of patients.[93] Within the first 24 hours post-operatively, massive blood flow shift to the grafted lung may result in pulmonary oedema.[93] In cases of severe graft dysfunction, inflammatory cells infiltrate the alveolar septum. Experimentally, in an attempt to curtail this event in rats, INO therapy has been investigated, but a minimum duration of treatment for 24 hours was found to be necessary.[93] Currently used methods in human patients include positive pressure ventilation, vasodilator therapy and inotropic drugs: these measures are, however, associated with barotrauma, haemodynamic instability, infection and bleeding.[90–93]

Summary

The unique anatomy of the lung affords a direct opportunity to investigate the role of NO in the pathophysiological processes of pulmonary vascular control and treatment. Many of the diseases that affect the pulmonary circulation are associated with raised PVR, impaired $V/\dot{Q}$ matching and diminished availability of NO derived from the endothelium, and INO has been successfully employed in their therapeutic management. The selectivity of INO arises both from its relative lack of systemic 'spillover' effect and its vasodilator effectiveness only in ventilated alveoli. This enables therapeutic advantage (decreased PAP and increased PaO_2) to be gained in diseases such as ARDS and PPH, though not in COPD. Questions currently remain about the relative efficacy of INO compared with alternative therapies and the possible long-term toxicity of the treatment.

References

1. Higenbottam T. Lung disease and pulmonary endothelial nitric oxide. *Exp Physiol* 1995; **80**: 855–864.
2. Borland CDR, Higenbottam TW. A simultaneous single breath measurement of pulmonary diffusing capacity with nitric oxide and carbon monoxide. *Eur Respir J* 1989; **2**; 56–63.
3. Cremona G, Higenbottam T, Takao M *et al.* Exhaled nitric oxide in isolated pig lung. *J Appl Physiol* 1995; **78**: 59–63.
4. Cremona G, Wood AM, Hall LW *et al.* Effects of inhibitors of nitric oxide release and action on vascular tone in isolated lungs of pig, sheep, dog and man. *J Physiol (Lond)* 1994; **481**: 185–195.
5. Barer G, Emery C, Stewart A *et al.* Endothelial control of the pulmonary circulation in normal and chronically hypoxic rats. *J Physiol (Lond)* 1993; **463**: 1–16.
6. Lundberg JON, Rinder J, Weitzberg E *et al.* Nasally exhaled nitric oxide in humans originates mainly in the paranasal sinuses. *Acta Physiol Scand* 1994; **152**: 431–432.
7. Alving K, Weitzberg E, Lundberg JM. Increased amount of nitric oxide in exhaled air of asthmatics. *Eur Respir J* 1993; **6**: 1368–1370.
8. Gerlach H, Rossaint R, Pappert D *et al.* Autoinhalation of nitric oxide after endogenous synthesis in nasopharynx. *Lancet* 1994; **343**: 518–519.

9. Lewandowski K, Busch T, Lohbrunner H *et al*. Low nitric oxide concentrations in exhaled gas and nasal airways of mammals without paranasal sinuses. *J Appl Physiol* 1998; **85**: 405–410.
10. Lundberg JON, Farkas-Szallasi T, Weitzberg E *et al*. High nitric oxide production in human paranasal sinuses. *Nature Med* 1995; **1**: 370–373.
11. Ferro CJ, Webb DJ. Endothelial dysfunction and hypertension. *Drugs* 1997; **53 (Suppl 1)**: 30–41.
12. Belvisi MG, Ward JK, Mitchell JA, Barnes PJ. Nitric oxide as a neurotransmitter in human airways. *Arch Int Pharmacodyn Ther* 1995; **329**: 97–110.
13. Shaul PW, Wells LB. Oxygen modulates nitric oxide production selectively in fetal pulmonary endothelial cells. *Am J Respir Cell Mol Biol* 1994; **11**: 432–438.
14. Johns RA, Linden JM, Peach MJ. Endothelium-dependent relaxation and cyclic GMP accumulation in rabbit pulmonary artery are selectively impaired by moderate hypoxia. *Circ Res* 1989; **65**: 1508–1515.
15. Crawley DE, Zhao L, Giembycz MA *et al*. Chronic hypoxia impairs soluble guanylyl cyclase-mediated pulmonary arterial relaxation in the rat. *Am J Physiol* 1992; **263**: L325–L332.
16. Hakim TS, Michel RP, Chang HK. Partitioning of pulmonary vascular resistance by arterial and venous occlusion. *J Appl Physiol* 1982; **52**: 710–715.
17. Liu SF, Crawley DE, Barnes PJ, Evans TW. Endothelium-derived relaxing factor inhibits hypoxic pulmonary vasoconstriction in rats. *Am Rev Respir Dis* 1991; **143**: 32–37.
18. Post JM, Hume JR, Archer SL, Weir EK. Direct role for potassium channel inhibition in hypoxic pulmonary vasoconstriction. *Am J Physiol* 1992; **262**: C882–C890.
19. Singh S, Evans TW. Nitric oxide, the biological mediator of the decade: fact or fiction? *Eur Respir J* 1997; **10**: 699–707.
20. Lindeborg DM, Kavanagh BP, Van Meurs K, Pearl RG. Inhaled nitric oxide does not alter the longitudinal distribution of pulmonary vascular resistance. *J Appl Physiol* 1995; **78**: 341–348.
21. Ignarro LJ. Biological actions and properties of endothelium-derived nitric oxide formed and released from artery and vein. *Circ Res* 1989; **65**: 1–21.
22. Rimar S, Gillis CN. Site of pulmonary vasodilation by inhaled nitric oxide in the perfused lung. *J Appl Physiol* 1995; **78**: 1745–1749.
23. Puybasset L, Rouby J-J, Mourgeon E *et al*. Factors influencing cardiopulmonary effects of inhaled nitric oxide in acute respiratory failure. *Am J Respir Crit Care Med* 1995; **152**: 318–328.
24. Ichinose F, Adrie C, Hurford WE, Zapol WM. Prolonged pulmonary vasodilator action of inhaled nitric oxide by zaprinast in awake lambs. *J Appl Physiol* 1995; **78**: 1288–1295.
25. Kinsella JP, Torielli F, Ziegler JW *et al*. Dipyridamole augmentation of response to nitric oxide [Letter]. *Lancet* 1995; **346**: 647–648.

26. Payen DM, Gatecel C, Plaisance P. Almitrine effect on nitric oxide inhalation in adult respiratory distress syndrome [Letter]. *Lancet* 1993; **341**: 1664.
27. Germann P, Pöschl G, Balassa A *et al.* Additional benefit of combined nitric oxide inhalational and gravitational therapy in patients with isolated ARDS [Abstract]. *Anesthesiology* 1994; **81**: A266.
28. Liu SF, Crawley DE, Evans TW, Barnes PJ. Endogenous nitric oxide modulates adrenergic neural vasoconstriction in guinea-pig pulmonary artery. *Br J Pharmacol* 1991; **104**: 565–569.
29. Lundberg JON, Lundberg JM, Settergren G *et al.* Nitric oxide, produced in the upper airways, may act in an 'aerocrine' fashion to enhance pulmonary oxygen uptake in humans. *Acta Physiol Scand* 1995; **155**: 467–468.
30. Frostell CG, Blomqvist H, Hedenstierna G *et al.* Inhaled nitric oxide selectively reverses human hypoxic pulmonary vasoconstriction without causing systemic vasodilation. *Anesthesiology* 1993; **78**: 427–435.
31. Yoshida K, Kasama K, Kitabatake M, Imai M. Biotransformation of nitric oxide, nitrite and nitrate. *Int Arch Occup Environ Health* 1983; **52**: 103–115.
32. Troncy E, Francœur M, Salazkin E *et al.* Systemic effect of inhaled nitric oxide: modification of renal hemodynamics in pigs. In: *The Biology of Nitric Oxide — Part 5*, Moncada S, Stamler J, Gross S, Higgs EA (Eds). Portland Press, London 1996: 319.
33. Stamler JS, Jaraki O, Osborne J *et al.* Nitric oxide circulates in mammalian plasma primarily as an *S*-nitroso adduct of serum albumin. *Proc Natl Acad Sci USA* 1992; **89**: 7674–7677.
34. Jia L, Bonaventura C, Bonaventura J, Stamler JS. *S*-nitrosohaemoglobin: a dynamic activity of blood involved in vascular control. *Nature* 1996; **280**: 221–226.
35. Olsen JS. Stopped-flow, rapid mixing measurements of ligand binding to hemoglobin and red cells. *Methods Enzymol* 1981; **76**: 631–651.
36. Samama CM, Diaby M, Fellahi J-L *et al.* Inhibition of platelet aggregation by inhaled nitric oxide in patients with acute respiratory distress syndrome. *Anesthesiology* 1995; **83**: 56–65.
37. Miller OI, Celermajer DS, Deanfield JE, Macrae DJ. Guidelines for the safe administration of inhaled nitric oxide. *Arch Dis Child* 1994; **70**: F47–F49.
38. Hugod C. Effect of exposure to 43 ppm nitric oxide and 3.6 ppm nitrogen dioxide on rabbit lung. A light and electron microscopic study. *Int Arch Occup Environ Health* 1979; **42**: 159–167.
39. Rossaint R, Falke KJ, Lopez F *et al.* Inhaled nitric oxide for the adult respiratory distress syndrome. *N Engl J Med* 1993; **328**: 399–405.
40. Lavoie A, Hall JB, Olson DM, Wylam ME. Life-threatening effects of discontinuing inhaled nitric oxide in severe respiratory failure. *Am J Respir Crit Care Med* 1996; **153**: 1985–1987.

41. Goldman AP, Haworth SG, Macrae DJ. Does inhaled nitric oxide suppress endogenous nitric oxide production? *J Thorac Cardiovasc Surg* 1996; **112**: 541–542.
42. Moncada S, Rees DD, Schulz R, Palmer RMJ. Development and mechanism of a specific supersensitivity to nitrovasodilators after inhibition of vascular nitric oxide synthesis *in vivo*. *Proc Natl Acad Sci USA* 1991; **88**: 2166–2170.
43. Rovira I, Chen TY, Winkler M *et al*. Effect of inhaled nitric oxide on pulmonary hemodynamics and gas exchange in an ovine model of ARDS. *J Appl Physiol* 1994; **76**: 345–355.
44. Ogura H, Saitoh D, Johnson AA *et al*. The effect of inhaled nitric oxide on pulmonary ventilation-perfusion matching following smoke inhalation injury. *J Trauma* 1994; **37**: 893–898.
45. Ogura H, Cioffi WG, Offner PJ *et al*. The effect of inhaled nitric oxide on pulmonary function after sepsis in a swine model. *Surgery* 1994; **116**: 313–321.
46. Fantini GA, Conte MS. Pulmonary failure following lower torso ischemia: clinical evidence for a remote effect of reperfusion injury. *Am Surg* 1995; **61**: 316–319.
47. Rossaint R, Gerlach H, Schmidt-Ruhnke H *et al*. Efficacy of inhaled nitric oxide in patients with severe ARDS. *Chest* 1995; **107**: 1107–1115.
48. Putensen C, Rasanen J, Lopez FA, Downs JB. Continuous positive airway pressure modulates effect of inhaled nitric oxide on the ventilation-perfusion distributions in canine lung injury. *Chest* 1994; **106**: 1563–1569.
49. Hillman ND, Meliones JN, Black DR *et al*. In acute lung injury inhaled nitric oxide improves ventilation-perfusion matching, pulmonary vascular mechanics, and transpulmonary vascular efficiency. *J Thorac Cardiovasc Surg* 1995; **110**: 593–600.
50. Young JD, Brampton WJ, Knighton JD, Finfer SR. Inhaled nitric oxide in acute respiratory failure in adults. *Br J Anaesth* 1994; **73**: 499–502.
51. Puybasset L, Stewart T, Rouby JJ *et al*. Inhaled nitric oxide reverses the increase in pulmonary vascular resistance induced by permissive hypercapnia in patients with acute respiratory distress syndrome. *Anesthesiology* 1994; **80**: 1254–1267.
52. Putensen C, Rasanen J, Downs JB. Effect of endogenous and inhaled nitric oxide on the ventilation-perfusion relationships in oleic-acid lung injury. *Am J Respir Crit Care Med* 1994; **150**: 330–336.
53. Mullahoo K, Doherty P. Inhaled nitric oxide therapy in neonates. *RT International* 1994; **Fall issue**: 39–45.
54. Roberts JD, Fineman JR, Morin FC *et al*. Inhaled nitric oxide and persistent pulmonary hypertension of the newborn. *N Engl J Med* 1997; **336**: 605–610.
55. Muller W, Kachel W, Lasch P *et al*. Inhaled nitric oxide for avoidance of extracorporeal membrane oxygenation in the treatment of severe persistent pulmonary hypertension of the newborn. *Intensive Care Med* 1996; **22**; 71–76.

56. Kinsella JP, McCurnin DC, Clark RH *et al.* Cardiac performance in ECMO candidates: echo-cardiographic predictors for ECMO. *J Pediat Surg* 1992; **27**: 44–47.
57. Bartlett RH, Gazzaniga AB, Toomasian J *et al.* Extracorporeal membrane oxygenation (ECMO) in neonatal respiratory failure: 100 cases. *Ann Surg* 1986; **204**: 236–245.
58. The Neonatal Inhaled Nitric Oxide Study Group. Inhaled nitric oxide in full-term and nearly full-term infants with hypoxic respiratory failure. *N Engl J Med* 1997; **336**: 597–604.
59. Tang SF, Miller OI. Low-dose inhaled nitric oxide for neonates with pulmonary hypertension. *J Paediatr Child Health* 1996; **32**: 419–423.
60. Kinsella JP, Neish SR, Shaffer E, Abman SH. Low dose inhalation nitric oxide in persistent pulmonary hypertension of the newborn. *Lancet* 1992; **340**: 819–820.
61. Roberts JD, Polaner DM, Lang P, Zapol WM. Inhaled nitric oxide in persistent pulmonary hypertension of the newborn. *Lancet* 1992; **340**: 818–819.
62. Davidson D, Barefield ES, Kattwinkel J *et al.* Inhaled nitric oxide for the early treatment of persistent pulmonary hypertension of the term newborn: a randomized, double-masked, placebo-controlled, dose-response, multicenter study. The I-NO/PPHN Study Group. *Pediatrics* 1998; **101**: 325–334.
63. Kinsella JP, Abman SH. Clinical pathophysiology of persistent pulmonary hypertension of the newborn and the role of inhaled nitric oxide therapy. *J Perinatol* 1996; **16**: S24–S27.
64. Dinh-Xuan AT, Higenbottam TW, Scott JP, Wallwork J. Primary pulmonary hypertension: diagnosis, medical and surgical treatment. *Respir Med* 1990; **84**: 189–197.
65. Rich S, Dantzker DR, Ayres SM *et al.* Primary pulmonary hypertension. A national prospective study. *Ann Intern Med* 1987; **107**: 216–223.
66. Loyd JE, Primm RK, Newman JH. Familial primary pulmonary hypertension: clinical patterns. *Am Rev Respir Dis* 1984; **129**: 194–197.
67. Packer M. Vasodilator therapy for primary pulmonary hypertension. Limitations and hazards. *Ann Intern Med* 1985; **103**: 258–270.
68. Weitzenblum E, Kessler R, Chaouat A. Hypertension artérielle pulmonaire d'origine iatrogène. *Rev Mal Respir* 1996; **13**: 133–139.
69. Tillman O, Speich R. Die primäre pulmonale Hypertonie. Klinik, Verlauf und prognostische Faktoren. *Schweiz Med Wochenschr* 1997; **127**: 923–934.
70. Bourdillon PDV, Oakley CM. Regression of primary pulmonary hypertension. *Br Heart J* 1976; **38**: 264–270.
71. Higenbottam TW. Primary pulmonary hypertension. *Br Med J* 1986; **293**: 1456–1457.
72. Channick RN, Newhart JW, Johnson FW *et al.* Pulsed delivery of inhaled nitric oxide to patients with primary pulmonary hypertension: an ambulatory delivery system and initial clinical tests. *Chest* 1996; **109**: 1545–1549.

73. Barbera JA, Roger N, Roca J *et al.* Worsening of pulmonary gas exchange with nitric oxide inhalation in chronic obstructive pulmonary disease. *Lancet* 1996; **347**: 436–440.
74. Voelkel NF, Tuder RM. Cellular and molecular mechanisms in the pathogenesis of severe pulmonary hypertension. *Eur Respir J* 1995; **8**: 2129–2138.
75. Giaid A, Saleh D. Reduced expression of endothelial nitric oxide synthase in the lungs of patients with pulmonary hypertension. *N Engl J Med* 1995; **333**: 214–221.
76. Riley MS, Pórszász J, Miranda J *et al.* Exhaled nitric oxide during exercise in primary pulmonary hypertension and pulmonary fibrosis. *Chest* 1997; **111**: 44–50.
77. Mason NA, Springall DR, Burke M *et al.* High expression of endothelial nitric oxide synthase in plexiform lesions of pulmonary hypertension. *J Pathol* 1998; **185**: 313–318.
78. Beghetti M, Habre W, Friedli B, Berner M. Continuous low dose inhaled nitric oxide for treatment of severe pulmonary hypertension after cardiac surgery in paediatric patients. *Br Heart J* 1995; **73**: 65–68.
79. Fineman JR, Wong J, Soifer SJ. Hyperoxia and alkalosis produce pulmonary vasodilation independent of endothelium-derived nitric oxide in newborn lambs. *Pediatr Res* 1993; **33**: 341–346.
80. Kavanagh BP, Mouchawar A, Goldsmith J, Pearl RG. Effects of inhaled NO and inhibition of endogenous NO synthesis in oxidant-induced acute lung injury. *J Appl Physiol* 1994; **76**: 1324–1329.
81. Wessel DL, Adatia I, Giglia TM *et al.* Use of inhaled nitric oxide and acetylcholine in the evaluation of pulmonary hypertension and endothelial function after cardiopulmonary bypass. *Circulation* 1993; **88**; 2128–2138.
82. Miller OI, Celermajer DS, Deanfield JE, Macrae DJ. Very low dose inhaled nitric oxide: a selective pulmonary vasodilator after operations for congenital heart disease. *J Thorac Cardiovasc Surg* 1994; **108**: 487–494.
83. Dreyfus G, Jebara VA, Guillemain R *et al.* Improved evaluation of pulmonary vascular resistance prior to heart transplantation. *Transplant Proc* 1989; **21**: 2559–2561.
84. Adatia I, Lillehei C, Arnold JH *et al.* Inhaled nitric oxide in the treatment of postoperative graft dysfunction after lung transplantation. *Ann Thorac Surg* 1994; **57**: 1311–1318.
85. Adatia I, Perry S, Landzberg M *et al.* Inhaled nitric oxide and hemodynamic evaluation of patients with pulmonary hypertension before transplantation. *J Am Coll Cardiol* 1995; **25**: 1656–1664.
86. Katayama Y, Hatanaka K, Hayashi T *et al.* Effects of inhaled nitric oxide in single lung transplantation in rats with monocrotaline-induced pulmonary hypertension. *J Heart Lung Transplant* 1995; **14**: 486–492.

87. Furchgott RF, Zawadzki JV. The obligatory role of endothelial cells in the relaxation of arterial smooth muscle by acetylcholine. *Nature* 1980; **288**: 373–376.
88. Addonizio LJ, Gersony WM, Robbins RC *et al.* Elevated pulmonary vascular resistance and cardiac transplantation. *Circulation* 1987; **76 (Suppl V)**: 52–55.
89. Griepp RB, Stinson EB, Dong E *et al.* Determinants of operative risk in human heart transplantation. *Am J Surg* 1971; **122**: 192–197.
90. Calhoon JH, Grover FL, Gibbons WJ *et al.* Single lung transplantation. Alternative indications and technique. *J Thorac Cardiovasc Surg* 1991; **101**: 816–825.
91. Armitage JM, Fricker FJ, Kurland G *et al.* Pediatric lung transplantation. The years 1985 to 1992 and the clinical trial of FK506. *J Thorac Cardiovasc Surg* 1993; **105**: 337–346.
92. Starnes VA, Lewiston NJ, Luikart H *et al.* Current trends in lung transplantation. Lobar transplantation and expanded use of single lungs. *J Thorac Cardiovasc Surg* 1992; **104**; 1060–1066.
93. Haydock DA, Trulock EP, Kaiser LR *et al.* Lung transplantation. Analysis of thirty-six consecutive procedures performed over a twelve-month period. *J Thorac Cardiovasc Surg* 1992; **103**: 329–340.

Chapter 21

The Role and Therapeutic Implications of Nitric Oxide in Systemic Hypertension

Pablo Forte and Nigel Benjamin

Introduction

The endothelium plays a vital role in the homeostasis of the cardiovascular system, and its secretory function is the most important factor regulating vascular tone. This endothelial action is exerted through the release of potent vasoactive substances that modulate the contractile activity of the vascular smooth muscle.[1] The best understood and probably the most significant of the endothelium-derived relaxing factors is NO, whose role in the vascular system has been intensively investigated in the last decade.

An abnormal endothelial L-arginine:NO pathway has been identified as an early event in a variety of cardiovascular diseases, including atherosclerosis and hypertension. This review will discuss the role of NO in blood pressure regulation and its involvement in the pathogenesis and development of essential hypertension. We will also examine potential therapeutic implications of NO-based pharmacotherapy in the treatment of this cardiovascular disease.

Physiological Role of Nitric Oxide in the Control of Blood Pressure

Regulation of Basal Vascular Tone

The demonstration that L-arginine is the physiological precursor for NO synthesis by the vascular endothelium[2] and the identification of specific inhibitors of this pathway (e.g. N^G-monomethyl-L-arginine, L-NMMA)[3] have allowed investigation of the role of NO in the regulation of blood pressure. In experimental animals, both short-term and chronic administration of NO synthase (NOS) inhibitors induce a time- and dose-dependent increase in blood pressure. This effect is accompanied by a decrease in urinary excretion of the NO oxidation products (nitrite and nitrate), indicating there is normally a continuous release of NO.[4] Similarly, in humans, local infusion of L-NMMA into the brachial artery leads to a dose-dependent fall in blood flow, showing that resistance vessels are in a constant state of NO-mediated vasodilatation.[5] In addition, studies in which NO inhibitors have been administered systemically have documented an increase in arterial blood pressure with a substantial rise in peripheral vascular resistance[6,7] (Figure 1). These findings suggest that under resting conditions the L-arginine:NO pathway is active in dilating smooth muscle at the level of resistance vessels, and that loss of NO synthesis leads to hypertension. The importance of basal endothelium-derived NO-mediated vasodilatation on blood pressure regulation is highlighted by studies in mice lacking the gene encoding for endothelial NO synthase (eNOS).[8] These rodents are markedly hypertensive, giving further evidence that basal NO generation from eNOS regulates blood pressure.

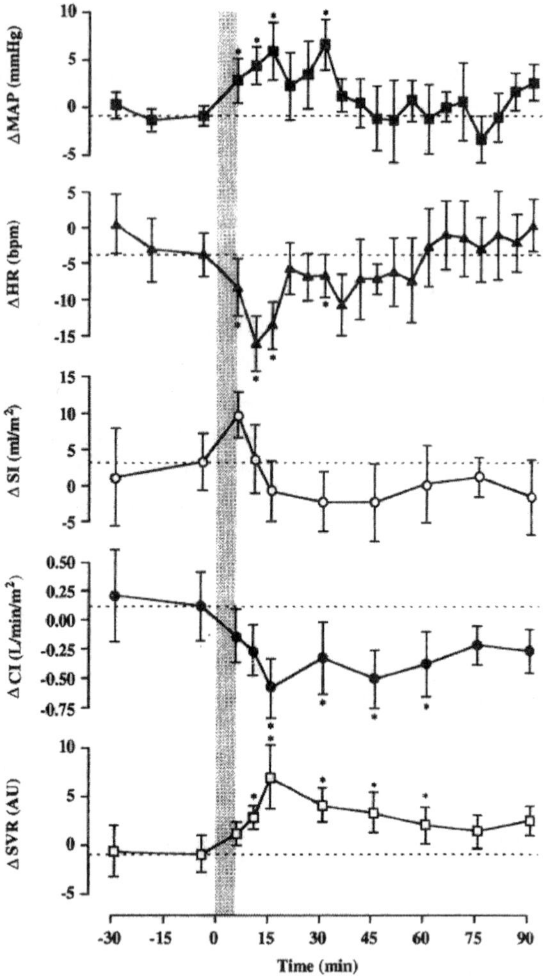

Figure 1. Effects of **L-NMMA** (shaded area) on differences from placebo. MAP, mean arterial pressure; **HR**, heart rate; **SI**, stroke index; **CI**, cardiac index; and **SVR**, systemic vascular resistance. Dashed line, last placebo-corrected pretreatment value for each index. *Significant difference ($p < 0.05$) from pre-dose. (Reproduced, with permission, from *J Hypertens* 1993; **11**: p1377)[6]

Volume Homeostasis

NO is an important factor in renal function. In the renal vasculature, the basal tone of glomerular and medullary microcirculation is regulated by endogenous NO. In experimental animals, acute inhibition of NOS is associated with a decrease in renal blood flow (RBF) and glomerular filtration rate (GFR), and increases both preglomerular and postglomerular vascular resistance.[9] In addition, the selective decrease of medullary blood flow by interstitial infusion of N^G-nitro-L-arginine methyl ester (L-NAME) into rat kidneys is associated with reductions of papillary blood flow, renal interstitial pressure and sodium excretion.[10] This demonstrates that NO is tonically active in the renal medullary circulation. In humans, inhibition of NO generation only caused modest renal vasoconstriction;[11] however, it substantially reduced GFR, indicating that NO has a greater effect on the afferent than the efferent arteriole. In addition, L-NMMA caused a significant decrease in fractional distal sodium excretion, implying an important physiological role for NO in the regulation of distal tubular sodium reabsorption. A more detailed discussion of the effects of NO on the renal circulation is given in Chapter 16.

NO also seems to be important for the renal response to volume expansion in rats. The natriuresis that follows acute increases in renal perfusion pressure is accompanied by augmented renal synthesis of NO, as suggested by increases in urinary excretion of nitrate.[12] Furthermore, inhibition of renal NOS by local, intrarenal infusion of arginine analogues produces a profound antidiuretic and antinatriuretic effect.[13] Similarly, acute sodium loading induces an increase in NO synthesis, and renal NOS inhibition blocks the homeostatic natriuresis that follows extracellular volume expansion.[14] The generation of NO in macula densa cells may also mediate a vasodilating component of the tubuloglomerular feedback response.[15] Collectively, these data demonstrate that NO plays an important role in the regulation of water/sodium excretion which is likely to influence blood pressure control.

Peripheral and Central Nervous System Control

Recent studies have suggested a possible NO-dependent role for the nervous system in the control of blood pressure. Low doses of intracisternal L-NAME markedly augmented renal nerve activity and slightly increased arterial blood pressure. In addition, both actions were blocked by cervical transection or by L-arginine administration.[16] Greenberg and colleagues reported a reciprocal control by noradrenergic vasoconstrictor and nitrergic vasodilator nerves in the arterial wall, whose disruption may contribute to the genesis of hypertension.[17] Recently, Sigmon and colleagues[18] reported that acute inhibition of neuronal NOS (nNOS) in mice decreases the hypertensive response to L-NAME. However, in animals lacking the gene encoding for nNOS, the systemic and renal haemodynamic responses to L-NAME were not different from controls. This suggests that nNOS may play a role in the acute regulation of blood pressure and nNOS-knock-out mice might develop compensatory mechanisms in the chronic control of blood pressure. Further studies are required to define the role of this NOS isoenzyme in the regulation of blood pressure and renal haemodynamics.

Pathophysiological Role of Nitric Oxide in Essential Hypertension

Because of the widespread involvement of NO in the physiological regulation of blood pressure, disturbances in its generation, release and action have been implicated in the pathogenesis and/or development of essential hypertension (Table 1).

Table 1. Role of the L-arginine-NO pathway in different models of hypertension.

Hypertension Model	Mechanisms
1. Animal models	
1.1. *Spontaneously hypertensive rat*	**Functionally ineffective NO**
	↑ production superoxide anion ↓ superoxide dismutase activity ↑ release of vasoconstrictore agents (prostaglandin H_2, thromboxane A_2) hypertrophy and fibrosis of vessels
	Decreased guanylate cyclase activity
1.2. *Dahl/Rapp salt-sensitive rat*	**Decreased production of NO**
	↓ constitutive and/or inducible NOS activity ↓ L-arginine availability, ↓ cofactors
	Decreased guanylate cyclase activity
2. Human hypertension	**Decreased basal production of NO**
	↓ NOS activity and/or ↓ cofactors ↓ transport of L-arginine, ↑ NOS inhibitors
	Decreased agonist-stimulated NO release
	(*not universal finding*)
3. Human pre-eclampsia	**Decreased NO production**
	↓ NOS activity and/or cofactors
	Increased NO production
	(*compensatory mechanism?*)

NOS = NO synthase

Classic Animal Models

Spontaneously hypertensive rat (*SHR*)

It appears that the endothelial production of NO is heterogeneously affected in this model of hypertension. In some vascular beds of hypertensive rats such as the mesenteric resistance arteries, the basal activity of constitutive NOS is augmented, as suggested by increases of nitrite + nitrate (the stable end-products of NO). This NOS appears to be functionally ineffective, however, as it is unable to increase the concentration of the second messenger guanosine 3',5'-cyclic monophosphate (cGMP) in the vascular smooth muscle cell to an appropriate level.[19] Therefore, it seems that additional events take place to prevent NO from accomplishing its normal vasorelaxant function. One mechanism involved is an increase in the degradation of NO, either by a rise in the production of oxidant molecules like superoxide anion or by a fall in superoxide dismutase (SOD) activity. Recently, Tschudi and colleagues demonstrated that the release of NO (measured using a porphyrinic microsensor) in the mesenteric vascular bed was unaffected in the SHR; however, there was increased degradation of NO by superoxide.[20] Furthermore, the intravenous administration of SOD decreased blood pressure in SHR, but not in normotensive animals.[21] Other possible mechanisms which could account for the blunted haemodynamic actions of NO are physical factors (e.g. hypertrophy and fibrosis of vessels), release of vasoconstrictor agents (e.g. prostaglandin H_2, thromboxane A_2) and the capacity of the vascular smooth muscle cells to respond to NO.

Conversely, in coronary and renal arteries of the SHR, the L-arginine:NO pathway does not seem to be impaired by hypertension. Tschudi and colleagues also showed that endothelium-dependent relaxations to acetylcholine (ACh) were comparable to normotensive WKY rats, at least up to the age of 15 weeks.[22] Furthermore, Kelm and colleagues demonstrated that the bradykinin-induced decrease in coronary vascular resistance and the parallel increase in NO release were similar in both rat strains.[23] Nava and colleagues showed that the medulla and heart of SHR displayed a higher Ca^{2+}-dependent NOS activity compared to that of WKY rats.[24] They concluded that the impaired renal medullary control of arterial pressure in this model of hypertension is not due to an impaired NO synthesis.

These data suggest that the endothelium, as well as the renal medulla and heart, respond to high blood pressure with increased NO production, probably representing a compensatory mechanism against hypertension. This is emphasized in the comparisons of SHR and L-NAME-induced hypertension in which the rate of mortality was markedly increased in the L-NAME treated rat, even when the degree of hypertension was similar in both groups.[25]

Dahl/Rapp salt-sensitive rat

In this hypertensive strain, there is an inability of the renal vasculature to dilate in response to increased salt intake. Studies have demonstrated that, compared to the Dahl/Rapp salt-resistant (DR) rat, its homologous salt-sensitive (DS) counterpart has little or no reduction in renal vascular resistance in response to high salt diet before an increased total peripheral resistance and hypertension develop.[26]

An abnormal L-arginine:NO pathway has been suggested to play an important role in the development of hypertension in this model of hypertension. Chen and Sanders reported that L-arginine, the substrate for NOS, increased urinary excretion of nitrate and simultaneously lowered blood pressure in DS rats fed on a high salt diet but not in the DR rat.[27] Furthermore, L-NMMA, a probe of NO activity, raised blood pressure more in the DR than in the DS rat. Similarly, Hu and Manning showed that the intravenous administration of L-arginine increased the whole body NO production of the DS rats. In addition, L-arginine prevented both the changes in the pressure-natriuresis relationship and the development of salt-induced hypertension in these rats.[28] Interestingly, this anti-hypertensive effect of L-arginine appears to be mediated through a mechanism involving the inducible NO synthase (iNOS), since dexamethasone, an inhibitor of this enzyme, completely abolished this effect.[29] This hypothesis has been strengthened by the work of Deng and Rapp, who showed that the locus for iNOS on chromosome 10 co-segregates with a gene determining blood pressure in this hypertensive strain.[30]

Alternatively, a recent study[31] has suggested that impairment in cGMP generation, the second messenger of NO, may play a primary role in the inability of the kidneys of DS rats to vasodilate in response to increased salt intake. This investigation found that the intravenous administration of atrial natriuretic peptide or nitroprusside, an NO donor, reduced renal vascular resistance in DR but not in DS rats fed on high salt diet. These results suggest that the L-arginine:NO pathway is likely to be involved in the regulation of blood pressure in Dahl rats fed on a high-salt diet, and the DS rat may be defective either in L-arginine availability or iNOS and/or in soluble guanylate cyclase (sGC) activity.

Human Hypertension

Inhibition of NOS has suggested that basal NO-mediated vasodilatation is abnormal in patients with hypertension. Calver and colleagues reported that the decrease in forearm blood flow induced by the vasoconstrictor L-NMMA was smaller in hypertensive than in normotensive subjects, with the response to L-NMMA correlating inversely with blood pressure.[32] Furthermore, we reported that after intravenous administration of L-$[^{15}N]_2$-arginine, the total urinary ^{15}N nitrate excretion was significantly lower in the hypertensive group compared with

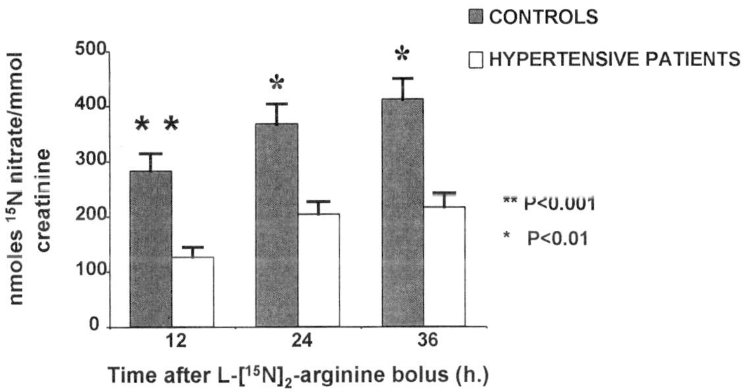

Figure 2. Mean (SEM) cumulative urinary excretion of ^{15}N nitrate after the administration of ^{15}N-labelled arginine in hypertensive patients and controls. (Reproduced with permission from: Forte P, Copland M, Smith LM *et al.*, Basal nitric oxide synthesis in essential hypertension. *Lancet* 1997; **349**: p839)

that in the control group[33] (Figure 2). In addition, urinary ^{15}N nitrate excretion correlated inversely with the blood pressure over the whole group of subjects studied. The reason for decreased synthesis is not clear, but could be explained by the following possibilities: (i) decreased NOS activity or a defect of any of its cofactors; (ii) abnormal entry of L-arginine into cells; and (iii) endogenous inhibitors of NOS (e.g. methylarginines). Regardless of the exact nature of the defect in the L-arginine:NO pathway, this abnormality could play an important role in the genesis or perpetuation of the hypertensive process, and thus contribute to the morbid consequences of the disease. Experimental data suggest that a reduction of NO synthesis leads to increased platelet aggregation and adhesion, increased monocyte chemotaxis, enhanced vasoconstrictor responses, and migration and proliferation of vascular smooth muscle cells.[34] These effects could strongly associate essential hypertension with vascular hypertrophy and occlusive vascular disease.

On the other hand, functional studies using muscarinic agonists to study agonist-stimulated NO release have reported conflicting results. Most studies have shown that endothelium-dependent vasorelaxation in response to ACh is reduced in patients with essential hypertension.[35,36] In a large study sample, however, we did not find any change in ACh-induced vasodilatation in the forearm arterial bed of patients with essential hypertension[37] (Figure 3). The reason for the difference in the results between these studies is not obvious, but the use of ACh to probe the L-arginine:NO pathway depends on the blood acetylcholinesterase activity,[38] forearm length,[39] race of the subjects, distribution of receptors and coupling of receptors to NOS. Furthermore, since ACh produces substantial vasodilatation even after infusion of L-NMMA,[40] the release of other substances (e.g. endothelium-derived hyperpolarizing factor, EDHF) could also explain these discrepancies.

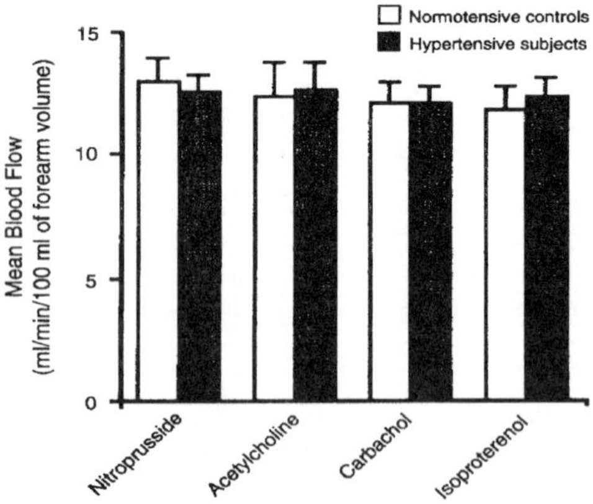

Figure 3. Mean (SEM) blood flow in the forearm during the infusion of nitroprusside, acetylcholine, carbachol and isoproterenol in 18 normotensive controls and 41 hypertensive subjects. (Reproduced with permission from: Cockcroft JR, Chowienczyk PJ, Benjamin N, Ritter JM, Preserved endothelium-dependent vasodilatation in patients with essential hypertension. *N Engl J Med* 1994; **330**: p1039)

These results suggest that although abnormal ACh-stimulated NO release is not a universal finding, basal levels of NO production appear to be diminished in patients with essential hypertension.

Human Pre-eclampsia

There is evidence that an augmented NO synthesis might contribute to the overall reduction in peripheral vascular resistance and the refractoriness to vasopressors in normal pregnancy. Previous studies in animals[41] and humans[42] have demonstrated increased plasma and urinary excretion of both cGMP and nitrate during normal pregnancy. In addition, N^G-nitro-L-arginine (a specific inhibitor of NO synthesis) annuls the blunted pressor response to angiotensin II, noradrenaline and vasopressin, and causes hypertension and fetal growth retardation in gravid rats.[43] Recently, Williams and colleagues showed that inhibition of NOS with L-NMMA reduced hand blood flow more during pregnancy than in the non-gravid state, and the L-NMMA response increased relative to that of noradrenaline as pregnancy progressed.[44] Furthermore, in studies of isolated perfused human placental cotyledons, Myatt and colleagues found that NO attenuated the actions of the vasoconstrictors endothelin-1 and the thromboxane mimetic U46619.[45]

It has been hypothesized that a deficiency in NO production may play an important role in the pathogenesis of pre-eclampsia in pregnancy. In pregnant rats, chronic blockade of NO synthesis produces a pre-eclampsia like syndrome, with sustained hypertension, proteinuria, thrombocytopoenia and intra-uterine growth retardation.[46] Attempts to extend this model of diminished NO synthesis to human pre-eclampsia have been limited. Studies have reported that the serum concentration and urinary excretion of nitrate, as well as umbilical venous plasma concentrations of cGMP, were significantly lower in women with pre-eclampsia compared with normal pregnant women.[47,48] In addition, McCarthy and colleagues found impaired endothelium-dependent relaxation to ACh in resistance arteries of subcutaneous fat from pre-eclamptic women.[49] The ability of perfused umbilical vessels to release NO in response to bradykinin was also reduced in pregnancy-induced hypertension.[50] These data suggest that defective NO production may be a possible cause for pre-eclampsia.

By contrast, other reports found an increased plasma concentration and urinary excretion of NO metabolites in women with pre-eclampsia, which correlated with changes in systolic blood pressure. Lyall and colleagues found significantly higher nitrate concentrations in umbilical venous serum of pre-eclamptic women than those of normal pregnant women.[51] Davidge and colleagues observed increased NO production and NOS activity in endothelial cells exposed to plasma from pre-eclamptic women.[52] However, the same authors reported in a subsequent paper[53] that, in the presence of shear stress, there were no differences in production rates of NO between cells exposed to plasma from either pre-eclamptic or normal pregnant women. These results support the hypothesis that increased NO production in this condition may be a compensatory response to improve blood flow, and limit adhesion and aggregation of platelets and monocytes to vascular endothelium.

As we can observe, there is no clear consensus as to whether the synthesis of NO is decreased or increased in women with pre-eclampsia. One explanation for these discrepancies may be associated with the relatively specificity of the methodology used to assess the production of NO. Indeed, in most of these studies, the measurements of plasma and urinary nitrate were confounded by the contribution of dietary nitrate. The half-life of nitrate in plasma is about 8 hours in man.[33,54] This means that to eliminate the effects of diet, the subjects must have a very low nitrate diet for 48 hours (about six half-lives) to clear plasma of 98.5% of exogenously-derived nitrate, and thus attain basal plasma concentration of this anion (see also Chapter 6). ACh causes dilatation by mechanisms other than *via* the generation of NO,[40] and cGMP is also the second messenger of other hormones such as atrial natriuretic peptide.[55] It would be of interest to determine the urinary excretion of ^{15}N-labelled nitrate production following intravenous administration of L-[guanidino-$^{15}N]_2$-arginine as an index of NO production in normal pregnancy and in women with pre-eclampsia. This technique overcomes all the limitations mentioned previously, since it is highly specific and the measurement of ^{15}N nitrate is independent of dietary sources.

It is clear that a large amount of work remains to be done before the role of NO in the pathogenesis of human pre-eclampsia is understood.

L-arginine:Nitric Oxide Pathway and the Pathogenesis of Essential Hypertension

Previous reports in animals and humans have shown an inverse correlation between the activity of the L-arginine:NO pathway and arterial blood pressure. However, whether the defect of the NO-dependent dilator system is a cause or a consequence of the increased blood pressure is as yet unclear. Although diminished release or efficacy of NO may lead directly to a rise in blood pressure, it is also possible that the vasoconstriction that accompanies hypertension may itself be associated with an abnormality in the L-arginine:NO pathway.

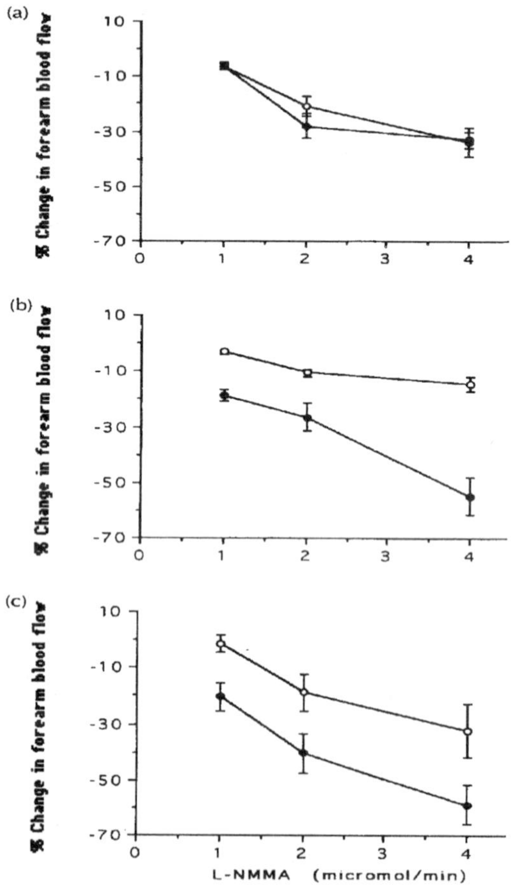

Figure 4. Percentage change in forearm blood flow to L-NMMA before (○) and after (●) 6 weeks of treatment with (a) placebo, (b) the ACE inhibitor enalapril, and (c) the Ca^{2+} antagonist amlodipine. (Reproduced with permission from: Lyons D, Webster J, Benjamin N, The effect of antihypertensive therapy on responsiveness to local intra-arterial N^G-monomethyl-L-arginine in patients with essential hypertension. *J Hypertens* 1994; **12:** p1050)

In two recent studies,[56,57] Taddei and colleagues found that the forearm vascular response to ACh was blunted in normotensive children of hypertensive patients as compared to normotensive subjects who did not have hypertensive parents, suggesting that an impairment of NO production precedes the onset of essential hypertension. We showed that platelets from untreated hypertensive patients exhibited a markedly reduced sensitivity to L-NMMA.[58] Moreover, Panza *et al.*[59] and Creager *et al.*[60] reported that decreased ACh-induced vasodilatation persisted in hypertensive patients after treatment with anti-hypertensive drugs that normalized blood pressure. These findings further support the view that impairment of NO-dependent relaxation may be the cause rather than the effect of elevated blood pressure. By contrast, both Calver and colleagues[61] and our own group[62] have shown that the forearm arterial responsiveness to L-NMMA in newly diagnosed patients with essential hypertension returned to normal following normalization of blood pressure achieved by standard anti-hypertensive drugs with differing modes of action (Figure 4).

It is thus evident that there is no clear agreement as to whether the defect in NO synthesis is a primary phenomenon or the result of long-standing hypertensive damage to endothelium. It would also be of interest to determine the changes in urinary excretion of ^{15}N-labelled nitrate after the administration of L-[^{15}N]$_2$-arginine in hypertensive patients treated with anti-hypertensive drugs.

Therapeutic Implications

Since reduced and/or impaired generation of NO is implicated in the development of essential hypertension, it may be desirable to mimic or enhance the physiological generation of NO. This may be achieved in several ways: using compounds that donate NO, stimulating receptors linked to the L-arginine:NO pathway, augmenting the action of endogenous NO (e.g. by increasing the levels of cGMP), or providing additional substrate (L-arginine) for its synthesis.

NO Donors

Vascular smooth muscle relaxation results from stimulation of cGMP production following the diffusion of NO into the smooth muscle cell.[63] Nitrosovasodilator compounds, such as nitroglycerin and sodium nitroprusside, are metabolized to NO inside the smooth muscle cell and cause vascular relaxation by increasing intracellular levels of cGMP (see also Chapter 4). There is increasing interest in the possibility that NO replacement therapy would be beneficial in hypertension. Recently, infusion of glyceryl trinitrate has been shown to improve uterine artery blood flow in pregnant women at risk from pre-eclampsia, whereas other vasodilators have only minor effects.[64] However, the use of these nitrovasodilators as effective therapeutic agents is dependent on electron-transfer activation before generation of the NO radical, and is limited by tolerance and toxicity.[65]

Compounds that directly deliver NO to vascular smooth muscle, without requiring enzymatic activation of a pro-drug, would be of important therapeutic value as vasodilators and as agents in countering the endothelial dysfunction known to be present in hypertension. Moreover, nitrovasodilators that target specific tissues (platelets or arterial smooth muscle) might have additional or alternative therapeutic indications to existing drugs. For example, MacAllister and colleagues reported that *S*-nitrosoglutathione was equipotent with nitroglycerin in the forearm arterial bed but approximately 8-fold less potent in dorsal hand veins.[66] In addition, Lees and colleagues reported that the infusion of *S*-nitrosoglutathione in women with pre-eclampsia reduced mean arterial blood pressure, platelet activation, and uterine artery resistance without compromising fetal Doppler indices.[67] However, much work remains to be done before we can determine the reality of their theoretical clinical potential and whether they may or may not be subject to the shortcomings, such as tolerance or toxicity, that compromise the use of certain currently available nitrosovasodilators.

Stimulation of Receptors Linked to the L-arginine:NO Pathway

Angiotensin converting enzyme inhibitors

The anti-hypertensive effect of angiotensin converting enzyme (ACE) inhibitors has been mainly attributed to diminished formation of angiotensin II. However, additional mechanisms appear to be involved in the hypotensive action of ACE inhibitors, because reports have shown that pre-treatment with a NOS inhibitor attenuated the acute blood pressure-lowering effect induced by both captopril and ramiprilat.[68] Indeed ACE inhibitors, besides preventing the generation of angiotensin II, act indirectly in this way also by preventing the breakdown of bradykinin, which stimulates bradykinin B_2 receptors on endothelial cells and activates the endothelial L-arginine:NO pathway.[69]

Chronic treatment with an ACE inhibitor not only decreases the ratio of plasma angiotensin II to angiotensin I, but also increases the plasma level of bradykinin.[70] In addition, this effect is accompanied by an increased production of cGMP, indicating the greater release of NO. Furthermore, we showed that local inhibition of converting enzyme within the human forearm increased the vasodilator response to intra-arterial bradykinin.[71] Similarly, in human coronary arteries, the ACE inhibitor enalapril enhanced the increase in local blood flow induced by bradykinin, while the B_2 receptor antagonist Hoe 140 prevented the response.[72] In addition, there was a reduction in flow-dependent dilatation after B_2 receptor blockade.

Besides this effect on NO production, bradykinin also exerts its vasodilatory action through the release of prostacyclin[73] and EDHF.[74] ACE inhibition may exhibit beneficial effects in addition to modulating the local concentration of bradykinin. Indeed, the enzyme is known to degrade potent vasodilator peptides such as substance P,[75] enkephalins,[76] neurotensin[77] and the C-natriuretic peptide.[78] Additionally, ACE inhibition is associated with significant elevation of angiotensin-(1–7), which enhances the release of NO.[79] Therefore, the contribution of these other biologically highly active mediators may also augment the endothelium-dependent relaxation produced by ACE inhibitors.

The effects of ACE inhibitors on endothelium-dependent vasodilatation have been explored in different models of hypertension. Lüscher and colleagues found that anti-hypertensive treatment with these drugs improved endothelium-dependent vasodilatation in Dahl salt-sensitive rats.[80] Clozel and colleagues reported that the ACE inhibitors cilazapril and captopril, but not hydralazine, improved endothelium-dependent relaxation in aortic rings of spontaneously hypertensive rats.[81] Furthermore, Hirooka and colleagues reported that administration of a single dose of captopril improves impaired endothelium-dependent vasodilatation (ACh-induced increase in forearm blood flow) in hypertensive patients.[82] However, ACE inhibitors were unable to improve the impaired endothelium-dependent vasodilatation in the forearm circulation of hypertensive patients after 8 weeks of treatment with captopril or enalapril.[83] One possibility is that ACE inhibitors may have been ineffective because the duration of therapy was too short. Indeed, Schiffrin and colleagues have reported that impaired endothelial function was improved after 1 and 2 years treatment with cilazapril.[84,85] However, whether these changes are attributed to increased production of NO, reduced generation of angiotensin II, increased concentration of EDHF or other mechanisms is a matter that remains to be elucidated.

Increasing the Release and/or Action of Endogenous NO

Calcium antagonists

Dihydropyridines have been introduced in cardiovascular pharmacology as Ca^{2+} channel antagonists with vasodilating properties that depend on an antagonism against L-type Ca^{2+} channels. Nevertheless, recent findings suggest that the vasorelaxation achieved with these compounds may also depend on endothelial function. Dhein and colleagues demonstrated that the vasodilating effects of four dihydropyridines (nitrendipine, nifedipine, nisoldipine, and nimodipine) on isolated mesenteric vascular bed were sensitive to L-NAME treatment, indicating a probable dependence on NO production.[86] Furthermore, Akopov reported that the vasodilator effects of isradipine and lacidipine on vessel segments with intact endothelium in the presence of L-NAME were significantly attenuated, whereas a cyclooxygenase inhibitor, indomethacin, did not significantly alter vasodilator responses of these Ca^{2+} antagonists.[87] However, the exact mechanism of how Ca^{2+} antagonists modulate

NO production is unclear. An enhancement of Ca^{2+} antagonist action by basally released NO is not likely to be mediated *via* an interaction with L-type Ca^{2+} channels, as endothelial cells from macrovascular arteries do not appear to express voltage-operated Ca^{2+} channels.[88] Recently, Salameh and colleagues suggested that nitrendipine elevates the intracellular Ca^{2+} in endothelial cells through activation of shear stress-activated cation-selective channels.[89]

Ca^{2+} antagonists have also been shown to facilitate the effects of NO in vascular smooth muscle cells. Kung and colleagues demonstrated that in vessels treated with a Ca^{2+} antagonist, endothelium-dependent relaxations and those due to the endothelium-independent vasodilator sodium nitroprusside were enhanced.[90] Moreover, Norman and colleagues reported that nifedipine and four other dihydropyridine Ca^{2+} entry blockers elicited vascular relaxation by selectively inhibiting cyclic nucleotide phosphodiesterase.[91]

These findings suggest that, in functionally intact vessels, vasorelaxation induced by 1,4-dihydropyridines may additionally be mediated by increasing the release and/or augmenting the action of NO. However, further studies are needed to determine the potential implications of these observations in endothelial dysfunction associated with human hypertension.

Inhibitors of cGMP phosphodiesterase

The actions of endogenous NO are generally mediated *via* stimulation of sGC and increased cGMP levels,[63] as described elsewhere in this volume. The intracellular levels of cGMP are regulated enzymatically by cyclases and/or by cGMP phosphodiesterase (PDE).[92] Since the half-life of cGMP is shortened by PDEs, the effect of cGMP may be enhanced or prolonged using selective inhibitors of these enzymes. PDEs are classified into seven major types; inhibition of PDE1 (by vinpocentine or SCH 51866) augments cGMP in vascular smooth muscle and causes vasorelaxation. The specificity of such compounds is restricted to a relatively narrow concentration range; however, it is probable that the development of more selective inhibitors of the cGMP PDE1 isoenzyme may constitute a potential therapeutic indication for treatment of essential hypertension.

Superoxide dismutase

The enzyme SOD catalyzes dismutation of superoxide anion radical ($.O_2$) to hydrogen peroxide and oxygen.[93] Since a rise in the production of oxidative radicals such as $.O_2$ increases the degradation of NO, the administration of this enzyme would prolong the half-life of NO. Indeed, Auch-Schwelk and colleagues showed that the intravenous administration of SOD decreased blood pressure in SHR.[21] However, Garcia and colleagues reported that exogenous intravascular administration of CuZn SOD did not improve the impaired endothelium-dependent vasodilatation in patients with essential hypertension.[94] It is plausible that CuZn SOD administration may not modify this anomaly, since it is not readily transported into the cytoplasm of vascular smooth muscle due to its large size and negative charge. Therefore, it is feasible that development of smaller molecules with high affinity for endothelial cell surfaces (i.e. SOD within liposomes) may provide future therapeutic applications.

L-arginine

If the availability of the L-arginine is considered as a rate-limiting factor for NO production *in vivo*, administration of L-arginine may be expected to result in increased NO formation. Long-term treatment of both SHR[95] and Dahl salt-sensitive rats[96] with L-arginine can prevent the development of hypertension. In both models, L-arginine treatment has been shown to shift the pressure-natriuresis relationship to lower pressures without altering total RBF or GFR. Furthermore, systemic infusion of L-arginine has been reported to have a hypotensive effect in certain patients with hypertension.[97] In addition, there was also an increase in the urinary excretion of nitrate and cGMP, suggesting that these effects may have been mediated through conversion of L-arginine to NO. However, the significance of these observations remains uncertain since D-arginine was not used to determine the stereo-specificity of the response. L-arginine is not only a substrate for the L-arginine:NO pathway; indeed, it is a powerful secretagogue of hormones and has complex actions that may be independent of the generation of NO.[98] Further studies are required to assess more clearly the role of L-arginine in the treatment of essential hypertension.

Summary

The widespread involvement of NO in the physiological regulation of the cardiovascular system suggests that disturbances in its generation, release or action could have important influence on the pathogenesis and/or development of essential hypertension. Nevertheless, the role of NO in hypertension is still controversial, and seems to vary depending on the stage of the disease and in the model studied. In spontaneously hypertensive rats, the production of NO appears to be increased but ineffective, probably because of increased inactivation or scavenging. In Dahl/Rapp salt-sensitive rats, the NO production and effector pathway seems to be defective, due either to impaired L-arginine availability or to reduced activity of iNOS and/or sGC. In human hypertension, the basal production of NO appears to be decreased; however, a decrease in ACh-stimulated NO activity is not a universal finding. In human pre-eclampsia, there is no clear consensus whether the rate of NO synthesis is decreased or increased. Most studies of essential hypertension have shown a correlation between the activity of the L-arginine:NO pathway and arterial blood pressure. Whether the defect in the NO-dependent dilator system is a cause or a consequence of increased blood pressure remains unclear, however.

The objective of NO-based pharmacotherapy is to reach appropriate homeostasis of NO metabolism in the target tissue where endogenous production of NO is abnormal. In essential hypertension, this may be achieved by increasing the production and action of NO through the use of ACE inhibitors and Ca^{2+} antagonists (1,4-dihydropyridines). Future prospects of NO-based pharmacotherapy may feature NO donors, inhibitors of cGMP phosphodiesterase, the administration of SOD enzyme and/or supplementation of the substrate of NO synthase, L-arginine.

References

1. Furchgott RF, Vanhoutte PM. Endothelium-derived relaxing and contracting factors. *FASEB J* 1989; **3**: 2007–2018.
2. Palmer RMJ, Ashton DS, Moncada S. Vascular endothelial cells synthesize nitric oxide from L-arginine. *Nature* 1988; **333**: 664–666.
3. Rees DD, Palmer RMJ, Schulz R *et al.* Characterization of three inhibitors of endothelial nitric oxide synthase *in vitro* and *in vivo. B J Pharmacol* 1990; **101**: 746–752.
4. Takahashi H, Hara K, Komiyama Y *et al.* Mechanism of hypertension induced by chronic inhibition of nitric oxide in rats. *Hypertens Res* 1995; **18**: 319–324.
5. Vallance P, Collier J, Moncada S. Effects of endothelium-derived nitric oxide on peripheral arteriolar tone in man. *Lancet* 1989; **2**: 997–1000.
6. Haynes WG, Noon JP, Walker BR, Webb DJ. Inhibition of nitric oxide synthesis increases blood pressure in healthy humans. *J Hypertens* 1993; **11**: 1375–1380.
7. Stamler JS, Loh E, Roddy MA *et al.* Nitric oxide regulates basal systemic and pulmonary vascular resistance in healthy humans. *Circulation* 1994; **89**: 2035–2040.
8. Huang PL, Huang Z, Mashimo H *et al.* Hypertension in mice lacking the gene for endothelial nitric oxide synthase. *Nature* 1995; **377**: 239–242.
9. Baylis C, Slangen B, Hussain S, Weaver C. Relationship between basal NO release and cyclooxygenase products in the normal rat kidney. *Am J Physiol* 1996; **271**: R1327–R1334.
10. Majid DS, Williams A, Kadowitz PJ, Navar LG. Renal responses to intra-arterial administration of nitric oxide donor in dogs. *Hypertension* 1993; **22**: 535–541.
11. Haynes WG, Hand MF, Dockrell ME *et al.* Physiological role in nitric oxide regulation of renal function in humans. *Am J Physiol* 1997; **272**: F364–F371.
12. Alberola A, Pinilla JM, Quesada T *et al.* Role of nitric oxide in mediating renal response to volume expansion. *Hypertension* 1992; **19**: 780–784.

13. Fenoy FJ, Ferrer P, Carbonell L, Garcia-Salom M. Role of nitric oxide on papillary blood flow and pressure natriuresis. *Hypertension* 1995; **25**: 408–414.
14. Shultz PJ, Tolins JP. Adaptation to increased dietary salt intake in the rat. Role of endogenous nitric oxide. *J Clin Invest* 1993; **91**: 642–650.
15. He XR, Greenberg SG, Briggs JP, Schnermann JB. Effect of nitric oxide on renin secretion. II. Studies in the perfused juxtaglomerular apparatus. *Am J Physiol* 1995; **268**: F953–F959.
16. Togashi H, Sakuma I, Yoshioka M *et al.* A central nervous system action of nitric oxide in blood pressure regulation. *J Pharmacol Exp Ther* 1992; **262**: 343–347.
17. Greenberg SS, Peevy K, Tanaka TP. Endothelium-derived and intraneuronal nitric oxide-dependent inhibition of norepinephrine efflux from sympathetic nerves by bradykinin. *Am J Hypertens* 1991; **4**: 464–467.
18. Sigmon DH, Carretero OA, Beierwaltes WH. Participation of neuronal nitric oxide synthase (nNOS) in blood pressure (BP) and renal hemodynamic regulation by nitric oxide in mice [Abstract]. *FASEB J* 1997; **11**: A266.
19. Nava E, Moreau P, Lüscher TF. Basal production of nitric oxide is increased, but inefficacious, in spontaneous hypertension [Abstract]. *Circulation* 1995; **92**: I–347.
20. Tschudi MR, Mesaros S, Lüscher TF, Malinski T. Direct in situ measurement of nitric oxide in mesenteric resistance arteries. Increased decomposition by superoxide in hypertension. *Hypertension* 1996; **27**: 32–35.
21. Auch-Schwelk W, Katusic ZS, Vanhoutte PM. Contractions to oxygen-derived free radicals are augmented in aorta of the spontaneously hypertensive rat. *Hypertension* 1989; **13**: 859–864.
22. Tschudi MR, Criscione L, Novosel D *et al.* Antihypertensive therapy augments endothelium-dependent relaxations in coronary arteries of spontaneously hypertensive rats. *Circulation* 1994; **89**: 2212–2218.
23. Kelm M, Feelisch M, Krebber T *et al.* Role of nitric oxide in the regulation of coronary vascular tone in hearts from hypertensive rats. Maintenance of nitric oxide forming capacity and increased basal production of nitric oxide. *Hypertension* 1995; **25**: 186–193.
24. Nava E, Llinas M, Gonzalez JD, Salazar JF. Nitric oxide synthase activity in renal cortex and medulla of normotensive and spontaneously hypertensive rats. *Am J Hypertens* 1996; **9**: 1236–1239.
25. Arnal JF, Battle T, Menard J, Michel JB. The vasodilatory effect of endogenous nitric oxide is a major counter-regulatory mechanism in the spontaneously hypertensive rat. *J Hypertens* 1993; **11**: 945–950.
26. Rapp JP. Dahl salt-susceptible and salt-resistant rats. *Hypertension* 1982; **4**: 753–763.
27. Chen, PY, Sanders PW. L-Arginine abrogates salt sensitive hypertension in Dahl/Rapp rats. *J Clin Invest* 1991; **88**: 1559–1567.

28. Hu L, Manning RD Jr. Role of nitric oxide in regulation of long-term pressure-natriuresis relationship in Dahl rats. *Am J Physiol* 1995; **268**: H2375–H2383.
29. Chen PY, Sanders PW. Role of nitric oxide synthesis in salt-sensitive hypertension in Dahl/Rapp rats. *Hypertension* 1993; **22**: 812–818.
30. Deng AY, Rapp JP. Locus for the inducible, but not a constitutive, nitric oxide synthase co-segregates with blood pressure in the Dahl salt-sensitive rat. *J Clin Invest* 1995; **95**: 2170–2177.
31. Simchon S, Manger W, Blumberg G *et al.* Impaired renal vasodilation and urinary cGMP excretion in Dahl salt-sensitive rats. *Hypertension* 1996; **27**: 653–657.
32. Calver A, Collier J, Moncada S, Vallance P. Effect of local intra-arterial N^G-monomethyl-L-arginine in patients with hypertension: the nitric oxide dilator mechanism appears abnormal. *J Hypertens* 1992; **10**: 1025–1031.
33. Forte P, Copland M, Smith LM *et al.* Basal nitric oxide synthesis in essential hypertension. *Lancet* 1997; **349**: 837–842.
34. Moncada S, Higgs EA. The L-arginine-nitric oxide pathway. *N Engl J Med* 1993; **329**: 2002–2012.
35. Linder L, Kiowski W, Buhler FR, Lüscher TF. Indirect evidence for release of endothelium-derived relaxing factor in human forearm circulation *in vivo*: blunted response in essential hypertension. *Circulation* 1990; **81**: 1762–1767.
36. Panza JA, Casino PR, Kilcoyne CM, Quyyumi AA. Role of endothelium-derived nitric oxide in the abnormal endothelium-dependent vascular relaxation of patients with essential hypertension. *Circulation* 1993; **87**: 1468–1474.
37. Cockcroft JR, Chowienczyk PJ, Benjamin N, Ritter JM. Preserved endothelium-dependent vasodilatation in patients with essential hypertension. *N Engl J Med* 1994; **330**: 1036–1040.
38. Duff F, Greenfield ADM, Shepherd JT, Thompson ID. A quantitative study of the response to acetylcholine and histamine of the blood vessels of the human hand and forearm. *J Physiol* (*Lond*) 1953; **120**: 160–170.
39. Chowienczyk PJ, Cockcroft JR, Ritter JM. Blood flow responses to intra-arterial acetylcholine in man: effects of basal flow and conduit vessel length. *Clin Sci* 1994; **87**: 45–51.
40. Chowienczyk PJ, Cockcroft JR, Ritter JM. Differential inhibition by N^G-monomethyl-L-arginine of vasodilator effects of acetylcholine and methacholine in human forearm vasculature. *Br J Pharmacol* 1993; **110**: 736–738.
41. Conrad KP, Joffe GM, Kruszyna H *et al.* Identification of increased nitric oxide biosynthesis during pregnancy in rats. *FASEB J* 1993; **7**: 566–571.
42. Begum S, Yamasaki M, Mochizuki M. The role of nitric oxide metabolites during pregnancy. *Kobe J Med Sci* 1996; **42**: 131–141.
43. Danielson LA, Conrad KP. Acute blockade of nitric oxide synthase inhibits renal vasodilation and hyperfiltration during pregnancy in chronically instrumented conscious rats. *J Clin Invest* 1995; **96**: 482–490.

44. Williams DJ, Vallance PJ, Neild GH *et al.* Nitric oxide-mediated vasodilation in human pregnancy. *Am J Physiol* 1997; **272**: H748–H752.
45. Myatt L, Brewer AS, Langdon G, Brockman DE. Attenuation of the vasoconstrictor effects of thromboxane and endothelin by nitric oxide in the human fetal-placental circulation. *Am J Obstet Gynecol* 1992; **166**: 224–230.
46. Molnar M, Suto T, Toth T, Hertelendy F. Prolonged blockade of nitric oxide synthesis in gravid rats produces sustained hypertension, proteinuria, thrombocytopenia, and intrauterine growth retardation. *Am J Obstet Gynecol* 1994; **170**: 1458–1466.
47. Seligman SP, Buyon JP, Clancy RM *et al.* The role of nitric oxide in the pathogenesis of preeclampsia. *Am J Obstet Gynecol.* 1994; **171**: 944–948.
48. Lauria MR, Standley CA, Sorokin Y *et al.* Brain natriuretic peptide and cyclic guanosine-3',5' monophosphate in pre-eclampsia. *J Matern Fetal Med.* 1996; **5**: 128–131.
49. McCarthy AL, Woolfson RG, Raju SK, Poston L. Abnormal endothelial cell function of resistance arteries from women with preeclampsia. *Am J Obstet Gynecol* 1993; **168**: 1323–1330.
50. Knock GA, Poston L. Bradykinin-mediated relaxation of isolated maternal resistance arteries in normal pregnancy and preeclampsia. *Am J Obstet Gynecol* 1996; **175**: 1668–1674.
51. Lyall F, Young A, Greer IA. Nitric oxide concentrations are increased in the fetoplacental circulation in preeclampsia. *Am J Obstet Gynecol* 1995; **173**: 714–718.
52. Davidge ST, Baker PN, Roberts JM. NOS expression is increased in endothelial cells exposed to plasma from women with preeclampsia. *Am J Physiol* 1995; **269**: H1106–H1112.
53. Baker PN, Stranko CP, Davidge ST *et al.* Mechanical stress eliminates the effects of plasma from patients with preeclampsia on endothelial cells. *Am J Obstet Gynecol* 1996; **174**: 730–736.
54. Jungersten L, Edlund A, Petersson AS, Wennmalm Å. Plasma nitrate as an index of nitric oxide formation in man: analyses of kinetics and confounding factors. *Clin Physiol* 1996; **16**: 369–379.
55. Hirata Y, Tomita M, Takada S, Yoshimi H. Vascular receptor binding activities and cyclic GMP responses by synthetic human and rat atrial natriuretic peptides (ANP) and receptor down-regulation by ANP. *Biochem Biophys Res Commun* 1985; **128**: 538–546.
56. Taddei S, Virdis A, Mattei P *et al.* Endothelium-dependent forearm vasodilation is reduced in normotensive subjects with familial history of hypertension. *J Cardiovasc Pharmacol* 1992; **20**: S193–S195.
57. Taddei S, Virdis A, Mattei P *et al.* Defective L-arginine-nitric oxide pathway in offspring of essential hypertensive patients. *Circulation* 1996; **94**: 1298–1303.
58. Cadwgan TM, Benjamin N. Evidence for altered platelet nitric oxide synthesis in essential hypertension. *J Hypertens* 1993; **11**: 417–420.

59. Panza JA, Quyyumi AA, Callahan TS, Epstein SE. Effect of antihypertensive treatment on endothelium-dependent vascular relaxation in patients with essential hypertension. *J Am Coll Cardiol* 1993; **21**: 1145–1151.
60. Creager MA, Roddy MA. Effect of captopril and enalapril on endothelial function in hypertensive patients. *Hypertension* 1994; **24**: 499–505.
61. Calver A, Collier J, Vallance P. Forearm blood flow responses to a nitric oxide synthase inhibitor in patients with treated essential hypertension. *Cardiovasc Res* 1994; **28**: 1720–1725.
62. Lyons D, Webster J, Benjamin N. The effect of antihypertensive therapy on responsiveness to local intra-arterial N^G-monomethyl-L-arginine in patients with essential hypertension. *J Hypertens* 1994; **12**: 1047–1052.
63. Waldman SA, Murad F. Biochemical mechanisms underlying vascular smooth muscle relaxation: the guanylate cyclase-cyclic GMP system. *J Cardiovasc Pharmacol* 1988; **12**: S115–S118.
64. Ramsay B, De Belder A, Campbell S *et al.* A nitric oxide donor improves uterine artery diastolic blood flow in normal early pregnancy and in women at high risk of pre-eclampsia. *Eur J Clin Invest* 1994; **24**: 76–78.
65. Morton WE. Occupational habituation to aliphatic nitrates and the withdrawal hazards of coronary disease and hypertension. *J Occup Med* 1977; **19**: 197–200.
66. MacAllister RJ, Calver AL, Riezebos J *et al.* Relative potency and arteriovenous selectivity of nitrovasodilators on human blood vessels: an insight into the targeting of nitric oxide delivery. *J Pharmacol Exp Ther* 1995; **273**: 154–160.
67. Lees C, Langford E, Brown AS *et al.* The effects of S-nitrosoglutathione on platelet activation, hypertension, and uterine and fetal Doppler in severe preeclampsia. *Obstet Gynecol* 1996; **88**: 14–19.
68. Holte HR, Bjornstad-Ostensen A, Berg T. The role of endogenous bradykinin in blood pressure homeostasis in spontaneously hypertensive rats. *Br J Pharmacol* 1996; **118**: 1925–1930.
69. Auch-Schwelk W, Duske E, Claus M *et al.* Endothelium-mediated vasodilation during ACE inhibition. *Eur Heart J* 1995; **16 (Suppl C)**: 59–65.
70. Wiemer G, Scholkens BA, Linz W. Endothelial protection by converting enzyme inhibitors. *Cardiol Res* 1994; **28**: 166–172.
71. Benjamin N, Cockcroft JR, Collier JG *et al.* Local inhibition of converting enzyme and vascular responses to angiotensin and bradykinin in the human forearm. *J Physiol (Lond)* 1989; **412**: 543–555.
72. Groves P, Kurz S, Just H, Drexler H. Role of endogenous bradykinin in human coronary vasomotor control. *Circulation* 1995; **92**: 3424–3430.
73. Wiemer G, Scholkens BA, Becker BH, Busse R. Ramiprilat enhances endothelial autacoid formation by inhibiting breakdown of endothelium-derived bradykinin. *Hypertension* 1991; **18**: 558–563.

74. Mombouli JV, Vanhoutte PM. Endothelium-derived hyperpolarizing factor(s) and the potentiation of kinins by converting enzyme inhibitors. *Am J Hypertens* 1995; **8**: 19S–27S.
75. Tan SA, Berk LS, Tan LG. Increased tachykinins substance P and neurokinin A, not bradykinin, cause the ACE inhibitor induced cough [Abstract]. *J Am Coll Card* 1997; **29 (Suppl A)**: 41A.
76. Melzig MF, Heder G. Dexamethasone-induced enhanced enkephalin degradation by angiotensin-converting enzyme on endothelial cells. *Pharmazie* 1995; **50**: 139–141.
77. Bauer B, Rose G, Dürr R, Ertl G. Angiotensin converting enzyme (ACE) inhibitor potentiates neurotensin effects in coronary circulation of the dog [Abstract]. *Circulation* 1990; **82**: III–454.
78. Seymour A, Mathers P, Abboa-Offei B. Potentiation of C-natriuretic peptide (CNP) by neutral endopeptidase (NEP) and angiotensin-converting enzyme (ACE) inhibitors [Abstract]. *FASEB J* 1995; **9**: A881.
79. Ping L, Chappell MC, Ferrario CM, Brosnihan KB. Angiotensin-(1–7) augments bradykinin-induced vasodilation by competing with ACE and releasing nitric oxide. *Hypertension* 1997; **29**: 394–400.
80. Lüscher TF, Vanhoutte PM, Raij L. Antihypertensive treatment normalizes decreased endothelium-dependent relaxations in rats with salt-induced hypertension. *Hypertension* 1987; **9**: III193–III197.
81. Clozel M, Kuhn H, Hefti F. Effects of angiotensin converting enzyme inhibitors and of hydralazine on endothelial function in hypertensive rats. *Hypertension* 1990; **16**: 532–540.
82. Hirooka Y, Imaizumi T, Masaki H *et al.* Captopril improves impaired endothelium-dependent vasodilation in hypertensive patients. *Hypertension* 1992; **20**: 175–180.
83. Creager MA, Roddy MA. Effect of captopril and enalapril on endothelial function in hypertensive patients. *Hypertension* 1994; **24**: 499–505.
84. Schiffrin EL, Deng LY, Larochelle P. Effects of a beta-blocker or a converting enzyme inhibitor on resistance arteries in essential hypertension. *Hypertension* 1994; **23**: 83–91.
85. Schiffrin EL, Deng LY. Comparison of effects of angiotensin I-converting enzyme inhibition and beta-blockade for 2 years on function of small arteries from hypertensive patients. *Hypertension* 1995; **25**: 699–703.
86. Dhein S, Zhao Y, Simsek S *et al.* Actions of 1,4-dihydropyridines in isolated mesenteric vascular beds. *J Cardiovasc Pharmacol* 1995; **26**: 784–791.
87. Akopov SE. Nitric oxide (EDRF) enhances the vasorelaxing effect of dihydropyridine calcium antagonists in isolated human middle cerebral arteries. *Methods Find Exp Clin Pharmacol* 1996; **18**: 101–104.
88. Bossu JL, Elhamdani A, Feltz A *et al.* Voltage-gated Ca entry in isolated bovine capillary endothelial cells: evidence of a new type of BAY K 8644-sensitive channel. *Pflügers Arch* 1992; **420**: 200–207.

89. Salameh A, Schomecker G, Breitkopf K *et al.* The effect of the calcium-antagonist nitrendipine on intracellular calcium concentration in endothelial cells. *Br J Pharmacol* 1996; **118**: 1899–1904.
90. Kung CF, Tschudi MR, Noll G *et al.* Differential effects of the calcium antagonist mibefradil in epicardial and intramyocardial coronary arteries. *J Cardiovasc Pharmacol* 1995; **26**: 312–318.
91. Norman JA, Ansell J, Phillips MA. Dihydropyridine Ca^{2+} entry blockers selectively inhibit peak I cAMP phosphodiesterase. *Eur J Pharmacol* 1983; **93**: 107–112.
92. Beavo JA, Conti M, Heaslip R. Multiple cyclic nucleotide phosphodiesterases. *Mol Pharmacol* 1994; **46**: 399–405.
93. Nakazono K, Watanabe N, Matsuno K *et al.* Does superoxide underlie the pathogenesis of hypertension? *Proc Natl Acad Sci USA* 1991; **88**: 10045–10048.
94. Garcia CE, Kilcoyne CM, Cardillo C *et al.* Effect of copper-zinc superoxide dismutase on endothelium-dependent vasodilation in patients with essential hypertension. *Hypertension* 1995; **26**: 863–868.
95. Ikenaga H, Suzuki H, Ishii N *et al.* Role of NO on pressure-natriuresis in Wistar-Kyoto and spontaneously hypertensive rats. *Kidney Int* 1993; **43**: 205–211.
96. Patel A, Layne S, Watts D, Kirchner KA. L-arginine administration normalizes pressure natriuresis in hypertensive Dahl rats. *Hypertension* 1993; **22**: 863–869.
97. Hishikawa K, Nakaki T, Suzuki H *et al.* Role of L-arginine-nitric oxide pathway in hypertension. *J Hypertens* 1993; **11**: 639–645.
98. Korbonits M, Trainer PJ, Fanciulli G *et al.* L-arginine is unlikely to exert neuroendocrine effects in humans *via* the generation of nitric oxide. *Eur J Endocrinol* 1996; **135**: 543–547.

Chapter 22

The Role of Nitric Oxide in Septic Shock and Other Inflammatory Conditions

T J Evans

Background

The presence of bacteria and other microorganisms within the bloodstream can trigger a complex series of events. Irrespective of the nature of the infecting organisms, however, the host response is remarkably similar, with the development of multi-organ failure, and in extreme cases profound hypotension, known as septic shock.[1] The characteristic of patients with this condition is that they have widespread vasodilatation, with a low systemic vascular resistance, and a poor response to conventional vasoconstrictors such as noradrenaline.[2,3] This in turn leads to hypotension, an inability to maintain adequate organ perfusion, and ultimately multi-organ failure and death.[4] In addition, there are effects on the myocardium. Initially, cardiac output increases in septic shock in an attempt to maintain organ perfusion.[5,6] However, the myocardium functions sub-optimally with a decreased ejection fraction, the output only being maintained by an increased stroke volume consequent to cardiac dilatation.[6] In the later stages of septic shock the heart may fail.[5] Although Gram-negative bacteria are the commonest cause of this condition, Gram-positive organisms are increasingly involved. In a small number of cases the syndrome develops secondary to a viral or fungal infection.[1] Patients who are admitted to hospital in septic shock have an extremely high mortality, usually in excess of 40%.[1,7,8]

One of the earliest observations on the production of NO metabolites by humans was that they were increased during an intercurrent infection.[9] This originated in research concerning the generation of inorganic oxides of nitrogen by mammalian cells, a process of potential importance in carcinogenesis.[9,10] Subsequent studies led to the realisation that the production of NO was the main source of reactive nitrogen intermediates in mammalian cells, and that one of the enzymes responsible for the production of NO, inducible NO synthase (iNOS), was strongly induced by pro-inflammatory cytokines and bacterial products such as lipopolysaccharide (LPS).[11] Hence, in any disease in which there was systemic production of pro-inflammatory cytokines and/or the presence of LPS and other bacterial products, the production of large quantities of the potent vasodilator NO from iNOS could lead to profound haemodynamic consequences. The haemodynamic changes of septic shock are exactly what would be predicted if there was widespread induction of iNOS in sepsis.

This chapter will: (a) summarize the various mediators that are important in sepsis; (b) focus on the evidence that stimulation of iNOS in sepsis does indeed result in the haemodynamic changes described above, and discuss the role that NO plays in the myocardium during sepsis; (c) consider the role of NO in other inflammatory conditions, such as cancer chemotherapy with cytokines. The role of NO in the haemodynamic changes associated with the endotoxaemia of cirrhosis is considered in Chapter 14.

Mediators of Sepsis

A large number of different mediators have been proposed to be important in producing the pathophysiological effects of sepsis.[4,12] The most important, however, are pro-inflammatory cytokines and bacterial products. Cytokines are polypeptides released usually from cells of the immune system which exert local powerful effects on nearby cells, acting to amplify and coordinate the immune response. The most important cytokine that is released in sepsis is tumour necrosis factor (TNF)-α, a product of macrophages which is released following stimulation with LPS and other microbial components.[1] Administration of TNF-α mimics many of the features of septic shock, with development of multi-organ failure and hypotension.[13] Levels of TNF-α are raised in patients with septic shock, and in a number of studies the levels achieved have been found to correlate with poor outcome.[14] Agents which neutralize TNF-α action are protective in a number of animal models of infection,[15] and such drugs are now undergoing clinical trials for use in sepsis and septic shock.[7,16]

TNF-α is not the only cytokine involved in the pathogenesis of septic shock. Interleukin (IL)-1, again produced from macrophages stimulated with LPS and other bacterial products, shares many features with TNF-α, including the induction of hypotension.[17] Interferon (IFN)-γ is another cytokine important in sepsis. It is produced largely from activated T cells, and acts principally to 'prime' the immune system to give a much enhanced response to other inflammatory cytokines such as TNF-α and IL-1.[18] Indeed, a key feature of all of these cytokines is that they show marked synergy one with another. Thus, any one agent on its own is much less potent than the combination of two or more cytokines.

The Gram-negative cell wall component LPS is an extremely potent activator of cytokine production. As mentioned above, Gram-positive organisms are of increasing importance in the development of septic shock. Gram-positive cell wall products such as lipoteichoic acid can mimic many of the effects of LPS.[19] In addition, Gram-positive organisms produce a variety of secreted exotoxins, some of which have the unusual property of 'superantigens'.[20] These are molecules that can cross link the MHC class II molecules on an antigen presenting cell with the T cell receptor of an adjacent lymphocyte. In distinction to ordinary antigens, which are only recognised by about 0.001% of circulating T cells, superantigens bind rather indiscriminately to T cells bearing a range of different V_β-segments. This can lead to activation of up to 30% of circulating T cells, with a massive outpouring of pro-inflammatory cytokines. This is probably the mechanism underlying the development of septic shock in staphylococcal toxic shock syndrome.[21]

The mixture of microbial products and pro-inflammatory cytokines described above is a potent stimulus to the induction of the iNOS isoform and consequent overproduction of NO. Moreover, both lipoteichoic acid and 'superantigens' derived from Gram-positive organisms can induce iNOS production in a variety of cells.[22,23]

Haemodynamic Effects of Nitric Oxide in Sepsis

Effects on Vessels

The characteristic feature of the hypotension associated with sepsis, *i.e.* septic shock, is that there is an abnormally low systemic vascular resistance, as described above. This contrasts with hypovolaemic shock, in which the peripheral vessels are intensely vasoconstricted, causing a high vascular resistance.[24] An additional feature of septic blood vessels is that they respond very poorly to powerful vasoconstrictors such as noradrenaline.[2] The peripheral vasculature is thus paralysed open, a so-called state of vasoplegia, which aptly summarizes the state of the vessels in sepsis. It has become clear that the major factor producing this effect in sepsis is the excess production of NO from iNOS induced directly in vascular smooth muscle. The evidence upon which this conclusion is based is considered below.

Animal studies — cells

NO in physiological solutions has a short half life (3–4 seconds), so that although NO is freely diffusible, its action is limited to only a few cell diameters from its site of synthesis,[25] unless bound to a 'carrier' molecule such as albumin or haemoglobin (see Chapters 2 and 3). Thus, to exert an effect on a particular cell, NO must originate either within that cell or closely adjacent to it. In the vessel wall, the endothelium lies immediately over the vascular smooth muscle, and release of NO from this site is a major determinant of resting vascular tone in man,[25] as discussed elsewhere in this book. In sepsis, the endothelium does not appear to be a major source of increased NO production. Although it can be induced to produce iNOS with suitable treatment, the amounts are low,[26] and in one study the normally stimulatory IFN-γ actually down-regulated iNOS production.[27]

NO production in vessels in sepsis is predominantly from vascular smooth muscle cells themselves. A number of studies have demonstrated that they are able to produce large quantities of NO following stimulation with pro-inflammatory cytokines and microbial products.[28,29] In common with iNOS synthesis in other cells, there is a lag of about 6 hours after stimulation before the appearance of significant NO production.[28,30] Once the enzyme is present, however, NO production continues at a high level.

Animal studies — blood vessels

Although it is clear that a variety of animals can produce iNOS in vascular smooth muscle cells maintained in culture, this does not necessarily imply that the same effect will be seen in an intact blood vessel. Indeed, in rats treated with endotoxin, a number of studies using immunochemical detection of iNOS failed to demonstrate significant production of iNOS within vascular smooth muscle.[31,32] Sensitivity of detection, however, is probably the reason for this failure, since in functional studies of vessels removed from endotoxin-treated rats, there is clearly an NO-dependent vasorelaxation that is largely independent of the presence of endothelium.[33,34] In addition, endotoxin alone is not the most potent stimulus for iNOS production.

It is also important to ascertain whether iNOS induction occurs in the pulmonary circulation as well as systemic vessels. *In vitro* studies using isolated pulmonary vascular smooth cells from rats have clearly shown that these cells can make iNOS on appropriate stimulation.[35] This induction of iNOS has been demonstrated in pulmonary vessels isolated from pigs treated with endotoxin,[36] suggesting that both pulmonary and systemic vascular smooth muscle can produce iNOS in whole animals.

Human studies

Initial experiments using human cells found that stimuli that gave good induction in rodent cells apparently failed to produce iNOS production in human cells, especially macrophages.[37] However, further studies have demonstrated unequivocally that human vascular smooth muscle cells and intact arteries can both produce iNOS after appropriate stimulation with a mixture of pro-inflammatory cytokines and/or LPS.[38–40] Indeed, in arteries isolated from humans with septic shock, there is an NO-dependent depressed response to noradrenaline.[41] Gram-positive products such as lipoteichoic acid can also induce iNOS in human vascular smooth muscle.[42]

Effects on Myocardium

The changes in myocardial function in sepsis are complex. Most patients with sepsis have a normal or increased cardiac output. However, there is a sub-population who have a reduced cardiac output.[5] In those patients with normal or increased cardiac output, there is often a considerable reduction in ejection fraction.[6] The cardiac output is maintained or increased by a compensatory dilatation of the ventricles, both on the right and the left side of the heart. These abnormalities return to normal within 7–10 days after sepsis has resolved. Clearly, it would be expected that in patients with cardiac dilatation this would contribute to the problems of maintaining blood pressure and organ perfusion. In fact no difference has been found in survivors and non-survivors of septic shock in their degree of myocardial depression.[6] While the exact role that the myocardial changes in sepsis play in determining outcome remain to be determined, it nevertheless seems likely that any therapy aimed at improving cardiac performance would be of benefit.

Pro-inflammatory cytokines are thought to play a crucial role in the myocardial changes seen in sepsis. NO has been shown to mediate the effects of these cytokines in many instances, but the exact role of NO in both immediate and delayed effects of cytokines on cardiac function is not yet clear. The first report, in 1992, of a direct effect of cytokines on cardiac contractility demonstrated that TNF-α produced a rapid (within 2–3 minutes) reduction in contractility of isolated hamster papillary muscle which was blocked by NOS inhibitors.[43] Such a rapid effect cannot be due to induction of iNOS and thus a direct effect on constitutive NOS activity, which is known to be present in cardiac myocytes, seems likely. However, a separate study on the immediate effects of TNF-α on adult cat myocytes found that negative inotropic effects were not NO-dependent.[44]

Other studies have analysed the effects of more prolonged incubations with cytokines or isolation of cardiac myocytes from endotoxin-treated animals. In adult endotoxin-treated rats, isolated cardiac myocytes have a reduced contractility that is NO-dependent.[45] Isolated heart preparations also show cytokine-inducible NOS which mediates reduced contractility.[46] The negative chronotropic effect of IL-1β on neonatal rat cardiac myocytes is also NO-dependent.[47] Further studies have shown that iNOS is induced in guinea-pig myocytes following cytokine stimulation, and accounts for the reduced contractility produced by these agents.[48] Studies in humans are necessarily limited, but the presence of iNOS within myocardial tissue isolated from humans dying of septic shock has been clearly demonstrated.[49]

It is clear, therefore, that inflammatory cytokines can induce iNOS in the heart, and that NO from this source produces a depression in cardiac performance. What is less certain is the role that NO from this source plays in the pathophysiology of sepsis. This is considered further below.

Indirect Effects

The central nervous system (CNS) exerts considerable control over tone of the vasculature and cardiac performance. Not only is there the neuronal constitutive isoform of NOS within the CNS, but after appropriate cytokine stimulation, iNOS is widely expressed within the CNS, both in neurones and glia.[50–52] Since pro-inflammatory cytokines are also expressed in brain in systemic inflammation, this raises the possibility that NO production in the CNS from iNOS may have pathophysiological effects. A number of studies have demonstrated that local injection of NOS inhibitors into the CNS produce increases in blood pressure and heart rate, suggesting that NO exerts central hypotensive and bradycardic effects.[53,54] Moreover, following injection of endotoxin in rats, there was a potentiation of the NO-dependent hypotensive and bradycardic effects deriving from the nucleus tractus solitarii.[53] We found in endotoxin-treated rats that there was strong antibody staining for iNOS in perivascular nerves within mesenteric blood vessels.[31] Taken together, these data suggest that there may be central indirect pathways utilising NO that affect blood pressure and cardiac output in sepsis. Definitive proof of the importance of this pathway in humans is lacking; clearly this is an area that warrants further research.

Effects of NO Production in the Whole Organism

The roles of NO in causing vasodilatation and myocardial depression in sepsis suggested that it might be of therapeutic benefit to inhibit NO production in this condition. The development of potent inhibitors of NOS has allowed this to be tested in both experimental animals and humans. Because of the pleiotropic effects of NO on the vasculature and myocardium, the net effect of NO production within the whole organism is complex. In particular, the overall result of a therapeutic intervention to prevent NO production during sepsis is difficult to predict. Only those aspects of NOS inhibition which affect haemodynamics will be considered here. However, NO is a potent microbicidal agent, and removal of this effect might be deleterious in the treatment of septic shock.[55,56] The data that have been obtained are often confusing, because of inherent differences between species, models of sepsis and presence or absence of an anaesthetic agent. In addition, the drugs that have been used for the most part to inhibit NOS are relatively non-selective, inhibiting all isoforms of NOS. However, some conclusions may be drawn. In addition, studies in mice homozygous for targeted disruptions of the iNOS gene allow further conclusions about the role of iNOS in sepsis to be drawn.

Animal studies — NOS inhibition

Rodent models of sepsis allow the study of relatively large numbers of animals, although correspondingly it is more difficult to obtain full haemodynamic data. In addition, unlike humans, rodents are extremely resistant to the effects of LPS, which is often used as a model of sepsis. A number of regimens are available to increase sensitivity, such as administration of the hepatotoxin D-galactosamine,[57,58] but actually bear little relation to the changes in human sepsis. A number of rodent models have shown that the non-selective NOS inhibitor N^G-monomethyl-L-arginine (L-NMMA) is able to reverse the hypotension produced by injection of LPS. The dose of the drug is critically important. In one study in rats, a dose of 10 mg/kg of L-NMMA 90 minutes after a bolus dose of LPS attenuated the fall in blood pressure and prevented death.[59] A dose of 300 mg/kg, however, produced rapid cardiovascular collapse and death of all animals. This was felt to reflect a need for a basal level of NO production to maintain blood flow to essential organs; complete NOS inhibition would lead to intense vasoconstriction and loss of essential organ perfusion. A variety of selective iNOS inhibitors are under development (see Chapter 7), and early reports suggest that these may be able to improve survival in rodent models of sepsis.[60–62]

A number of studies of NOS inhibitors in models of sepsis have been performed in large animals, such as dogs, sheep, and pigs. These are well summarized in recent reviews of NO and septic shock.[63,64] A number of common themes emerge. In virtually all cases, the administration of the NOS inhibitor increased the mean arterial pressure of the affected animal mainly by increasing the systemic vascular resistance. In many cases there was a rise in pulmonary vascular resistance, and in almost all cases a decrease in cardiac index. In some cases, the drug had a clear adverse effect on survival, although few studies addressed this directly, and the data available do not allow firm conclusions to be drawn.

A number of features of these different models need to be pointed out. One drug, amino-L-arginine, had clear neurotoxic effects in dogs, inducing epileptic fits.[65] This does not appear to be a problem with L-NMMA. In addition, unlike nitro-L-arginine, which is a less effective inhibitor of iNOS compared to endothelial NOS, L-NMMA is equipotent for both NOS isoforms;[66] see also Chapter 7. In general, most studies have used an intravenous injection of endotoxin to mimic the changes of septic shock. Endotoxin produces different haemodynamic effects depending on the exact nature of the model. Dogs, for example, tend to have a depressed cardiac index following endotoxin administration.[65,67] Sheep and pigs tend to increase their cardiac index following injection of endotoxin, more in keeping with the majority of humans with septic shock.[68,69] However, the degree to which animals are resuscitated with fluids during the experimental period may also alter the haemodynamic response. Thus, in one study in pigs treated with an endotoxin infusion of 15 μg/kg per hour for three hours and receiving 0.14 ml fluid/kg per minute, cardiac index dropped by 50% over a three hour period.[70] In another study, pigs receiving 200 μg/kg of endotoxin for 20 minutes, but resuscitated with 1 ml/kg per minute of fluid, showed a 50% rise in cardiac index after 2 hours, which then declined to baseline levels 4 hours after endotoxin treatment.[68] The presence or absence of an anaesthetic agent will also affect the haemodynamic responses observed. Clearly, these differences may alter individual trial results. A further consideration is the timing of drug administration. iNOS requires a period of about 6 hours for induction,[25] so that drug dosage before that time may be deleterious because of effects predominantly on endothelial NOS which is required to maintain vascular patency to essential organs. Indeed, there are reports of a time-dependency of the effects of NOS inhibitor administration following endotoxin challenge.[71]

The mechanism of the depression of cardiac output produced by NOS inhibition is not clear. Studies on cells have shown that NO inhibits cardiac contractility, as discussed above. Thus, NOS inhibition should produce a rise, not a fall, in cardiac output. The observed effect is probably due to the other haemodynamic changes produced by NOS inhibition. The rise in systemic and particularly pulmonary resistance will place an additional burden on the myocardium, and this may be exacerbated by coronary vasoconstriction and reduction in coronary artery perfusion, leading to reduction in cardiac output.

As mentioned above, most of the animal studies showed that NOS inhibition produced elevation in pulmonary artery pressure in the animals that received endotoxin.[67,68] In untreated control animals, however, there was little effect of NOS inhibition on pulmonary artery pressure. This suggests that under normal conditions basal NO production contributes little to pulmonary arterial pressure, but that in endotoxaemia, NO production is an important mechanism to limit its increase (see also Chapter 8). Clearly, rises in pulmonary artery pressure are potentially deleterious, increasing the work of the right heart. One possible approach to limiting these deleterious rises in pulmonary artery pressure is to administer NO by inhalation (see also Chapter 20). This will produce vasorelaxation in the pulmonary circulation but will have no systemic effect. A number of studies have addressed this issue. In a porcine model of shock, this combined approach was very successful, with little increase in pulmonary arterial pressure, and correspondingly a significant improvement in cardiac output and increase in blood pressure.[70] Survival was also dramatically improved. Another study found the same improvements in pulmonary artery pressure, but little effect on cardiac output.[72] This approach would warrant further study.

An indirect means of antagonising the effects of NO on vascular smooth muscle is to inhibit the production of cGMP by soluble guanylate cyclase, as this is the main effector of NO action. One way in which this can be achieved is through the drug methylene blue. This agent has been shown to be beneficial in animal models of sepsis,[73] but as its effects are rather non-specific and certainly not confined to inhibiting the action of NO, it will not be considered further here.

Animal studies — NOS knockout models

An alternative way of analysing the role of NO in the pathophysiology of sepsis is to study mice which are homozygous for targeted disruptions in the gene for iNOS.[74] Three groups have independently studied such animals. One group found that these mice were resistant to the hypotensive effects of LPS when anaesthetized, and that in animals which were not anaesthetized the reduction in LPS-induced hypotension was much less marked. Sensitized animals showed no protection from the lethal effects of endotoxin.[75] Another research group, however, found no protection against endotoxin-induced death in non-anaesthetized NOS knockout animals, though blood pressure was not recorded.[76] A third study found that non-anaesthetized knockout mice were protected against endotoxin treatment, but the control wild-type population was from a different mouse strain, making a true comparison between wild-type and mutant animals impossible.[77] These studies do support the view that NO from iNOS is important in the hypotension and mortality resulting from endotoxin treatment, but also suggest that there are iNOS-independent mechanisms of death in endotoxin shock.

Human studies

The first report of the administration of a NOS inhibitor, L-NMMA, to humans with septic shock was in 1991.[78] This suggested that the drug did improve blood pressure. A subsequent study by the same group in 12 patients with septic shock extended these findings. This demonstrated that L-NMMA produced a dose-dependent rise in mean arterial pressure and systemic vascular resistance, but with a decrease in cardiac output and rise in pulmonary vascular resistance.[79] These findings are very similar to many of the observations on experimental animals treated with this drug, as described above. A similar study had very comparable findings, with rises in systemic vascular resistance and blood pressure but a fall in cardiac output.[80] Interestingly, these drug-induced changes were largely reversed by L-arginine, the substrate for NOS. In both cases, no adverse results of treatment were seen.

Of some concern is a report of two patients who received L-NMMA as adjunctive treatment for septic shock and who died suddenly of acute left ventricular failure.[81] Both patients received the drug for longer than 24 hours. It is obviously difficult to make any firm conclusions from such a report. More encouragingly, published results of ongoing trials of NOS inhibitors in sepsis show that L-NMMA, even when given at doses as high as 20 mg/kg per hour for 8 hours, causes no apparent adverse effects.[82] Moreover, results from a double-blind phase II trial of L-NMMA in 312 patients with sepsis have recently been presented.[83] Broadly, these confirm that the drug (given at a median dose of 5 mg/kg per hour for 24 hours) elevates blood pressure in septic shock, with a reduced need for concurrent vasopressor support. A further trial, enrolling a greater number of patients and designed to produce outcome data, is now planned. Such trials are of course the only way in which the ultimate utility of inhibiting NOS during sepsis can be properly evaluated.

Nitric Oxide in Other Inflammatory Conditions

NO is involved in the pathophysiology of a number of other inflammatory conditions, such as inflammatory bowel disease,[84] rheumatoid arthritis[85] and acute pancreatitis.[86] Of these three specific conditions, however, the profound haemodynamic changes of septic shock are a feature only of acute pancreatitis.[86] Chemotherapy with IL-2 for cancer can result in a shock-like syndrome, with hypotension and capillary leak. Induction of iNOS plays an important role in the development of this syndrome, and trials of L-NMMA to prevent its development have been performed. These have shown beneficial effects of the drug on blood pressure, with no evidence of toxicity.[64] As in the use of this drug in sepsis, depression of cardiac output was noted at the highest doses. NO may not always be responsible for the adverse haemodynamic consequences of cancer chemotherapy with cytokines. One study of the shock-like syndrome induced by TNF-α therapy found no evidence for the involvement of NO.[87]

Conclusions

There is no doubt that NO plays an important role in the haemodynamic changes of sepsis. It is one of the factors involved in the hypotension of septic shock and probably is also responsible for some of the myocardial depression of this condition. The source of NO in sepsis is from induction of the iNOS isoform. The use of drugs which inhibit NOS has clearly demonstrated the potential benefit of targeting this pathway in sepsis, and clinical studies to date are encouraging. Definitive proof of their benefit remains to be determined, however. One problem with existing agents is that they are relatively non-specific, inhibiting both the constitutive and inducible NOS isoforms. Selective inhibition of iNOS might allow the excess NO production of sepsis to be abrogated, while maintaining the basal NO output required for essential organ perfusion. Another approach may be to combine NOS inhibition with inhaled NO therapy, to minimize the adverse effects of NOS inhibition on pulmonary artery pressure and cardiac output. Future clinical studies should allow a definitive answer to the role that NOS inhibition will play in the treatment of patients with septic shock.

References

1. Glauser MP, Zanetti G, Baumgartner J-D, Cohen J. Septic shock: pathogenesis. *Lancet* 1991; **338**: 732–736.
2. Thijs LG, Schneider AJ, Groeneveld AB. The haemodynamics of septic shock. *Intens Care Med* 1990; **16 (Suppl 3)**: S182–S186.
3. Fink MP. Just say no to NO., yes? *Crit Care Med* 1996; **24**: 1772–1774.
4. Evans TJ, Krausz T. Pathogenesis and pathology of shock. In: *Recent Advances in Histopathology 16*, Anthony PP, MacSween RNM (Eds). Churchill Livingstone, Edinburgh 1994: 21–47.
5. Nishijima H, Weil MH, Shubin H, Cavanilles J. Hemodynamic and metabolic studies on shock associated with gram negative bacteremia. *Medicine* 1973; **52**: 287–294.
6. Parker MM, Shelhamer JH, Bacharach SL *et al.* Profound but reversible myocardial depression in patients with septic shock. *Ann Intern Med* 1984; **100**: 483–490.

7. Cohen J, Carlet J. INTERSEPT: an international, multicenter, placebo-controlled trial of monoclonal antibody to human tumor necrosis factor-α in patients with sepsis. International Sepsis Trial Study Group. *Crit Care Med* 1996; **24**: 1431–1440.
8. Abraham E, Wunderink R, Silverman H *et al.* for the TNF-α MAb Sepsis Study Group. Efficacy and safety of monoclonal antibody to human tumor necrosis factor α in patients with sepsis syndrome. A randomized, controlled, double-blind, multicenter clinical trial. *J Am Med Assoc* 1995; **273**: 934–941.
9. Green LC, Ruiz de Luzuriaga K, Wagner DA *et al.* Nitrate biosynthesis in man. *Proc Natl Acad Sci USA* 1981; **78**: 7764–7768.
10. Green LC, Tannenbaum SR, Goldman P. Nitrate synthesis in the germfree and conventional rat. *Science* 1981; **212**: 56–58.
11. Nathan C. Nitric oxide as a secretory product of mammalian cells. *FASEB J* 1992; **6**: 3051–3064.
12. Evans TJ, Cohen J. Mediators: nitric oxide and other toxic oxygen species. *Curr Topic Microbiol Immunol* 1996; **216**: 189–207.
13. Tracey KJ, Beutler B, Lowry SF *et al.* Shock and tissue injury induced by recombinant human cachectin. *Science* 1986; **234**: 470–474.
14. Waage A, Halstensen A, Espevik T. Association between tumor necrosis factor in serum and fatal outcome in patients with meningococcal disease. *Lancet* 1987; **i**: 355–357.
15. Silva AT, Bayston KF, Cohen J. Prophylactic and therapeutic effects of a monoclonal antibody to tumour necrosis factor-α in experimental Gram-negative shock. *J Infect Dis* 1990; **162**: 421–427.
16. Reinhart K, Wiegand-Lohnert C, Grimminger F *et al.* Assessment of the safety and efficacy of the monoclonal anti-tumor necrosis factor antibody-fragment, MAK 195F, in patients with sepsis and septic shock: a multicenter, randomized, placebo-controlled, dose-ranging study. *Crit Care Med* 1996; **24**: 733–742.
17. Dinarello C. Interleukin-1 and interleukin-1 antagonism. *Blood* 1991; **8**: 1627–1652.
18. Billiau A. Gamma-interferon: the match that lights the fire? *Immunol Today* 1988; **9**: 37–40.
19. Keller R, Fischer W, Keist R, Bassetti S. Macrophage response to bacteria: induction of marked secretory and cellular activities by lipoteichoic acids. *Infect Immun* 1992; **60**: 3664–3672.
20. Sékaly R-P. Introduction: bacterial superantigens. *Sem Immunol* 1993; **5**: 1–2.
21. Todd JK. Staphylococcal toxin syndromes. *Ann Rev Med* 1985; **36**: 337–347.
22. De Kimpe SJ, Kengatharan M, Thiemermann C, Vane JR. The cell wall components peptidoglycan and lipoteichoic acid from *Staphylococcus aureus* act in synergy to cause shock and organ failure. *Proc Natl Acad Sci USA* 1995; **92**: 10359–10363.

23. Zembowicz A, Vane JR. Induction of nitric oxide synthase activity by toxic shock syndrome toxin 1 in a macrophage-monocyte cell line. *Proc Natl Acad Sci USA* 1992; **89**: 2051–2055.
24. Bond RF, Johnson G III. Vascular adrenergic interactions during hemorrhagic shock. *Fed Proc* 1985; **44**: 281–289.
25. Moncada S, Palmer RMJ, Higgs EA. Nitric oxide: physiology, pathophysiology, and pharmacology. *Pharmacol Rev* 1991; **43**: 109–142.
26. Lamas S, Michel T, Brenner BM, Marsden PA. Nitric oxide synthesis in endothelial cells: evidence for a pathway inducible by TNF-α. *Am J Physiol* 1991; **261**: C634–C641.
27. Lamas S, Michel T, Collins T *et al.* Effects of interferon-γ on nitric oxide synthase activity and endothelin production by vascular endothelial cells. *J Clin Invest* 1992; **90**: 879–887.
28. Busse R, Mülsch A. Induction of nitric oxide synthase by cytokines in vascular smooth muscle cells. *FEBS Lett* 1990; **275**: 87–90.
29. Vanhoutte PM. Inducible nitric oxide synthase and vascular smooth muscle. *Jpn J Pharmacol* 1992; **58 Suppl 2**: 192P–199P.
30. Geng Y-J, Hansson GK, Holme E. Interferon-γ and tumor necrosis factor synergize to induce nitric oxide production and inhibit mitochondrial respiration in vascular smooth muscle cells. *Circ Res* 1992; **71**: 1268–1276.
31. Buttery LDK, Evans TJ, Springall DR *et al.* Immunochemical localization of inducible nitric oxide synthase in endotoxin-treated rats. *Lab Invest* 1994; **71**: 755–764.
32. Sato K, Miyakawa K, Takeya M *et al.* Immunohistochemical expression of inducible nitric oxide synthase (iNOS) in reversible endotoxic shock studied by a novel monoclonal antibody against rat iNOS. *J Leukoc Biol* 1995; **57**: 36–44.
33. Julou-Schaeffer G, Gray GA, Fleming I *et al.* Loss of vascular responsiveness induced by endotoxin involves L-arginine pathway. *Am J Physiol* 1990; **259**: H1038–H1043.
34. Schneider F, Bucher B, Schott C *et al.* Effect of bacterial lipopolysaccharide on function of rat small femoral arteries. *Am J Physiol* 1994; **266**: H191–H198.
35. Nakayama DK, Geller DA, Lowenstein CJ *et al.* Cytokines and lipopolysaccharide induce nitric oxide synthase in cultured rat pulmonary artery smooth muscle. *Am J Respir Cell Mol Biol* 1992; **7**: 471–476.
36. Villamor E, Pérez-Vizcaíno F, Ruiz T *et al.* Group B *Streptococcus* and *E. coli* LPS-induced NO-dependent hyporesponsiveness to noradrenaline in isolated intrapulmonary arteries of neonatal piglets. *Br J Pharmacol* 1995; **115**: 261–266.
37. Denis M. Human monocytes/macrophages: NO or no NO? *J Leukoc Biol* 1994; **55**: 682–684.
38. MacNaul KL, Hutchinson NI. Differential expression of iNOS and cNOS mRNA in human vascular smooth muscle cells and endothelial cells under normal

and inflammatory conditions. *Biochem Biophys Res Commun* 1993; **196**: 1330–1334.
39. Thorin-Trescases N, Hamilton CA, Reid JL *et al.* Inducible L-arginine/nitric oxide pathway in human internal mammary artery and saphenous vein. *Am J Physiol* 1995; **268**: H1122–H1132.
40. Berrazueta JR, Salas E, Amado JA *et al.* Induction of nitric oxide synthase in human mammary arteries *in vitro. Eur J Pharmacol* 1994; **251**: 303–305.
41. Tsuneyoshi I, Kanmura Y, Yoshimura N. Nitric oxide as a mediator of reduced arterial responsiveness in septic patients. *Crit Care Med* 1996; **24**: 1083–1086.
42. Tsuneyoshi I, Kanmura Y, Yoshimura N. Lipoteichoic acid from *Staphylococcus aureus* depresses contractile function of human arteries *in vitro* due to the induction of nitric oxide synthase. *Anesth Analg* 1996; **82**: 948–953.
43. Finkel MS, Oddis CV, Jacob TD *et al.* Negative inotropic effects of cytokines on the heart mediated by nitric oxide. *Science* 1992; **257**: 387–389.
44. Yokoyama T, Vaca L, Rossen RD *et al.* Cellular basis for the negative inotropic effects of tumor necrosis factor-α in the adult mammalian heart. *J Clin Invest* 1993; **92**: 2303–2012.
45. Brady AJB, Poole-Wilson PA, Harding SE, Warren JB. Nitric oxide production within cardiac myocytes reduces their contractility in endotoxemia. *Am J Physiol* 1992; **263**: H1963–H1966.
46. Stein B, Frank P, Schmitz W *et al.* Endotoxin and cytokines induce direct cardiodepressive effects in mammalian cardiomyocytes *via* induction of nitric oxide synthase. *J Mol Cell Cardiol* 1996; **28**: 1631–1639.
47. Roberts AB, Vodovotz Y, Roche NS *et al.* Role of nitric oxide in antagonistic effects of transforming growth factor-β and interleukin-1β on the beating rate of cultured cardiac myocytes. *Mol Endocrinol* 1992; **6**: 1921–1930.
48. Schulz R, Panas DL, Catena R *et al.* The role of nitric oxide in cardiac depression induced by interleukin-1β and tumour necrosis factor-α. *Br J Pharmacol* 1995; **114**: 27–34.
49. Thoenes M, Förstermann U, Tracey WR *et al.* Expression of inducible nitric oxide synthase in failing and non-failing human heart. *J Mol Cell Cardiol* 1996; **28**: 165–169.
50. Minc Golomb D, Tsarfaty I, Schwartz JP. Expression of inducible nitric oxide synthase by neurones following exposure to endotoxin and cytokine. *Br J Pharmacol* 1994; **112**: 720–722.
51. Wallace MN, Bisland SK. NADPH-diaphorase activity in activated astrocytes represents inducible nitric oxide synthase. *Neuroscience* 1994; **59**: 905–919.
52. Galea E, Reis DJ, Feinstein DL. Cloning and expression of inducible nitric oxide synthase from rat astrocytes. *J Neurosci Res* 1994; **37**: 406–414.
53. Tseng CJ, Liu HY, Lin HC *et al.* Cardiovascular effects of nitric oxide in the brain stem nuclei of rats. *Hypertension* 1996; **27**: 36–42.

54. el Karib AO, Sheng J, Betz AL, Malvin RL. The central effects of a nitric oxide synthase inhibitor (N^{ω}-nitro-L-arginine) on blood pressure and plasma renin. *Clin Exp Hypertens* 1993; **15**: 819–832.
55. Nathan C, Hibbs JB Jr. Role of nitric oxide synthesis in macrophage antimicrobial activity. *Curr Opin Immunol* 1991; **3**: 65–70.
56. Granger DL, Hibbs JB Jr. High-output nitric oxide: weapon against infection? *Trends Microbiol* 1996; **4**: 46–47.
57. Galanos C, Freudenburg MA, Reutter W. Galactosamine-induced sensitization to the lethal effects of endotoxins. *Proc Natl Acad Sci USA* 1979; **76**: 5939–5943.
58. Freundenberg MA, Galanos C. Tumour necrosis factor alpha mediates lethal activity of killed Gram-negative and Gram-positive bacteria in D-galactosamine-treated mice. *Infect Immun* 1991; **59**: 2110–2115.
59. Nava E, Palmer RMJ, Moncada S. The role of nitric oxide in endotoxic shock: effects of N^G-monomethyl-L-arginine. *J Cardiovasc Pharmacol* 1992; **20 (Suppl 12)**: S132–S134.
60. Marletta MA. Approaches toward selective inhibition of nitric oxide synthase. *J Med Chem* 1995; **37**: 1899–1907.
61. Southan GJ, Szabó C, Thiemermann C. Isothioureas: potent inhibitors of nitric oxide synthases with variable isoform selectivity. *Br J Pharmacol* 1995; **114**: 510–516.
62. Szabó C, Southan GJ, Thiemermann C. Beneficial effects and improved survival in rodent models of septic shock with *S*-methylisothiourea sulfate, a potent and selective inhibitor of inducible nitric oxide synthase. *Proc Natl Acad Sci USA* 1994; **91**: 12472–12476.
63. Cobb JP, Danner RL. Nitric oxide and septic shock. *J Am Med Assoc* 1996; **275**: 1192–1196.
64. Kilbourn RG, Szabó C, Traber DL. Beneficial versus detrimental effects of nitric oxide synthase inhibitors in circulatory shock: lessons learned from experimental and clinical studies. *Shock* 1997; **7**: 235–246.
65. Cobb JP, Natanson C, Hoffman WD *et al.* N^{ω}-amino-L-arginine, an inhibitor of nitric oxide synthase, raises vascular resistance but increases mortality rates in awake canines challenged with endotoxin. *J Exp Med* 1992; **176**: 1175–1182.
66. Gross SS, Stuehr DJ, Aisaka K *et al.* Macrophage and endothelial cell nitric oxide synthesis: cell-type selective inhibition by N^G-aminoarginine, N^G-nitroarginine and N^G-methylarginine. *Biochem Biophys Res Commun* 1990; **170**: 96–103.
67. Cobb JP, Natanson C, Quezado ZMN *et al.* Differential hemodynamic effects of L-NMMA in endotoxemic and normal dogs. *Am J Physiol* 1995; **268**: H1634–H1642.
68. Robertson FM, Offner PJ, Ciceri DP *et al.* Detrimental hemodynamic effects of nitric oxide synthase inhibition in septic shock. *Arch Surg* 1994; **129**: 149–155.

69. Meyer J, Lentz CW, Stothert JC Jr *et al.* Effects of nitric oxide synthesis inhibition in hyperdynamic endotoxemia. *Crit Care Med* 1994; **22**: 306–312.
70. Klemm P, Thiemermann C, Winklmaier G *et al.* Effects of nitric oxide synthase inhibition combined with nitric oxide inhalation in a porcine model of endotoxin shock. *Br J Pharmacol* 1995; **114**: 363–368.
71. Laszlo F, Whittle BJR, Moncada S. Time-dependent enhancement or inhibition of endotoxin-induced vascular injury in rat intestine by nitric oxide synthase inhibitors. *Br J Pharmacol* 1994; **111**: 1309–1315.
72. Weitzberg E, Rudehill A, Modin A, Lundberg JM. Effect of combined nitric oxide inhalation and N^G-nitro-L-arginine infusion in porcine endotoxin shock. *Crit Care Med* 1995; **23**: 909–918.
73. Keaney JF Jr, Puyana JC, Francis S *et al.* Methylene blue reverses endotoxin-induced hypotension. *Circ Res* 1994; **74**: 1121–1125.
74. Gross SS, Kilbourn RG, Griffith OW. NO in septic shock: good, bad or ugly? Learning from iNOS knockouts. *Trends Microbiol* 1996; **4**: 47–49.
75. MacMicking JD, Nathan C, Hom G *et al.* Altered responses to bacterial infection and endotoxic shock in mice laking inducible nitric oxide synthase. *Cell* 1995; **81**: 641–650.
76. Laubach VE, Shesely EG, Smithies O, Sherman PA. Mice lacking inducible nitric oxide synthase are not resistant to lipolysaccharide-induced death. *Proc Natl Acad Sci USA* 1995; **92**: 10688–10692.
77. Wei X-Q, Charles I, Smith A *et al.* Altered immune responses in mice lacking inducible nitric oxide synthase. *Nature* 1995; **375**: 408–411.
78. Petros A, Bennett D, Vallance P. Effect of nitric oxide synthase inhibitors on hypotension in patients with septic shock. *Lancet* 1991; **338**: 1557–1559.
79. Petros A, Lamb G, Leone A *et al.* Effects of a nitric oxide synthase inhibitor in humans with septic shock. *Cardiovasc Res* 1994; **28**: 34–39.
80. Lorente JA, Landín L, De Pablo R *et al.* L-arginine pathway in the sepsis syndrome. *Crit Care Med* 1993; **21**: 1287–1295.
81. Locker GJ, Burgmann H, Staudinger T *et al.* Fatal cardiac side effects of L-NMMA? [Letter]. *Crit Care Med* 1996; **24**: 1930.
82. Grover R, Zaccardelli D, Colice G *et al.* The cardiovascular effects of 546C88 in human septic shock [Abstract]. *Int Care Med* 1995; **21 (Suppl 1)**: S20.
83. Anzueto A, Lodato R, Lorente J *et al.* for the International 546C88 Septic Shock Group. Multicenter placebo-controlled, double-blind trial of the nitric oxide synthase inhibitor 546C88 in patients with septic shock: acute hemodynamic effects [Abstract]. *Am J Respir Crit Care Med* 1997; **155**: A263.
84. Singer II, Kawka DW, Scott S *et al.* Expression of inducible nitric oxide synthase and nitrotyrosine in colonic epithelium in inflammatory bowel disease. *Gastroenterology* 1996; **111**: 871–885.
85. Kaur H, Halliwell B. Evidence for nitric oxide-mediated oxidative damage in chronic inflammation. Nitrotyrosine in serum and synovial fluid from rheumatoid patients. *FEBS Lett* 1994; **350**: 9–12.

86. Al-Mufti RA, Williamson RCN, Mathie RT. Increased nitric oxide activity in a rat model of acute pancreatitis. *Gut* 1998; **43**: 564–570.
87. Zwaveling JH, Maring JK, Moshage H *et al.* Role of nitric oxide in recombinant tumor necrosis factor-α-induced circulatory shock: a study in patients treated for cancer with isolated limb perfusion. *Crit Care Med* 1996; **24**: 1806–1810.

Index